INTRODUCTION TO DYNAMICS AND CONTROL

INTRODUCTION TO DYNAMICS AND CONTROL

Leonard Meirovitch

College of Engineering
Virginia Polytechnic Institute and State University
Blacksburg, Virginia

John Wiley & Sons

New York Chichester Brisbane Toronto Singapore

Library of Congress Cataloging in Publication Data:

Meirovitch, Leonard.
Introduction to dynamics and control.

Bibliography: p.
Includes index.
1. Dynamics. 2. Control theory. I. Title.
QA845.M45 1985 620.1'04 84-20938
ISBN 0-471-87074-9

Printed in the United States of America

10 9 8 7 6 5 4 3 2 1

To Jo Anne

Preface

The only permanent thing in engineering education is change. One area of engineering education that has experienced substantial change in recent years is dynamics. Of course, dynamics is one of the oldest branches of physics, with its development as a science beginning with Galileo, close to four centuries ago. However, the study of dynamics of particles and rigid bodies as an engineering subject is not nearly so old, having been introduced as standard fare in engineering curricula much more recently, perhaps only after World War II. An even-later addition to engineering curricula is the study of vibrations, which can be ragarded as the part of dynamics concerned with the motion of elastic systems. In dynamics and vibrations the emphasis is on analysis, which implies the determination of the response to given excitations, and any design amounts to changing the system parameters so as to produce satisfactory response. This type of design tends to be very unsatisfactory at times, as the improvement in performance achieved by changing the parameters is very limited. Quite recently, a new approach to system design has been gaining increasing acceptance. The approach consists of designing forces that, when applied to the system, bring about a satisfactory response. This approach, known as control, is rapidly becoming a ubiquitous part of the engineering curriculum. To be sure, the idea of control is as old as ancient Egypt, but control theory is relatively new.

Ideally, a student graduated in aerospace engineering, mechanical engineering and engineering science should be familiar with dynamics, vibrations, and control. Perhaps because of chronological order of development and tradition, the subjects of dynamics of particles and rigid bodies, vibrations, and control are being taught in the above curricula in three separate courses, and a good case can be made for continuing to do so. This is certainly true if the interest lies in only one or two of the three subjects. A problem arises when all three subjects are of interest and there is no room in the curriculum for three courses in dynamics of systems, in which case a painful choice must be made as to the course to omit. A solution to this problem may be to teach the three subjects on an integrated basis. From a philosophical point of view, the most persuasive argument in favor of an integrated approach is that dynamics, vibrations, and control really belong together, and studying them together is likely to enhance the understanding of all three. From a pedagogical point of view, teaching the three subjects on an integrated basis can result in greater efficiency, achieved through the elimination of duplicated material. In this regard, we should point out that various principles of dynamics are ordinarily covered in courses in dynamics as well as in courses in vibrations. Moreover, many concepts and mathematical techniques used in vibrations are also used in control. Finally, before one can design controls for dynamical systems, it is often necessary to derive the system equations of motion, which is in the domain of dynamics. It is conceivable that, by eliminating the duplication, the same material can be taught in two courses instead of three.

This book covers the subjects of dynamics, vibrations, and control on an integrated basis, a feature characterizing a number of other texts on dynamics of systems. The main difference between this and other texts on the subject lies in the selection of the material, with this text placing heavier emphasis on the fundamentals of dynamics. The book regards dynamics, vibrations, and control as essential parts of the more general area of dynamics of systems, and the material is structured so as to make the text suitable for courses on these subjects taught separately or on an integrated basis. The material is intended for courses ranging from the junior level to the senior level. To help with the tailoring of the material to a given course, a chapter-by-chapter review of the material follows.

Chapter 1 presents concepts and mathematical techniques from linear system theory that are basic to the study of vibrations and control. The relation between the excitation and response is discussed in a general way, without going into the derivation of the system equations. Then, a variety of functions of special interest in vibrations and control are introduced. The chapter concludes with the development of the state equations, a form particularly suited for the computation of the system response, The material in this chapter can be covered in a course on the junior level. The chapter can be omitted if the interest lies in a course strictly on the dynamics of particles and rigid bodies.

Chapter 2 is devoted to kinematics. This the place to begin a course on dynamics of particles and rigid bodies. For courses on vibrations and control, there is no need for so much kinematical detail, and the material is recommended as review only.

Chapter 3 is concerned with the dynamics of a single particle. This is a very fundamental chapter in which Newton's laws and certain principles of dynamics are introduced, and the first six sections should be included in any junior-level course on the dynamics of systems. The last two sections are concerned with applications from orbital mechanics and should be considered only for a course on the dynamics of particles.

Chapter 4 treats the response of low-order linear systems. It introduces many concepts and mathematical techniques of fundamental importance in the study of vibrations and control. The material is suitable for a junior course.

Chapter 5 extends the ideas of Chapter 3 to systems of particles. Once again, this is a very basic chapter, and the first six sections should be included in any junior-level course on the dynamics of systems. The last two sections are primarily of interest in an aerospace curriculum.

Chapter 6 is concerned with rigid-body dynamics. The first nine sections contain fundamental material and are recommended for any course in dynamics of systems at the junior level. The last section is more advanced and is suitable primarily for a senior-level course on dynamics.

Chapter 7 presents elements of Lagrangian mechanics, the object being the efficient derivation of the equations of motion. The material is intended for a senior-level course on dynamics and vibrations.

Chapter 8 deals with the vibration of multi-degree-of-freedom systems. The material is suitable for a course in vibrations and to a lesser extent for a course in control, the latter only if the interest lies in multi-input, multi-output elastic systems. This is senior-level material.

Chapter 9 presents fundamental concepts of system stability, such as equilibrium points and the motion characteristics in the neighborhood of such points. The material is basic to dynamics, vibrations, and control. For the most part, this is senior-level material.

Chapter 10 is concerned with modern computational techniques for the response of both linear and nonlinear systems. It introduces concepts such as the transition matrix, and it presents an introduction to discrete-time systems, an approach permitting the evaluation of the response on a digital computer. Moreover, it discusses techniques for the computation of the response of nonlinear systems. The material is intended for senior courses on vibrations and control.

Chapter 11 is the one in which all three disciplines, dynamics, vibrations, and control, come together. It presents the basic concepts and techniques of classical control, and it provides a glimpse into modern control. The material ranges from junior level to senior level, although most material can be covered in junior courses.

The book concludes with an appendix on the Laplace transformation, which is of interest to both vibrations and control.

The author wishes to acknowledge the help he received from colleagues and students. Thanks are due to Professor Leonard R. Anderson, Professor Haim Baruh, Mr. Jeffrey K. Bennighot, Mr. Debasish Ghosh, Mr. Mark A. Norris, Professor Hayrani Öz, Mr. Roger D. Quinn and Professor Terence A. Weisshaar. Special thanks are due to Professor Frederick H. Lutze, who made many valuable suggestions. Last, but not least, the author wishes to express his appreciation to Bill Stenquist for his efficient handling of the project and to Mrs Norma B. Guynn for her excellent job of typing the manuscript.

Leonard Meirovitch

Contents

INTRODUCTION TO DYNAMICS AND CONTROL

CHAPTER 1
Concepts From Linear System Theory

1.1 INTRODUCTION

Broadly speaking, linear system theory is concerned with the behavior of linear systems, and in particular with the relation between excitation and response. The theory is of fundamental importance to the study of vibrations and control. It includes selected topics from the theory of ordinary differential equations and, for higher-order systems, certain topics from matrix theory also. In this chapter, we are concerned primarily with general relations between excitation and response.

The relation between the excitation and response can be carried out in the frequency domain or in the time domain. Frequency-domain analysis is used when the excitation is harmonic, and time-domain analysis is used in the case of arbitrary excitation. For the most part, the interest lies in dynamical systems described by ordinary differential equations with constant coefficients. The analysis of such systems can be carried out conveniently by means of the Laplace transformation. Quite often, the integration of the system differential equations can be performed more efficiently if the equations are of first order. But, equations of higher order can be transformed into first-order equations by expressing them in terms of state variables. In this chapter, we introduce all these topics.

The derivation of the system equations represents a study in itself, and indeed such a study is the subject of the following chapters. In this chapter, no derivations of the system equations are carried out, and the discussion is in terms of a generic system. Moreover, the relation between the excitation and response is treated in a general way, with a more detailed analysis to be presented later in the text, after the differential equations for a variety of systems have been derived. The purpose of this chapter is to introduce certain definitions and concepts that will prove extremely useful in the analysis and design of dynamical systems.

1.2 CONCEPTS FROM SYSTEM ANALYSIS

A *system* is defined as an assemblage of parts or components acting together as a whole. When acted on by a given *excitation*, the system exhibits a certain *response*. *Dynamics* is the study of this cause-and-effect relation.

We distinguish between *physical systems* and *engineering systems*. The former can be broadly defined as systems that can be found in nature and the latter as systems that are the product of man. For example, the moon is a physical system, but an artificial satellite is an engineering system.

Dynamic analysis entails several stages. The first stage is to identify the system to be studied. Because in most cases a system is very complex, it is often necessary to develop a simple idealized model of the system, referred to as a *physical model*. To develop such a model, one must make almost invariably certain simplifying assumptions. The expectation is that the simplified model will be able to simulate the behavior of the actual system reasonably well. The next stage is to develop a *mathematical model*. In this stage, the various components of the physical system are identified, and their excitation–response characteristics are determined on the basis of appropriate physical laws. Then, the mathematical model is derived in the form of *system equations*, where the equations must be such that the various components are guaranteed to act as a system and not as individual entities. System equations describe system behavior in the sense that they provide the relation between response and excitation in implicit form. The equations have most frequently the form of *differential equations* and less frequently the form of *integral equations*. At times, they can have the form of *difference equations*. The third stage is to solve the system equations and interpret the solution. If the system behavior, as inferred from the solution, matches the observed behavior reasonably well, then the model can be regarded as being representative of the actual system and can be used to predict its behavior under different excitations. Otherwise, further refinement of the model is necessary. At this point, it should be noted that the simplifying assumptions made in deriving the mathematical model may hold only in a given range, so that the model should be accepted as representative of the system only in this range.

The dynamic analysis outlined above can be performed on a physical system or an engineering system. Generally, an engineering system involves another task, namely, *design*. For example, if the excitation is given and there is a need for a system exhibiting a desired response, and if no such system exists in natural state, then one may wish to design one. Design involves virtually all phases of dynamic analysis. The only difference is that in design the analysis can be repeated many times. Indeed, if the designed system does not perform as desired, then changes must be made until it does. It should be clear that a design is not unique, and there may be several designs giving satisfactory performance.

In system analysis terminology, systems are often referred to as *plants*, or *processes*. Moreover, the excitation is known as the *input signal* and the response as the *output signal*, or simply input and output. The cause-and-effect relationship can be shown schematically in the form of the *block diagram* of Fig. 1.1. As an illustration, we can envision the vertical free flight of a rocket (Fig. 1.2). The system is the rocket itself. The excitation consists of the gravity and drag forces, both

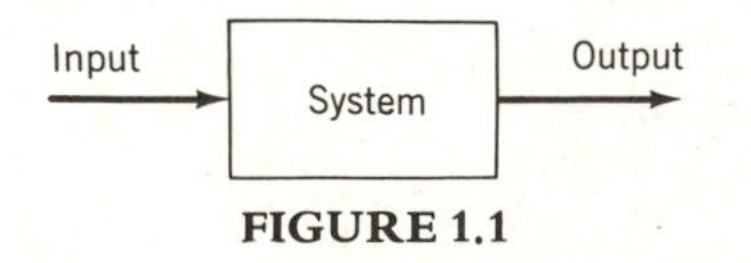

FIGURE 1.1

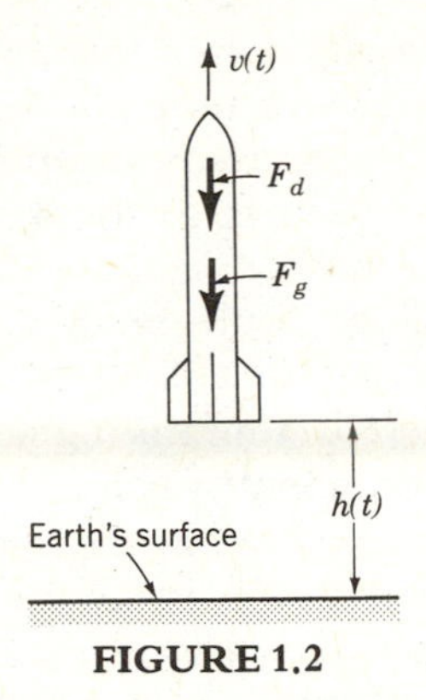

FIGURE 1.2

acting in a direction opposing the motion. As response, we can consider either the altitude $h(t)$, or the velocity $v(t)=\dot{h}(t)$, or both the altitude and the velocity. The system shown in Fig. 1.1, or Fig. 1.2, is said to be *uncontrolled* because there are no forces applied by design. The only forces acting on the system are those arising naturally.

If the system is to exhibit a desired output, then one must select a certain input and subject the system to the selected input through a *controller*, as shown in Fig. 1.3. If the input is essentially predetermined and is not influenced by the output, then the control system is an *open-loop* system. The reason for this term will become evident shortly. An example of such a system is a rocket in powered flight, as shown in Fig. 1.4. In addition to the gravity and drag forces, the rocket is acted upon by the thrust T. The time history of the thrust is designed in advance so as to fulfill certain mission objectives, such as reaching a given altitude at a given time, but is not influenced by the current altitude or velocity.

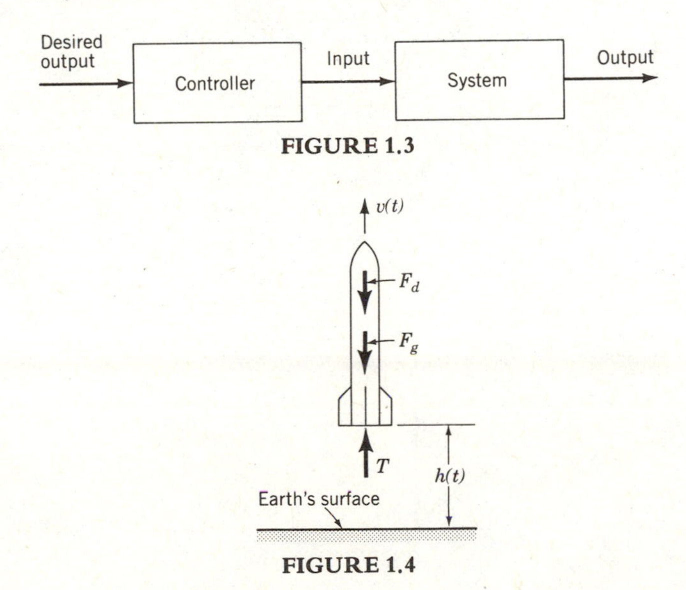

FIGURE 1.3

FIGURE 1.4

In the above discussion, the implication was that the input affects the output, but that the output has no influence on the input. Yet, in many cases the input does depend on the output, either inadvertently or by design. For example, it is conceivable that unknown or unforeseen factors may actually prevent a system from meeting the desired objectives, in which case one may wish to consider the actual output in implementing a change in the input such that the objectives are met. Such a system is depicted in the more elaborate block diagram of Fig. 1.5. In addition to the controller, there is a measuring device and a comparing device. The measuring device senses the output, which is then fed to the comparing device. The difference between the desired and the measured output is called the *output error*. Based on this error, the controller issues the necessary input commands to reduce the error to zero. Such a control system is known as a *closed-loop* system, which explains the term "open loop" introduced earlier. As an illustration, let us consider once again the rocket in powered flight, but this time there is an accelerometer on board (Fig. 1.6). As output, we shall consider now the velocity instead of the altitude. The sensed acceleration is integrated once to yield the output velocity. The difference between the desired and measured velocity can be used to calculate the necessary change in the engine thrust to reduce the difference to zero.

A more common example of a closed-loop control system is the automobile. The driver acts as sensor, comparing device, and controller. Indeed, the driver takes continuous measurements by looking at the road, compares the sightings with the desired path, and issues commands to the vehicle through the steering wheel, for direction, and through the accelerator or brakes, for acceleration or deceleration.

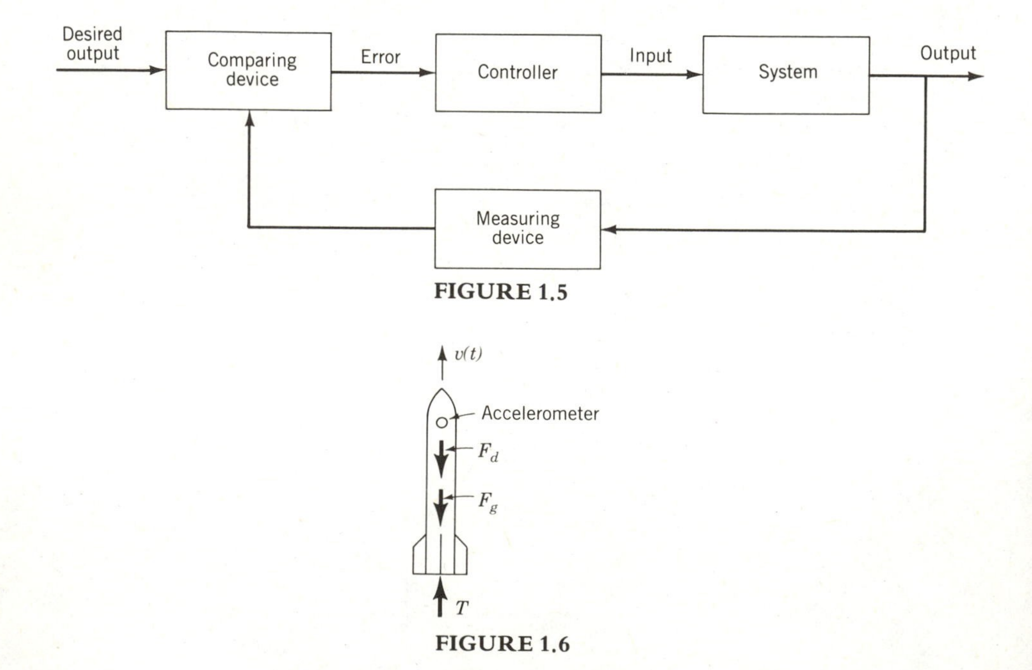

FIGURE 1.5

FIGURE 1.6

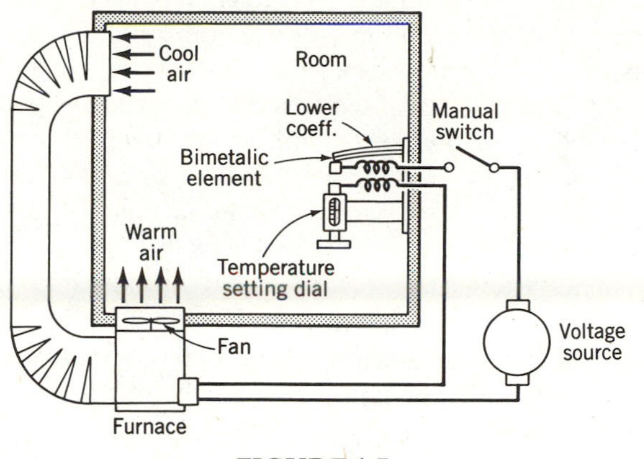

FIGURE 1.7

Another example of a closed-loop control system is a heating system using a thermostat to regulate the temperature (Fig. 1.7). The thermostat includes a bimetallic element, i.e., an element consisting of two metallic plates of different coefficients of thermal expansion welded together. For low temperatures the tip of the element closes a switch, actuating an electrical circuit that causes the furnace to burn and some fans to push the warm air into the room. As the air in the room becomes warmer, the different metals expand at different rates causing the bimetallic element to bend. At a given temperature, the tip of the element bends away, thus opening the circuit and causing the furnace to stop burning and the fans to stop running. In the case of the automobile a human being was inserted in the control loop, whereas in the case of the furnace there was no human being, so that the latter represents *automatic control.*

In this text, we shall discuss both uncontrolled and controlled dynamical systems.

1.3 CLASSIFICATION OF SYSTEMS. SUPERPOSITION PRINCIPLE

In developing a mathematical model for a system, it is necessary to identify the system components and determine their excitation–response characteristics. These characteristics are governed by given physical laws and are described in terms of the so-called *system parameters.* These parameters can be divided into two broad classes: parameters that are not functions of the spatial coordinates and parameters that are. Those in the first class are known as *lumped parameters* and those in the second class as *distributed parameters.* In the case of lumped-parameter systems the excitation and response are functions of *time alone,* and in the case of distributed-parameter systems they are functions of both *spatial coordinates* and *time.* The behavior of lumped-parameter systems is governed by *ordinary differential equations* and that of distributed-parameter systems by *partial differen-*

tial equations. An example of the first is a set of rigid disks mounted on a massless shaft, and an example of the second is a thin membrane. In this text, we shall be concerned exclusively with lumped-parameter systems. In the mathematical formulation, i.e., in the differential equations, the system parameters appear in the form of coefficients. If the coefficients do not depend on time, i.e., if they are constant, then the system is said to be *time invariant.* If the coefficients do depend on time, then the system is known as *time varying.* In this text, we confine ourselves to systems with constant coefficients.

One property of a system that has profound implications in mathematical analysis is *linearity.* To introduce the concept, it will prove convenient to express the input–output relation in the form of a block diagram. Figure 1.8 shows this diagram, in which $r(t)$ denotes the input and $c(t)$ the output. For this system, let us consider two pairs of input–output relations, $r_1(t)$, $c_1(t)$ and $r_2(t)$, $c_2(t)$, as shown in Figs. 1.9*a* and 1.9*b*. Then, for the same system, let us consider an input $r_3(t)$ (Fig. 1.9*c*) in the form of the *linear combination*

$$r_3(t)=\alpha_1 r_1(t)+\alpha_2 r_2(t) \tag{1.1}$$

where α_1 and α_2 are constant scalars. If the output $c_3(t)$ represents a linear combination of the same form, namely,

$$c_3(t)=\alpha_1 c_1(t)+\alpha_2 c_2(t) \tag{1.2}$$

then *the system is linear.* On the other hand, if

$$c_3(t)\neq\alpha_1 c_1(t)+\alpha_2 c_2(t) \tag{1.3}$$

the system is nonlinear.

Equation (1.2) represents one of the most powerful statements in system analysis.

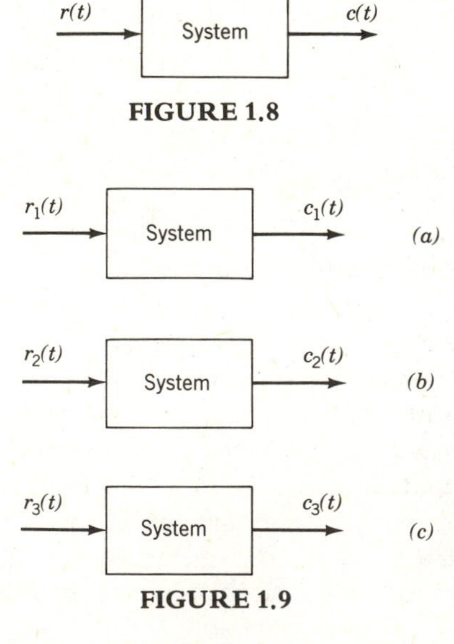

FIGURE 1.8

FIGURE 1.9

In words, the statement implies that *for a linear system responses to different excitations can be obtained separately and then combined linearly*. This statement is known as the *principle of superposition*, and it represents the most fundamental principle of linear system theory.

Both linear and nonlinear systems will be discussed here, although the emphasis will be on linear systems. Later in this chapter, we examine ways of recognizing linear and nonlinear systems.

1.4 SYSTEM DIFFERENTIAL EQUATION

Let us consider a generic system described by the ordinary differential equation of order n

$$a_0 \frac{d^n c(t)}{dt^n} + a_1 \frac{d^{n-1} c(t)}{dt^{n-1}} + \cdots + a_{n-1} \frac{dc(t)}{dt} + a_n c(t) = r(t) \tag{1.4}$$

where $c(t)$ is the output, $r(t)$ is the input, and the a_i $(i=0, 1, \ldots, n)$ are coefficients representing the system parameters. Note that the order of the differential equation is determined by the order of the highest derivative.

Equation (1.4) states that the output is related to the input through a differential equation. The system can be demonstrated to be linear. Indeed, considering the two input–output pairs, $r_1(t)$, $c_1(t)$ and $r_2(t)$, $c_2(t)$, we can write

$$a_0 \frac{d^n c_1(t)}{dt^n} + a_1 \frac{d^{n-1} c_1(t)}{dt^{n-1}} + \cdots + a_{n-1} \frac{dc_1(t)}{dt} + a_n c_1(t) = r_1(t) \tag{1.5a}$$

$$a_0 \frac{d^n c_2(t)}{dt^n} + a_1 \frac{d^{n-1} c_2(t)}{dt^{n-1}} + \cdots + a_{n-1} \frac{dc_2(t)}{dt} + a_n c_2(t) = r_2(t) \tag{1.5b}$$

Then, if we let the input have the form of the linear combination

$$r_3(t) = \alpha_1 r_1(t) + \alpha_2 r_2(t) \tag{1.6}$$

where α_1 and α_2 are two constant scalars, we can use Eqs. (1.5) to write

$$\begin{aligned}
r_3(t) &= \alpha_1 \left[a_0 \frac{d^n c_1(t)}{dt^n} + a_1 \frac{d^{n-1} c_1(t)}{dt^{n-1}} + \cdots + a_{n-1} \frac{dc_1(t)}{dt} + a_n c_1(t) \right] \\
&\quad + \alpha_2 \left[a_0 \frac{d^n c_2(t)}{dt^n} + a_1 \frac{d^{n-1} c_2(t)}{dt^{n-1}} + \cdots + a_{n-1} \frac{dc_2(t)}{dt} + a_n c_2(t) \right] \\
&= a_0 \left[\alpha_1 \frac{d^n c_1(t)}{dt^n} + \alpha_2 \frac{d^n c_2(t)}{dt^n} \right] + a_1 \left[\alpha_1 \frac{d^{n-1} c_1(t)}{dt^{n-1}} + \alpha_2 \frac{d^{n-1} c_2(t)}{dt^{n-1}} \right] + \cdots \\
&\quad + a_{n-1} \left[\alpha_1 \frac{dc_1(t)}{dt} + \alpha_2 \frac{d^n c_2(t)}{dt^n} \right] + a_n [\alpha_1 c_1(t) + \alpha_2 c_2(t)] \\
&= a_0 \frac{d^n}{dt^n} [\alpha_1 c_1(t) + \alpha_2 c_2(t)] + a_1 \frac{d^{n-1}}{dt^{n-1}} [\alpha_1 c_1(t) + \alpha_2 c_2(t)] + \cdots \\
&\quad + a_{n-1} \frac{d}{dt} [\alpha_1 c_1(t) + \alpha_2 c_2(t)] + a_n [\alpha_1 c_1(t) + \alpha_2 c_2(t)]
\end{aligned} \tag{1.7}$$

from which we conclude that the output to $r_3(t)$ is

$$c_3(t)=\alpha_1 c_1(t)+\alpha_2 c_2(t) \tag{1.8}$$

which represents a linear combination of the same form as $r_3(t)$, Eq. (1.6). Hence *the system is linear*. It can be verified that in general *a system is linear if the dependent coordinate describing the system behavior and its time derivatives appear to the first power only in the differential equation.* Note that this also rules out products of the dependent coordinate and its time derivatives. Of course, some derivatives or even the coordinate itself can be absent. The above statement is true whether the coefficients a_i are constant or time dependent. The only restriction on a_i is that they do not depend on $c(t)$.

Equation (1.4) represents an input–output relation for the system. We shall study the solution of various special cases of Eq. (1.4) later in this text. At this point, it may prove of interest to present Eq. (1.4) in the context of the block diagram of Fig. 1.1. To this end, we write Eq. (1.4) in the symbolic form

$$\left(a_0 \frac{d^n}{dt^n}+a_1 \frac{d^{n-1}}{dt^{n-1}}+\cdots+a_{n-1} \frac{d}{dt}+a_n\right) c(t)=r(t) \tag{1.9}$$

where the left side is said to have an *operator form*. To elaborate on this idea, we can write Eq. (1.9) in the more compact form

$$D(t)c(t)=r(t) \tag{1.10}$$

where

$$D(t)=a_0 \frac{d^n}{dt^n}+a_1 \frac{d^{n-1}}{dt^{n-1}}+\cdots+a_{n-1} \frac{d}{dt}+a_n \tag{1.11}$$

represents a *linear homogeneous differential operator* and contains all the dynamic characteristics of the system, i.e., all the system parameters multiplying the appropriate derivatives or unity, where unity can be regarded as the derivative of zero order. The relation between the input and output can be represented by the block diagram of Fig. 1.10, which implies that the input and output are related through a differential expression. Although the idea is interesting, Fig. 1.10 gives no clues as to how to produce a solution of the differential equation, Eq. (1.4). Later in this chapter, we shall derive analogous algebraic relations, which will go a long way toward pointing the way to a solution.

Before we proceed with the discussion of the response of systems, it may prove of interest to give an example of a nonlinear system and examine the circumstances under which this particular system can be treated as linear. To this end, let us consider the second-order system

$$a_0 \frac{d^2c(t)}{dt^2}+a_2 \sin c(t)=0 \tag{1.12}$$

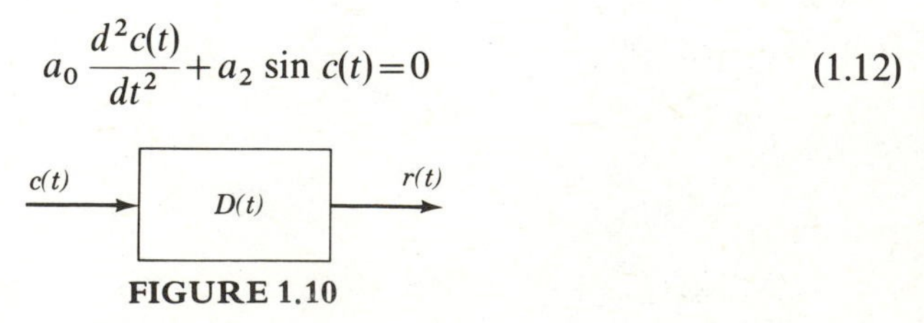

FIGURE 1.10

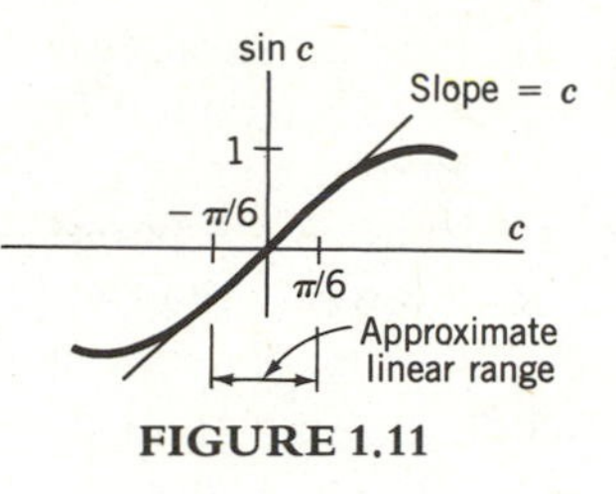

FIGURE 1.11

Equation (1.12) is nonlinear because $\sin c$ is a nonlinear function of c. For the system to be linear, the term $\sin c$ must be replaced by c. With reference to Fig. 1.11, which is a plot of the ordinary sine function, we note that there is a region in the neighborhood of the origin in which the function $\sin c$ can be replaced by the tangent to the curve, which has a slope equal to c. It can be verified that $\sin c$ can be approximated by c up to $c=\pi/6$ with an error of less than 5%. Hence, if this accuracy is sufficient, then one can regard $|c|<\pi/6$ as an *approximate linear range* for Eq. (1.12). Of course, if higher accuracy is desired, then this range must be reduced. The conclusion is that *the system described by Eq. (1.12) acts like a linear system for sufficiently small values of c*. We shall see later in this text that Eq. (1.12) describes the motion of a simple pendulum.

1.5 SYSTEM RESPONSE

To obtain the system response, more often than not one must solve a differential equation. For discrete systems, this is an ordinary differential equation, as shown in Section 1.4. Moreover, for linear time-invariant systems, the differential equation is linear with constant coefficients, where the coefficients represent the system parameters.

The solution of differential equations consists of two parts: the *homogeneous solution* and the *particular solution*. The homogeneous solution corresponds to the case in which the external excitation is zero. Hence, the homogeneous solution can be attributed entirely to initial excitation. On the other hand, the particular solution is the part of the response caused entirely by the external excitation, with the initial conditions assumed to be equal to zero. Of course, the homogeneous and the particular solutions complement one another. Indeed, for linear systems one can invoke the superposition principle and combine the homogeneous and the particular solutions linearly to obtain the total response. The above classification is somewhat artificial because initial conditions are generally produced by external factors, but the classification is helpful in the analysis of linear systems.

The nature of the response depends on the excitation, in addition to the dynamic characteristics of the system itself. In this regard, it is convenient to distinguish between *steady-state response* and *transient response*. In general, steady-state response is one in which the system achieves a certain type of equilibrium, such as a constant response or a response that repeats itself ad infinitum, without approaching zero or without growing indefinitely with time. In describing the steady-state response, time becomes an incidental factor. In fact, quite often the steady-state response can be obtained from the total response by letting t approach infinity. On the other hand, the transient response depends strongly on time. Broadly speaking, steady-state response occurs in the case of constant, harmonic, or periodic excitation, and transient response occurs in the case of initial excitation and in the case of external excitation other than the ones just mentioned. This external excitation is often called transient excitation.

The nature of the excitation affects not only the response but also the choice of methods for determining the response. Indeed, the methods for obtaining a steady-state solution are different from those for obtaining a transient solution. In the case of steady-state harmonic excitation, the indicated method of solution is harmonic balance. The response is known as frequency response, and the investigation of the response characteristics is carried out in the frequency domain. On the other hand, in the case of transient excitation, the Laplace transform method appears eminently suited. The investigation of the response to transient excitation is usually carried out in the time domain. In this text, we consider both frequency-domain and time-domain techniques.

1.6 IMPEDANCE FUNCTION AND FREQUENCY RESPONSE

We showed in Section 1.4 that the differential equation governing the behavior of a linear system can be written in the symbolic form

$$Dc(t)=r(t) \tag{1.13}$$

where D is a linear homogeneous differential operator. The input–output relation described by Eq. (1.13) was expressed schematically by the block diagram of Fig. 1.10. Equation (1.13) can be regarded as a general input–output relation, valid for any linear system. It indicates that the excitation $r(t)$ can be obtained by operating on the response $c(t)$ with a given differential operator $D=D(t)$, where $D(t)$ differs from system to system. The operator $D(t)$ contains in a compact form all the dynamic characteristics of the system.

In analysis, one is interested typically in determining the response to a given excitation, which amounts to solving the differential equation (1.13) and obtaining $c(t)$ for a given $r(t)$. This can be expressed symbolically in the form

$$c(t)=D^{-1}r(t) \tag{1.14}$$

where the operator D^{-1} can be interpreted as being the reciprocal of the operator D. The input–output relation given by Eq. (1.14) is shown schematically in Fig. 1.12.

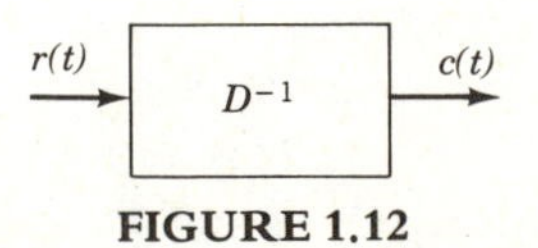

FIGURE 1.12

Because obtaining $c(t)$ from Eq. (1.13) requires integrating the equation, the operator D^{-1} represents an *integral operator*, so that Eq. (1.14) represents the integration process in a nice compact form, and the same can be said about the block diagram of Fig. 1.12. Although the block diagram of Fig. 1.12 displays the input–output relation in a more natural way than the block diagram of Fig. 1.10, we are no closer to a solution than we were before. Indeed, we merely traded a differential operator for an integral operator, and neither one of the operators is very helpful in producing a solution.

Equations (1.13) and (1.14) represent input–output relations in the time domain, and they are symbolic in nature since by themselves they do not lead to the system response. Our interest lies in an input–output relation that not only is simple and compact but also permits the derivation of the response. Once again, we wish to distinguish between the steady-state case and the transient case. In particular, we wish to distinguish between harmonic excitation and arbitrary excitation.

Let us consider Eq. (1.13) and assume that the excitation is harmonic, or

$$r(t)=r_0 \cos \omega t \tag{1.15}$$

where ω is known as the *excitation frequency*, or the *driving frequency*. Note that ω has units of radians per second (rad/s). The classical approach is to insert Eq. (1.15) into Eq. (1.13), to assume a solution in the form

$$c(t)=C_1 \sin \omega t + C_2 \cos \omega t \tag{1.16}$$

and to impose harmonic balance to determine the coefficients C_1 and C_2, i.e., to equate the coefficients of $\sin \omega t$ and $\cos \omega t$ on both sides of the equation. Clearly, the same approach can be used if the excitation is $r(t)=r_0 \sin \omega t$.

Instead of working with trigonometric functions, we choose to work with exponential functions, where the exponent is the imaginary number $i\omega t$, $i=\sqrt{-1}$. At first glance, this may appear as an unnecessary complication, as it involves use of complex algebra. It turns out, however, that the use of complex algebra simplifies the derivation of the solution. Hence, let us consider an excitation of the form

$$r(t)=r_0 e^{i\omega t} \tag{1.17}$$

Before proceeding with the formal solution, it will prove useful to examine the nature of the exponential function $e^{i\omega t}$. To this end, let us expand $e^{i\omega t}$ in the Taylor series

$$\begin{aligned} e^{i\omega t} &= 1 + i\omega t + \frac{1}{2!}(i\omega t)^2 + \frac{1}{3!}(i\omega t)^3 + \frac{1}{4!}(i\omega t)^4 + \frac{1}{5!}(i\omega t)^5 + \cdots \\ &= 1 - \frac{1}{2!}(\omega t)^2 + \frac{1}{4!}(\omega t)^4 - \cdots + i\left[\omega t - \frac{1}{3!}(\omega t)^3 + \frac{1}{5!}(\omega t)^5 - \cdots\right] \end{aligned} \tag{1.18}$$

Recognizing that

$$\begin{aligned} 1 - \frac{1}{2!}(\omega t)^2 + \frac{1}{4!}(\omega t)^4 - \cdots &= \cos \omega t \\ \omega t - \frac{1}{3!}(\omega t)^3 + \frac{1}{5!}(\omega t)^5 - \cdots &= \sin \omega t \end{aligned} \tag{1.19}$$

we conclude that

$$e^{i\omega t} = \cos \omega t + i \sin \omega t \tag{1.20}$$

Equation (1.20) can be represented in the complex plane, as shown in Fig. 1.13. Hence, $e^{i\omega t}$ can be interpreted as a complex vector of unit magnitude making an angle ωt with respect to the real axis, so that

$$\text{Re } e^{i\omega t} = \cos \omega t, \qquad \text{Im } e^{i\omega t} = \sin \omega t \tag{1.21}$$

which implies that $\cos \omega t$ and $\sin \omega t$ are the projections of $e^{i\omega t}$ on the real axis and imaginary axis, respectively. The angle ωt increases linearly with time, so that as time unfolds, the vector $e^{i\omega t}$ rotates in the complex plane with the angular velocity ω in the counterclockwise sense. Clearly, the notation of Eq. (1.17) permits us to pursue the solutions to $r_0 \cos \omega t$ and $r_0 \sin \omega t$ simultaneously.

Next, let us assume a solution of Eq. (1.13), with $r(t)$ given by Eq. (1.17), in the form

$$c(t) = C(i\omega)e^{i\omega t} \tag{1.22}$$

where $C(i\omega)$ is in general a complex expression. Observing that

$$\frac{d^r}{dt^r} e^{i\omega t} = (i\omega)^r e^{i\omega t}, \qquad r = 1, 2, \ldots, n \tag{1.23}$$

we can write

$$D(t)e^{i\omega t} = Z(i\omega)e^{i\omega t} \tag{1.24}$$

where $D(t)$ is the differential operator introduced earlier and $Z(i\omega)$ is a mere complex function. Hence, inserting Eqs. (1.17), (1.22), and (1.24) into Eq. (1.13), we obtain

$$Z(i\omega)C(i\omega)e^{i\omega t} = r_0 e^{i\omega t} \tag{1.25}$$

From Eq. (1.25), we can write

$$C(i\omega) = \frac{r_0}{Z(i\omega)} \tag{1.26}$$

In obtaining Eq. (1.26), we simply equated the coefficient of $e^{i\omega t}$ on both sides of Eq. (1.25), which is the complex version of the determination of the coefficients C_1 and C_2 in Eq. (1.16). Note, however, that the complex algebra approach yields the response not only to $r(t) = r_0 \cos \omega t$ but also to $r(t) = r_0 \sin \omega t$. Indeed, inserting

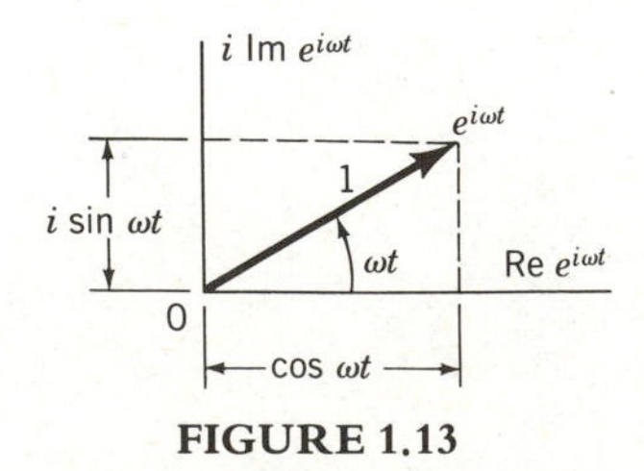

FIGURE 1.13

Eq. (1.26) into Eq. (1.22), we obtain

$$c(t)=\frac{r_0}{Z(i\omega)}\,e^{i\omega t} \tag{1.27}$$

which represents the response to the harmonic excitation in the form of Eq. (1.17). It is a steady-state response. We note that by retaining the real part of $c(t)$ in Eq. (1.27) we obtain the response to $r_0 \cos \omega t$ and that by retaining the imaginary part we obtain the response to $r_0 \sin \omega t$.

Equation (1.27) can be rewritten

$$c(t)=G(i\omega)r_0 e^{i\omega t} \tag{1.28}$$

where

$$G(i\omega)=\frac{1}{Z(i\omega)} \tag{1.29}$$

The input–output relation (1.28) for the case of harmonic excitation is displayed schematically in the block diagram of Fig. 1.14. It will prove of interest to contrast the block diagram of Fig. 1.14 with that of Fig. 1.12. Whereas in Fig. 1.12 multiplication by D^{-1} implies a symbolic operation, in Fig. 1.14 it involves an actual multiplication by $G(i\omega)$, so that Fig. 1.14 represents a genuine block diagram. The function $Z(i\omega)$ represents a complex algebraic expression containing all the information concerning the dynamic characteristics of the system, much in the same way that the differential operator $D(t)$ does. The clear advantage of $Z(i\omega)$ over $D(t)$ is that $Z(i\omega)$ is a mere algebraic expression with a clearly defined reciprocal $G(i\omega)$. Hence, the operation described in the block diagram of Fig. 1.14 is very simple, and it represents a process whereby the solution is obtained. The function $Z(i\omega)$ relating the response to the excitation for simple harmonic excitation is known as the *system impedance*, and the reciprocal $G(i\omega)$ is known as the *system admittance*, or the *frequency response*. The frequency response plays a very important role in vibrations and control.

Example 1.1

Derive the impedance and frequency response for a second-order system.

Letting $n=2$ in Eq. (1.4), the system differential equation is

$$a_0\frac{d^2c(t)}{dt^2}+a_1\frac{dc(t)}{dt}+a_2c(t)=r(t) \tag{a}$$

so that, from Eq. (1.11), the differential operator $D(t)$ is

$$D(t)=a_0\frac{d^2}{dt^2}+a_1\frac{d}{dt}+a_2 \tag{b}$$

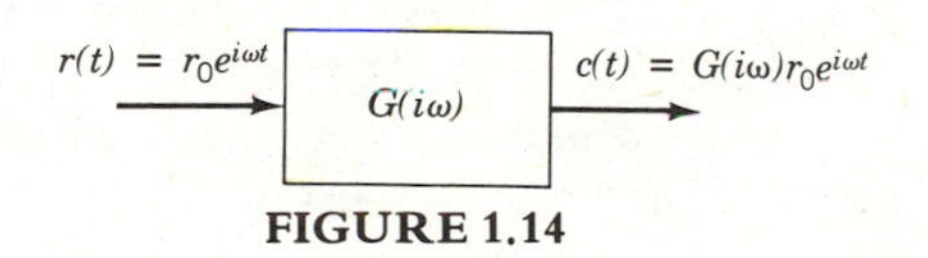

FIGURE 1.14

Hence, using Eqs. (1.23), we can write

$$D(t)e^{i\omega t}=\left(a_0\frac{d^2}{dt^2}+a_1\frac{d}{dt}+a_2\right)e^{i\omega t}=[a_0(i\omega)^2+a_1 i\omega+a_2]e^{i\omega t} \tag{c}$$

Comparing Eq. (c) with Eq. (1.24), we conclude that the impedance function is

$$Z(i\omega)=a_0(i\omega)^2+a_1 i\omega+a_2 \tag{d}$$

From Eq. (1.29), the frequency response is simply

$$G(i\omega)=\frac{1}{Z(i\omega)}=\frac{1}{a_0(i\omega)^2+a_1 i\omega+a_2} \tag{e}$$

1.7 TRANSFER FUNCTION

The simple relation between input and output expressed by Eq. (1.28) is valid only for the case of harmonic excitation. The question arises whether such a simple and compact input–output relation also exists in the more general case of transient excitation. The answer is a qualified yes, in the sense that such a relation does indeed exist, but it is not in the time domain. It is in a complex domain, known as the Laplace domain, and it can be derived by means of the Laplace transformation method (see Appendix).

Let us define the Laplace transform of the input by

$$R(s)=\mathscr{L}r(t)=\int_0^\infty e^{-st}r(t)\,dt \tag{1.30}$$

and the Laplace transform of the output by

$$C(s)=\mathscr{L}c(t)=\int_0^\infty e^{-st}c(t)\,dt \tag{1.31}$$

where s is a subsidiary variable, generally a complex quantity. The variable s defines the Laplace domain mentioned above. We shall show later that, Laplace transforming both sides of Eq. (1.13), we obtain a compact input–output relation of the type

$$C(s)=G(s)R(s) \tag{1.32}$$

where $G(s)$ is an algebraic expression known as the *system transfer function.* One can think of $G(s)$ as the Laplace transform of $D^{-1}(t)$, but it is perhaps better to think of $G(s)$ as the reciprocal of the Laplace transform of $D(t)$. The relation between the transformed input $R(s)$ and the transformed output $C(s)$ is shown schematically in Fig. 1.15. But, unlike Fig. 1.12, which expresses a mere symbolic relation, Fig. 1.15 simply states that *the transformed output* $C(s)$ *can be obtained by multiplying the transformed input* $R(s)$ *by the transfer function* $G(s)$, so that Fig. 1.15 represents a genuine block diagram in a manner analogous to Fig. 1.14. Hence, by working in the s-domain instead of the time domain, the block diagram is no longer a

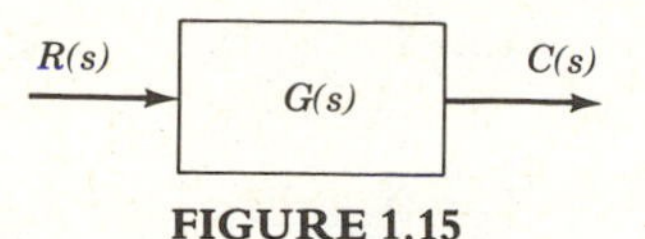

FIGURE 1.15

symbolic way of describing the input–output relation, but is now one with significant practical applications as far as the solution of the system differential equation is concerned.

The system output is obtained by performing an inverse transformation from the Laplace domain to the time domain. Considering Eq. (1.32), we can write the inversion process in the form

$$c(t)=\mathscr{L}^{-1}C(s)=\mathscr{L}^{-1}G(s)R(s) \tag{1.33}$$

The inverse transformation $\mathscr{L}^{-1}$ is defined as a line integral in the complex s-plane. In many applications, however, line integrations are not really necessary, and the inverse of a function can be found readily in tables of Laplace transform pairs, such as the one at the end of the Appendix (Section A.7). In other applications, the function itself cannot be found in tables, but it can be decomposed into a linear combination of simpler functions whose transforms are in tables. This decomposition can be carried out by the method of partial fractions, as discussed in the Appendix.

There are several functions of particular importance in vibrations and in control. In vibrations they are important because they can be used to synthesize a large variety of functions, thus permitting the calculation of the response to complicated excitations with relative ease. In control they are important because they can be used as reference inputs, permitting the evaluation of the performance of control designs. These functions are the unit impulse, the unit step function, and the unit ramp function, and they belong to the class of singularity functions. We discuss singularity functions in the next section and the response to singularity functions in subsequent sections.

Example 1.2

Derive the transfer function of the system of Example 1.1. Then establish the relation between the transfer function and the frequency response function.

To obtain the transfer function, we must first take the Laplace transform Eq. (a) of Example 1.1. From Section A.2, we conclude that for zero initial conditions the Laplace transform of the rth derivative of $c(t)$ is

$$\mathscr{L}\frac{d^r c(t)}{dt^r}=\int_0^\infty e^{-st}\frac{d^r c(t)}{dt^r}\,dt=s^r C(s), \qquad r=1,2,\ldots,n \tag{a}$$

where $C(s)$ is the Laplace transform of $c(t)$, $C(s)=\mathscr{L}c(t)$. Transforming both sides of Eq. (a) of Example 1.1, we obtain

$$(a_0 s^2+a_1 s+a_2)C(s)=R(s) \tag{b}$$

where $R(s)$ is the Laplace transform of $r(t)$, $R(s)=\mathscr{L}r(t)$. From Eq. (b), we can write

$$C(s)=\frac{1}{a_0s^2+a_1s+a_2}R(s) \tag{c}$$

so that, comparing Eq. (c) with Eq. (1.32), we conclude that the desired transfer function is

$$G(s)=\frac{1}{a_0s^2+a_1s+a_2} \tag{d}$$

To establish the relation between the transfer function and the frequency response function, we compare Eq. (d) with Eq. (e) of Example 1.1. The obvious conclusion is that *the transfer function is simply the frequency response with iω replaced by s, and vice versa.* This statement is valid for any linear system with constant coefficients.

1.8 SINGULARITY FUNCTIONS

The singularity functions form the basis for the time-domain analysis of linear systems. The class of singularity functions is characterized by the fact that every singularity function and all its derivatives are continuous functions of time, except at a given value of time. They are also characterized by the fact that the singularity functions can be obtained from one another by successive integration or differentiation.

One of the most important and commonly encountered singularity function is the *unit impulse*, or the *Dirac delta function.* The mathematical definition of the unit impulse is

$$\delta(t-a)=0, \qquad t\neq a \tag{1.34a}$$

$$\int_{-\infty}^{\infty}\delta(t-a)\,dt=1 \tag{1.34b}$$

The function is displayed in Fig. 1.16. From Fig. 1.16, we conclude that the Dirac delta function can be regarded as the result of a limiting process. In particular, the function is zero everywhere, except over a small time interval ε in the neighborhood of $t=a$. Over that interval the amplitude is very large and equal to $1/\varepsilon$. As $\varepsilon\to 0$, the

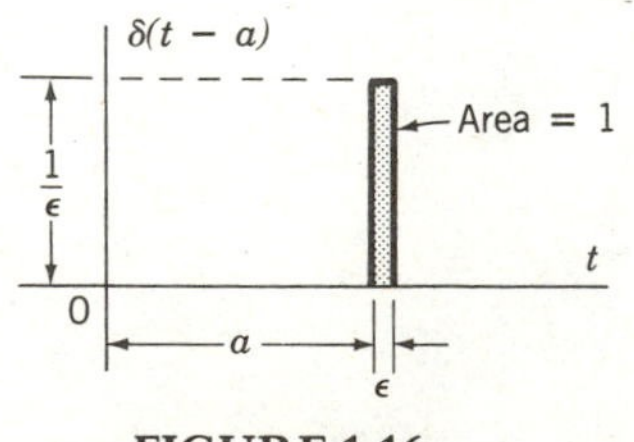

FIGURE 1.16

time interval approaches zero and the amplitude tends to infinity, but in such a way that the area under the curve remains constant and equal to unity, which explains the term "unit impulse." It is clear that the units of $\delta(t-a)$ are inverse seconds (s^{-1}).

It is obvious that the unit impulse can be used to represent impulsive excitations. What may not be so obvious is that arbitrary excitations can be regarded as superpositions of impulses, a fact of considerable importance in vibrations and in control. This subject is discussed in Section 1.11.

The Dirac delta function possesses a property that is extremely useful in the evaluation of integrals involving the delta function. Indeed, invoking the mean-value theorem and considering Eq. (1.34b), we can write

$$\int_{-\infty}^{\infty} f(t)\,\delta(t-a)\,dt = f(a)\int_{-\infty}^{\infty} \delta(t-a)\,dt = f(a) \tag{1.35}$$

so that the integral of a function $f(t)$ weighted by the Dirac delta function acting at $t=a$ is simply the function $f(t)$ evaluated at $t=a$. The above property is sometimes referred to as the *sampling property*. Note that the limits of integration need not be $-\infty$ and ∞, but they must be such that they bracket the point $t=a$.

Perhaps of equal importance to the unit impulse is the *unit step function*, defined mathematically as

$$u(t-a) = \begin{cases} 0 & \text{for} \quad t<a \\ 1 & \text{for} \quad t>a \end{cases} \tag{1.36}$$

We note that the unit step function is dimensionless. The unit step function is depicted in Fig. 1.17. Consistent with a statement made in the beginning of this section, the unit step function is closely related to the unit impulse. Indeed, it is easy to verify that

$$u(t-a) = \int_{-\infty}^{t} \delta(\zeta-a)\,d\zeta \tag{1.37}$$

or the unit step function is the integral of the unit impulse. Conversely,

$$\delta(t-a) = \frac{du(t-a)}{dt} \tag{1.38}$$

or the unit impulse is the time derivative of the unit step function.

The effect of multiplying a function $f(t)$ by $u(t-a)$ is to annihilate the portion of $f(t)$ corresponding to $t<a$ and to leave unaffected the portion corresponding to $t>a$. Figure 1.18 shows an example in which $f(t)=\sin\omega t$ and $a=0$. The property

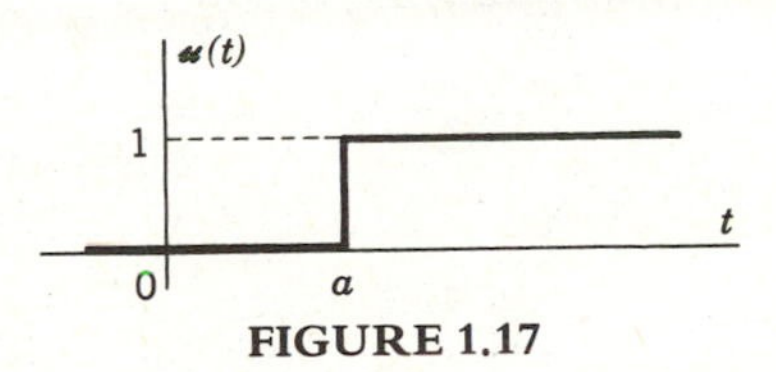

FIGURE 1.17

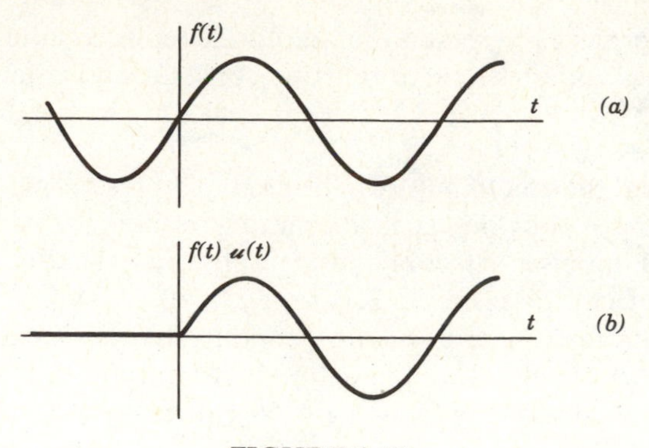

FIGURE 1.18

of the unit step function just described can be used on occasions to simplify certain expressions. For example, the function shown in Fig. 1.18*b* can be described mathematically as

$$f(t)=\begin{cases}\sin \omega t & \text{for} \quad t>0 \\ 0 & \text{for} \quad t<0\end{cases} \tag{1.39}$$

or in the more compact form

$$f(t)=\sin \omega t\, u(t) \tag{1.40}$$

The unit step function is also useful in synthesizing given functions. As an example, the rectangular pulse displayed in Fig. 1.19 can be expressed

$$f(t)=f_0\left[u\left(t+\frac{T}{2}\right)-u\left(t-\frac{T}{2}\right)\right] \tag{1.41}$$

Another function of interest is the *unit ramp function*, defined as

$$r(t-a)=\begin{cases}t-a & \text{for} \quad t>a \\ 0 & \text{for} \quad t<a\end{cases} \tag{1.42}$$

or more compactly as

$$r(t-a)=(t-a)u(t-a) \tag{1.43}$$

The unit of the unit ramp function is the second (s). The unit ramp function is

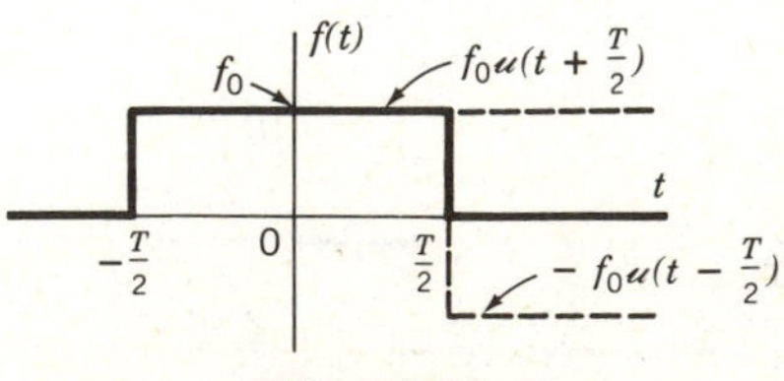

FIGURE 1.19

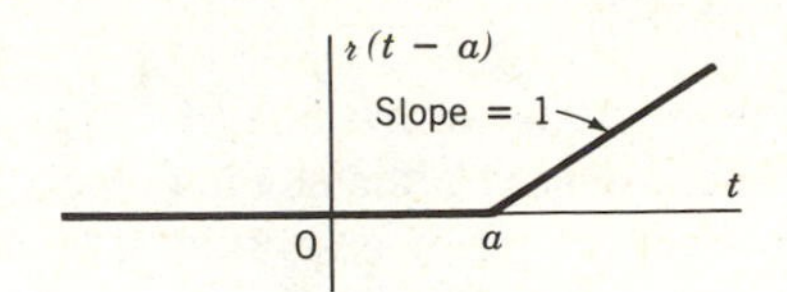

FIGURE 1.20

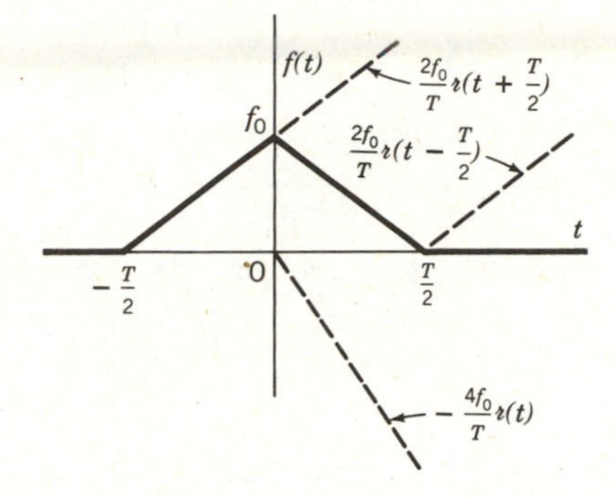

FIGURE 1.21

exhibited in Fig. 1.20. Clearly, the unit ramp function is the integral of the unit step function, or

$$r(t-a)=\int_{-\infty}^{t} u(\zeta-a)\,d\zeta \tag{1.44}$$

and, conversely, the unit step function is the time derivative of the unit ramp function

$$u(t-a)=\frac{dr(t-a)}{dt} \tag{1.45}$$

As with the unit step function, the unit ramp function can also be used to synthesize certain functions. For example, the triangular pulse shown in Fig. 1.21 can be expressed mathematically in the compact form

$$f(t)=\frac{2f_0}{T}\left[r\left(t+\frac{T}{2}\right)-2r(t)+r\left(t-\frac{T}{2}\right)\right] \tag{1.46}$$

The response to singularity functions is of considerable interest in vibrations and in control. In general, one would expect that the discontinuous nature of some of the singularity functions is likely to cause difficulties in the derivation of the system response. However, no particular difficulty is encountered in deriving the response by the Laplace transform method. We shall have ample opportunity throughout this text to verify this statement.

1.9 IMPULSE RESPONSE

An important concept in linear system analysis is the impulse response. Indeed, inputs in the form of an impulse are used often as reference inputs in vibrations to check the system response. Perhaps more important is the fact that arbitrary functions can be synthesized by means of impulses of varying amplitude applied at different times.

The impulse response $g(t)$ is defined as the response of a system to a unit impulse applied at $t=0$, with the initial conditions being equal to zero.* The block diagram of Fig. 1.22 shows the relation between the unit impulse and the impulse response in schematic form. The impulse response contains in integrated form the same information as that contained by the differential operator $D(t)$ defining the system, or that contained by the transfer function $G(s)$. Hence, we must conclude that the impulse response $g(t)$ and the transfer function $G(s)$ are related. Of course, we recognize that $g(t)$ is in the time domain and $G(s)$ in the s-domain. To establish the relation between $g(t)$ and $G(s)$, we consider the Laplace transform of the unit impulse applied at $t=0$ as follows:

$$\Delta(s)=\mathscr{L}\,\delta(t)=\int_0^\infty e^{-st}\,\delta(t)\,dt=e^{-st}\Big|_{t=0}\int_0^\infty \delta(t)\,dt=1 \tag{1.47}$$

where use has been made of Eqs. (1.34b) and (1.35). Letting $c(t)=g(t)$ and $R(s)=\Delta(s)=1$ in Eq. (1.33), we obtain simply

$$g(t)=\mathscr{L}^{-1}G(s) \tag{1.48}$$

or, *the impulse response is equal to the inverse Laplace transform of the transfer function.* Hence, *the impulse response and the transfer function represent a Laplace transform pair.* This provides ample justification for denoting the impulse response by $g(t)$ instead of $h(t)$.

Example 1.3

Derive the impulse response for the system of Example 1.1 for the case in which $a_1=0$, $a_2=a_0\omega^2$.

The transfer function for the system was derived in Example 1.2. Using the given

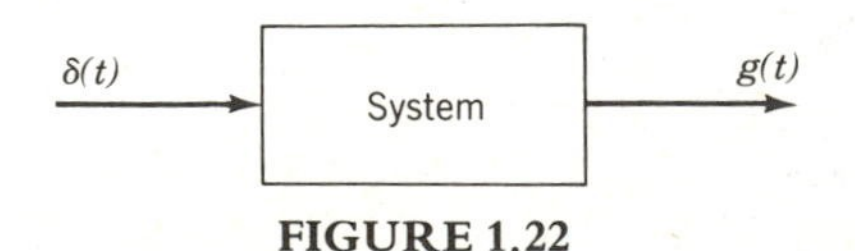

FIGURE 1.22

*In the technical literature, the impulse response is denoted frequently by $h(t)$. The notation used here has many advantages over the traditional one.

values of the coefficients, we can write

$$G(s)=\frac{1}{a_0(s^2+\omega^2)} \tag{a}$$

Hence, from the table of Laplace transforms in Section A.7, we obtain simply

$$g(t)=\mathscr{L}^{-1}\frac{1}{a_0(s^2+\omega^2)}=\frac{1}{a_0\omega}\sin\omega t \tag{b}$$

and, because by definition $g(t)$ is zero for $t<0$, the impulse response should be written more correctly in the form

$$g(t)=\frac{1}{a_0\omega}\sin\omega t\, u(t) \tag{c}$$

1.10 STEP RESPONSE

Another important concept in linear system analysis is the step response. The unit step function is used frequently in control as a reference input for the purpose of evaluating the system performance. Step functions of infinitesimal amplitude and different starting times can be used to synthesize arbitrary functions, but this fact will not be used in this text.

The step response $\mathscr{s}(t)$ is defined as the response of a system to a unit step function applied at $t=0$, with the initial conditions being equal to zero. The corresponding block diagram is shown in Fig. 1.23. To derive the step response, we make use of Eq. (1.33) once again. To this end, we calculate the Laplace transform of the unit step function applied at $t=0$ as follows:

$$U(s)=\mathscr{L}u(t)=\int_0^\infty e^{-st}u(t)\,dt=\int_0^\infty e^{-st}\,dt=\left.\frac{e^{-st}}{-s}\right|_0^\infty=\frac{1}{s} \tag{1.49}$$

Hence, letting $c(t)=\mathscr{s}(t)$ and $R(s)=U(s)$ in Eq. (1.33) and using Eq. (1.49), we can write

$$\mathscr{s}(t)=\mathscr{L}^{-1}\frac{G(s)}{s} \tag{1.50}$$

Example 1.4

Derive the step response for the first-order system

$$a_0\frac{dc(t)}{dt}+a_1c(t)=r(t) \tag{a}$$

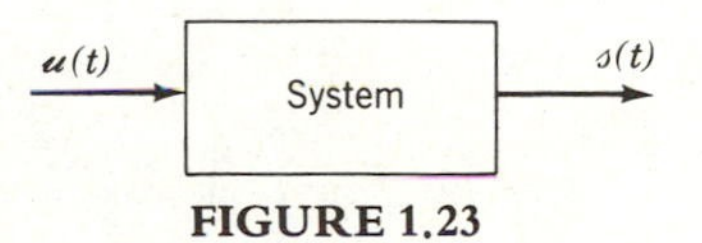

FIGURE 1.23

It is easy to verify that the transfer function is

$$G(s)=\frac{1}{a_0}\frac{1}{s+\alpha}, \qquad \alpha=\frac{a_1}{a_0} \tag{b}$$

so that, using Eq. (1.50), the step response is

$$\mathscr{s}(t)=\frac{1}{a_0}\mathscr{L}^{-1}\frac{1}{s(s+\alpha)} \tag{c}$$

Next, consider the partial fractions expansion

$$\frac{1}{s(s+\alpha)}=\frac{1}{\alpha}\left(\frac{1}{s}-\frac{1}{s+\alpha}\right) \tag{d}$$

Inserting Eq. (d) into Eq. (c) and recalling the second of Eqs. (b), we can write

$$\mathscr{s}(t)=\frac{1}{a_1}\mathscr{L}^{-1}\left(\frac{1}{s}-\frac{1}{s+\alpha}\right) \tag{e}$$

But, from Section A.7,

$$\mathscr{L}^{-1}\frac{1}{s}=\mathscr{u}(t), \qquad \mathscr{L}^{-1}\frac{1}{s+\alpha}=e^{-\alpha t}\mathscr{u}(t) \tag{f}$$

so that, introducing Eqs. (f) into Eq. (e), we obtain the step response

$$\mathscr{s}(t)=\frac{1}{a_1}(1-e^{-\alpha t})\mathscr{u}(t) \tag{g}$$

Example 1.5

Obtain the response of the system of Example 1.4 to an excitation in the form of a rectangular pulse similar to that shown in Fig. 1.19.

From Eq. (1.41), we can write the input in the form

$$r(t)=r_0\left[\mathscr{u}\left(t+\frac{T}{2}\right)-\mathscr{u}\left(t-\frac{T}{2}\right)\right] \tag{a}$$

where $\mathscr{u}(t+T/2)$ is a unit step function applied at $t=-T/2$ and $\mathscr{u}(t-T/2)$ is a unit step function applied at $t=T/2$. Because the input is the difference between two step functions, the output must be the difference between two corresponding step responses. Hence, using Eq. (g) of Example 1.4, we can write the response

$$\begin{aligned}c(t)&=r_0\left[\mathscr{s}\left(t+\frac{T}{2}\right)-\mathscr{s}\left(t-\frac{T}{2}\right)\right]\\&=\frac{r_0}{a_1}\left\{[1-e^{-\alpha(t+T/2)}]\mathscr{u}\left(t+\frac{T}{2}\right)-[1-e^{-\alpha(t-T/2)}]\mathscr{u}\left(t-\frac{T}{2}\right)\right\}\end{aligned} \tag{b}$$

1.11 RESPONSE TO ARBITRARY EXCITATION. THE CONVOLUTION INTEGRAL

In Sections 1.9 and 1.10, we showed how to obtain the response to impulse and step functions or simple combinations of step functions. The question remains as to how to produce the response to arbitrary excitations. We propose to address this question in two ways. The first approach consists of regarding the arbitrary excitation as a superposition of impulses of varying amplitude and time of application. The second approach is simply a follow-up on the Laplace inversion process given by Eq. (1.33). Of course, both approaches lead to the same result.

Figure 1.24 shows an arbitrary input $r(t)$. The shaded thin rectangle represents an impulse of magnitude $r(\tau)\,\Delta\tau$ applied at $t=\tau$. Hence, the input $r(t)$ can be regarded as a superposition of such impulses, or

$$r(t)=\sum r(\tau)\,\Delta\tau\,\delta(t-\tau) \tag{1.51}$$

But, for linear systems, if the magnitude of the input is increased by a given amount, the magnitude of the output increases in the same proportion. Hence, with reference to the block diagram of Fig. 1.22, the response to the impulse $r(\tau)\,\Delta\tau\,\delta(t-\tau)$ is

$$\Delta c(t,\tau)\cong r(\tau)\,\Delta\tau\,g(t-\tau) \tag{1.52}$$

where $g(t-\tau)$ is the impulse response shifted by the time interval $t=\tau$. This input–output relation is shown schematically in the block diagram of Fig. 1.25. According to Eq. (1.51), the input can be represented by a linear combination of impulses. It follows that, for linear systems, the output can be represented by a similar linear combination of impulse responses. Hence, using Eq. (1.52), we can write the response to $r(t)$ in the form

$$c(t)\cong\sum \Delta c(t,\tau)=\sum r(\tau)\,\Delta\tau\,g(t-\tau) \tag{1.53}$$

In the limit, as $\Delta\tau\to 0$, the summation process becomes an integration process,

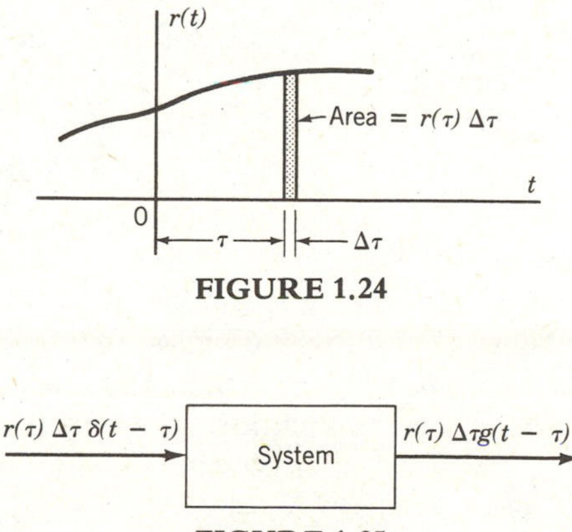

FIGURE 1.24

FIGURE 1.25

so that

$$c(t)=\int_{-\infty}^{\infty} r(\tau)g(t-\tau)\,d\tau \tag{1.54}$$

The right side of Eq. (1.54) is known as the *convolution integral*, or the *superposition integral*.

According to the integral in Eq. (1.54), the input is a function of τ and the impulse response is a function of $t-\tau$; that is, the impulse response is shifted. It turns out that the same result is obtained if the input is shifted instead of the impulse response. To show this, let us consider the change of variables

$$t-\tau=\sigma, \qquad \tau=t-\sigma, \qquad d\tau=-d\sigma \tag{1.55}$$

and the corresponding change in the integration limits

$$\tau=-\infty\rightarrow\sigma=\infty, \qquad \tau=\infty\rightarrow\sigma=-\infty \tag{1.56}$$

Inserting Eqs. (1.55) and (1.56) into Eq. (1.54), we can write

$$c(t)=\int_{\infty}^{-\infty} r(t-\sigma)g(\sigma)(-d\sigma)=\int_{-\infty}^{\infty} r(t-\sigma)g(\sigma)\,d\sigma \tag{1.57}$$

But, because σ is merely a dummy variable, we can replace it by τ, so that Eqs. (1.54) and (1.57) can be combined into

$$c(t)=\int_{-\infty}^{\infty} r(\tau)g(t-\tau)\,d\tau=\int_{-\infty}^{\infty} r(t-\tau)g(\tau)\,d\tau \tag{1.58}$$

Hence, the convolution integral is symmetric in $r(t)$ and $g(t)$ in the sense that it does not matter which of the two functions is shifted.

The convolution integrals in Eq. (1.58) have infinite limits, which can cause difficulties in the evaluation of the integrals. Such difficulties can be obviated if the infinite limits can be replaced by finite ones. In this regard, we note that in general the interest lies in inputs $r(t)$ defined only for $t>0$; that is, $r(t)=0$ for $t<0$. Moreover, we observe that $g(t)$ is zero for negative argument. Hence, considering inputs defined for $t>0$ only, we can adjust the limits of integration in Eq. (1.58) so that the convolution integrals become

$$c(t)=\int_{0}^{t} r(\tau)g(t-\tau)\,d\tau=\int_{0}^{t} r(t-\tau)g(\tau)\,d\tau \tag{1.59}$$

The question remains as to which of the two functions to shift, $r(\tau)$ or $g(\tau)$. As a rule of thumb, it is advisable to shift the simpler of the two functions.

The approach to the derivation of the convolution integral described above is based on physical considerations to some extent. The second approach is entirely mathematical and based on the convolution theorem (Section A.6). According to the convolution theorem, if a Laplace transform $F(s)$ is the product of two Laplace transforms $F_1(s)$ and $F_2(s)$, i.e., if

$$F(s)=F_1(s)F_2(s) \tag{1.60}$$

then the inverse transformation of $F(s)$ has the form

$$f(t)=\mathscr{L}^{-1}F(s)=\mathscr{L}^{-1}F_1(s)F_2(s)$$
$$=f_1(t)*f_2(t)=\int_0^t f_1(\tau)f_2(t-\tau)\,d\tau=\int_0^t f_1(t-\tau)f_2(\tau)\,d\tau \tag{1.61}$$

where

$$f_1(t)=\mathscr{L}^{-1}F_1(s), \qquad f_2(t)=\mathscr{L}^{-1}F_2(s) \tag{1.62}$$

are the inverse Laplace transforms of $F_1(s)$ and $F_2(s)$, respectively. The notation $f_1(t)*f_2(t)$ is merely symbolic and implies not the product but the so-called convolution of the two functions $f_1(t)$ and $f_2(t)$. The integrals on the right side of Eq. (1.61) can be identified as convolution integrals.

But, according to Eq. (1.32), the transformed output $C(s)$ can be written in the form of the product

$$C(s)=G(s)R(s) \tag{1.63}$$

where $G(s)$ is the transfer function, which is equal to the transformed impulse response, and $R(s)$ is the transformed input. Hence, using the convolution theorem, Eq. (1.61), we find that the output has the form

$$c(t)=g(t)*r(t)=\int_0^t g(\tau)r(t-\tau)\,d\tau=\int_0^t g(t-\tau)r(\tau)\,d\tau \tag{1.64}$$

in which

$$c(t)=\mathscr{L}^{-1}C(s), \qquad g(t)=\mathscr{L}^{-1}G(s), \qquad r(t)=\mathscr{L}^{-1}R(s) \tag{1.65}$$

are the output, impulse response, and input, respectively. Equation (1.64) is clearly identical to Eq. (1.59).

Example 1.6

Show that the step response is the integral of the impulse response.

We shall prove the proposition by means of the convolution integral. Letting $r(t)=\mathscr{u}(t)$ and $c(t)=\mathscr{s}(t)$, we can use either Eq. (1.59) or Eq. (1.64) to write the step response in the form

$$\mathscr{s}(t)=\int_0^t g(\tau)\mathscr{u}(t-\tau)\,d\tau \tag{a}$$

But, by the definition of the unit step function,

$$\mathscr{u}(t-\tau)=\begin{cases}1 & \text{for } t>\tau \\ 0 & \text{for } t<\tau\end{cases} \tag{b}$$

Reading the inequalities in Eq. (b) in reversed order, we conclude that $\mathscr{u}(t-\tau)$ is equal to unity over the interval of integration and equal to zero for $\tau>t$. Hence $\mathscr{u}(t-\tau)$ can be omitted from the integrand without affecting the value of the integral, so that

$$\mathscr{s}(t) = \int_0^t g(\tau)\, d\tau \tag{c}$$

which is the desired result.

1.12 STATE EQUATIONS

In Section 1.4, we expressed the relation between the input and output in the form of an nth-order linear differential equation, Eq. (1.4). Moreover, in Section 1.11 we derived the solution of such a linear differential equation for any arbitrary input in terms of a convolution integral. The convolution integral involves the impulse response, and for relatively large n the impulse response may be difficult to obtain. Another approach to the solution of Eq. (1.4) is to transform it into a set of n simultaneous first-order differential equations. The solution of the set of first-order equations also involves a convolution integral, but it does not require the impulse response.

To derive the set of first-order equations, let us consider the transformation

$$\begin{aligned}
c(t) &= x_1(t) \\
\frac{dc(t)}{dt} &= \frac{dx_1(t)}{dt} = x_2(t) \\
\frac{d^2c(t)}{dt^2} &= \frac{dx_2(t)}{dt} = x_3(t) \\
&\vdots \\
\frac{d^{n-1}c(t)}{dt^{n-1}} &= \frac{dx_{n-1}(t)}{dt} = x_n(t) \\
\frac{d^n c(t)}{dt^n} &= \frac{dx_n(t)}{dt}
\end{aligned} \tag{1.66}$$

and rewrite Eq. (1.4) in the form

$$\frac{d^n c(t)}{dt^n} = -\frac{a_1}{a_0}\frac{d^{n-1}c(t)}{dt^{n-1}} - \frac{a_2}{a_0}\frac{d^{n-2}c(t)}{dt^{n-2}} - \cdots - \frac{a_{n-1}}{a_0}\frac{dc(t)}{dt} - \frac{a_n}{a_0}c(t) + \frac{1}{a_0}r(t) \tag{1.67}$$

Combining Eqs. (1.66) and (1.67), we obtain the desired set of first-order equations

$$\begin{aligned}
\dot{x}_1(t) &= x_2(t) \\
\dot{x}_2(t) &= x_3(t) \\
&\vdots \\
\dot{x}_{n-1}(t) &= x_n(t) \\
\dot{x}_n(t) &= -\frac{a_n}{a_0}x_1(t) - \frac{a_{n-1}}{a_0}x_2(t) - \cdots - \frac{a_2}{a_0}x_{n-1}(t) - \frac{a_1}{a_0}x_n(t) + \frac{1}{a_0}r(t)
\end{aligned} \tag{1.68}$$

where the overdots denote derivatives with respect to time. The variables $x_1(t)$, $x_2(t), \ldots, x_n(t)$ are known as *state variables* and Eqs. (1.68) as *state equations.*

The state equations can be expressed in matrix form. To this end, we introduce the n-dimensional *state vector*

$$\mathbf{x}(t) = [x_1(t) \quad x_2(t) \quad \cdots \quad x_n(t)]^{\mathrm{T}} \tag{1.69}$$

where the superscript T denotes the transpose of the quantity in the brackets, as well as the $n \times n$ coefficient matrix

$$A = \begin{bmatrix} 0 & 1 & 0 & 0 & \cdots & 0 & 0 \\ 0 & 0 & 1 & 0 & \cdots & 0 & 0 \\ 0 & 0 & 0 & 1 & \cdots & 0 & 0 \\ \vdots & \vdots & \vdots & \vdots & & \vdots & \vdots \\ 0 & 0 & 0 & 0 & \cdots & 0 & 1 \\ -\dfrac{a_n}{a_0} & -\dfrac{a_{n-1}}{a_0} & -\dfrac{a_{n-2}}{a_0} & -\dfrac{a_{n-3}}{a_0} & \cdots & -\dfrac{a_2}{a_0} & -\dfrac{a_1}{a_0} \end{bmatrix} \tag{1.70}$$

and the n-dimensional coefficient vector

$$\mathbf{b} = [0 \quad 0 \quad 0 \quad \cdots \quad 0 \quad 1/a_0]^{\mathrm{T}} \tag{1.71}$$

Then, Eqs. (1.68) can be written in the compact form

$$\dot{\mathbf{x}}(t) = A\mathbf{x}(t) + \mathbf{b}r(t) \tag{1.72}$$

Equations of this type are encountered frequently in control. The solution of such equations is discussed in Chapter 10.

PROBLEMS

1.1 Give one example of each of the following: (1) an uncontrolled system (2) an open-loop controlled system, and (3) a closed-loop controlled system.

1.2 Determine whether the system described by the differential equation

$$t^2 \frac{d^2c}{dt^2} + t\frac{dc}{dt} + (t^2 - m^2)c = 0$$

where m is an integer, is linear or nonlinear.

1.3 Repeat Problem 1.2 for the differential equation

$$\frac{d^2c}{dt^2} + (a - bc^2)c = 0$$

where a and b are constant scalars.

1.4 Repeat Problem 1.2 for the differential equation

$$\frac{d^2c}{dt^2}+(a+b\cos 2t)c=0$$

where a and b are constant scalars.

1.5 Repeat Problem 1.2 for the differential equation

$$\frac{d^2c}{dt^2}+(a+b\cos 2c)c=0$$

where a and b are constant scalars.

1.6 A system is described by the differential equation

$$\frac{dc}{dt}+2c=3\cos 2t$$

Determine the following: (1) the system impedance, (2) the frequency response, and (3) the steady-state response.

1.7 Derive the transfer function for the system described by the differential equation

$$\frac{d^2c}{dt^2}+3\frac{dc}{dt}+2c=\frac{dr}{dt}+3r$$

1.8 Derive the impulse response of the system described by the differential equation

$$\frac{d^2c}{dt^2}+3\frac{dc}{dt}+2c=r$$

1.9 Derive the step response of the system of Problem 1.8.

1.10 The ramp response is defined as the response of a system to a unit ramp function applied at $t=0$, with the initial conditions being equal to zero. Derive the ramp response of the system of Example 1.3 by using the Laplace transformation method.

1.11 Solve Problem 1.10 by the convolution integral.

1.12 Derive the response of the system of Example 1.3 to the trapezoidal pulse shown in Fig. 1.26.

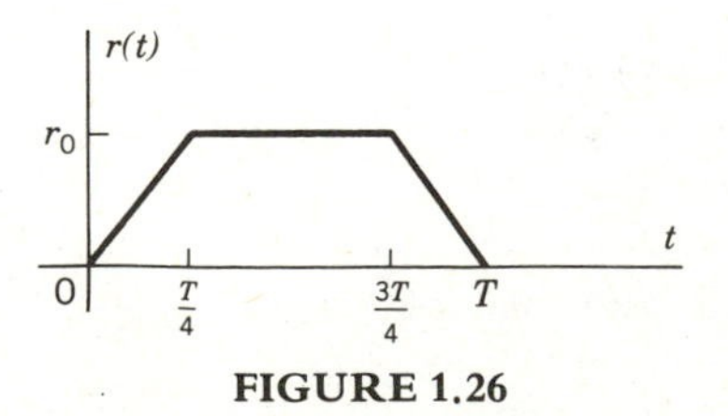

FIGURE 1.26

1.13 Derive the state equations for the system described by

$$\frac{d^3c}{dt^3}+6\frac{d^2c}{dt^2}+11\frac{dc}{dt}+6c=r$$

and then express the equations in matrix form.

CHAPTER 2
Kinematics

2.1 INTRODUCTION

The motion of a mechanical system is governed by Newton's second law, which relates the forces acting on the system to its acceleration. A description of the acceleration requires a reference frame. In applying Newton's second law, it is necessary to use an inertial reference frame, that is, a reference frame fixed in an inertial space (Section 3.1). Quite often, however, it is more convenient to describe the motion in terms of moving reference frames. This is true in the case of rotating bodies, in which case rotating reference frames are more suitable than inertial frames.

It becomes clear from the above brief discussion that the description of the motion requires more effort than the description of the forces. In fact, the description of the motion represents a study in itself. The study of the motion of a body without regard to the forces causing the motion is known as *kinematics*. It is perhaps appropriate to think of kinematics as the geometry of motion. Because of its importance in the derivation of the differential equations governing the motion of mechanical systems by Newtonian mechanics, this chapter is devoted entirely to kinematics. The material is fundamental to the study of the dynamics of systems of particles and rigid bodies.

2.2 MOTION RELATIVE TO A FIXED REFERENCE FRAME. CARTESIAN COMPONENTS

Let us consider the motion of a particle P along curve C in a three-dimensional space (Fig. 2.1). To describe the motion, it is convenient to introduce a reference frame in the form of a rectangular set of axes xyz with the origin at the fixed point 0.

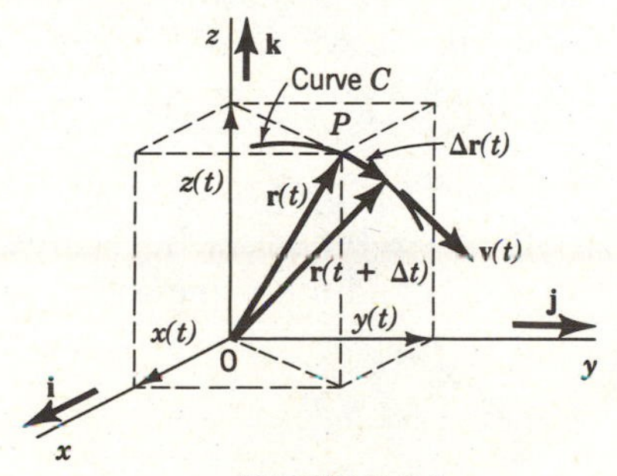

FIGURE 2.1

The axes xyz are assumed to maintain a fixed orientation in space. The position of the particle P in space is defined at any time t by the three *cartesian coordinates* $x(t)$, $y(t)$, and $z(t)$. These coordinates can be regarded as the components of the *position vector*, or *radius vector* $\mathbf{r}(t)$. An expression relating $\mathbf{r}(t)$ and its components can be provided by introducing the vectors $\mathbf{i}$, $\mathbf{j}$, and $\mathbf{k}$, where the vectors have unit magnitude and directions along the axes x, y, and z, respectively. Such vectors are referred to as *unit vectors*. Because both the magnitude and direction of each of the vectors are constant, these are constant unit vectors. The position vector $\mathbf{r}(t)$ can then be expressed in terms of the components $x(t)$, $y(t)$, $z(t)$ and the unit vectors $\mathbf{i}$, $\mathbf{j}$, $\mathbf{k}$ as follows:

$$\mathbf{r}(t) = x(t)\mathbf{i} + y(t)\mathbf{j} + z(t)\mathbf{k} \tag{2.1}$$

and we note that Eq. (2.1) is merely a convenient way of writing the three components of the position vector simultaneously. The position has units of length, which in SI units* is the meter (m).

The *velocity* of the particle P in space is defined as the time rate of change of the position. The velocity has in general three components also, so that it too can be regarded as a vector. Denoting the *velocity vector* of P by $\mathbf{v}(t)$ and using rules of differential calculus, we can refer to Fig. 2.1 and write

$$\mathbf{v}(t) = \lim_{\Delta t \to 0} \frac{\mathbf{r}(t+\Delta t) - \mathbf{r}(t)}{\Delta t} = \lim_{\Delta t \to 0} \frac{\Delta \mathbf{r}(t)}{\Delta t} = \frac{d\mathbf{r}(t)}{dt} = \dot{\mathbf{r}}(t) \tag{2.2}$$

where the overdot denotes differentiation with respect to time. The velocity has units meters per second (m/s). Introducing Eq. (2.1) into Eq. (2.2) and recalling that the unit vectors $\mathbf{i}$, $\mathbf{j}$, and $\mathbf{k}$ are constant, we obtain

$$\begin{aligned}\mathbf{v}(t) &= \frac{dx(t)}{dt}\mathbf{i} + \frac{dy(t)}{dt}\mathbf{j} + \frac{dz(t)}{dt}\mathbf{k} = \dot{x}(t)\mathbf{i} + \dot{y}(t)\mathbf{j} + \dot{z}(t)\mathbf{k} \\ &= v_x(t)\mathbf{i} + v_y(t)\mathbf{j} + v_z(t)\mathbf{k}\end{aligned} \tag{2.3}$$

where

$$v_x(t) = \dot{x}(t), \qquad v_y(t) = \dot{y}(t), \qquad v_z(t) = \dot{z}(t) \tag{2.4}$$

are the cartesian components of the velocity vector. From Fig. 2.1, we observe that as $\Delta t \to 0$ the increment $\Delta \mathbf{r}(t)$ in the position vector corresponding to the time increment Δt aligns itself with the curve C and becomes the differential $d\mathbf{r}(t)$. Hence, from Eq. (2.2), we conclude that *the velocity vector* $\mathbf{v}(t)$ *is tangent to the curve* C *at all times.*

The *acceleration* of the particle P is defined as the time rate of change of the velocity. As with the velocity, the acceleration can be regarded as a vector. Denoting the *acceleration vector* of P by $\mathbf{a}(t)$ and following the pattern established for the

*The acronym derives from the French name for the system of units, namely, *Système International d'Unités.*

velocity vector, we can write the acceleration vector in terms of cartesian components in the form

$$\mathbf{a}(t)=\dot{\mathbf{v}}(t)=\ddot{\mathbf{r}}(t)=a_x(t)\mathbf{i}+a_y(t)\mathbf{j}+a_z(t)\mathbf{k} \tag{2.5}$$

where

$$a_x(t)=\dot{v}_x(t)=\ddot{x}(t), \qquad a_y(t)=\dot{v}_y(t)=\ddot{y}(t), \qquad a_z(t)=\dot{v}_z(t)=\ddot{z}(t) \tag{2.6}$$

in which double overdots denote second derivatives with respect to time. The acceleration has units meters per second squared (m/s^2).

There are two special cases of motion of particular interest, namely, rectilinear motion and planar motion.

i. Rectilinear Motion

Rectilinear motion implies motion along a straight line. Because in this case there is only one component of motion, we can dispense with the vector notation and describe the motion in terms of scalar quantities. Denoting the line along which the motion takes place by s and the distance of the particle P from the fixed origin 0 by $s(t)$ (Fig. 2.2), the velocity of P can be written simply

$$v(t)=\frac{ds(t)}{dt}=\dot{s}(t) \tag{2.7}$$

and the acceleration of P

$$a(t)=\frac{dv(t)}{dt}=\frac{d^2s(t)}{dt^2}=\ddot{s}(t) \tag{2.8}$$

In Eqs. (2.7) and (2.8), the distance $s(t)$, the velocity $v(t)$, and the acceleration $a(t)$ are regarded as explicit functions of the time t. Using Eqs. (2.7) and (2.8), we can derive a relation among s, v, and a in which the time t is only implicit. To this end, let us use the chain rule for differentiation and write

$$a=\frac{dv}{dt}=\frac{dv}{ds}\frac{ds}{dt}=\frac{dv}{ds}v \tag{2.9}$$

Equation (2.9) can be rewritten in the form

$$a\,ds=v\,dv \tag{2.10}$$

which is the desired result.

Integrating Eq. (2.10) between the points $s=s_1, v=v_1$ and $s=s_2, v=v_2$, we obtain

$$\int_{s_1}^{s_2} a\,ds=\int_{v_1}^{v_2} v\,dv=\frac{1}{2}v^2\Big|_{v_1}^{v_2}=\frac{1}{2}(v_2^2-v_1^2) \tag{2.11}$$

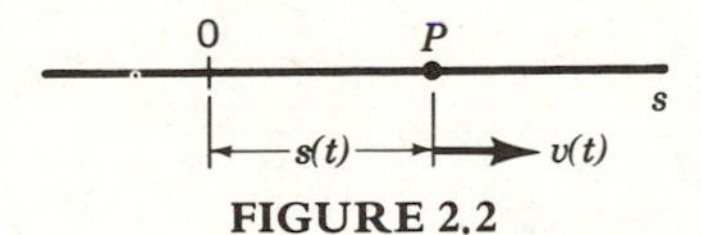

FIGURE 2.2

Example 2.1
An automobile starting from rest travels with the constant acceleration a_0 for 6 s (Fig. 2.3). Determine the value a_0 given that the automobile has reached the velocity $v_f=108$ kilometers per hour (km/h) at the end of the 6 s. What is the distance traveled by the automobile?

Integrating Eq. (2.8), we obtain

$$\int_0^v dv=\int_0^t a\,dt=a_0\int_0^t dt \tag{a}$$

or

$$v(t)=a_0 t \tag{b}$$

Letting $t=t_f=6$ s, $v(t)=v(t_f)=v_f=108$ km/h in Eq. (b), we obtain

$$a_0=\frac{v_f}{t_f}=\frac{108\times 1000}{3600}\frac{1}{6}=5 \quad \text{m/s}^2 \tag{c}$$

To determine the distance traveled, we integrate Eq. (2.7), with the result

$$\int_0^s ds=\int_0^t v\,dt=\int_0^t a_0 t\,dt \tag{d}$$

so that

$$s=\tfrac{1}{2}a_0 t^2 \tag{e}$$

Letting $t=t_f=6$ s in Eq. (e) and using Eq. (c), we obtain

$$s_f=\tfrac{1}{2}5\times 6^2=90 \quad \text{m} \tag{f}$$

Example 2.2
Determine the distance traveled by the automobile of Example 2.1 by using Eq. (2.11).

Letting $a=a_0=\text{const}$, $s_1=0$, $s_2=s_f$, $v_1=0$, $v_2=v_f$ in Eq. (2.11), we obtain

$$a_0 s_f=\tfrac{1}{2}v_f^2 \tag{a}$$

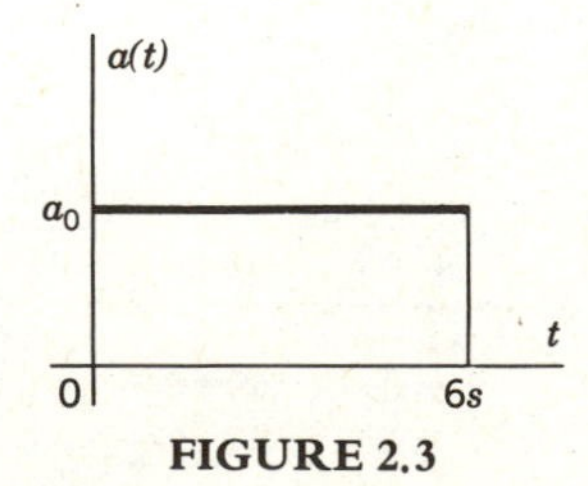

FIGURE 2.3

or

$$s_f = \frac{1}{2}\frac{v_f^2}{a_0} = \frac{1}{2}\left(\frac{108 \times 1000}{3600}\right)^2 \frac{1}{5} = 90 \quad \text{m} \tag{b}$$

which is the same result as that obtained in Example 2.1.

ii. Planar Motion. Trajectories

Let us consider a particle traveling in the plane xz with the constant acceleration

$$\mathbf{a} = \ddot{z}\mathbf{k} = -g\mathbf{k} \tag{2.12}$$

after being propelled with the initial velocity $\mathbf{v}_0$ from the origin 0 (Fig. 2.4), where g is the acceleration due to gravity. We propose to calculate the trajectory described by the particle in the plane xz. To this end, we first determine the motion as a function of time and then eliminate the explicit dependence of the motion on the time t.

Multiplying Eq. (2.5) by dt, integrating with respect to time, and considering Eq. (2.12), we obtain

$$\int_{\mathbf{v}_0}^{\mathbf{v}} d\mathbf{v} = \int_0^t \mathbf{a}\, dt = -gt\mathbf{k} \tag{2.13}$$

or

$$\mathbf{v} = \mathbf{v}_0 - gt\mathbf{k} \tag{2.14}$$

Multiplying Eq. (2.2) by dt, integrating with respect to time, and considering Eq. (2.14), we can write

$$\int_0^{\mathbf{r}} d\mathbf{r} = \int_0^t \mathbf{v}\, dt = \int_0^t (\mathbf{v}_0 - gt\mathbf{k})\, dt = \mathbf{v}_0 t - \tfrac{1}{2}gt^2\mathbf{k} \tag{2.15}$$

or

$$\mathbf{r} = \mathbf{v}_0 t - \tfrac{1}{2}gt^2\mathbf{k} \tag{2.16}$$

Letting v_0 be the magnitude of the initial velocity vector $\mathbf{v}_0$ and α the angle between $\mathbf{v}_0$ and the x axis, we can write the vector in terms of the cartesian components

$$\mathbf{v}_0 = v_0 \cos\alpha\, \mathbf{i} + v_0 \sin\alpha\, \mathbf{k} \tag{2.17}$$

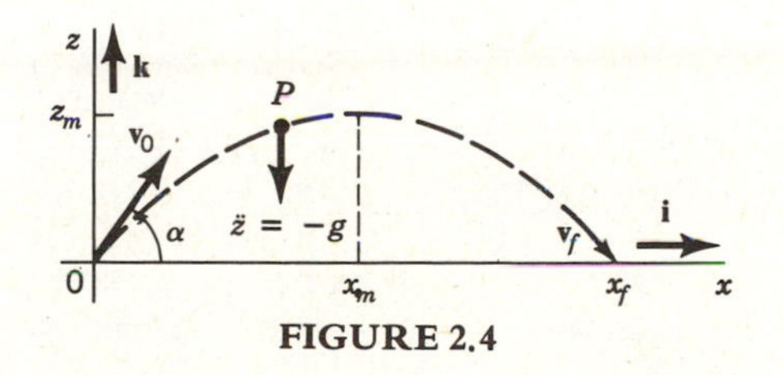

FIGURE 2.4

which permits us to express the position vector **r** in terms of Cartesian components as follows:

$$x=v_0 t \cos\alpha, \qquad z=v_0 t \sin\alpha-\tfrac{1}{2}gt^2 \tag{2.18}$$

Equations (2.18) give the position of the particle as an explicit function of time.

To derive the trajectory equation, let us solve the first of Eqs. (2.18) for the time t. The result is

$$t=\frac{x}{v_0\cos\alpha} \tag{2.19}$$

Inserting Eq. (2.19) into the second of Eqs. (2.18), we obtain

$$z=x\tan\alpha-\frac{1}{2}g\frac{x^2}{v_0^2\cos^2\alpha} \tag{2.20}$$

which is the *trajectory equation.* It represents a parabola. To determine its shape in more detail, we calculate the point at which the particle reaches its maximum altitude. From calculus, this point is characterized by zero slope, or

$$\frac{dz}{dx}=0 \tag{2.21}$$

Inserting Eq. (2.20) into Eq. (2.21), we can write

$$\frac{dz}{dx}=\tan\alpha-g\frac{x}{v_0^2\cos^2\alpha}=0 \tag{2.22}$$

Denoting by $x=x_m$ the distance along the x axis corresponding to the maximum altitude $z=z_m$, we obtain from Eq. (2.22)

$$x_m=\frac{v_0^2}{g}\sin\alpha\cos\alpha \tag{2.23}$$

Letting $x=x_m$ and $z=z_m$ in Eq. (2.20) and using Eq. (2.23), we obtain the maximum altitude

$$z_m=\frac{1}{2}\frac{v_0^2}{g}\sin^2\alpha \tag{2.24}$$

The trajectory is symmetric with respect to the vertical through $x=x_m$, as shown in Fig. 2.4. We conclude from Fig. 2.4 that the particle hits the ground again at $x_f=2x_m$. On the basis of geometric considerations, we also conclude from Fig. 2.4 that the impact velocity is

$$\mathbf{v}_f=v_0\cos\alpha\,\mathbf{i}-v_0\sin\alpha\,\mathbf{k} \tag{2.25}$$

2.3 PLANAR MOTION IN TERMS OF CURVILINEAR COORDINATES

There are instances in which the use of coordinates other than cartesian is advisable. This is often the case when the motion of the particle is not rectilinear, but follows a curved path. In such cases the use of curvilinear coordinates is likely to be more advantageous. There are two important sets of curvilinear coordinates in common use for planar motion: (i) radial and transverse coordinates and (ii) tangential and normal coordinates. In this section, we derive the velocity and acceleration expressions for both sets of coordinates.

i. Radial and Transverse Coordinates

The radial and transverse coordinates are commonly known as polar coordinates. Let us consider a particle P traveling along curve C, as shown in Fig. 2.5, and define the radial axis r as the one coinciding at all times with the direction of the radius vector $\mathbf{r}(t)$ from the origin 0 to the point P. The transverse axis θ is normal to the radial axis, as shown in Fig. 2.5. To describe the motion, it is convenient to introduce the unit vectors $\mathbf{u}_r(t)$ and $\mathbf{u}_\theta(t)$ in the radial and transverse directions, respectively. However, unlike the unit vectors $\mathbf{i}$, $\mathbf{j}$, and $\mathbf{k}$ of Section 2.2, the unit vectors $\mathbf{u}_r(t)$ and $\mathbf{u}_\theta(t)$ are not constant, but depend on time. This can be explained easily by observing that the radius vector $\mathbf{r}(t)$ changes directions continuously as the particle P moves along the curve C. Because the unit vector $\mathbf{u}_r(t)$ is aligned with $\mathbf{r}(t)$, $\mathbf{u}_r(t)$ also changes directions continuously. Hence, although the unit vector $\mathbf{u}_r(t)$ has constant magnitude (equal to unity), the mere fact that it changes directions makes it a time-dependent unit vector. Because $\mathbf{u}_\theta(t)$ is normal to $\mathbf{u}_r(t)$, $\mathbf{u}_\theta(t)$ is also a time-dependent unit vector.

Denoting the magnitude of the radius vector $\mathbf{r}(t)$ by $r(t)$, we can express the vector $\mathbf{r}(t)$ in the form

$$\mathbf{r}(t) = r(t)\mathbf{u}_r(t) \tag{2.26}$$

Following the pattern used to derive Eq. (2.2), the velocity $\mathbf{v}(t)$ of P is simply

$$\mathbf{v}(t) = \dot{\mathbf{r}}(t) = \dot{r}(t)\mathbf{u}_r(t) + r(t)\dot{\mathbf{u}}_r(t) \tag{2.27}$$

Equation (2.27) involves the time derivative $\dot{\mathbf{u}}_r(t)$ of the unit vector $\mathbf{u}_r(t)$. Before we can complete the expression of $\mathbf{v}(t)$ in terms of radial and transverse components, it is necessary to derive an expression for $\dot{\mathbf{u}}_r(t)$. We use this opportunity to evaluate $\dot{\mathbf{u}}_\theta(t)$ as well. From Fig. 2.6, we note that at the end of the time increment Δt the vector $\mathbf{u}_r(t)$ becomes the vector $\mathbf{u}_r(t+\Delta t)$, the difference between the two vectors

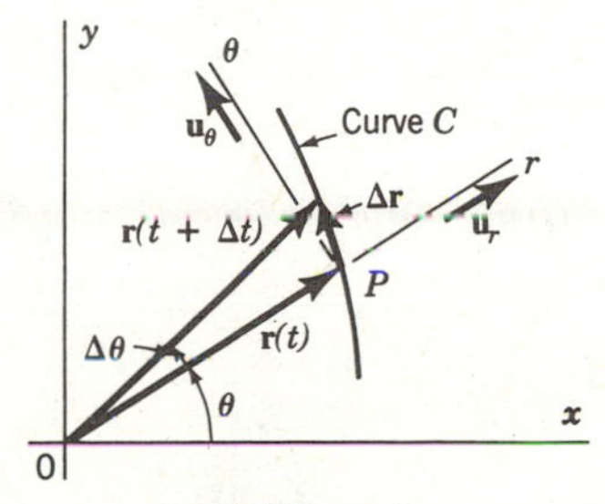

FIGURE 2.5

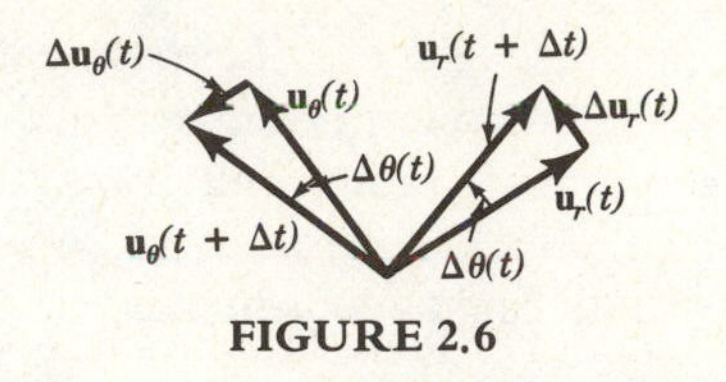

FIGURE 2.6

being $\Delta\mathbf{u}_r(t)$. For small increments Δt, the magnitude of $\Delta\mathbf{u}_r(t)$ is equal to the magnitude of $\mathbf{u}_r(t)$ multiplied by the correspondingly small angle $\Delta\theta(t)$, and the direction of $\Delta\mathbf{u}_r(t)$ is parallel to that of $\mathbf{u}_\theta(t)$. By analogy, the difference between the vectors $\mathbf{u}_\theta(t+\Delta t)$ and $\mathbf{u}_\theta(t)$ is the increment $\Delta\mathbf{u}_\theta(t)$, whose magnitude is equal to the magnitude of $\mathbf{u}_\theta(t)$ multiplied by $\Delta\theta(t)$, and its direction is opposite to that of $\mathbf{u}_r(t)$. Because $\mathbf{u}_r(t)$ and $\mathbf{u}_\theta(r)$ are unit vectors, we can write

$$\begin{aligned}\Delta\mathbf{u}_r(t)&=\mathbf{u}_r(t+\Delta t)-\mathbf{u}_r(t)=1\times\Delta\theta(t)\,\mathbf{u}_\theta(t)\\ \Delta\mathbf{u}_\theta(t)&=\mathbf{u}_\theta(t+\Delta t)-\mathbf{u}_\theta(t)=1\times\Delta\theta(t)[-\mathbf{u}_r(t)]\end{aligned}\tag{2.28}$$

Hence, dividing Eqs. (2.28) by Δt and taking the limit, we obtain the time derivatives of $\mathbf{u}_r(t)$ and $\mathbf{u}_\theta(t)$ in the form

$$\begin{aligned}\dot{\mathbf{u}}_r(t)&=\lim_{\Delta t\to 0}\frac{\Delta\mathbf{u}_r(t)}{\Delta t}=\lim_{\Delta t\to 0}\frac{\Delta\theta(t)}{\Delta t}\,\mathbf{u}_\theta(t)=\dot{\theta}(t)\mathbf{u}_\theta(t)\\ \dot{\mathbf{u}}_\theta(t)&=\lim_{\Delta t\to 0}\frac{\Delta\mathbf{u}_\theta(t)}{\Delta t}=\lim_{\Delta t\to 0}\frac{\Delta\theta(t)}{\Delta t}[-\mathbf{u}_r(t)]=-\dot{\theta}(t)\mathbf{u}_r(t)\end{aligned}\tag{2.29}$$

where $\dot{\theta}(t)$ is the angular rate of change of the radius vector $\mathbf{r}(t)$ as its tip moves along the curve C. Introducing the first of Eqs. (2.29) into Eq. (2.27), we obtain

$$\mathbf{v}(t)=\dot{r}\mathbf{u}_r+r\dot{\theta}\mathbf{u}_\theta=v_r\mathbf{u}_r+v_\theta\mathbf{u}_\theta\tag{2.30}$$

where

$$v_r=\dot{r},\qquad v_\theta=r\dot{\theta}\tag{2.31}$$

are the *radial and transverse components of the velocity vector*, respectively.

The acceleration of point P is obtained by simply taking the derivative of the velocity vector. Hence, using Eq. (2.30) and considering Eqs. (2.29), we can write

$$\begin{aligned}\mathbf{a}(t)=\dot{\mathbf{v}}(t)&=\ddot{r}\mathbf{u}_r+\dot{r}\dot{\mathbf{u}}_r+\dot{r}\dot{\theta}\mathbf{u}_\theta+r\ddot{\theta}\mathbf{u}_\theta+r\dot{\theta}\,\dot{\mathbf{u}}_\theta\\ &=(\ddot{r}-r\dot{\theta}^2)\mathbf{u}_r+(r\ddot{\theta}+2\dot{r}(\dot{\theta})\mathbf{u}_\theta=a_r\mathbf{u}_r+a_\theta\mathbf{u}_\theta\end{aligned}\tag{2.32}$$

where

$$a_r=\ddot{r}-r\dot{\theta}^2,\qquad a_\theta=r\ddot{\theta}+2\dot{r}\dot{\theta}\tag{2.33}$$

are the *radial and transverse components of the acceleration vector*, respectively.

It should be noted that by adding the coordinate z to the polar coordinates r and θ, we obtain the *cylindrical coordinates* r, θ, z. The velocity and acceleration vectors can be expressed in terms of cylindrical coordinates by simply adding $v_z\mathbf{k}=\dot{z}\mathbf{k}$ and $a_z\mathbf{k}=\ddot{z}\mathbf{k}$ to Eqs. (2.30) and (2.32), respectively.

Example 2.3
A cyclist enters the semicircular track of radius $R=50$ m shown in Fig. 2.7 with the velocity $v_A=15$ m/s, decelerates circumferentially at a uniform rate, and exits with the velocity $v_C=10$ m/s. Calculate the circumferential deceleration, the time it takes to complete the semicircle, and the velocity at point B.

Let us denote the magnitude of the circumferential deceleration by c, so that

$$a_\theta=-c=R\ddot{\theta} \tag{a}$$

Integrating Eq. (a) with respect to time, we obtain

$$v_\theta=R\dot{\theta}=v_A+\int_0^t a_\theta\,dt=v_A-ct \tag{b}$$

Letting $t=t_f$, $v_\theta=v_C$ in Eq. (b), we can write

$$ct_f=v_A-v_C=15-10=5 \quad \text{m/s} \tag{c}$$

Integration of Eq. (b) with respect to time yields

$$R\theta=v_At-\tfrac{1}{2}ct^2 \tag{d}$$

Introducing $\theta_f=\pi$, $t=t_f$ and $ct_f=5$ in Eq. (d) and solving for t_f, we obtain

$$t_f=\frac{R\theta_f}{v_A-\frac{1}{2}ct_f}=\frac{50\pi}{15-\frac{1}{2}\times 5}=4\pi \quad \text{s} \tag{e}$$

so that, from Eqs. (c) and (e),

$$c=\frac{5}{4\pi} \quad \text{m/s}^2 \tag{f}$$

To obtain v_B, we make use of Eq. (2.11), recall that $a_\theta=\text{const}$, and write

$$a_\theta s_{AB}=\tfrac{1}{2}(v_B^2-v_A^2) \tag{g}$$

so that

$$\begin{aligned}v_B&=\sqrt{v_A^2+2a_\theta s_{AB}}=\sqrt{v_A^2-2cR\pi/2}=\sqrt{15^2-2(5/4\pi)(50\pi/2)}\\&=\sqrt{225-62.5}=12.75 \quad \text{m/s}\end{aligned} \tag{h}$$

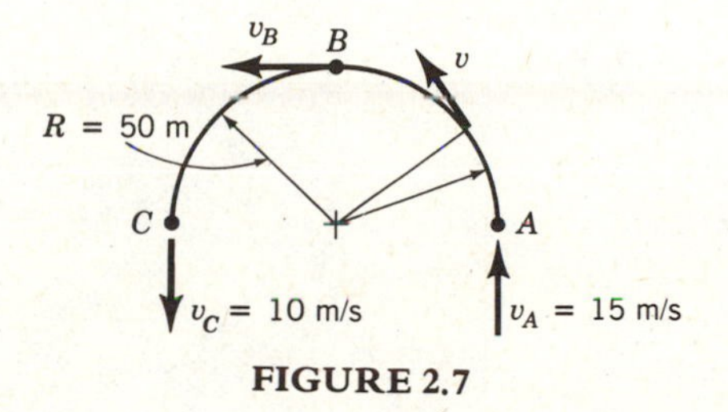

FIGURE 2.7

ii. Tangential and Normal Coordinates

Let us define the tangential and normal coordinates as a set of axes tangent and normal to curve C at the point P and time t, as shown in Fig. 2.8. The point 0′ denotes the center of curvature of the curve C and ρ is the radius of curvature corresponding to the instantaneous position of the point P on the curve. As P moves along C, both the center of curvature and the radius of curvature change, except when the curve C is a circle. If Δs is the distance along C traveled by the particle P during the time increment Δt, we can write

$$\Delta \mathbf{r} \cong \Delta s\, \mathbf{u}_t \tag{2.34}$$

Hence, the velocity is simply

$$\mathbf{v} = \lim_{\Delta t \to 0} \frac{\Delta \mathbf{r}}{\Delta t} = \lim_{\Delta t \to 0} \frac{\Delta s}{\Delta t}\, \mathbf{u}_t = \dot{s}\mathbf{u}_t \tag{2.35}$$

where $\dot{s}$ is the magnitude of the velocity vector. As expected, the velocity vector has only a tangential component, or the *velocity vector is tangent to the curve C at all times*. The acceleration vector is obtained by differentiating Eq. (2.35) with respect to time. The result is

$$\mathbf{a} = \ddot{s}\mathbf{u}_t + \dot{s}\dot{\mathbf{u}}_t \tag{2.36}$$

But, from Fig. 2.9, we can write

$$\dot{\mathbf{u}}_t = \lim_{\Delta t \to 0} \frac{\mathbf{u}_t(t+\Delta t) - \mathbf{u}_t(t)}{\Delta t} = \lim_{\Delta t \to 0} \frac{\Delta \mathbf{u}_t(t)}{\Delta t} = \lim_{\Delta t \to 0} \frac{1 \times \Delta\phi \mathbf{u}_n(t)}{\Delta t} = \dot{\phi}\mathbf{u}_n \tag{2.37}$$

where $\dot{\phi}$ is the angular rate of change of the vector $\boldsymbol{\rho}$ from O′ to P as its tip moves

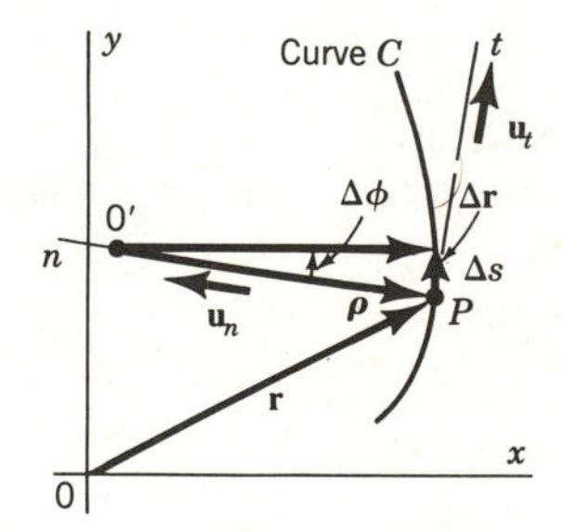

FIGURE 2.8

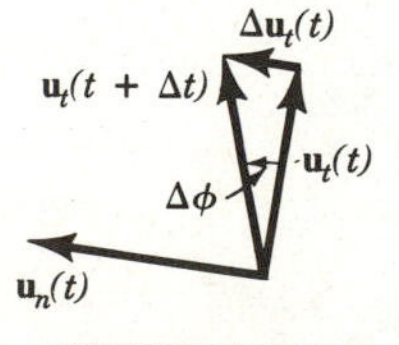

FIGURE 2.9

along C. Observing that

$$\Delta s = \rho\, \Delta\phi \tag{2.38}$$

where ρ is the magnitude of $\boldsymbol{\rho}$, we obtain

$$\dot{\phi} = \lim_{\Delta t \to 0} \frac{\Delta\phi}{\Delta t} = \lim_{\Delta t \to 0} \frac{\Delta\phi}{\Delta s}\frac{\Delta s}{\Delta t} = \lim_{\Delta t \to 0} \frac{1}{\rho}\frac{\Delta s}{\Delta t} = \frac{\dot{s}}{\rho} \tag{2.39}$$

Hence, introducing Eqs. (2.37) and (2.39) into Eq. (2.36), we have

$$\mathbf{a} = \ddot{s}\mathbf{u}_t + \frac{\dot{s}^2}{\rho}\mathbf{u}_n = a_t\mathbf{u}_t + a_n\mathbf{u}_n \tag{2.40}$$

where

$$a_t = \ddot{s}, \qquad a_n = \frac{\dot{s}^2}{\rho} \tag{2.41}$$

are the *tangential and normal components of the acceleration vector*, and we observe that the normal component is directed toward the center of curvature.

Example 2.4

Consider the parabolic trajectory of Fig. 2.4 and calculate the radius of curvature at the top of the parabola.

From the second of Eqs. (2.41), the radius of curvature is

$$\rho = \frac{\dot{s}^2}{a_n} = \frac{v_t^2}{a_n} \tag{a}$$

At the top of the parabola, we have

$$a_n = g, \qquad v_t = v_0 \cos\alpha \tag{b}$$

so that

$$\rho = \frac{v_0^2 \cos^2\alpha}{g} \tag{c}$$

2.4 MOVING REFERENCE FRAMES

In Sections 2.2 and 2.3, we described the motion of a particle in terms of a fixed reference frame. Yet, on many occasions it is more convenient to use moving reference frames. Hence, let us consider a reference frame xyz moving relative to the fixed reference frame XYZ (Fig. 2.10). We distinguish several cases: (i) the frame xyz translates relative to XYZ, (ii) the frame xyz rotates relative to XYZ, and (iii) the frame xyz both translates and rotates relative to XYZ. In all three cases, we can express the position of a point P relative to XYZ system in the form

$$\mathbf{R} = \mathbf{r}_A + \mathbf{r}_{AP} \tag{2.42}$$

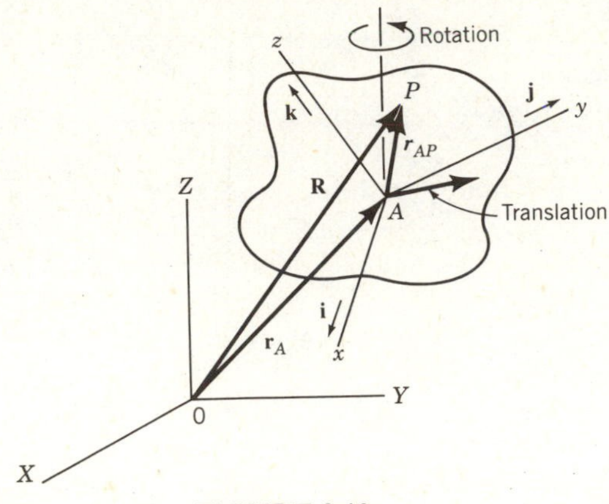

FIGURE 2.10

where $\mathbf{r}_A$ is the radius vector from the origin 0 of the fixed axes XYZ to the origin A of the moving axes xyz and $\mathbf{r}_{AP}$ is the radius vector from A to P, where the latter represents the position of P relative to A. The velocity of P relative to an inertial space is the time derivative of $\mathbf{R}$, or

$$\mathbf{v}=\dot{\mathbf{R}}=\mathbf{v}_A+\mathbf{v}_{AP} \tag{2.43}$$

where $\mathbf{v}_A=\dot{\mathbf{r}}_A$ is the velocity of point A and $\mathbf{v}_{AP}=\dot{\mathbf{r}}_{AP}$ is the velocity of P relative to A. Similarly, the acceleration of P relative to an inertial space is the time derivative of $\mathbf{v}$, or

$$\mathbf{a}=\dot{\mathbf{v}}=\ddot{\mathbf{R}}=\mathbf{a}_A+\mathbf{a}_{AP} \tag{2.44}$$

where $\mathbf{a}_A=\dot{\mathbf{v}}_A=\ddot{\mathbf{r}}_A$ is the acceleration of point A and $\mathbf{a}_{AP}=\dot{\mathbf{v}}_{AP}=\ddot{\mathbf{r}}_{AP}$ is the acceleration of P relative to A.

The evaluation of the velocity $\mathbf{v}_A$ and acceleration $\mathbf{a}_A$ of point A follows the same pattern as that established in Sections 2.2 and 2.3 for the motion of a particle relative to a fixed reference frame. On the other hand, the evaluation of the velocity $\mathbf{v}_{AP}$ and acceleration $\mathbf{a}_{AP}$ of point P relative to A depends on the case considered. If the frame xyz merely translates relative to the frame XYZ, then the calculation of $\mathbf{v}_{AP}$ and $\mathbf{a}_{AP}$ is similar to the calculation of $\mathbf{v}_A$ and $\mathbf{a}_A$, respectively. However, if the frame xyz rotates relative to the frame XYZ, then the evaluation of $\mathbf{v}_{AP}$ and $\mathbf{a}_{AP}$ becomes more involved and, in fact, requires the introduction of new concepts. In the following, the three cases indicated above are discussed separately.

i. The Frame *xyz* Translates Relative to *XYZ*

In this case it is convenient to let axes x, y, and z be parallel to axes X, Y, and Z, respectively. Because axes x, y, and z are in pure translation, they do not change directions in space, so that the unit vectors $\mathbf{i}$, $\mathbf{j}$, and $\mathbf{k}$ (Fig. 2.10) remain constant. The evaluation of the velocity and acceleration by means of Eqs. (2.43) and (2.44), respectively, represent mere vector additions, which implies simple addition of the corresponding components.

ii. The Frame *xyz* Rotates Relative to *XYZ*

In this section, we confine ourselves to the case in which *point P is fixed in the frame xyz*, so that *the motion of P relative to A is due entirely to the rotation of the frame xyz*. The case in which P moves relative to the frame xyz is discussed later in this chapter. The motion of P relative to A due to the rotation of the frame xyz can be best visualized by imagining that the system xyz is embedded in a rigid body, where a rigid body is defined as a body such that the distance between any two of its points is constant. Because the triad xyz is embedded in the rigid body, the rotation of the frame is identical to that of the rigid body.

Before we discuss the rotational motion of the triad xyz, it is necessary to introduce the concept of *angular velocity*. Let us assume that the rigid body is rotating instantaneously about a given axis AB, and consider a point P at a distance $\rho = |\boldsymbol{\rho}|$ from point C on axis AB, where C is the intersection between axis AB and a plane normal to AB and containing the point P (Fig. 2.11). In the time increment Δt the vector $\boldsymbol{\rho}$ from C to P sweeps an angle $\Delta\theta$ in a plane normal to AB. Hence, taking the limit, we can write

$$\lim_{\Delta t \to 0} \frac{\Delta\theta}{\Delta t} = \dot{\theta} \tag{2.45}$$

or the vector $\boldsymbol{\rho}$ rotates in a plane normal to the axis AB with the angular rate $\dot{\theta}$. Because the vector $\boldsymbol{\rho}$ is embedded in the rigid body, we conclude that the rigid body, and hence the triad xyz, rotates about axis AB at the same rate $\dot{\theta}$. Next, we propose to represent this angular rate as a vector. As Fig. 2.11 indicates, such a vector, denoted $\boldsymbol{\omega}$, is directed along the axis AB. Clearly, its magnitude must

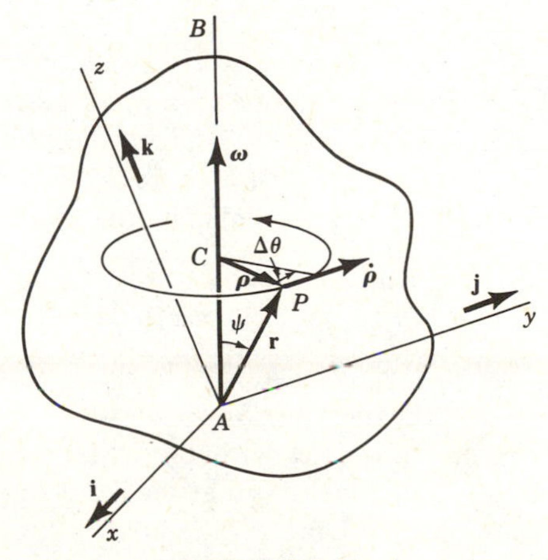

FIGURE 2.11

be $\dot{\theta}$, or

$$|\omega| = \dot{\theta} \tag{2.46}$$

If $\boldsymbol{\omega}$ satisfies the rules of operations of vectors, then it can be used to represent the *angular velocity vector* of the rigid body, or of the frame xyz. The angular velocity has units radians per second (rad/s).

To verify the above proposition, let us consider the rate of change $\dot{\boldsymbol{\rho}}$ of the vector $\boldsymbol{\rho}$ due to the rotation of the body. We observe from Fig. 2.11 that the tip P of the vector $\boldsymbol{\rho}$ describes a circle of radius ρ, so that $\dot{\boldsymbol{\rho}}$ is tangent to the circle at P, and hence it is normal to the plane defined by the vectors $\boldsymbol{\omega}$ and $\boldsymbol{\rho}$. In the time increment Δt, the vector $\boldsymbol{\rho}$ undergoes the change in magnitude

$$\Delta\rho = \rho\,\Delta\theta \tag{2.47}$$

so that the rate of change of the magnitude of the vector $\boldsymbol{\rho}$ is

$$\dot{\rho} = \lim_{\Delta t \to 0} \frac{\Delta\rho}{\Delta t} = \lim_{\Delta t \to 0} \rho \frac{\Delta\theta}{\Delta t} = \rho\dot{\theta} \tag{2.48}$$

From vector analysis, however, it can be easily verified that the vector $\dot{\boldsymbol{\rho}}$ can be expressed as the cross product, or vector product,

$$\dot{\boldsymbol{\rho}} = \boldsymbol{\omega} \times \boldsymbol{\rho} \tag{2.49}$$

as both the direction and magnitude of $\dot{\boldsymbol{\rho}}$ are according to the definition of a vector product. Hence, the *angular velocity of the body, and of axes xyz, can be represented by a vector* $\boldsymbol{\omega}$ whose direction is along the axis of rotation AB.

Equation (2.49) can be generalized by observing that

$$\rho = r \sin\psi \tag{2.50}$$

where r is the magnitude of the vector $\mathbf{r}$ and ψ is the angle between AB and $\mathbf{r}$. Hence the rate of change $\dot{\mathbf{r}}$ of the vector $\mathbf{r}$ can be written as the vector product

$$\dot{\mathbf{r}} = \boldsymbol{\omega} \times \mathbf{r} \tag{2.51}$$

Equation (2.51) gives the time derivative of a vector constant in magnitude and embedded in a set of axes rotating relative to an inertial space with the angular velocity $\boldsymbol{\omega}$. The equation is valid for any vector, regardless of the physical meaning.

As pointed out earlier, the velocity of point P relative to point A is due entirely to the rotation of the frame xyz. Hence, replacing the vector $\mathbf{r}$ in Eq. (2.51) by the radius vector $\mathbf{r}_{AP}$, we obtain

$$\mathbf{v}_{AP} = \dot{\mathbf{r}}_{AP} = \boldsymbol{\omega} \times \mathbf{r}_{AP} \tag{2.52}$$

and, because point A is at rest, $\mathbf{v}_{AP}$ is also the absolute velocity of P, $\mathbf{v} = \mathbf{v}_{AP}$.

We demonstrated above that the angular velocity of a rotating set of axes can be represented by a vector coinciding with the instantaneous axis of rotation. The implication is that at a later time the rotation can take place about a different axis. Moreover, the rotation rate can be different, so that the angular velocity vector $\boldsymbol{\omega}$ can undergo changes both in direction and magnitude. We refer to the time rate of change of $\boldsymbol{\omega}$ as the *angular acceleration*

$$\boldsymbol{\alpha}=\dot{\boldsymbol{\omega}} \tag{2.53}$$

The angular acceleration vector $\boldsymbol{\alpha}$ is shown in Fig. 2.12. The angular acceleration has units radians per second squared (rad/s^2).

The acceleration of point P relative to A can be obtained from Eq. (2.52) by simply taking the time derivative, or

$$\mathbf{a}_{AP}=\dot{\mathbf{v}}_{AP}=\dot{\boldsymbol{\omega}}\times \mathbf{r}_{AP}+\boldsymbol{\omega}\times \dot{\mathbf{r}}_{AP} \tag{2.54}$$

so that, using Eqs. (2.52) and (2.53), we obtain

$$\mathbf{a}_{AP}=\boldsymbol{\alpha}\times \mathbf{r}_{AP}+\boldsymbol{\omega}\times(\boldsymbol{\omega}\times \mathbf{r}_{AP}) \tag{2.55}$$

and, because point A is unaccelerated, $\mathbf{a}_{AP}$ is also the absolute acceleration of P, $\mathbf{a}=\mathbf{a}_{AP}$.

iii. The Frame *xyz* Translates and Rotates Relative to *XYZ*

When the origin A of the rotating frame xyz is not fixed, but moves relative to the inertial frame XYZ with the velocity $\mathbf{v}_A$ and acceleration $\mathbf{a}_A$, then the absolute velocity and acceleration of P are given by Eqs. (2.43) and (2.44), respectively, in which $\mathbf{v}_{AP}$ is given by Eq. (2.52) and $\mathbf{a}_{AP}$ by (2.55). Hence, the absolute velocity of P is

$$\mathbf{v}=\mathbf{v}_A+\mathbf{v}_{AP}=\mathbf{v}_A+\boldsymbol{\omega}\times \mathbf{r}_{AP} \tag{2.56}$$

and the absolute acceleration of P is

$$\mathbf{a}=\mathbf{a}_A+\mathbf{a}_{AP}=\mathbf{a}_A+\mathbf{a}\times \mathbf{r}_{AP}+\boldsymbol{\omega}\times(\boldsymbol{\omega}\times \mathbf{r}_{AP}) \tag{2.57}$$

At times, it is advisable to use more than one rotating reference frame. Let us consider the case in which the reference frame $x_1y_1z_1$ rotates with the angular

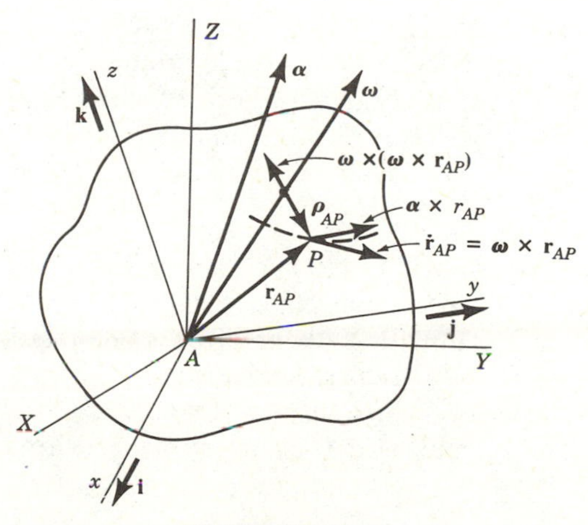

FIGURE 2.12

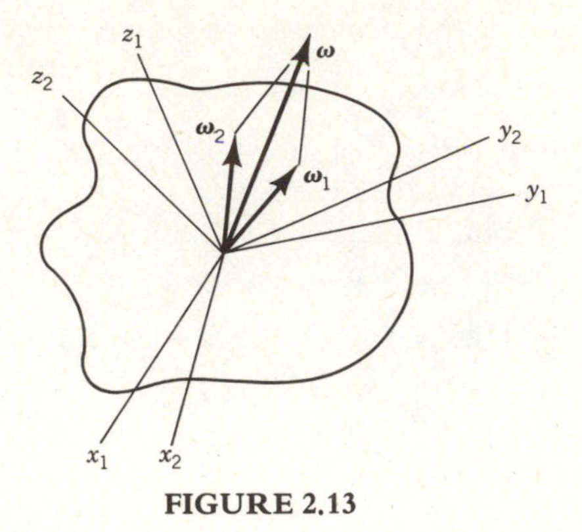

FIGURE 2.13

velocity ω_1 and the reference frame $x_2y_2z_2$ rotates relative to the frame $x_1y_1z_1$ with the angular velocity ω_2 (Fig. 2.13). Then, the angular velocity of the frame $x_2y_2z_2$ is simply

$$\omega=\omega_1+\omega_2 \tag{2.58}$$

The question arises as to the angular acceleration of the frame $x_2y_2z_2$. If we assume that the vector ω is expressed in terms of components along the frame $x_1y_1z_1$, the angular acceleration consists of two parts: the first due to the change in the components of ω relative to the frame $x_1y_1z_1$ and the second due to the fact that ω is expressed in terms of components along a rotating frame. Denoting the first part by α' and recognizing that the second part can be obtained from Eq. (2.51) by replacing $\mathbf{r}$ by ω and ω by ω_1, we can write

$$\alpha=\dot{\omega}=\alpha'+\omega_1\times\omega=\alpha'+\omega_1\times\omega_2 \tag{2.59}$$

The interesting aspect of Eq. (2.59) is that the frame $x_2y_2z_2$ has an angular acceleration even when both ω_1 and ω_2 are constant. The explanation lies in the fact that the constancy of ω_2 implies only that the frame $x_2y_2z_2$ rotates uniformly relative to the frame $x_1y_1z_1$. But, as the frame $x_1y_1z_1$ rotates with the angular velocity ω_1, it causes an acceleration of the frame $x_2y_2z_2$ equal to $\omega_1\times\omega_2$, which is due entirely to a rate of change in the direction of ω_2.

***Example* 2.5**

An airplane flies directly north at 1000 km/h, and it almost collides with another airplane. To the pilot of the first airplane it appears that the second airplane is flying at 800 km/h in a southwest path, at a 210° angle relative to the x-axis (Fig. 2.14). Determine the velocity of the second airplane.

Denoting the velocity of the first airplane by $\mathbf{v}_A$, the velocity of the second airplane by $\mathbf{v}_B$, and the velocity of the second airplane relative to the first by $\mathbf{v}_{AB}$, we can write

$$\mathbf{v}_B=\mathbf{v}_A+\mathbf{v}_{AB} \tag{a}$$

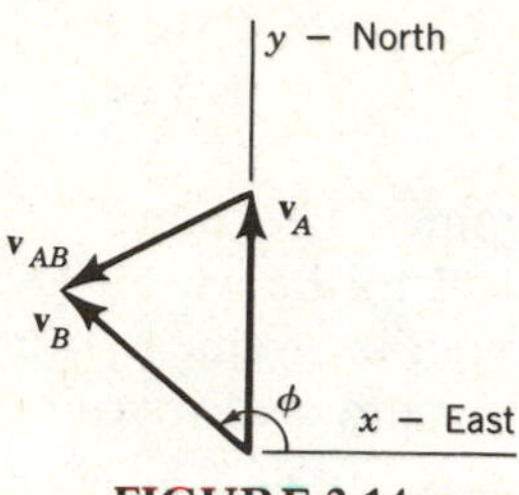

FIGURE 2.14

where, from Fig. 2.14,

$$\mathbf{v}_A = 1000 \quad \text{km/h} \tag{b}$$

and

$$\mathbf{v}_{AB} = 800(\sin 210°\mathbf{i} + \cos 210°\mathbf{j}) = -693\mathbf{i} - 400\mathbf{j} \quad \text{km/h} \tag{c}$$

Hence,

$$\mathbf{v} = 1000\mathbf{j} - 693\mathbf{i} - 400\mathbf{j} = -693\mathbf{i} + 600\mathbf{j} \tag{d}$$

so that the second airplane is flying with the speed

$$v = [(-693)^2 + 600^2]^{1/2} \cong 917 \quad \text{km/h} \tag{e}$$

at an angle

$$\phi = \tan^{-1} \frac{600}{-693} = 139.11° \tag{f}$$

as shown in Fig. 2.14.

Example 2.6
A bicycle travels on a circular track of radius R with the circumferential velocity v_A and acceleration a_A. Figure 2.15 shows one of the bicycle wheels. Determine the

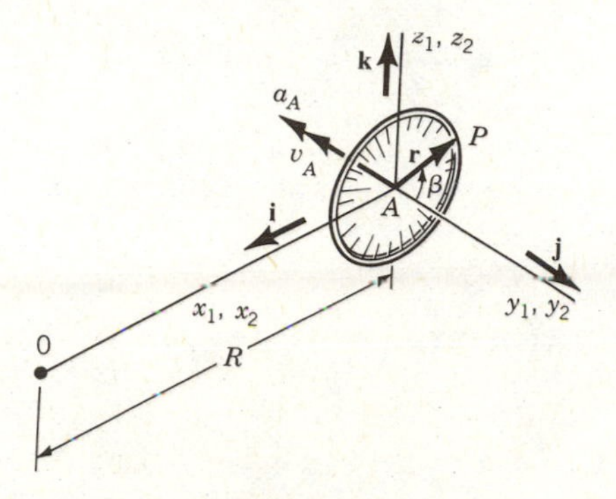

FIGURE 2.15

velocity $\mathbf{v}$ and acceleration $\mathbf{a}$ of a point P on the tire when the radius from the center of the wheel to the point P makes an angle β with respect to the horizontal plane. The radius of the wheel is r.

We consider two reference frames. The frame $x_1y_1z_1$ is attached to the bicycle, and the frame $x_2y_2z_2$ is attached to the wheel. The two frames are assumed to coincide instantaneously, although the frame $x_2y_2z_2$ rotates relative to the frame $x_1y_1z_1$. The velocity and acceleration of P are calculated according to Eqs. (2.56) and (2.57), where $\boldsymbol{\omega}$ and $\boldsymbol{\alpha}$ are given by Eqs. (2.58) and (2.59), respectively. In the first place, the velocity of the wheel center is simply

$$\mathbf{v}_A = -v_A \mathbf{j} \tag{a}$$

and the angular velocity of the frame $x_1y_1z_1$ is

$$\boldsymbol{\omega}_1 = \frac{v_A}{R}\mathbf{k} \tag{b}$$

Moreover, the angular velocity of the frame $x_2y_2z_2$ relative to $x_1y_1z_1$ is

$$\boldsymbol{\omega}_2 = \frac{v_A}{r}\mathbf{i} \tag{c}$$

so that the absolute angular velocity of $x_2y_2z_2$ is

$$\boldsymbol{\omega} = \boldsymbol{\omega}_1 + \boldsymbol{\omega}_2 = \frac{v_A}{R}\mathbf{k} + \frac{v_A}{r}\mathbf{i} \tag{d}$$

The radius vector from A to P is

$$\mathbf{r}_{AP} = r(\cos\beta\,\mathbf{j} + \sin\beta\,\mathbf{k}) \tag{e}$$

Hence, the velocity of P is

$$\begin{aligned}\mathbf{v} &= \mathbf{v}_A + \boldsymbol{\omega}\times\mathbf{r}_{AP} = -v_A\mathbf{j} + \left(\frac{v_A}{r}\mathbf{i} + \frac{v_A}{R}\mathbf{k}\right)\times r(\cos\beta\,\mathbf{j} + \sin\beta\,\mathbf{k}) \\ &= -\frac{v_A r}{R}\cos\beta\,\mathbf{i} - v_A(1+\sin\beta)\mathbf{j} + v_A\cos\beta\,\mathbf{k}\end{aligned} \tag{f}$$

The acceleration of A has two components: a tangential component due to the acceleration of the bicycle along the track and a normal component due to motion along a curvilinear track. Hence,

$$\begin{aligned}\mathbf{a}_A &= -a_A\mathbf{j} + \boldsymbol{\omega}_1\times(\boldsymbol{\omega}_1\times\mathbf{r}_{OA}) = -a_A\mathbf{j} + \frac{v_A}{R}\mathbf{k}\times\left[\frac{v_A}{R}\mathbf{k}\times(-R)\mathbf{i}\right] \\ &= \frac{v_A^2}{R}\mathbf{i} - a_A\mathbf{j}\end{aligned} \tag{g}$$

Using Eq. (2.59), the angular acceleration of the frame $x_2y_2z_2$ is

$$\boldsymbol{\alpha}=\dot{\boldsymbol{\omega}}_1'+\dot{\boldsymbol{\omega}}_2'+\boldsymbol{\omega}_1\times\boldsymbol{\omega}_2=\frac{a_A}{R}\mathbf{k}+\frac{a_A}{r}\mathbf{i}+\frac{v_A}{R}\mathbf{k}\times\frac{v_A}{r}\mathbf{i}$$

$$=\frac{a_A}{r}\mathbf{i}+\frac{v_A^2}{Rr}\mathbf{j}+\frac{a_A}{R}\mathbf{k} \tag{h}$$

Hence, the acceleration of P is

$$\mathbf{a}=\mathbf{a}_A+\boldsymbol{\alpha}\times\mathbf{r}_{AP}+\boldsymbol{\omega}\times(\boldsymbol{\omega}\times\mathbf{r}_{AP})$$

$$=\frac{v_A^2}{R}\mathbf{i}-a_A\mathbf{j}+\left(\frac{a_A}{r}\mathbf{i}+\frac{v_A^2}{Rr}\mathbf{j}+\frac{a_A}{R}\mathbf{k}\right)\times r(\cos\beta\,\mathbf{j}+\sin\beta\,\mathbf{k})$$

$$+\left(\frac{v_A}{R}\mathbf{i}+\frac{v_A}{R}\mathbf{k}\right)\times\left[\left(\frac{v_A}{r}\mathbf{i}+\frac{v_A}{R}\mathbf{k}\right)\times r(\cos\beta\,\mathbf{j}+\sin\beta\,\mathbf{k})\right]$$

$$=\left[\frac{v_A^2}{R}(1+2\sin\beta)-\frac{a_Ar}{R}\cos\beta\right]\mathbf{i}-\left[\left(\frac{1}{r}+\frac{r}{R^2}\right)v_A^2\cos\beta+a_A\sin\beta\right]\mathbf{j}$$

$$-\left(\frac{v_A^2}{r}\sin\beta-a_A\cos\beta\right)\mathbf{k} \tag{i}$$

2.5 PLANAR MOTION OF RIGID BODIES

In Section 2.4, we derived expressions for the general three-dimensional motion of rigid bodies. In a large number of applications, however, the motion is confined to two dimensions, so that it appears desirable to simplify the expressions for the velocity and acceleration, Eqs. (2.56) and (2.57), respectively. To this end we consider the case in which the body moves in the xy-plane, as shown in Fig. 2.16. We first express the motion in terms of cartesian components. Hence, we write

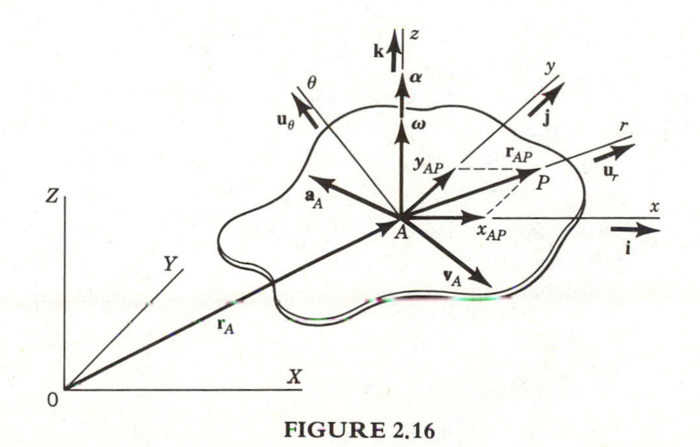

FIGURE 2.16

the radius vector $\mathbf{r}_{AP}$ in the simple form

$$\mathbf{r}_{AP} = x_{AP}\mathbf{i} + y_{AP}\mathbf{j} \tag{2.60}$$

where the various quantities are defined in Fig. 2.16. Planar motion implies that

$$\boldsymbol{\omega} = \omega\mathbf{k}, \qquad \boldsymbol{\alpha} = \alpha\mathbf{k} \tag{2.61}$$

where ω and α are the magnitude of the angular velocity and acceleration, respectively. Introducing Eqs. (2.60) and (2.61) into Eq. (2.56), we obtain the velocity

$$\mathbf{v} = \mathbf{v}_A + \boldsymbol{\omega} \times \mathbf{r}_{AP} = \mathbf{v}_A + \omega\mathbf{k} \times (x_{AP}\mathbf{i} + y_{AP}\mathbf{j}) = \mathbf{v}_A - \omega(y_{AP}\mathbf{i} - x_{AP}\mathbf{j}) \tag{2.62}$$

Moreover, inserting Eqs. (2.60) and (2.61) into Eqs. (2.57), we obtain the acceleration

$$\begin{aligned} \mathbf{a} &= \mathbf{a}_A + \boldsymbol{\alpha} \times \mathbf{r}_{AP} + \boldsymbol{\omega} \times (\boldsymbol{\omega} \times \mathbf{r}_{AP}) = \mathbf{a}_A + \alpha\mathbf{k} \times (x_{AP}\mathbf{i} + y_{AP}\mathbf{j}) + \omega\mathbf{k} \times [\omega\mathbf{k} \times (x_{AP}\mathbf{i} + y_{AP}\mathbf{j})] \\ &= \mathbf{a}_A - \alpha(y_{AP}\mathbf{i} - x_{AP}\mathbf{j}) - \omega^2(x_{AP}\mathbf{i} + y_{AP}\mathbf{j}) = \mathbf{a}_A - \alpha(y_{AP}\mathbf{i} - x_{AP}\mathbf{j}) - \omega^2\mathbf{r}_{AP} \end{aligned} \tag{2.63}$$

At times, it is more convenient to express the motion in terms of radial and transverse components. To this end, we write the radius vector in the form

$$\mathbf{r}_{AP} = r_{AP}\mathbf{u}_r \tag{2.64}$$

where $\mathbf{u}_r$ is the unit vector in the radial direction. Then, it is easy to verify that the velocity vector is

$$\mathbf{v} = \mathbf{v}_A + \omega r_{AP}\mathbf{u}_\theta \tag{2.65}$$

and the acceleration vector is

$$\mathbf{a} = \mathbf{a}_A + \alpha r_{AP}\mathbf{u}_\theta - \omega^2\mathbf{r}_{AP} = \mathbf{a}_A + \alpha r_{AP}\mathbf{u}_\theta - \omega^2 r_{AP}\mathbf{u}_r \tag{2.66}$$

where $\mathbf{u}_\theta$ is the unit vector in the transverse direction.

Equation (2.62) regards the velocity of an arbitrary point P in a rigid body as consisting of two terms, the first representing the velocity of translation of a reference point A and the second representing the velocity due to rotation about A. However, there exists a point C such that the velocity of P can be regarded instantaneously as due entirely to rotation about C. It follows that the point C is instantaneously at rest. For this reason, the point C is called the *instantaneous center of rotation*. Point C does not necessarily lie inside the body. If the velocity vector is known in full (i.e., both the magnitude and direction are known) and if the angular velocity of the body is given, then the instantaneous center can be determined by a graphical procedure. Figure 2.17 shows the rigid body and the velocity vector $\mathbf{v}$ associated with point P. Using radial and transverse coordinates, the velocity vector can be written in the form

$$\mathbf{v} = \boldsymbol{\omega} \times \mathbf{r} = \omega\mathbf{k} \times r\mathbf{u}_r = \omega r\mathbf{u}_\theta = v\mathbf{u}_\theta \tag{2.67}$$

where

$$v = \omega r \tag{2.68}$$

is the magnitude of the velocity vector. Of course, the radius vector $\mathbf{r}$ must be normal to the velocity vector $\mathbf{v}$. From Eq. (2.68), we obtain the magnitude of the

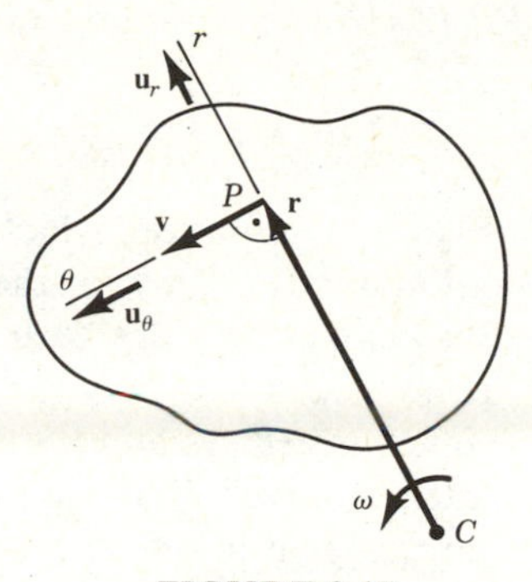

FIGURE 2.17

radius vector by writing simply

$$r = \frac{v}{\omega} \tag{2.69}$$

Figure 2.17 shows the instantaneous center C for the case in which the angular velocity $\boldsymbol{\omega}$ is in the counterclockwise sense.

A more common case is that in which the velocity vector $\mathbf{v}_1$ of one point in the body is known in full and the direction of the velocity vector $\mathbf{v}_2$ of a different point is also known. Then, the instantaneous center C can be determined by the graphical procedure depicted in Fig. 2.18. The velocity vectors satisfy the equations

$$\mathbf{v}_1 = \boldsymbol{\omega} \times \mathbf{r}_1, \qquad \mathbf{v}_2 = \boldsymbol{\omega} \times r_2 \tag{2.70a, b}$$

so that the magnitudes of $\mathbf{v}_1$ and $\mathbf{v}_2$ satisfy

$$v_1 = \omega r_1, \qquad v_2 = \omega r_2 \tag{2.71a, b}$$

where r_1 and r_2 can be determined from Fig. 2.18. Then, from Eq. (2.71a), the angular velocity magnitude is simply

$$\omega = \frac{v_1}{r_1} \tag{2.72}$$

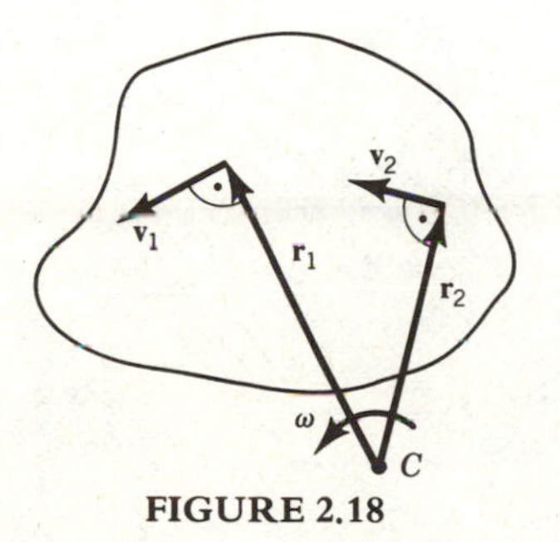

FIGURE 2.18

and, inserting Eq. (2.72) into Eq. (2.71b), we obtain the magnitude of $\mathbf{v}_2$ in the form

$$v_2 = v_1 \frac{r_1}{r_2} \tag{2.73}$$

Point C is an instantaneous center of rotation, which implies that the motion can be interpreted as pure rotation about C only for that particular instant. At a different time, the instantaneous center coincides with a different point. As the motion of the body unfolds, the point C traces a curve in the inertial space representing the locus of the instantaneous centers and known as the *space centrode.* At the same time, the point C traces a curve relative to the body and known as the *body centrode.* The two curves are tangent to one another at any given instant, as shown in Fig. 2.19, and the motion can be interpreted geometrically as the rolling of the body centrode on the space centrode.

The concept of instantaneous center can be used at times to solve kinematical problems.

Example 2.7

The planar framework shown in Fig. 2.20 consists of three rigid bars hinged at the points A, B, C and D. The rigid AB rotates with the angular velocity $\omega_{AB}=5$ rad/s and angular acceleration $\alpha_{AB}=2$ rad/s^2. Calculate the velocity $\mathbf{v}_C$ and acceleration $\mathbf{a}_C$ of point C.

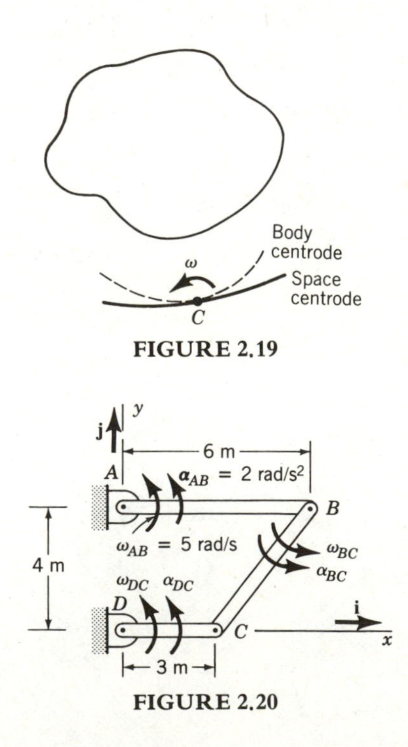

FIGURE 2.19

FIGURE 2.20

To solve the problem, we calculate the velocity and acceleration of point C by regarding point C in two ways: (1) as belonging to bar BC and (2) as belonging to bar DC. This procedure yields a sufficient number of equations to solve for all the unknown quantities, including $\mathbf{v}_C$ and $\mathbf{a}_C$.

The given quantities can be written in vector form in terms of cartesian components as follows:

$$\mathbf{r}_{AB}=6\mathbf{i} \quad \text{m}, \qquad \boldsymbol{\omega}_{AB}=\omega_{AB}\mathbf{k}=5\mathbf{k} \quad \text{rad/s},$$
$$\boldsymbol{\alpha}_{AB}=\alpha_{AB}\mathbf{k}=2\mathbf{k} \quad \text{rad/s}^2, \qquad \mathbf{r}_{BC}=-3\mathbf{i}-4\mathbf{j} \quad \text{m}, \qquad \mathbf{r}_{DC}=3\mathbf{i} \quad \text{m} \tag{a}$$

where, from Fig. 2.20, the notation is obvious. First, we calculate $\mathbf{v}_C$ and $\mathbf{a}_C$ by regarding C as belonging to bar BC. Before we can do that, we must calculate $\mathbf{v}_B$ and $\mathbf{a}_B$. Considering Eq. (2.62) with $\mathbf{v}_A=\mathbf{0}$ and $P=B$, the velocity of point B is

$$\mathbf{v}_B=\boldsymbol{\omega}_{AB}\times\mathbf{r}_{AB}=-\omega_{AB}(y_{AB}\mathbf{i}-x_{AB}\mathbf{j})=\omega_{AB}x_{AB}\mathbf{j}=5\times 6\mathbf{j}=30\mathbf{j} \quad \text{m/s} \tag{b}$$

and considering Eq. (2.63) with $\mathbf{a}_A=\mathbf{0}$ the acceleration of point B is

$$\begin{aligned}\mathbf{a}_B&=\boldsymbol{\alpha}_{AB}\times\mathbf{r}_{AB}+\boldsymbol{\omega}_{AB}\times(\boldsymbol{\omega}_{AB}\times\mathbf{r}_{AB})=-\alpha_{AB}(y_{AB}\mathbf{i}-x_{AB}\mathbf{j})-\omega_{AB}^2\mathbf{r}_{AB}\\ &=-2\times 6\mathbf{j}-5^2\times 6\mathbf{i}=-150\mathbf{i}+12\mathbf{j} \quad \text{m/s}^2\end{aligned} \tag{c}$$

Similarly, using Eqs. (2.62) and (2.63) with B and C replacing A and P, respectively, we can write

$$\begin{aligned}\mathbf{v}_C&=\mathbf{v}_B+\boldsymbol{\omega}_{BC}\times\mathbf{r}_{BC}=30\mathbf{j}-\omega_{BC}(y_{BC}\mathbf{i}-x_{BC}\mathbf{j})\\ &=4\omega_{BC}\mathbf{i}+(30-3\omega_{BC})\mathbf{j} \quad \text{m/s}\end{aligned} \tag{d}$$

$$\begin{aligned}\mathbf{a}_C&=\mathbf{a}_B+\boldsymbol{\alpha}_{BC}\times\mathbf{r}_{BC}+\boldsymbol{\omega}_{BC}\times(\boldsymbol{\omega}_{BC}\times\mathbf{r}_{BC})\\ &=-150\mathbf{i}+12\mathbf{j}-\alpha_{BC}(y_{BC}\mathbf{i}-x_{BC}\mathbf{j})-\omega_{BC}^2\mathbf{r}_{BC}\\ &=(-150+4\alpha_{BC}+3\omega_{BC}^2)\mathbf{i}+(12-3\alpha_{BC}+4\omega_{BC}^2)\mathbf{j} \quad \text{m/s}^2\end{aligned} \tag{e}$$

Next, we regard point C as belonging to bar DC. The velocity of point C is simply

$$\mathbf{v}_C=\boldsymbol{\omega}_{DC}\times\mathbf{r}_{DC}=\omega_{DC}x_{DC}\mathbf{j}=3\omega_{DC}\mathbf{j} \quad \text{m/s} \tag{f}$$

and the acceleration of C is

$$\begin{aligned}\mathbf{a}_C&=\boldsymbol{\alpha}_{DC}\times\mathbf{r}_{DC}+\boldsymbol{\omega}_{DC}\times(\boldsymbol{\omega}_{DC}\times\mathbf{r}_{DC})\\ &=\alpha_{DC}x_{DC}\mathbf{j}-\omega_{DC}^2\mathbf{r}_{DC}=-3\omega_{DC}^2\mathbf{i}+3\alpha_{DC}\mathbf{j} \quad \text{m/s}^2\end{aligned} \tag{g}$$

Comparing the components of $\mathbf{v}_C$ in Eqs. (d) and (f), we conclude that

$$\omega_{BC}=0, \qquad \omega_{DC}=10 \quad \text{rad/s} \tag{h}$$

so that

$$\mathbf{v}_C=30\mathbf{j} \quad \text{m/s} \tag{i}$$

Similarly, comparing the components of $\mathbf{a}_C$ in Eqs. (e) and (g), we obtain

$$\alpha_{BC}=-37.5 \quad \text{rad/s}^2, \qquad \alpha_{DC}=-33.5 \quad \text{rad/s}^2 \tag{j}$$

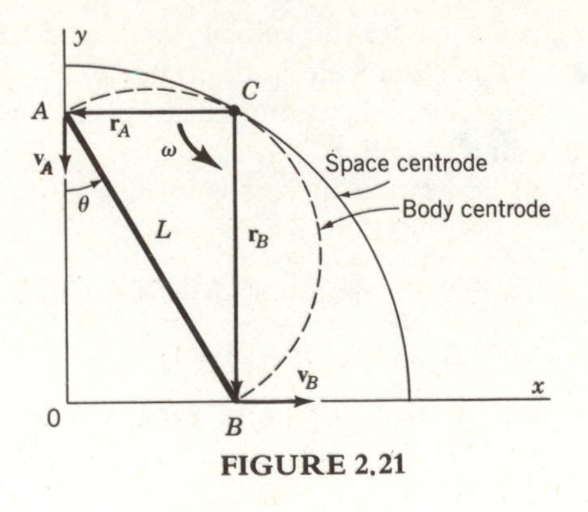

FIGURE 2.21

so that

$$\mathbf{a}_C = -300\mathbf{i} - 100.5\mathbf{j} \quad \text{m/s}^2 \tag{k}$$

Example 2.8

A rod of length $L = 10$ m slides with the end A touching the wall and the end B touching the floor (Fig. 2.21). If the velocity of A has the magnitude $v_A = 50$ m/s when the angle between the rod and the wall is $\theta = 30°$, determine the angular velocity of the rod and the velocity of B by using the concept of instantaneous center. Then, plot the body and space centrodes.

The instantaneous center lies at the intersection of the normal to the wall at A and the normal to the floor at B, as shown in Fig. 2.21. From Fig. 2.21, we obtain

$$r_A = L \sin\theta = 10 \times 0.5 = 5 \quad \text{m}, \qquad r_B = L\cos\theta = 10 \times 0.866 = 8.66 \quad \text{m} \tag{a}$$

Then, the angular velocity is simply

$$\omega = \frac{v_A}{r_A} = \frac{50}{5} = 10 \quad \text{rad/s} \tag{b}$$

in the counterclockwise sense and the magnitude of the velocity vector $\mathbf{v}_B$ is

$$v_B = \omega r_B = 10 \times 8.66 = 86.6 \quad \text{m/s} \tag{c}$$

Of course, the velocity is in the x-direction, as shown in Fig. 2.21.

From Fig. 2.21, we observe that the points A, 0, B, and C are the corners of a rectangle with the diagonals equal to L. Hence, C is always at a distance L from 0, so that the space centrode represents one quarter of a circle of radius L and with the center at 0. On the other hand, the vectors $\mathbf{r}_A$ and $\mathbf{r}_B$ make a 90° angle at C. From geometry, the locus of the points representing the corner of a right triangle corresponding to the right angle is a semicircle with the diameter coinciding with the side of the triangle opposing the right angle. This semicircle is the body centrode. The space and body centrodes are shown in Fig. 2.21 in solid and dashed lines, respectively. It is clear from Fig. 2.21 that, as the rod slides, the body centrode rolls on the space centrode.

2.6 GENERAL CASE OF MOTION

Let us return to the system of Section 2.4 and consider the case in which the particle P is no longer at rest relative to the moving frame xyz, but can move relative to that frame. To treat this case, it is advisable to derive first an expression for the time derivative of a vector with time-dependent magnitude and embedded in a rotating reference frame. Considering an arbitrary vector $\mathbf{r}$ and referring to Fig. 2.11, we can write

$$\mathbf{r}=x\mathbf{i}+y\mathbf{j}+z\mathbf{k} \tag{2.74}$$

where x, y, and z are the cartesian components of the vector and $\mathbf{i}$, $\mathbf{j}$, and $\mathbf{k}$ are unit vectors along these axes. In the case at hand, the components x, y, and z of the vector $\mathbf{r}$ are not constant, and, of course, the unit vectors $\mathbf{i}$, $\mathbf{j}$, and $\mathbf{k}$ are not constant either, as they rotate with the same angular velocity $\boldsymbol{\omega}$ as the moving frame. Hence, the time derivative of the vector $\mathbf{r}$ can be written in the form

$$\dot{\mathbf{r}}=\dot{x}\mathbf{i}+x\dot{\mathbf{i}}+\dot{y}\mathbf{j}+y\dot{\mathbf{j}}+\dot{z}\mathbf{k}+z\dot{\mathbf{k}} \tag{2.75}$$

where the time derivative of the unit vectors $\mathbf{i}$, $\mathbf{j}$, and $\mathbf{k}$ can be obtained from Eq. (2.51), as the derivative of any vector embedded in a rotating reference frame. Hence, replacing $\mathbf{r}$ by $\mathbf{i}$, $\mathbf{j}$, and $\mathbf{k}$ in Eq. (2.51) in sequence, we obtain

$$\dot{\mathbf{i}}=\boldsymbol{\omega}\times\mathbf{i}, \qquad \dot{\mathbf{j}}=\boldsymbol{\omega}\times\mathbf{j}, \qquad \dot{\mathbf{k}}=\boldsymbol{\omega}\times\mathbf{k} \tag{2.76}$$

Inserting Eqs. (2.76) into Eq. (2.75), we obtain the time derivative of $\mathbf{r}$

$$\dot{\mathbf{r}}=\dot{x}\mathbf{i}+\dot{y}\mathbf{j}+\dot{z}\mathbf{k}+\boldsymbol{\omega}\times(x\mathbf{i}+y\mathbf{j}+z\mathbf{k})=\dot{\mathbf{r}}'+\boldsymbol{\omega}\times\mathbf{r} \tag{2.77}$$

where

$$\dot{\mathbf{r}}'=\dot{x}\mathbf{i}+\dot{y}\mathbf{j}+\dot{z}\mathbf{k} \tag{2.78}$$

can be identified as the time rate of change of $\mathbf{r}$ regarding the reference frame xyz as inertial.

Equation (2.77) can be used to derive the velocity and acceleration of point P. From Fig. 2.22, the position vector of point P has the form

$$\mathbf{R}=\mathbf{r}_A+\mathbf{r}_{AP} \tag{2.79}$$

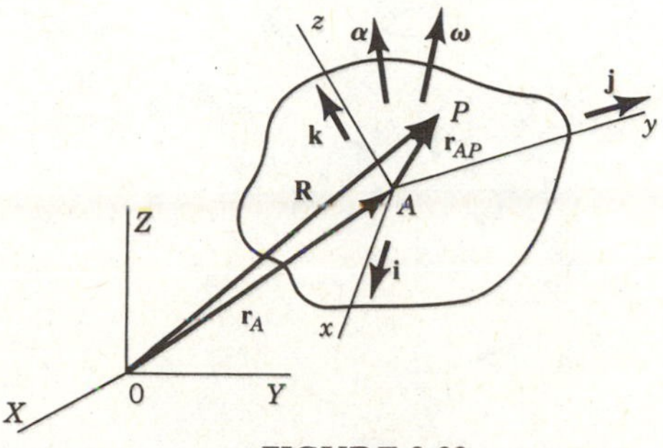

FIGURE 2.22

so that, using Eq. (2.77), we can write the absolute velocity of P as follows:

$$\mathbf{v}=\dot{\mathbf{R}}=\dot{\mathbf{r}}_A+\dot{\mathbf{r}}_{AP}=\mathbf{v}_A+\mathbf{v}'_{AP}+\boldsymbol{\omega}\times\mathbf{r}_{AP} \tag{2.80}$$

where $\mathbf{v}_A$ is the velocity of the origin A relative to the inertial space,

$$\mathbf{v}'_{AP}=\dot{x}_{AP}\mathbf{i}+\dot{y}_{AP}\mathbf{j}+\dot{z}_{AP}\mathbf{k} \tag{2.81}$$

is the velocity of P relative to the moving frame xyz, in which x_{AP}, y_{AP}, and z_{AP} are the cartesian components of $\mathbf{r}_{AP}$, and $\boldsymbol{\omega}\times\mathbf{r}_{AP}$ is the velocity of P due entirely to the rotation of the frame xyz. Similarly, from Eq. (2.80), we write the absolute acceleration of P in the form

$$\begin{aligned}\mathbf{a}=\dot{\mathbf{v}}&=\dot{\mathbf{v}}_A+\frac{d}{dt}(\mathbf{v}'_{AP})+\dot{\boldsymbol{\omega}}\times\mathbf{r}_{AP}+\boldsymbol{\omega}\times\dot{\mathbf{r}}_{AP}\\&=\mathbf{a}_A+\mathbf{a}'_{AP}+\boldsymbol{\omega}\times\mathbf{v}'_{AP}+\boldsymbol{\alpha}\times\mathbf{r}_{AP}+\boldsymbol{\omega}\times(\mathbf{v}'_{AP}+\boldsymbol{\omega}\times\mathbf{r}_{AP})\\&=\mathbf{a}_A+\mathbf{a}'_{AP}+2\boldsymbol{\omega}\times\mathbf{v}'_{AP}+\boldsymbol{\alpha}\times\mathbf{r}_{AP}+\boldsymbol{\omega}\times(\boldsymbol{\omega}\times\mathbf{r}_{AP})\end{aligned} \tag{2.82}$$

where $\mathbf{a}_A=\dot{\mathbf{v}}_A$ is the acceleration of A relative to the inertial space,

$$\mathbf{a}'_{AP}=\ddot{x}_{AP}\mathbf{i}+\ddot{y}_{AP}\mathbf{j}+\ddot{z}_{AP}\mathbf{k} \tag{2.83}$$

is the acceleration of P relative to the rotating frame xyz, $2\boldsymbol{\omega}\times\mathbf{v}'_{AP}$ is the so-called *Coriolis acceleration*, and $\boldsymbol{\alpha}\times\mathbf{r}_{AP}+\boldsymbol{\omega}\times(\boldsymbol{\omega}\times\mathbf{r}_{AP})$ is the acceleration of P due entirely to the rotation of the frame xyz, in which $\boldsymbol{\alpha}=\dot{\boldsymbol{\omega}}$ is the angular acceleration of the frame.

Example 2.9

An automobile travels north at the constant speed v relative to the ground. The earth rotates with the angular velocity $\boldsymbol{\Omega}$ relative to the inertial space, where $\boldsymbol{\Omega}$ can be assumed to be constant. Ignore the motion of the earth's center 0 relative to the inertial space, and calculate the velocity and acceleration of the automobile when the automobile is at a latitude λ.

Let us consider a reference frame xyz, with x pointing south, y pointing east, and z pointing to the zenith (Fig. 2.23). Then, recognizing that point P coincides

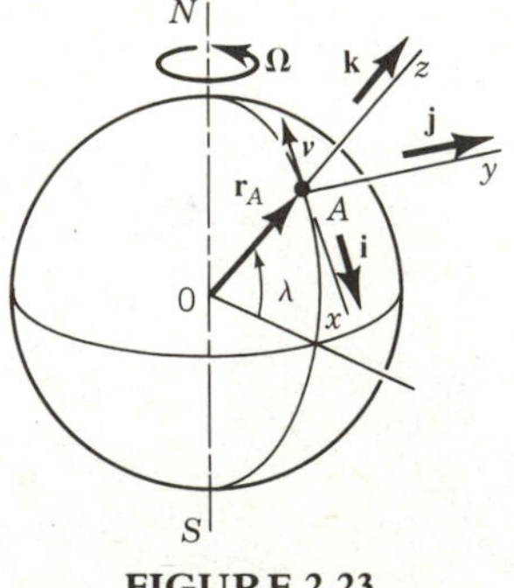

FIGURE 2.23

with point A, and that the automobile travels on a circular path of radius r_A, we can write

$$\mathbf{r}_A = r_A\mathbf{k}, \qquad \boldsymbol{\omega} = \Omega(-\cos\lambda\,\mathbf{i} + \sin\lambda\,\mathbf{k}),$$
$$\mathbf{r}_{AP} = \mathbf{0}, \qquad \mathbf{v}'_{AP} = -v\mathbf{i}, \qquad \mathbf{a}'_{AP} = -\frac{v^2}{r_A}\mathbf{k} \tag{a}$$

where r_A and Ω are the magnitudes of $\mathbf{r}_A$ and $\boldsymbol{\Omega}$, respectively. From Eq. (2.80), we can write the velocity of the automobile

$$\mathbf{v} = \mathbf{v}_A + \mathbf{v}'_{AP} = \mathbf{v}_A - v\mathbf{i} \tag{b}$$

where the velocity of the origin A is

$$\mathbf{v}_A = \boldsymbol{\omega} \times \mathbf{r}_A = \Omega(-\cos\lambda\,\mathbf{i} + \sin\lambda\,\mathbf{k}) \times r_A\mathbf{k} = \Omega r_A \cos\lambda\,\mathbf{j} \tag{c}$$

Hence,

$$\mathbf{v} = -v\mathbf{i} + \Omega r_A \cos\lambda\,\mathbf{j} \tag{d}$$

Using Eq. (2.82), we can write the acceleration of the automobile in the form

$$\begin{aligned}\mathbf{a} &= \mathbf{a}_A + \mathbf{a}'_{AP} + 2\boldsymbol{\omega} \times \mathbf{v}'_{AP} = \boldsymbol{\omega} \times (\boldsymbol{\omega} \times \mathbf{r}_A) + \mathbf{a}'_{AP} + 2\boldsymbol{\omega} \times \mathbf{v}'_{AP} \\ &= \Omega(-\cos\lambda\,\mathbf{i} + \sin\lambda\,\mathbf{k}) \times \Omega r_A \cos\lambda\,\mathbf{j} - \frac{v^2}{r_A}\mathbf{k} + 2\Omega(-\cos\lambda\,\mathbf{i} + \sin\lambda\,\mathbf{k}) \times (-v\mathbf{i}) \\ &= -\Omega^2 r_A \sin\lambda\cos\lambda\,\mathbf{i} - 2\Omega v \sin\lambda\,\mathbf{j} - \left(\frac{v^2}{r_A} + \Omega^2 r_A \cos^2\lambda\right)\mathbf{k}\end{aligned} \tag{e}$$

Example 2.10

A sprinkler rotates with the constant angular velocity ω, as shown in Fig. 2.24. The pipe is in the form of an arc of a circle of radius of curvature ρ. Determine the velocity and acceleration of a particle of water as it leaves the pipe with the constant velocity $\dot{s}$ relative to the pipe.

Let us introduce a reference frame xyz, with x and y as shown in Fig. 2.24 and with z perpendicular to x and y, and let $\mathbf{i}$, $\mathbf{j}$, and $\mathbf{k}$ be unit vectors along these axes. In addition, let t and n be tangential and normal directions with associated unit vectors $\mathbf{u}_t$ and $\mathbf{u}_n$, where $\mathbf{u}_t$ makes a 30° angle relative to $\mathbf{i}$. Note that point A

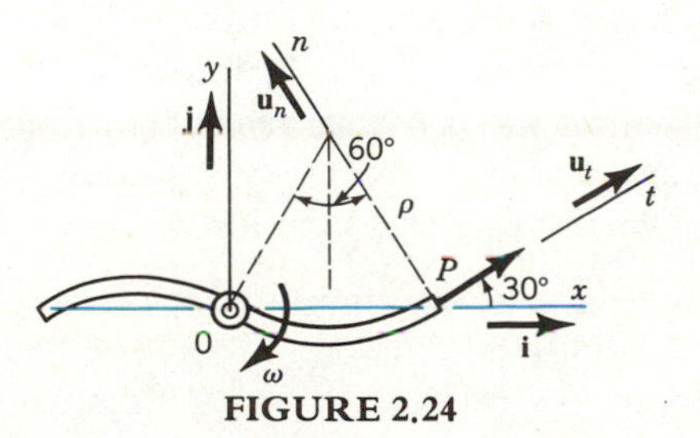

FIGURE 2.24

coincides with point 0. In terms of this notation, we have

$$\mathbf{r}_{AP}=\rho\mathbf{i}, \qquad \boldsymbol{\omega}=-\omega\mathbf{k}, \qquad \mathbf{v}'_{AP}=v_t\mathbf{u}_t=\dot{s}\mathbf{u}_t=\frac{\dot{s}}{2}(\sqrt{3}\mathbf{i}+\mathbf{j}) \tag{a}$$

Then, using Eq. (2.80) and recognizing that $\mathbf{v}_A=\mathbf{0}$, we can write

$$\begin{aligned}\mathbf{v}&=\mathbf{v}'_{AP}+\boldsymbol{\omega}\times\mathbf{r}_{AP}=\frac{\dot{s}}{2}(\sqrt{3}\mathbf{i}+\mathbf{j})+(-\omega\mathbf{k})\times\rho\mathbf{i}\\&=\frac{\sqrt{3}}{2}\dot{s}\mathbf{i}+\left(\frac{\dot{s}}{2}-\omega\rho\right)\mathbf{j}\end{aligned} \tag{b}$$

To calculate the acceleration, we recognize that, although $\dot{s}$ is constant in magnitude, we have the relative acceleration

$$\mathbf{a}'_{AP}=a_n\mathbf{u}_n=\frac{\dot{s}^2}{\rho}\mathbf{u}_n=\frac{\dot{s}^2}{2\rho}(-\mathbf{i}+\sqrt{3}\mathbf{j}) \tag{c}$$

Moreover, $\mathbf{a}_A=\dot{\boldsymbol{\omega}}=\mathbf{0}$, so that Eq. (2.82) yields

$$\begin{aligned}\mathbf{a}&=\mathbf{a}'_{AP}+2\boldsymbol{\omega}\times\mathbf{v}'_{AP}+\boldsymbol{\omega}\times(\boldsymbol{\omega}\times\mathbf{r}_{AP})\\&=\frac{\dot{s}^2}{2\rho}(-\mathbf{i}+\sqrt{3}\mathbf{j})+2(-\omega)\mathbf{k}\times\frac{\dot{s}}{2}(\sqrt{3}\mathbf{i}+\mathbf{j})+(-\omega\mathbf{k})\times(-\omega\rho\mathbf{j})\\&=\left(-\frac{\dot{s}^2}{2\rho}+\omega\dot{s}-\omega^2\rho\right)\mathbf{i}+\left(\frac{\sqrt{3}\dot{s}^2}{2\rho}-\sqrt{3}\omega\dot{s}\right)\mathbf{j}\end{aligned} \tag{d}$$

PROBLEMS

2.1 An automobile moves according to the formula $s(t)=-0.05t^3+2t^2+100$, where s is the distance in meters and t is the time in seconds. Determine the following: (1) the initial acceleration, (2) the velocity reached in 10 s, (3) the maximum velocity reached and the time when this occurs, and (4) the distance at $t=20$ s.

2.2 Plot the distance, velocity, and acceleration as functions of time for the automobile of Problem 2.1 for $0<t<20$ s.

2.3 An automobile is traveling at 72 km/h when the driver observes that a traffic light 240 m ahead is turning red. The driver knows that the traffic light is timed to stay red for 20 s. Determine the following: (1) the uniform deceleration that would permit the driver to reach the traffic light just as it turns green and (2) the speed of the automobile as it reaches the traffic light. Derive an expression for the distance as a function of time.

2.4 Water is dripping from a leaking faucet at constant intervals of time. At the instant a drop is ready to fall, the preceding drop has already fallen 1.25 m. Determine the following: (1) the interval of time between the beginning of free fall of two succeeding water drops and (2) the distance between the two drops as a function of time.

2.5 Rocks leave the chute shown in Fig. 2.25 with the velocity $v=5$ m/s. The rocks are to drop in a basket 10 m below. At what distance B should the basket be placed? Determine the velocity and the angle with respect to the horizontal at which the rocks enter the basket.

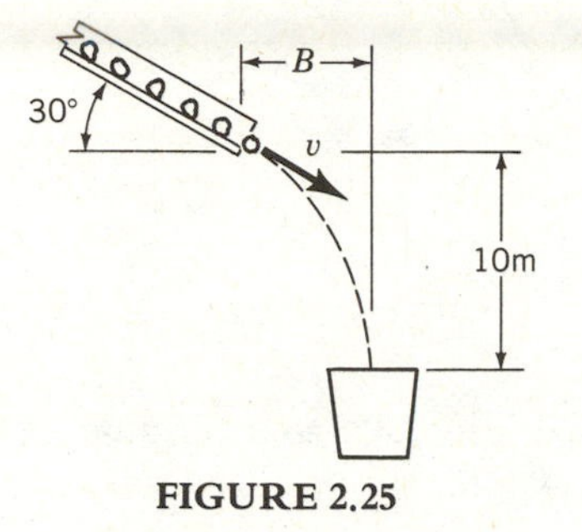

FIGURE 2.25

2.6 A particle travels in a resisting medium with the acceleration $a=-0.01v^2$. If the particle begins its motion with the velocity $v_0=15$ m/s, determine the distance traveled by the time the velocity falls to $v=10$ m/s.

2.7 A particle moves with the acceleration $a=-cs$, where c is a constant and s is the distance. Determine an equation relating the velocity v and the distance s, and plot the function $v=v(s)$ for the initial values $s_0=0$, $v_0=2$, and the constant $c=4$.

2.8 A satellite moves around the earth in a circular orbit of radius $R=6600$ km. Assuming that the radial acceleration (due to the attraction of the earth) is equal to $0.94g$, where g is the acceleration due to gravity at sea level, determine the velocity of the satellite and the orbital period (the time necessary for one complete revolution).

2.9 The orbit of a satellite moving around the earth is given by the expression

$$r=\frac{p}{1+e\cos\theta}$$

where r is the magnitude of the radius vector from the center of the earth to the satellite and θ is the angle between a given reference direction and the radius vector. Knowing that the satellite motion satisfies the equation

$$r^2\dot{\theta}=h=\text{const}$$

determine the velocity and acceleration of the satellite as a function of θ.

2.10 A cam rotates with the constant angular velocity ω, as shown in Fig. 2.26. The follower is constrained so as to move vertically only. Assuming that the roller is always in contact with the cam, determine the velocity and acceleration of the follower if the shape of the cam is given by

$$r = e - b\cos\theta, \qquad e > b$$

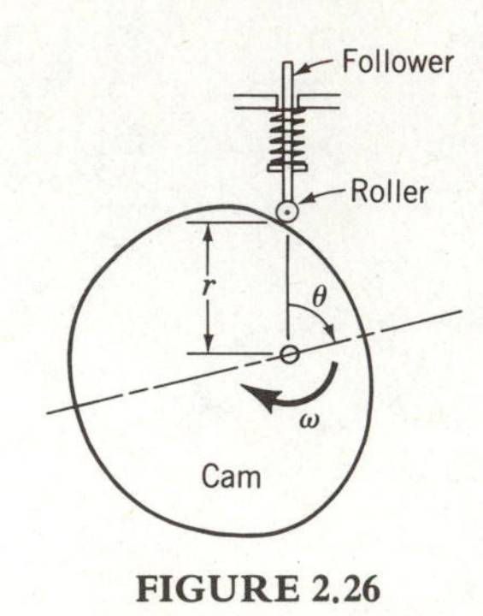

FIGURE 2.26

2.11 Consider the particle of Fig. 2.4, and determine the radius of curvature of the trajectory immediately after the particle leaves point 0.

2.12 Determine the radius of curvature of the trajectory described by the rocks of Problem 2.5 at the time they enter the basket.

2.13 During a rainstorm the raindrops have been observed from a side window of an automobile stopped at a red light to fall with the velocity of 12 m/s at an angle of 10° with respect to the vertical. After the light has changed, the automobile resumed its motion and reached a given cruising speed at which time the raindrops were observed from the same side window to fall at an angle of 40° with respect to the vertical. Determine the velocity of the automobile and of the raindrops relative to the automobile.

2.14 A boy on a bicycle traveling at 5 m/s wants to throw a ball into an open truck directly ahead of him and traveling at 10 m/s on a street making a 90° angle with the street on which he is traveling. If he can throw the ball with the velocity of 20 m/s relative to the bicycle, determine the direction in which he must throw the ball and the velocity of the ball relative to the ground.

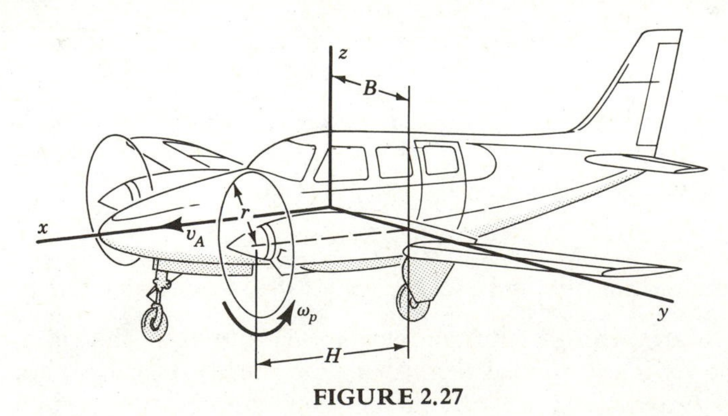

FIGURE 2.27

2.15 The propeller airplane shown in Fig. 2.27 is taxiing in a straight line with the uniform velocity v_A while the propeller rotates with the angular velocity ω_p. Determine the velocity and acceleration of a point at the tip of the propeller blade when the blade is horizontal.

2.16 Repeat Problem 2.15 for the case in which the airplane makes a turn of radius R to the right.

2.17 The gun turret shown in Fig. 2.28 rotates with the angular velocity ω_t and angular acceleration α_t when the gun barrel is being raised with the angular velocity ω_b. Determine the velocity and acceleration of a point at the end of the barrel when the barrel makes an arbitrary angle θ with respect to the horizontal.

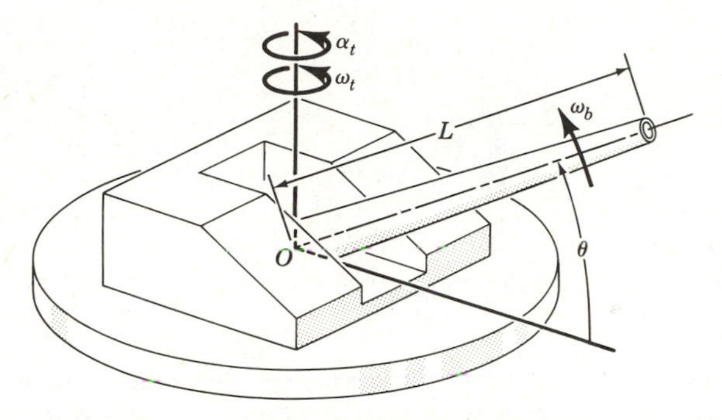

FIGURE 2.28

2.18 A disk of radius r rolls without slip inside a circular cylinder of radius R, as shown in Fig. 2.29. Determine the acceleration of point A and the velocity and acceleration of point B. Express the results in terms of $\dot{\theta}$ and $\ddot{\theta}$.

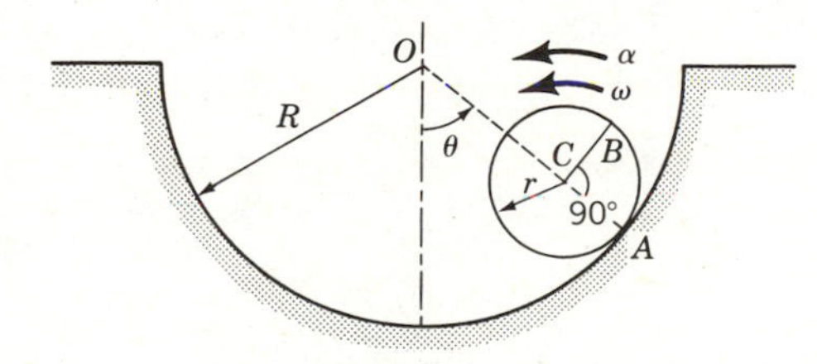

FIGURE 2.29

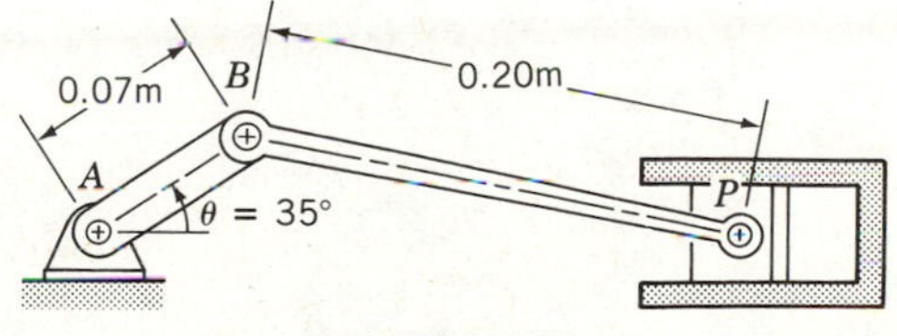

FIGURE 2.30

2.19 The crank AB of the automobile engine shown in Fig. 2.30 rotates counterclockwise with the constant angular velocity of 3000 rotations per minute (rpm). Determine the following: (1) the angular velocity of the connecting rod BP, (2) the velocity of the piston P, (3) the angular acceleration of the connecting rod, and (4) the acceleration of the piston.

2.20 In the position shown in Fig. 2.31, bar AB rotates counterclockwise with the constant angular velocity $\omega_{AB} = 10$ rad/s. Determine the following: (1) the angular velocity of bar BC, (2) the velocity of point C, (3) the angular velocity of bar CD, (4) the angular acceleration of bar BC, (5) the angular acceleration of bar CD, and (6) the acceleration of point C.

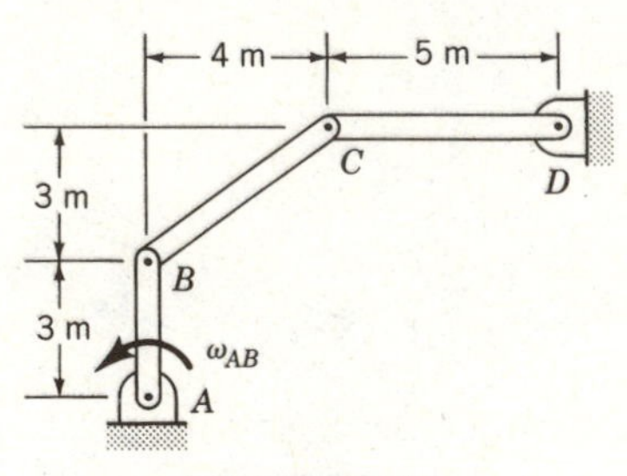

FIGURE 2.31

2.21 Solve the problem of Example 2.6 by using only the rotating reference frame xyz and by regarding the motion of P as motion relative to that frame.

2.22 Solve Problem 2.16 by using only one reference frame, namely, one attached to the airplane.

2.23 Solve Problem 2.17 by using only one reference frame, namely, one attached to the turret.

2.24 A river is flowing east with the constant speed of 3 km/h. Determine the acceleration of a drop of water. The latitude of the river is $\lambda = 30°$.

CHAPTER 3
Dynamics of a Particle

3.1 INTRODUCTION

In Chapter 2, we considered the motion of a particle without regard to the forces causing the motion. In this chapter, we propose to relate the motion to the forces causing it. This relation is commonly known as the equations of motion.

The beginnings of dynamics can be traced to the development of geometry by the ancient Greeks, such as Euclid and Pythagoras. Indeed, Euclidean geometry was later accepted as the framework for Newtonian mechanics. The Greeks were not successful in explaining the motion of bodies, however. For example, Aristotle believed that wherever there is motion there must be a force, although he was puzzled by the ability of bodies to move even in the absence of forces. The first to place the study of dynamics on a sound scientific foundation was Galileo. His *law of inertia* states that force causes a change in velocity but that no force is necessary to maintain motion in which the magnitude and direction of the velocity do not change, thus anticipating Newton's first law and perhaps the second law as well. Galileo postulated the existence of an *inertial space* or *Galilean reference frame*, where an inertial system is either at rest or translating with uniform velocity relative to a fixed space. It was Newton, however, who formulated the laws of motion in a clear and concise manner. Moreover, Newton's correct interpretation of Kepler's planetary laws resulted in his *law of gravitation*. Based on observations of the motion of planets made by Tycho Brahe, Kepler enunciated three laws of planetary motion. These laws were strictly geometric in nature, and it remained for Newton to give them physical content by demonstrating that the motion of planets is governed by the so-called *inverse square law*. Of course, the same law governs the motion of man-made satellites.

Newton's laws dominated mechanics through most of the nineteenth century, although in the latter part of the nineteenth century cracks began to appear. These discrepancies were finally resolved by Einstein's special and general relativity theories. The incidents in which Newtonian mechanics fails to provide the correct answers are not very common in engineering practice, so that Newtonian mechanics can be accepted with confidence, unless relativistic effects are present. Relativistic effects become important when the bodies under consideration move with velocities approaching in magnitude the speed of light. Cases in which relativistic mechanics must be invoked lie beyond the scope of this text.

3.2 NEWTON'S LAWS

Newton's laws were formulated for a single particle and can be extended to systems of particles and rigid bodies. There are three laws of motion, although the first law is merely a special case of the second law. Together with Newton's gravitational law, they form the basis for the so-called *Newtonian mechanics.* Newton referred to his laws as *axioms.*

Newtonian mechanics postulates the existence of *inertial systems of reference,* that is, systems of reference that are either at rest or moving with uniform velocity relative to a fixed reference frame. The motion of any particle is measured relative to such an inertial system and is said to be *absolute.* Newton's laws can be stated as follows:

First Law. *If there are no forces acting on a particle, then the particle will move in a straight line with constant velocity.*

A particle is defined as an idealization of a material body whose dimensions are very small when compared with the distance to other bodies, so that in essence a particle can be regarded as a point mass. Denoting the *resultant force vector* by $\mathbf{F}$ and the *absolute velocity vector* by $\mathbf{v}$, where we recall that an absolute velocity is measured relative to an inertial frame of reference, we can state the first law mathematically as

$$\text{If} \quad \mathbf{F}=\mathbf{0}, \quad \text{then} \quad \mathbf{v}=\text{const} \tag{3.1}$$

Second Law. *A particle acted on by a force moves so that the force vector is equal to the time rate of change of the linear momentum vector.*

The *linear momentum vector,* denoted $\mathbf{p}$, is defined as the product of the mass m of the particle and the absolute velocity vector $\mathbf{v}$, or

$$\mathbf{p}=m\mathbf{v} \tag{3.2}$$

Then, the second law can be stated mathematically as

$$\mathbf{F}=\frac{d\mathbf{p}}{dt}=\frac{d}{dt}(m\mathbf{v}) \tag{3.3}$$

In SI units, the unit of mass is the kilogram (kg) and the unit of force is the newton (N). The kilogram is a basic unit and the newton is a derived unit and is such that $1\text{N}=1\ \text{kg}\cdot\text{m/s}^2$. The mass of the particle is defined as a positive quantity whose value does not depend on time, so that Eq. (3.3) can be rewritten in the familiar form

$$\mathbf{F}=m\mathbf{a} \tag{3.4}$$

where

$$\mathbf{a}=d\mathbf{v}/dt \tag{3.5}$$

is the *absolute acceleration* of m. Clearly, the first law is a special case of the second law. Equation (3.4) represents the *equations of motion* for a particle.

Third Law. *When two particles exert forces on one another, the forces act along the line joining the particles, and they are equal in magnitude but opposite in directions.*

This law is also known as the *law of action and reaction.* Denoting by $\mathbf{F}_{12}$ the force exerted by particle m_2 on particle m_1, and vice versa, the third law can be stated mathematically as

$$\mathbf{F}_{12} = -\mathbf{F}_{21} \tag{3.6}$$

where the vectors $\mathbf{F}_{12}$ and $\mathbf{F}_{21}$ are *collinear* (Fig. 3.1). Because of this, there are no moments acting on the particles m_1 and m_2. There are cases, however, when moments do act on particles, in which cases the forces are no longer collinear. This is the case when there are electromagnetic forces between moving particles. Such exceptions to the third law are very rare indeed and will be excluded from further discussion.

In addition to the three laws of motion, Newton formulated the *law of universal gravitation.* The law states that gravity produces mutual attraction forces of the type shown in Fig. 3.1 and that the magnitude of these forces is given by

$$F(r) = \frac{Gm_1 m_2}{r^2} \tag{3.7}$$

where r is the distance between the two particles and G is the *universal gravitational constant.* Later in this chapter we shall show how to establish the value of G. The gravitational law, Eq. (3.7), is commonly known as the *inverse square law.*

Newton's second law relates the forces acting on a particle with the acceleration of the particle. Hence, if some of these quantities are given, then one can use Newton's second law to solve for the remaining quantities. Of course, the number of unknowns must be equal to the number of equations of motion. An indispensable tool in the application of Newton's second law is the *free-body diagram,* a diagram containing all the forces acting on the particle. In drawing a free-body diagram, one must isolate the particle from any other particles. If the particle is attached to a massless member, such as a spring or a string, then in isolating the particle any force internal to the system, such as the force in the spring or string, becomes external to the system and must be treated as such. The concept of free-body diagram is equally useful for rigid bodies. We shall have ample opportunity to use the concept in this text.

Example 3.1

A stuntman rides a motorcycle inside a vertical cylindrical wall of radius $R = 15$ m (Fig. 3.2*a*). If the minimum velocity required to perform the stunt is $v = 54$ km/h,

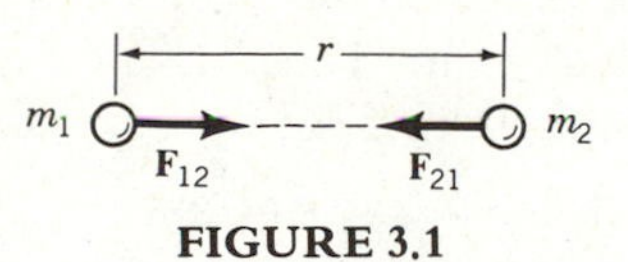

FIGURE 3.1

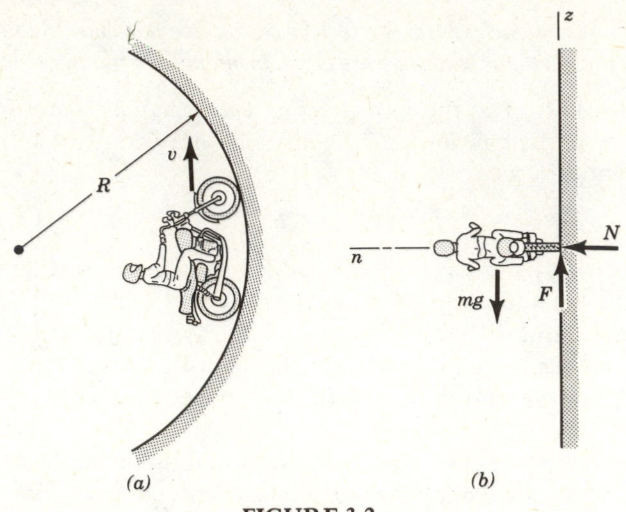

FIGURE 3.2

determine the coefficient of friction μ between the tires and the wall. Treat the system as a particle.

The equations of motion in the normal and vertical directions (Fig. 3.2*b*) are

$$F_n = N = ma_n$$
$$F_z = F - mg = ma_z = 0 \tag{a}$$

where

$$F = \mu N \tag{b}$$

From Eqs. (2.41), the acceleration component in the normal direction is

$$a_n = \frac{v^2}{R} \tag{c}$$

Solving Eqs. (a), (b), and (c), we obtain

$$\mu = \frac{mg}{N} = \frac{gR}{v^2} = \frac{9.81 \times 15}{(54 \times 1000/3600)^2} = 0.654 \tag{d}$$

3.3 INTEGRATION OF THE EQUATIONS OF MOTION

Let us consider the particle m shown in Fig. 3.3 and assume that the rectangular axes xyz represent an inertial frame. Then, the absolute position of m can be expressed in terms of the cartesian components x, y, and z by the radius vector

$$\mathbf{r} = x\mathbf{i} + y\mathbf{j} + z\mathbf{k} \tag{3.8}$$

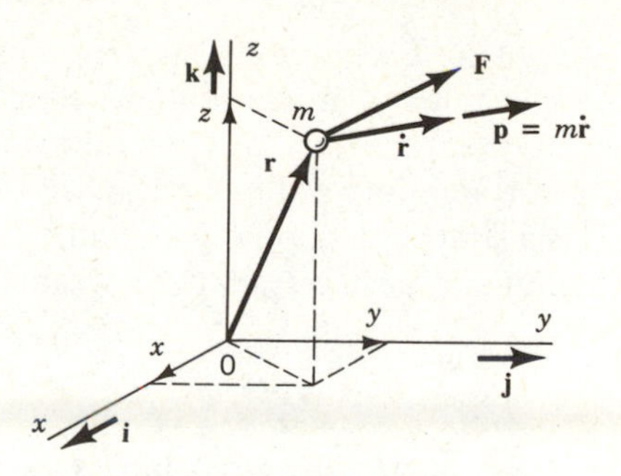

FIGURE 3.3

where **i**, **j**, and **k** are associated constant unit vectors. From Section 2.2, we can write the velocity and acceleration of m in the form

$$\mathbf{v} = \dot{\mathbf{r}} = v_x \mathbf{i} + v_y \mathbf{j} + v_z \mathbf{k} \tag{3.9a}$$

$$\mathbf{a} = \ddot{\mathbf{r}} = a_x \mathbf{i} + a_y \mathbf{j} + a_z \mathbf{k} \tag{3.9b}$$

where

$$v_x = \dot{x}, \qquad v_y = \dot{y}, \qquad v_z = \dot{z} \tag{3.10a}$$

$$a_x = \ddot{x}, \qquad a_y = \ddot{y}, \qquad a_z = \ddot{z} \tag{3.10b}$$

are the cartesian components of the velocity and acceleration vectors, respectively.

The force **F** depends in general on the position **r**, the velocity **v**, and possibly the time t, or

$$\mathbf{F} = \mathbf{F}(\mathbf{r}, \mathbf{v}, t) \tag{3.11}$$

The force vector can be expressed as follows:

$$\mathbf{F} = F_x \mathbf{i} + F_y \mathbf{j} + F_z \mathbf{k} \tag{3.12}$$

where

$$F_x = F_x(x, y, z, \dot{x}, \dot{y}, \dot{z}, t), \quad F_y = F_y(x, y, z, \dot{x}, \dot{y}, \dot{z}, t), \quad F_z = F_z(x, y, z, \dot{x}, \dot{y}, \dot{z}, t) \tag{3.13}$$

are the cartesian components of the vector.

The motion of m is governed by Newton's second law, as expressed by Eq. (3.4). Introducing Eqs. (3.9b), (3.10b), (3.12), and (3.13) into Eq. (3.4), we can write the three cartesian components of the equations of motion in the form

$$m\ddot{x} = F_x(x, y, z, \dot{x}, \dot{y}, \dot{z}, t) \tag{3.14a}$$

$$m\ddot{y} = F_y(x, y, z, \dot{x}, \dot{y}, \dot{z}, t) \tag{3.14b}$$

$$m\ddot{z} = F_z(x, y, z, \dot{x}, \dot{y}, \dot{z}, t) \tag{3.14c}$$

Equations (3.14) represent a set of three second-order ordinary differential equations of motion. At least in principle, they can be integrated to determine the three components $x(t)$, $y(t)$, and $z(t)$ of the motion. This requires the explicit form of the components F_x, F_y, and F_z of the force vector $\mathbf{F}$, as well as six conditions permitting the evaluation of the six constants of integration involved. The six conditions are normally the initial displacements $x(0)$, $y(0)$, $z(0)$ and the initial velocities $\dot{x}(0)$, $\dot{y}(0)$, $\dot{z}(0)$. In practice, however, straightforward integration of the equations of motion is possible only when $\mathbf{F}$ is a simple function of $\mathbf{r}$, $\mathbf{v}$, and t, or in some special cases. In more complicated cases, it may be necessary to integrate the equations of motion numerically.

Example 3.2

Let us consider a particle m moving in the vicinity of a point 0 on the earth's surface (Fig. 3.4*a*) and determine the motion of the particle if the initial conditions are $x(0)=z(0)=0$, $\dot{x}(0)=v_{x0}$, and $\dot{z}(0)=v_{z0}$.

We assume that the motion involves distances that are much smaller than the radius of the earth. Letting $m_1=m$, $m_2=M$, and $r\cong R$ in Eq. (3.7), where M and R are the mass and the radius of the earth, respectively, the force vector can be written in the form (Fig. 3.4*b*)

$$\mathbf{F}=-\frac{mMG}{R^2}\mathbf{k}=-mg\mathbf{k}=\text{const} \tag{a}$$

where $g=GM/R^2$ is the gravitational acceleration. Equation (a) implies that in

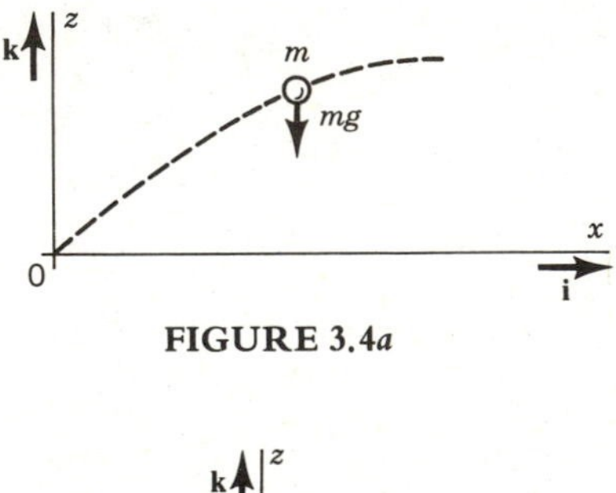

FIGURE 3.4*a*

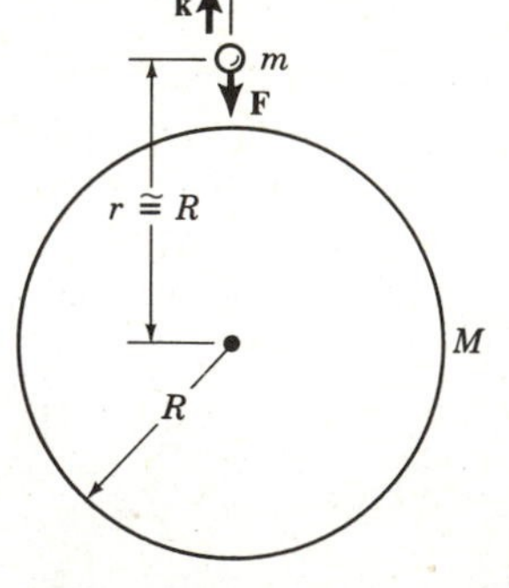

FIGURE 3.4*b*

the neighborhood of a given point on the earth's surface the gravitational force field can be assumed to be uniform.

Equations (3.14) in conjunction with Eq. (a) yield the differential equations of motion

$$m\ddot{x}=0, \qquad m\ddot{z}=-mg \tag{b}$$

Integrating Eqs. (b) twice with respect to time and considering the initial conditions, we obtain

$$x=v_{x0}t, \qquad z=v_{z0}t-\tfrac{1}{2}gt^2 \tag{c}$$

This problem has already been solved in Section 2.2 by regarding it simply as a kinematical problem. Of course, the solution is the same, as can be observed from Eqs. (2.18), where we recognize that $v_0 \cos\alpha=v_{x0}$ and $v_0 \sin\alpha=v_{z0}$. Eliminating t from Eqs. (c), one can derive the trajectory equation, that is, an equation relating x and z alone, with the time t acting as an implicit parameter. The trajectory equation has the form

$$z=x\frac{v_{z0}}{v_{x0}}-\frac{1}{2}g\frac{x^2}{v_{x0}^2} \tag{d}$$

and can be verified to be identical to Eq. (2.20) in Section 2.2.

Example 3.3

A rock of mass m is thrown vertically upward from the earth's surface with the initial velocity $\mathbf{v}_0$. Assume that there is a force proportional to the velocity resisting the motion (Fig. 3.5), and derive an expression for the velocity of the rock when it hits the ground upon its return.

Because the motion is one dimensional, we can dispense with the vector notation and write the differential equation of motion

$$\sum F_z=-mg-c\dot{z}=m\ddot{z} \tag{a}$$

where c is a proportionality constant. Equation (a) can be rewritten as

$$m\ddot{z}+\dot{c}z=-mg \tag{b}$$

Equation (b) is of second order, so that its solution requires two initial conditions.

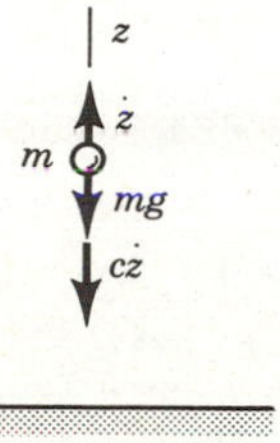

FIGURE 3.5

These conditions are

$$z(0)=0, \qquad \dot{z}(0)=v_0 \tag{c}$$

The solution of Eq. (b) can be obtained conveniently by the Laplace transformation method. Transforming both sides of Eq. (b), we obtain

$$m[s^2Z(s)-\dot{z}(0)-sz(0)]+c[sZ(s)-z(0)]=-\frac{mg}{s} \tag{d}$$

where $Z(s)$ is the Laplace transform of $z(t)$. Considering Eqs. (c), we can solve Eq. (d) for $Z(s)$ as follows:

$$Z(s)=\frac{v_0}{s(s+c/m)}-\frac{g}{s^2(s+c/m)} \tag{e}$$

The solution of Eq. (b) is obtained by carrying out the inverse Laplace transformation on Eq. (e). To this end, we recall from Example 1.4 that

$$\mathscr{L}^{-1}\frac{1}{s(s+c/m)}=\frac{m}{c}(1-e^{-ct/m}) \tag{f}$$

Moreover, let us consider the partial fractions expansion

$$\frac{1}{s^2(s+c/m)}=\frac{A+Bs}{s^2}+\frac{C}{s+c/m}=\frac{(A+Bs)(s+c/m)+Cs^2}{s^2(s+c/m)} \tag{g}$$

Equating the coefficients of s^2 and s at the numerator of the last fraction in Eq. (g) to zero and the remaining term to 1, we obtain the equations

$$B+C=0, \qquad A+Bc/m=0, \qquad Ac/m=1 \tag{h}$$

which have the solution

$$A=m/c, \qquad B=-C=-(m/c)^2 \tag{i}$$

so that the partial fractions expansion has the form

$$\frac{1}{s^2(s+c/m)}=\frac{m}{c}\frac{1}{s^2}-\left(\frac{m}{c}\right)^2\left(\frac{1}{s}-\frac{1}{s+c/m}\right) \tag{j}$$

Using the table of Laplace transform pairs in Section A.7, we can write

$$\mathscr{L}^{-1}\frac{1}{s^2(s+c/m)}=\frac{m}{c}t-\left(\frac{m}{c}\right)^2(1-e^{-ct/m}) \tag{k}$$

so that, considering Eqs. (f) and (k), we conclude that the inverse transformation of Eq. (e), and hence the solution of Eq. (b), is

$$\begin{aligned} z(t)&=\frac{mv_0}{c}(1-e^{-ct/m})-\frac{mg}{c}\left[t-\frac{m}{c}(1-e^{-ct/m})\right] \\ &=-\frac{m}{c}\left[gt-\left(v_0+\frac{mg}{c}\right)(1-e^{-ct/m})\right] \end{aligned} \tag{l}$$

To obtain the velocity of the rock when it strikes the ground, we must first produce an expression for the velocity. Differentiating Eq. (1) with respect to time, we obtain simply

$$\dot{z}(t) = -\frac{mg}{c} + \left(v_0 + \frac{mg}{c}\right) e^{-ct/m} \tag{m}$$

The rock strikes the ground at the instant $t = t_f$, where t_f denotes the final time, at which time $z(t_f) = 0$. Hence, from Eq. (1), t_f is the solution of the equation

$$gt_f - \left(v_0 + \frac{mg}{c}\right)(1 - e^{-ct_f/m}) = 0 \tag{n}$$

so that, letting $\dot{z}(t_f) = v_f$ in Eq. (m), where v_f is the final velocity of the rock, and considering Eq. (n), we obtain

$$v_f = v_0 - gt_f \tag{o}$$

where t_f is the solution of Eq. (n).

In general, the rock increases its velocity continuously as it falls toward the ground. However, under certain circumstances, the rock can reach constant velocity before it strikes the ground. Indeed, if the ratio c/m is relatively large, then for a certain value of time such that $ct \gg m$, we conclude from Eq. (m) that the velocity reaches the constant value

$$\dot{z} = -\frac{mg}{c} = \text{const} \tag{p}$$

Because the velocity is constant for any subsequent time, it must be the same velocity with which the rock strikes the ground, or

$$v_f = -\frac{mg}{c} \tag{q}$$

The negative sign can be explained by the fact that, when the rock strikes the ground, its velocity is directed along the negative direction of the z-axis.

3.4 IMPULSE AND MOMENTUM

The motion of a particle can be determined by integrating the equations of motion, as shown in Section 3.3. This task can be made easier at times by one of several principles based on Newton's second law. One such principle is the impulse-momentum principle, which is simply an integral with respect to time of Newton's second law.

The *linear impulse vector* is defined as the time integral of the force vector between two distinct times. Letting these times be t_1 and t_2, we can write the mathematical definition of the linear impulse vector

$$\hat{\mathbf{F}} = \int_{t_1}^{t_2} \mathbf{F}\, dt \tag{3.15}$$

where $\hat{\mathbf{F}}$ has unit newton·seconds. Multiplying Eq. (3.3) by dt and integrating between the times t_1 and t_2, we obtain

$$\hat{\mathbf{F}} = \int_{t_1}^{t_2} \mathbf{F}\, dt = \int_{t_1}^{t_2} \frac{d\mathbf{p}}{dt}\, dt = \mathbf{p}_2 - \mathbf{p}_1 = m\mathbf{v}_2 - m\mathbf{v}_1 \tag{3.16}$$

where

$$\mathbf{p}_i = m\mathbf{v}_i = m\mathbf{v}(t_i), \qquad i = 1, 2 \tag{3.17}$$

is the linear momentum vector at $t = t_i$. Letting

$$\Delta\mathbf{p} = \mathbf{p}_2 - \mathbf{p}_1 \tag{3.18}$$

be the change in the linear momentum vector between the times t_1 and t_2, we can rewrite Eq. (3.16) in the form

$$\hat{\mathbf{F}} = \Delta\mathbf{p} \tag{3.19}$$

or, *the linear impulse vector corresponding to the times t_1 and t_2 is equal to the change in the linear momentum vector between the same two instants.* It should be stressed that both the impulse and the linear momentum are vector quantities. Equation (3.19) is sometimes referred to as a first integral of motion with respect to time.

If there are no forces acting on the particle ($\mathbf{F} = \mathbf{0}$), Eq. (3.16) yields

$$\mathbf{p}_2 = \mathbf{p}_1 = \mathbf{p} = \text{const} \tag{3.20}$$

or *in the absence of forces the linear momentum does not change.* Equation (3.20) is the mathematical statement of the principle of *conservation of linear momentum.* It is also a restatement of Newton's first law.

Example 3.4

A ball of mass $m = 10^{-2}$ kg approaches a rigid wall with the velocity $v_1 = 40$ m/s normal to the wall (Fig. 3.6*a*), hits the rigid wall (Fig. 3.6*b*), and returns with the velocity $v_2 = 30$ m/s also normal to the wall (Fig. 3.6*c*). It is known that the force between the ball and the wall during impact is a triangular function of time (Fig. 3.7) and that the ball is in contact with the wall for a time interval $\Delta t = 0.01$ s. Calculate the maximum force on the ball.

The motion is one dimensional, so that, choosing the return direction as positive, we can write the impulse-momentum relation

$$\tfrac{1}{2}\mathbf{F}_{\max}\,\Delta t = mv_2 - (-mv_1) = m(v_2 + v_1) \tag{a}$$

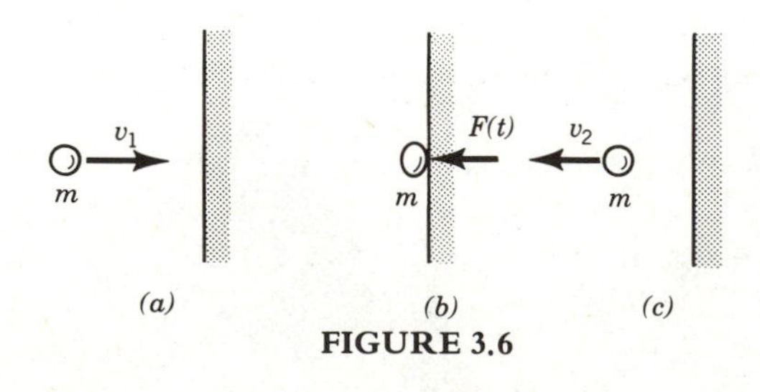

FIGURE 3.6

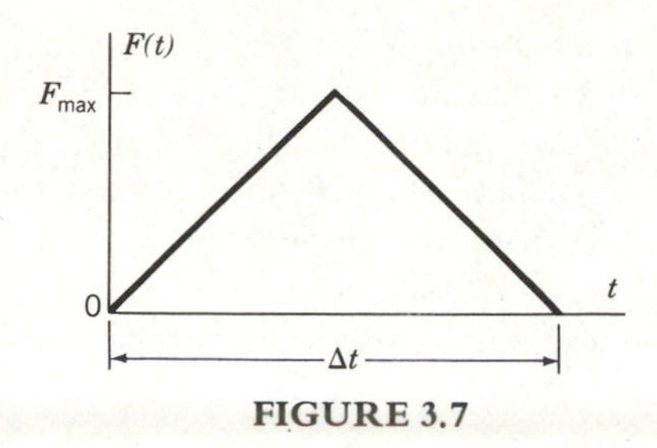

FIGURE 3.7

or

$$F_{max}=\frac{2m(v_2+v_1)}{\Delta t}=\frac{2\times 10^{-2}(40+30)}{0.01}=140 \quad \text{N} \tag{b}$$

3.5 MOMENT OF A FORCE AND ANGULAR MOMENTUM ABOUT A FIXED POINT

Let us consider a particle of mass m moving under the action of a force $\mathbf{F}$. We shall denote the position of m relative to the origin 0 of the inertial frame xyz by $\mathbf{r}$ and the absolute velocity of m by $\dot{\mathbf{r}}$ (Fig. 3.3). The *moment of momentum* or *angular momentum* of m with respect to point 0 is defined as the moment of the linear momentum $\mathbf{p}$ about 0 and is represented mathematically by the cross product (vector product) of the radius vector $\mathbf{r}$ and the linear momentum vector $\mathbf{p}$. Denoting the angular momentum about 0 by $\mathbf{H}_0$, we can write

$$\mathbf{H}_0=\mathbf{r}\times\mathbf{p}=\mathbf{r}\times m\dot{\mathbf{r}} \tag{3.21}$$

Let us consider now the rate of change of $\mathbf{H}_0$. Assuming that m is constant, we have

$$\dot{\mathbf{H}}_0=\dot{\mathbf{r}}\times m\dot{\mathbf{r}}+\mathbf{r}\times m\ddot{\mathbf{r}}=\mathbf{r}\times m\ddot{\mathbf{r}} \tag{3.22}$$

where $\dot{\mathbf{r}}\times m\dot{\mathbf{r}}=m(\dot{\mathbf{r}}\times\dot{\mathbf{r}})=\mathbf{0}$. By Newton's second law, however,

$$m\ddot{\mathbf{r}}=\mathbf{F} \tag{3.23}$$

Moreover, recognizing that

$$\mathbf{r}\times\mathbf{F}=\mathbf{M}_0 \tag{3.24}$$

is the moment of the force $\mathbf{F}$ about point 0, we obtain

$$\mathbf{M}_0=\dot{\mathbf{H}}_0 \tag{3.25}$$

or, *the moment of a force about a fixed point* 0 *is equal to the time rate of change of the moment of momentum about* 0.

By definition, the cross product of two vectors is a vector normal to the plane defined by the two vectors involved in the product. Hence, $\mathbf{H}_0$ is a vector normal to both $\mathbf{r}$ and $\dot{\mathbf{r}}$. On the other hand, $\mathbf{M}_0$ is a vector normal to $\mathbf{r}$ and $\mathbf{F}$. Because in general $\mathbf{r}$, $\dot{\mathbf{r}}$, and $\mathbf{F}$ are not in the same plane, $\mathbf{M}_0$ and $\mathbf{H}_0$ are not in the same direction. However, from Eq. (3.25) we conclude that $\mathbf{M}_0$ and $\dot{\mathbf{H}}_0$ are in the same direction. In the special case in which the motion is planar, the vectors $\mathbf{M}_0$, $\mathbf{H}_0$, and $\dot{\mathbf{H}}_0$ are all in the same direction, namely, in the direction normal to the plane of motion.

Next, let us define the angular impulse about 0 between the times t_1 and t_2 as the vector

$$\hat{\mathbf{M}}_0=\int_{t_1}^{t_2}\mathbf{M}_0\,dt \tag{3.26}$$

where $\hat{\mathbf{M}}_0$ has the unit newton·meter·seconds. Multiplying Eq. (3.25) by dt and integrating with respect to time between t_1 and t_2, we obtain

$$\hat{\mathbf{M}}_0=\int_{t_1}^{t_2}\mathbf{M}_0\,dt=\int_{t_1}^{t_2}\frac{d\mathbf{H}_0}{dt}\,dt=\mathbf{H}_{02}-\mathbf{H}_{01} \tag{3.27}$$

where

$$\mathbf{H}_{0i}=\mathbf{H}_0(t_i)=\mathbf{r}(t_i)\times m\dot{\mathbf{r}}(t_i)=\mathbf{r}(t_i)\times m\mathbf{v}(t_i),\qquad i=1,2 \tag{3.28}$$

is the angular momentum at $t=t_i$. Letting

$$\Delta\mathbf{H}_0=\mathbf{H}_{02}-\mathbf{H}_{01} \tag{3.29}$$

be the change in the angular momentum vector between the times t_1 and t_2, Eq. (3.27) becomes

$$\hat{\mathbf{M}}_0=\Delta\mathbf{H}_0 \tag{3.30}$$

or, *the angular impulse vector about* 0 *between the times* t_1 *and* t_2 *is equal to the change in the angular momentum vector about* 0 *between the same two instants.*

If there are no torques about 0, $\mathbf{M}_0=\mathbf{0}$, Eq. (3.27) yields

$$\mathbf{H}_{02}=\mathbf{H}_{01}=\mathbf{H}_0=\text{const} \tag{3.31}$$

or, *in the absence of torques about* 0 *the angular momentum about* 0 *is constant,* which is the statement of the principle of *conservation of angular momentum.* Note that the conservation of the angular momentum does not require that the force be zero. Indeed, the angular momentum is conserved if the force passes through 0. Such a force is known as a *central force* and plays an important role in the motion of planets and satellites.

Example 3.5

A particle of mass m moves on a smooth surface while attached to a string rotating with the angular velocity ω. At the same time, the string is being pulled through a small hole, as shown in Fig. 3.8. Let the distance between the hole and the particle when the angular velocity is ω_1 be equal to R, and calculate the angular velocity ω_2 when the distance is reduced to $R/2$.

The only force acting on the mass m is the tension in the string. Because this

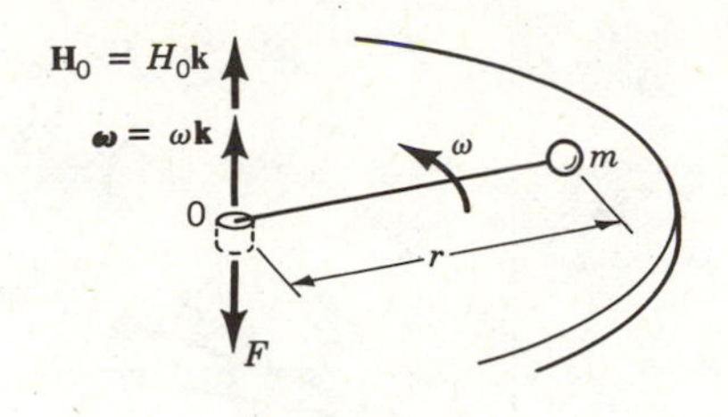

FIGURE 3.8

force goes through the center of rotation 0, as shown in Fig. 3.8, there is no torque about 0, so that the angular momentum about 0 is conserved. This being a planar problem, we can dispense with the vector notation and write

$$H_0 = mr^2\omega = \text{const} \tag{a}$$

Letting $r=R$, $\omega=\omega_1$ and $r=R/2$, $\omega=\omega_2$ in Eq. (a), in sequence, we obtain

$$mR^2\omega_1 = m(R/2)^2\omega_2 \tag{b}$$

from which it follows that

$$\omega_2 = 4\omega_1 \tag{c}$$

3.6 WORK AND ENERGY

Let us consider a particle m moving along a curve s under the action of a given force $\mathbf{F}$ (Fig. 3.9). By definition, the *increment of work* corresponding to the displacement of m from position $\mathbf{r}$ to position $\mathbf{r}+d\mathbf{r}$ is given by the dot product (scalar product)

$$dW = \mathbf{F}\cdot d\mathbf{r} \tag{3.32}$$

Clearly, dW is a scalar quantity. By Newton's second law, however, we have $\mathbf{F}=m\ddot{\mathbf{r}}$ and from kinematics $d\mathbf{r}=\dot{\mathbf{r}}\,dt$, so that

$$dW = m\ddot{\mathbf{r}}\cdot\dot{\mathbf{r}}\,dt = m\frac{d\dot{\mathbf{r}}}{dt}\cdot\dot{\mathbf{r}}\,dt = m\dot{\mathbf{r}}\cdot d\dot{\mathbf{r}} = d(\tfrac{1}{2}m\dot{\mathbf{r}}\cdot\dot{\mathbf{r}}) \tag{3.33}$$

But the *kinetic energy* of the particle m is defined as

$$T = \tfrac{1}{2}m\dot{\mathbf{r}}\cdot\dot{\mathbf{r}} \tag{3.34}$$

and we note that T is a scalar function. Hence, inserting Eq. (3.34) into Eq. (3.33), we obtain

$$dW = dT \tag{3.35}$$

Next, let us consider the work performed by $\mathbf{F}$ in moving the particle m from position $\mathbf{r}_1$ to position $\mathbf{r}_2$ (Fig. 3.9). Considering Eqs. (3.32) and (3.35), we can write

$$W_{1-2} = \int_{\mathbf{r}_1}^{\mathbf{r}_2} \mathbf{F}\cdot d\mathbf{r} = \int_{T_1}^{T_2} dT = T_2 - T_1 \tag{3.36}$$

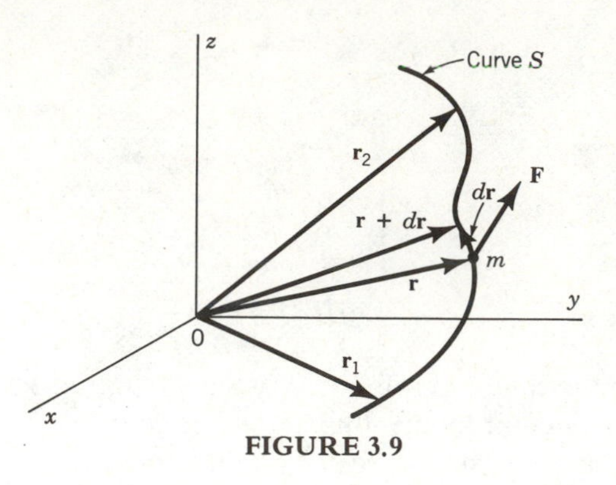

FIGURE 3.9

so that *the work performed by the force* **F** *in moving the particle m from position* $\mathbf{r}_1$ *to position* $\mathbf{r}_2$ *is equal to the change in the kinetic energy from* T_1 *to* T_2.

There exists one class of forces for which *the work performed depends only on the terminal positions* $\mathbf{r}_1$ *and* $\mathbf{r}_2$, and not on the path taken to travel from $\mathbf{r}_1$ to $\mathbf{r}_2$. Considering two distinct paths, I and II, as shown in Fig. 3.10, we conclude that the above statement implies that

$$\underset{\text{path I}}{\int_{\mathbf{r}_1}^{\mathbf{r}_2}} \mathbf{F}\cdot d\mathbf{r} = \underset{\text{path II}}{\int_{\mathbf{r}_1}^{\mathbf{r}_2}} \mathbf{F}\cdot d\mathbf{r} \tag{3.37}$$

Equation (3.37) yields

$$\underset{\text{path I}}{\int_{\mathbf{r}_1}^{\mathbf{r}_2}} \mathbf{F}\cdot d\mathbf{r} - \underset{\text{path II}}{\int_{\mathbf{r}_1}^{\mathbf{r}_2}} \mathbf{F}\cdot d\mathbf{r} = \underset{\text{path I}}{\int_{\mathbf{r}_1}^{\mathbf{r}_2}} \mathbf{F}\cdot d\mathbf{r} + \underset{\text{path II}}{\int_{\mathbf{r}_2}^{\mathbf{r}_1}} \mathbf{F}\cdot d\mathbf{r} = \oint \mathbf{F}\cdot d\mathbf{r} = 0 \tag{3.38}$$

where $\oint$ denotes an integral over a closed path. Hence, the statement that the work performed does not depend on the path is equivalent to the statement that *the work performed in traveling over a closed path* (starting at a given point and returning to the same point) *is zero.* Forces for which the above statements are true for all possible paths are said to be *conservative* and will be identified by the subscript c.

Consider now a conservative force $\mathbf{F}_c$, choose a path passing through the

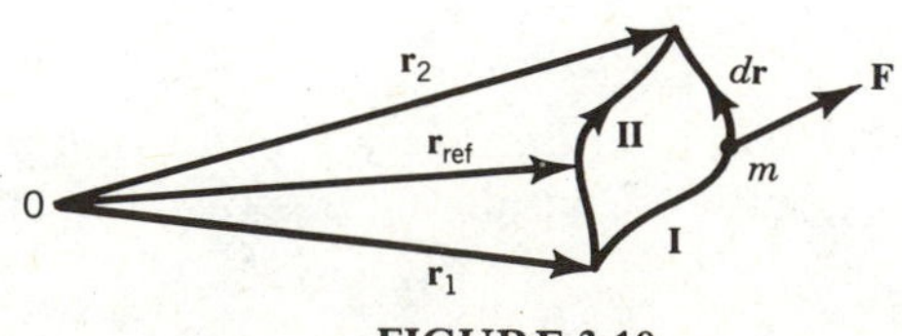

FIGURE 3.10

reference position $\mathbf{r}_{\text{ref}}$ (Fig. 3.10) and write

$$W_{1-2c} = \int_{\mathbf{r}_1}^{\mathbf{r}_2} \mathbf{F}_c \cdot d\mathbf{r} = \int_{\mathbf{r}_1}^{\mathbf{r}_{\text{ref}}} \mathbf{F}_c \cdot d\mathbf{r} + \int_{\mathbf{r}_{\text{ref}}}^{\mathbf{r}_2} \mathbf{F}_c \cdot d\mathbf{r}$$
$$= \int_{\mathbf{r}_1}^{\mathbf{r}_{\text{ref}}} \mathbf{F}_c \cdot d\mathbf{r} - \int_{\mathbf{r}_2}^{\mathbf{r}_{\text{ref}}} \mathbf{F}_c \cdot d\mathbf{r} \tag{3.39}$$

Then, defining the *potential energy* as *the work performed by a conservative force in moving a particle from the position* $\mathbf{r}$ *to the reference position* $\mathbf{r}_{\text{ref}}$, or

$$V(\mathbf{r}) = \int_{\mathbf{r}}^{\mathbf{r}_{\text{ref}}} \mathbf{F}_c \cdot d\mathbf{r} \tag{3.40}$$

where we note that V is a scalar function, and using definition (3.40), we can rewrite Eq. (3.39)

$$W_{1-2c} = V_1 - V_2 = -(V_2 - V_1) \tag{3.41}$$

where

$$V_i = V(\mathbf{r}_i), \qquad i = 1, 2 \tag{3.42}$$

Equation (3.41) states that *the work performed by a conservative force in moving a particle from* $\mathbf{r}_1$ *to* $\mathbf{r}_2$ *is equal to the negative of the change in the potential energy from* V_1 *to* V_2. Note that, because our interest lies mainly in changes in the potential energy rather than in the potential energy itself, the reference position is arbitrary, i.e., it can be selected at will. It should be pointed out that, in contrast with the potential energy, the kinetic energy represents an absolute quantity, because it is expressed in terms of velocities relative to an inertial space.

In general, we can distinguish between conservative and nonconservative forces, so that the work can be separated accordingly, or

$$W_{1-2} = W_{1-2c} + W_{1-2nc} \tag{3.43}$$

where

$$W_{1-2nc} = \int_{\mathbf{r}_1}^{\mathbf{r}_2} \mathbf{F}_{nc} \cdot d\mathbf{r} \tag{3.44}$$

in which W_{1-2nc} is the nonconservative work and $\mathbf{F}_{nc}$ the nonconservative force. Inserting Eqs. (3.36) and (3.41) into Eq. (3.43), we obtain

$$T_2 - T_1 = -(V_2 - V_1) + W_{1-2nc} \tag{3.45}$$

Introducing the *total energy* as *the sum of the kinetic energy and potential energy*

$$E = T + V \tag{3.46}$$

we can rearrange Eq. (3.45) into

$$W_{1-2nc} = E_2 - E_1 \tag{3.47}$$

or, *the work performed by the nonconservative force* $\mathbf{F}_{nc}$ *in moving a particle from* $\mathbf{r}_1$ *to* $\mathbf{r}_2$ *is equal to the change in the total energy from* E_1 *to* E_2. The work–energy relation in any of its forms, Eq. (3.36), or Eq. (3.41), or Eq. (3.47), is sometimes referred to as a first integral of motion with respect to displacement.

Equation (3.41) can also be written in the form

$$dW_c = -dV \tag{3.48}$$

and Eq. (3.43) in the form

$$dW = dW_c + dW_{nc} \tag{3.49}$$

so that, introducing Eqs. (3.35) and (3.48) into Eq. (3.49) and recalling Eq. (3.46), we obtain

$$dW_{nc} = dE \tag{3.50}$$

But, from Eq. (3.44), we have

$$dW_{nc} = \mathbf{F}_{nc} \cdot d\mathbf{r} \tag{3.51}$$

so that

$$\mathbf{F}_{nc} \cdot d\mathbf{r} = dE \tag{3.52}$$

Dividing both sides of Eq. (3.52) by dt, we obtain

$$\mathbf{F}_{nc} \cdot \dot{\mathbf{r}} = \dot{E} \tag{3.53}$$

But, for any general force $\mathbf{F}$, the expression

$$P = \mathbf{F} \cdot \dot{\mathbf{r}} \tag{3.54}$$

represents the rate of work performed by the force $\mathbf{F}$ and is known as the *power*. Hence, Eq. (3.53) states that the power associated with the nonconservative force $\mathbf{F}_{nc}$ is equal to the time rate of change of the total energy E. Note that nonconservative forces can either add energy to a system, as in the case of applied forces, or dissipate energy, as in the case of damping forces.

In the case in which there are no nonconservative forces ($\mathbf{F}_{nc} = \mathbf{0}$), Eq. (3.53) yields

$$E = \text{const} \tag{3.55}$$

or, *in the absence of nonconservative forces the total energy remains constant*. This statement is known as the principle of *conservation of energy*. Conservative forces are characterized by the fact that they depend on the position $\mathbf{r}$ alone and not on the velocity $\dot{\mathbf{r}}$ or time t.

Example 3.6

A particle m is originally at rest at point 1 when it begins to slide on a smooth surface (Fig. 3.11). Calculate the velocity v_2 with which the particle leaves the slide.

The only forces acting on m are the reaction force $\mathbf{N}$ exerted by the surface and the gravitational force. The reaction force is normal to the path, so that it performs no work. The gravitational force can be regarded as conservative (although it is not necessary to do so), so that we can solve the problem by using the conservation of energy principle, Eq. (3.55).

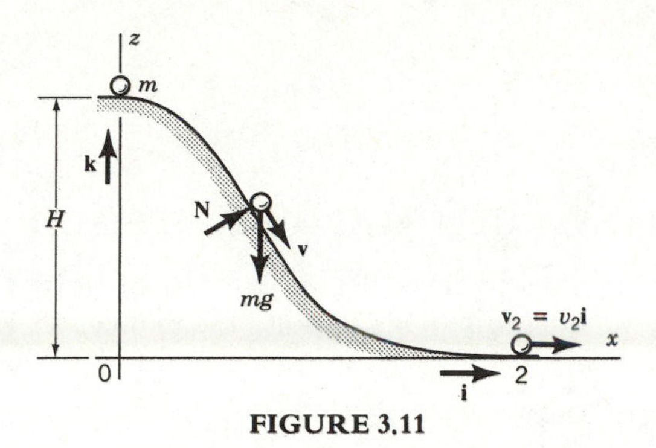

FIGURE 3.11

Letting point 1 be the reference position, Eq. (3.40) yields the potential energy in position 2 in the form

$$V_2 = V(\mathbf{r}_2) = \int_{\mathbf{r}_2}^{\mathbf{r}_1} (-mg\mathbf{k})\cdot d\mathbf{r} = \int_{x_2,0}^{0,H} (-mg\mathbf{k})\cdot(dx\mathbf{i}+dz\mathbf{k})$$

$$= -mg\int_0^H dz = -mgH \tag{a}$$

so that the potential energy depends only on the weight mg and the difference in heights H between the reference position 1 and point 2. Of course, because point 1 represents the reference position, $V_1 = 0$. In the reference position the particle is at rest, so that $T_1 = 0$. On the other hand, at point 2 the kinetic energy is

$$T_2 = \tfrac{1}{2}mv_2^2 \tag{b}$$

Hence, using Eq. (3.50), we can write

$$E = T_1 + V_1 = T_2 + V_2 \tag{c}$$

or

$$0 = -mgH + \tfrac{1}{2}mv_2^2 \tag{d}$$

which yields the exit velocity

$$v_2 = \sqrt{2gH} \tag{e}$$

Example 3.7

Consider the particle m of Example 3.5 and determine the work performed in moving the particle from the position $r = R$ to the position $r = R/2$.

Because the particle moves on a horizontal table, there is no change in the potential energy, so that Eq. (3.47) yields

$$W_{1-2\text{nc}} = T_2 - T_1 \tag{a}$$

where, recalling from Example 3.5 that $\omega_2 = 4\omega_1$,

$$\begin{aligned} T_1 &= \tfrac{1}{2}mv_1^2 = \tfrac{1}{2}mR^2\omega_1^2 \\ T_2 &= \tfrac{1}{2}mv_2^2 = \tfrac{1}{2}m[(R/2)\omega_2)]^2 = 2mR^2\omega_1^2 \end{aligned} \tag{b}$$

Hence,

$$W_{1-2\text{nc}} = \tfrac{3}{2} m R^2 \omega_1^2 \tag{c}$$

3.7 MOTION IN A CENTRAL-FORCE FIELD

Let us consider a particle m moving under the influence of a force $\mathbf{F}$ passing through a fixed point 0 at all times (Fig. 3.12). Because the moment of $\mathbf{F}$ about 0 is zero, the angular momentum about 0 is conserved (Section 3.5), so that

$$\mathbf{H}_0 = \mathbf{r} \times m\mathbf{v} = \text{const} \tag{3.56}$$

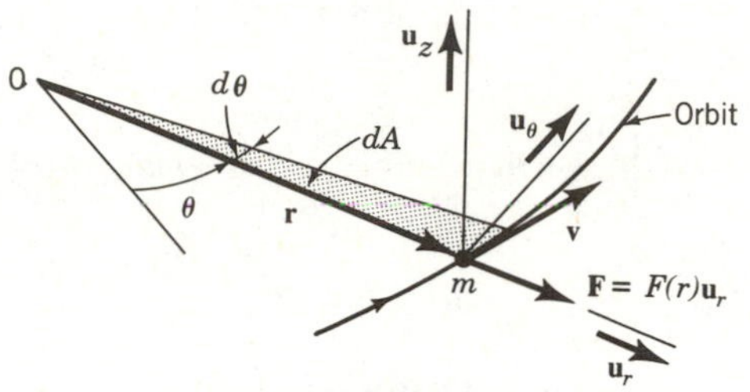

FIGURE 3.12

The constancy of $\mathbf{H}_0$ implies that *the magnitude H_0 and the direction in space of the angular momentum vector $\mathbf{H}_0$ are both constant*. Moreover, because the vector $\mathbf{H}_0$ is normal to both the radius vector $\mathbf{r}$ and the velocity vector $\mathbf{v}$, it follows that *the motion takes place in a plane fixed in space*. It will prove convenient to refer the motion to a set of cylindrical axes defined by the unit vectors $\mathbf{u}_r$, $\mathbf{u}_\theta$, and $\mathbf{u}_z$, where $\mathbf{u}_r$ is in the radial direction, $\mathbf{u}_\theta$ is in the transverse direction and in the plane of motion, and $\mathbf{u}_z$ is in the direction normal to the plane of motion. Whereas $\mathbf{u}_z$ is a constant unit vector, $\mathbf{u}_r$ and $\mathbf{u}_\theta$ are not constant because their directions change with the position of m. In essence, we propose to describe the motion in terms of the polar coordinates r and θ, which depend on the instantaneous position of the particle m.

Before proceeding with the investigation of the motion, we recall from Section 2.3 that the absolute velocity vector has the expression

$$\mathbf{v} = v_r \mathbf{u}_r + \mathbf{v}_\theta \mathbf{u}_\theta \tag{3.57}$$

where

$$v_r = \dot{\mathbf{r}}, \qquad v_\theta = r\dot{\theta} \tag{3.58}$$

are the radial and transverse components of the velocity vector $\mathbf{v}$, in which $\mathbf{r} = r\mathbf{u}_r$ is the radius vector. Moreover, the absolute acceleration has the form

$$\mathbf{a} = a_r \mathbf{u}_r + a_\theta \mathbf{u}_\theta \tag{3.59}$$

where

$$a_r = \ddot{r} - r\dot{\theta}^2, \qquad a_\theta = r\ddot{\theta} + 2\dot{r}\dot{\theta} \tag{3.60}$$

are the radial and transverse components of the acceleration vector **a**.

The constancy of the angular momentum, Eq. (3.56), can be given an interesting geometric interpretation. Indeed, inserting Eqs. (3.57) and (3.58) into Eq. (3.56), we obtain

$$\mathbf{H}_0 = \mathbf{r} \times m\mathbf{v} = r\mathbf{u}_r \times m(\dot{r}\mathbf{u}_r + r\dot{\theta}\mathbf{u}_\theta) = mr^2\dot{\theta}\mathbf{u}_z = \text{const} \tag{3.61}$$

where we recognized that $\mathbf{u}_r \times \mathbf{u}_r = \mathbf{0}$ and that $\mathbf{u}_r \times \mathbf{u}_\theta = \mathbf{u}_z$. We have already established that the direction of $\mathbf{H}_0$, which coincides with the direction of the unit vector $\mathbf{u}_z$, is fixed in space. On the other hand, we observe that the triangular differential element of area dA shown in Fig. 3.12 can be written as

$$dA = \tfrac{1}{2}(r)(rd\theta) = \tfrac{1}{2}r^2\, d\theta \tag{3.62}$$

Dividing both sides of Eq. (3.62) by dt and considering Eq. (3.61), we obtain

$$\dot{A} = \tfrac{1}{2}r^2\dot{\theta} = H_0/2m = \text{const} \tag{3.63}$$

where

$$H_0 = mr^2\dot{\theta} = \text{const} \tag{3.64}$$

is the magnitude of the angular momentum vector $\mathbf{H}_0$. Equation (3.63) represents the mathematical statement of *Kepler's second law* for planetary motion: *Every planet moves in such a way that its radius vector sweeps over equal areas in equal times.*

Next, let us turn our attention to the equations of motion. Letting the central force have the general form

$$\mathbf{F} = F_r\mathbf{u}_r + F_\theta\mathbf{u}_\theta = F_r\mathbf{u}_r \tag{3.65}$$

the radial and transverse components of Newton's second law can be written as

$$ma_r = F_r, \qquad ma_\theta = F_\theta = 0 \tag{3.66}$$

Introducing Eqs. (3.60) into Eqs. (3.66), we obtain

$$m(\ddot{r} - r\dot{\theta}^2) = F_r \tag{3.67a}$$

$$m(r\ddot{\theta} + 2\dot{r}\dot{\theta}) = 0 \tag{3.67b}$$

Equation (3.67b) can be rewritten as

$$\frac{m}{r}\frac{d}{dt}(r^2\dot{\theta}) = 0 \tag{3.68}$$

which, upon integration, leads to a reaffirmation of Kepler's second law.

The above results, including Kepler's second law, were reached solely on the assumption of a force passing through the fixed center 0 and without specifying the explicit functional dependence of the central force.

3.8 THE INVERSE SQUARE LAW. ORBITS OF PLANETS AND SATELLITES

In Section 3.7, we discussed the motion of a particle m under a central force. A force of particular importance is that given by Newton's inverse square law. Strictly speaking, the force is along the line joining two moving particles. If the mass of one of the particles is considerably larger than the mass of the other, then the motion can be regarded as that of the small particle moving around the massive particle. This is certainly the case with many planets and satellites. Moreover, if we further assume that the massive particle is at rest in an inertial space, then the motion of the small particle can be regarded as taking place about a fixed point.

Let us consider a particle m moving around a massive particle M under the action of a force according to Newton's gravitational law, Eq. (3.7). Letting $m_1 = m$ and $m_2 = M$ in Eq. (3.7), introducing the notation

$$GM = \mu \tag{3.69}$$

and dividing through by m, we can write Eq. (3.67a) in the form

$$\ddot{r} - r\dot{\theta}^2 = \frac{F_r}{m} = -\frac{\mu}{r^2} \tag{3.70}$$

where the minus sign on the right side of the equation can be explained by the fact that the force is in the opposite direction to that shown in Fig. 3.12. Moreover, Kepler's second law can be rewritten in the form

$$r^2\dot{\theta} = h = \text{const} \tag{3.71}$$

where the constant h can be identified as the angular momentum per unit mass, $h = H_0/m$, and is sometimes referred to as *specific angular momentum.*

The object now is to eliminate the time dependence from Eq. (3.70) and obtain an equation in terms of the polar coordinates r and θ alone. This equation, in which the time does not appear explicitly, is known as the *orbit equation.* As it turns out, to derive the orbit equation, we do not work with the radial distance r but with its reciprocal. Hence, let us consider Eq. (3.71) and write

$$\dot{r} = \frac{dr}{dt} = \frac{dr}{d\theta}\frac{d\theta}{dt} = \frac{h}{r^2}\frac{dr}{d\theta} = -h\frac{d}{d\theta}\left(\frac{1}{r}\right) = -h\frac{du}{d\theta} \tag{3.72}$$

where u is the reciprocal of r, or

$$u = 1/r \tag{3.73}$$

Moreover, we have

$$\ddot{r} = \frac{d\dot{r}}{dt} = \frac{d\dot{r}}{d\theta}\frac{d\theta}{dt} = \frac{h}{r^2}\frac{d\dot{r}}{d\theta} = -h^2u^2\frac{d^2u}{d\theta^2} \tag{3.74}$$

Introducing Eqs. (3.71), (3.73), and (3.74) into Eq. (3.70), and dividing through by $-h^2u^2$, we obtain the second-order differential equation

$$\frac{d^2u}{d\theta^2} + u = \frac{\mu}{h^2} \tag{3.75}$$

which does not contain the time explicitly. The complete solution of Eq. (3.75) is

$$u = \frac{\mu}{h^2} + C\cos(\theta - \theta_0) \tag{3.76}$$

and it represents the orbit equation. The first term on the right side of Eq. (3.76) is constant and is recognized as the particular solution; the second term is harmonic and can be identified as the homogeneous solution, in which C and θ_0 are the two constants of integration. There are several types of orbits possible, with the type of orbit depending on the constants of integration, and in particular on C. The constant θ_0 turns out to be of minor importance. In the following, we shall determine the relation between the type of orbit and the constant C.

Physically, the orbit is determined by the initial conditions, and hence by the energy of the orbiting body. Because for bodies in the solar system there is virtually no energy dissipation, the orbits of planets will remain the same. Actually, this statement is only approximately true, as there are always perturbing forces present. In the case of artificial earth satellites, the energy dissipation is negligible if the orbits are sufficiently high to avoid atmospheric drag. Note that Eqs. (3.70) and (3.71) imply no energy dissipation. Because there is no force component in the transverse direction, the *potential energy per unit mass* can be written

$$V(r) = \int_r^{r_0} \left(-\frac{\mu}{\xi^2}\right) d\xi = \frac{\mu}{\xi}\bigg|_r^{r_0} = \frac{\mu}{r_0} - \frac{\mu}{r} \tag{3.77}$$

where we replaced r on the right side of Eq. (3.70) by ξ, a dummy variable of integration. We observe that the term μ/r_0 is constant. As pointed out earlier, in comparing the potential energy associated with two different positions, the constant is immaterial. Hence, in such cases it will prove convenient to choose as reference position $r_0 = \infty$, which has the advantage that it renders this constant zero. We shall return to this subject later in this section. Using Eqs. (3.58), we can write the *kinetic energy per unit mass*

$$T = \tfrac{1}{2}(v_r^2 + v_\theta^2) = \tfrac{1}{2}(\dot{r}^2 + r^2\dot{\theta}^2) \tag{3.78}$$

so that, letting $r_0 = \infty$ in Eq. (3.77), the *total energy per unit mass* becomes

$$E = T + V = \tfrac{1}{2}(\dot{r}^2 + r^2\dot{\theta}^2) - \mu/r \tag{3.79}$$

Inserting Eqs. (3.71)–(3.73) into Eq. (3.79) and considering solution (3.76), we obtain

$$\begin{aligned} E &= \frac{1}{2}\left[\left(-h\frac{du}{d\theta}\right)^2 + (hu)^2\right] - \mu u = \frac{h^2}{2}\left\{[-C\sin(\theta-\theta_0)]^2 \right. \\ &\quad \left. + \left[\frac{\mu}{h^2} + C\cos(\theta-\theta_0)\right]^2\right\} - \frac{\mu^2}{h^2} - \mu C\cos(\theta-\theta_0) \\ &= \frac{h^2}{2}\left[C^2 - \left(\frac{\mu}{h^2}\right)^2\right] \end{aligned} \tag{3.80}$$

from which it follows that

$$C=\frac{\mu}{h^2}\sqrt{1+\frac{2Eh^2}{\mu^2}}=\frac{\mu}{h^2}\,e \tag{3.81}$$

where

$$e=\sqrt{1+\frac{2Eh^2}{\mu^2}} \tag{3.82}$$

Introducing Eq. (3.81) into Eq. (3.76) and recalling Eq. (3.73), we obtain the orbit equation in the form

$$r=\frac{h^2}{\mu}\frac{1}{1+e\cos(\theta-\theta_0)} \tag{3.83}$$

which represents geometrically the equation of a *conic section* in polar form with the *focus* at the center 0. The shape of the orbit is determined by the parameters h^2/μ and e, where the latter is known as the orbit *eccentricity*. The type of conic section depends mathematically on the eccentricity e and physically on the total energy E. In the following, we present a brief discussion of the various types of orbits possible.

There are two classes of orbits: closed and open. Among the closed orbits we find the *circle* and the *ellipse*, and among the open orbits the *parabola* and *hyperbola*. A point characterized by $dr/d\theta=0$ is known as an *apsis*, and it represents either the point on the orbit closest to the focus or that farthest from the focus. A circle has an infinity of *apsides*, because at every point of a circle $dr/d\theta=0$, which is consistent with the fact that all points on a circle are equidistant from the center. An ellipse has two apsides. On the other hand, open orbits such as the parabola and hyperbola have only one apsis. The shorter apsis is called *pericenter*, and the longer one is called *apocenter*. For orbits with the focus at the earth's center they are known as *perigee* and *apogee*, and for orbits with the focus at the sun's center they are known as *perihelion* and *aphelion*, respectively. The angle θ_0 can be identified as the angle from a given reference line to the pericenter. If θ is measured from the pericenter, then $\theta_0=0$. In future discussion, we shall assume that this is the case. Introducing the notation

$$h^2/\mu=p \tag{3.84}$$

where p is known as the *semi-latus rectum*, the orbit equation can be written in the simple form

$$r=\frac{p}{1+e\cos\theta} \tag{3.85}$$

Note that the semi-latus rectum is the radial distance corresponding to $\theta=\pi/2$. The various possible orbits, together with the eccentricity e and the total energy E, are listed below:

1. Circle: $e=0$, $E=-\mu^2/2h^2$.
2. Ellipse: $0<e<1$, $-\mu^2/2h^2<E<0$.

3. Parabola: $e=1$, $E=0$.

4. Hyperbola: $e>1$, $E>0$.

The fact that the total energy can be negative has no particular physical significance and is only a reflection of choosing the reference position r_0 for the potential energy as infinity. As pointed out earlier, this is an acceptable choice if the energy thus calculated is used only to compare energy requirements for various orbits and not as an absolute energy requirement. Indeed, if one is interested in the energy necessary to place an earth satellite in orbit, then the reference position for the potential energy must be the surface of the earth, at a distance R from the earth's center. This adds the constant μ/R to the total energy, Eq. (3.79), so that E is no longer negative.

Observing that $e=0$ and $e=1$ are merely two values out of an entire range of values of e, we conclude that circles and parabolas are really limiting cases of ellipses and hyperbolas, respectively. We shall discuss the various orbits separately.

1. *Circle:* $e=0$, $E=-\mu^2/2h^2$. For zero eccentricity, Eq. (3.85) yields the orbit equation $r=r_c=p=\text{const}$, where r_c is the radius of the circular orbit. In a circular orbit, the gravitational force and the centrifugal force are in balance, so that the radial velocity component is zero and the transverse component is simply equal to the circular velocity v_c. The circular velocity can be obtained from Eq. (3.79). Indeed, letting $\dot{r}=0$, $r\dot{\theta}=v_c$, and $E=-\mu^2/2h^2$ in Eq. (3.79), we obtain

$$\frac{1}{2}v_c^2-\frac{\mu}{r_c}=-\frac{\mu^2}{2h^2} \tag{3.86}$$

But, from Eqs. (3.84) and (3.85), we conclude that

$$h^2=\mu r_c \tag{3.87}$$

so that, inserting Eq. (3.87) into Eq. (3.86), we obtain the circular velocity

$$v_c=\sqrt{\mu/r_c} \tag{3.88}$$

Defining the period T as the time necessary for one complete revolution, $\theta=2\pi$, integrating $dt=(r_c/v_c)\,d\theta$ over one period and using Eq. (3.88), we obtain

$$T=\frac{2\pi r_c}{v_c}=2\pi\frac{r_c^{3/2}}{\mu^{1/2}} \tag{3.89}$$

2. *Ellipse:* $0<e<1$, $-\mu^2/2h^2<E<0$. Because this is a closed orbit, the motion is periodic. Denoting by r_p and r_a the pericenter and apocenter, respectively, and letting $\theta=0$ and $\theta=\pi$ in Eq. (3.75), we obtain

$$r_p=\frac{p}{1+e\cos 0}=\frac{p}{1+e} \tag{3.90a}$$

$$r_a=\frac{p}{1+e\cos\pi}=\frac{p}{1-e} \tag{3.90b}$$

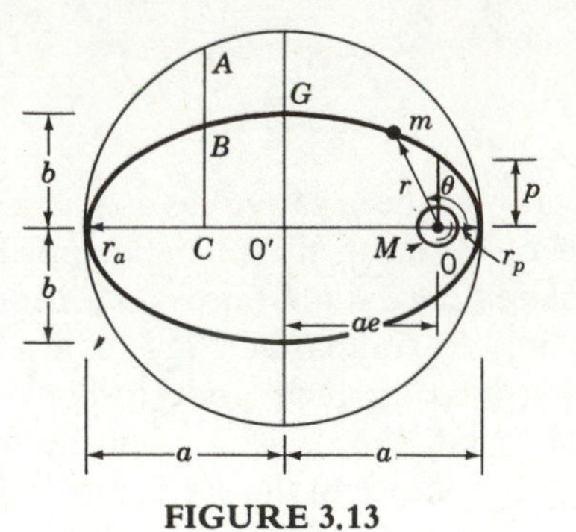

FIGURE 3.13

A typical ellipse is shown in Fig. 3.13, in which $r_{\mathrm{p}}, r_{\mathrm{a}}, e$, and p are identified. Another quantity of interest is the *semimajor axis* given by

$$a=\tfrac{1}{2}(r_{\mathrm{p}}+r_{\mathrm{a}}) \tag{3.91}$$

Using Eqs. (3.90), the semimajor axis is simply

$$a=\frac{1}{2}\left(\frac{p}{1+e}+\frac{p}{1-e}\right)=\frac{p}{1-e^2} \tag{3.92}$$

Moreover, recalling Eqs. (3.82) and (3.84), we obtain

$$a=\frac{h^2}{\mu}\left(-\frac{\mu^2}{2Eh^2}\right)=-\frac{\mu}{2E} \tag{3.93}$$

To construct the ellipse, we recall from analytic geometry that the ellipse is defined as the locus of points for which the sum of the distances from the two foci of the ellipse is constant and equal to $2a$. A simpler way of constructing the ellipse is by recognizing that the ellipse and the circle are related. Indeed, the relation between the ellipse and the circle is defined by the ratio (see Fig. 3.13)

$$\overline{BC}/\overline{AC}=b/a=\text{const} \tag{3.94}$$

where b is the ellipse *semiminor axis*. If we consider Pythagoras's theorem for the triangle $00'G$, we conclude that the semiminor axis is

$$b=\sqrt{a^2-(ae)^2}=a\sqrt{1-e^2} \tag{3.95}$$

so that

$$\overline{BC}/\overline{AC}=\sqrt{1-e^2} \tag{3.96}$$

The importance of the ellipse in planetary motion is underscored by *Kepler's first law* which states: *Each planet revolves in an elliptic orbit about the sun at one focus.*

A quantity of particular interest is the orbital period T, which can be obtained from Eq. (3.71) in the form

$$T=\int_0^{2\pi}\frac{r^2}{h}\,d\theta \tag{3.97}$$

Introducing Eqs. (3.84), (3.85), and (3.92) into Eq. (3.97), we can write

$$T=\int_0^{2\pi}\frac{1}{h}\frac{p^2\,d\theta}{(1+e\cos\theta)^2}=\frac{2a^{3/2}(1-e^2)^{3/2}}{\mu^{1/2}}\int_0^{\pi}\frac{d\theta}{(1+e\cos\theta)^2}$$
$$=\frac{2a^{3/2}(1-e^2)^{3/2}}{\mu^{1/2}}\left[\frac{-\sin\theta}{1+e\cos\theta}+\frac{2}{(1-e)^{1/2}}\tan^{-1}\frac{(1-e)^{1/2}\tan\theta/2}{1+e}\right]_0^{\pi}$$
$$=2\pi\frac{a^{3/2}}{\mu^{1/2}} \tag{3.98}$$

Equation (3.98) is the mathematical statement of *Kepler's third law: The squares of the periodic times of the planets are proportional to the cubes of the semimajor axes of the ellipses.* As we shall see later, this law is only approximately valid.

Finally, we wish to calculate the velocity at any point on the ellipse. From Eqs. (3.77), (3.79), and (3.93), we obtain simply

$$v^2=(\dot{r}^2+r^2\dot{\theta}^2)=2(E-V)=\mu\left(\frac{2}{r}-\frac{1}{a}\right) \tag{3.99}$$

Clearly, the highest velocity is at the pericenter and the lowest at the apocenter.

Note that Eqs. (3.96) and (3.97) are also valid for the circle if one simply replaces both a and r by r_c.

An interesting question arises as to what happens if a satellite expected to be placed in a circular earth orbit fails to achieve the circular velocity v_c, as given by Eq. (3.88). In this case, the satellite goes into a subcircular elliptic orbit, with the launching point playing the role of apogee instead of the expected perigee. Of course, if $r=r_a$ and $v=v_a$ are not large enough, the orbit will intersect the earth's surface (which is another way of saying that the satellite will crash or will burn up because of atmospheric friction).

3. *Parabola:* $e=1$, $E=0$. The parabola is the open orbit (Fig. 3.14) requiring the least amount of energy. We shall not go into the geometry of the parabola, but concentrate our attention on the velocity required to achieve parabolic

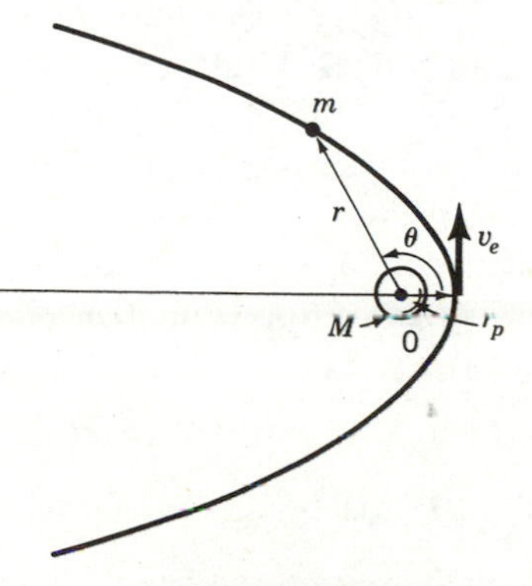

FIGURE 3.14

orbit. Recalling that for a parabolic orbit $E=0$, we can use Eq. (3.79) to write

$$\frac{1}{2}v^2-\frac{\mu}{r}=0 \tag{3.100}$$

where $v=\sqrt{\dot{r}^2+r^2\dot{\theta}^2}$ is the magnitude of the velocity vector. Hence, from Eq. (3.100), the velocity in a parabolic orbit is

$$v=\sqrt{2\mu/r} \tag{3.101}$$

It is obvious from Eq. (3.101) that the velocity tends to zero as the satellite approaches infinity. Denoting the velocity at the pericenter r_p by v_e, Eq. (3.101) yields

$$v_e=\sqrt{2\mu/r_p} \tag{3.102}$$

Because this is the velocity necessary to escape the gravitational pull of the massive particle M, v_e is referred to as the *escape velocity*, which explains the subscript e. Clearly, v_e is the minimum velocity for which an open orbit is achieved. In the case of an earth satellite, it is the velocity that must be imparted to the satellite at perigee to escape the gravitational pull of the earth.

Note that very elongated (high eccentricity) ellipses can be mistaken for parabolas, which had been the case with the orbit of the comet Halley. The orbit is an elongated ellipse with the sun at one focus and having a period of close to 76 years.

4. *Hyperbola:* $e>1$, $E>0$. If the velocity at the pericenter is larger than v_e, then the resulting orbit is a hyperbola. Because in this case the energy E is higher than the minimum required for an open orbit, the hyperbola is known as a high-energy orbit. Letting $r_0=\infty$ in Eq. (3.77), we observe that $V\to 0$ as $r\to\infty$. But, because $E>0$, which implies that $T>0$, there will be some residual velocity as $r\to\infty$. The velocity at infinity is called the *hyperbolic excess velocity* and characterizes the energy of a hyperbolic orbit.

Example 3.8

A satellite is launched in an earth orbit. At injection in orbit, the satellite is at an altitude of $H=2\times10^5$ m, and its velocity is parallel to the earth surface and has the magnitude $v=8\times10^3$ m/s. Determine the following: (1) the orbit eccentricity, (2) the semimajor axis, and (3) the orbital period. Use the following data: $M=5.9774\times10^{24}$ kg, $G=6.6685\times10^{-11}$ m^3/kg·s^2 and $R=6.3781\times10^6$ m, where M is the mass of the earth, G is the universal gravitational constant, and R is the radius of the earth.

Because the velocity at injection is parallel to the earth surface, the injection point corresponds to the perigee. (Note that the velocity is sufficiently high to achieve orbit.) Hence, we have

$$\begin{aligned} r_p&=R+H=(6.3781+0.2)10^6\text{ m}=6.5781\times10^6\text{ m}\\ v_p&=v=8\times10^3\text{ m/s}\end{aligned} \tag{a}$$

To calculate the eccentricity e, we observe from Eq. (3.82) that we need the total energy E, the specific angular momentum h, and $\mu = GM$. First, let us calculate

$$\mu = GM = 6.6685 \times 10^{-11} \times 5.9774 \times 10^{24} = 3.9860 \times 10^{14} \quad \mathrm{m^3/s^2} \tag{b}$$

Because $t=0$ at perigee, from Eq. (3.79), we can write

$$E = \frac{1}{2}v_p^2 - \frac{\mu}{r_p} = \frac{1}{2} \times 8^2 \times 10^6 - \frac{3.9860 \times 10^{14}}{6.5781 \times 10^6}$$
$$= -2.8595 \times 10^7 \quad \mathrm{m^2/s^2} \tag{c}$$

Moreover, the specific angular momentum is

$$h = r_p v_p = 6.5781 \times 10^6 \times 8 \times 10^3 = 5.2625 \times 10^{10} \quad \mathrm{m^2/s} \tag{d}$$

Using Eq. (3.82), we obtain

$$e = \sqrt{1 + \frac{2Eh^2}{\mu^2}} = \sqrt{1 - \frac{2 \times 2.8595 \times 10^7 \times (5.2625 \times 10^{10})^2}{(3.9860 \times 10^{14})^2}}$$
$$= 0.0561 \tag{e}$$

from which we conclude that the orbit is elliptical.

The semimajor axis can be calculated by means of Eq. (3.93). The result is

$$a = -\frac{\mu}{2E} = \frac{3.9860 \times 10^{14}}{2 \times 2.8595 \times 10^7} = 6.9697 \times 10^6 \quad \mathrm{m} \tag{f}$$

Finally, from Eq. (3.98) the orbital period is

$$T = 2\pi \sqrt{\frac{a^3}{\mu}} = 2\pi \sqrt{\frac{(6.9697 \times 10^6)^3}{3.9860 \times 10^{14}}}$$
$$= 5.7907 \times 10^3 \quad \mathrm{s} = 1.6085 \quad \mathrm{h} \tag{g}$$

PROBLEMS

3.1 A bullet of mass 10^{-2} kg leaves the gun barrel with a velocity of 0.8 km/s. If the firing is known to create a force on the bullet having a half-sine form of amplitude 7500 N, determine the duration of the force.

3.2 A pendulum consists of a rigid rod of length L and a bob of mass m (Fig. 3.15). The plane of the pendulum is made to rotate with the constant angular velocity Ω about a vertical axis through 0. Determine the forces on the bob and the torque on the rotating shaft. Note that the angle θ can vary with time, $\theta = \theta(t)$.

3.3 Determine the energy dissipation in the system of Example 3.4.

3.4 The bob of a pendulum of mass $m = 5$ kg and length $L = 2$ m is released from rest in a position defined by the angle $\theta_0 = 60°$ with respect to the vertical. Assuming that the string is inextensible, (1) determine the tension in the

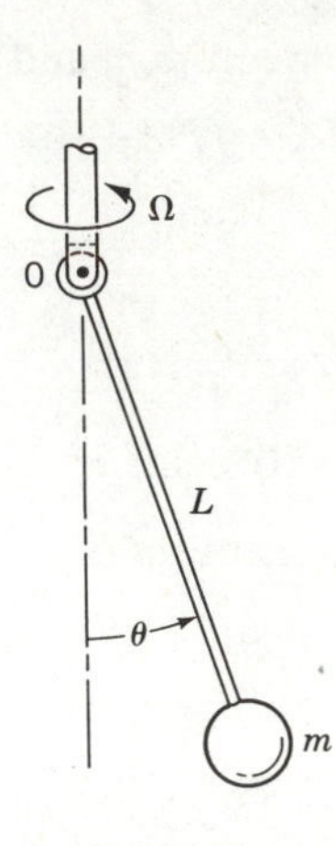

FIGURE 3.15

string when the bob is in the positions $\theta_1 = 30°$ and $\theta_2 = 0°$ and (2) calculate the angular impulse about the point of support between the times $t = t_0 = 0$ and $t = t_2$, where t_2 is the time corresponding to $\theta = \theta_2$.

3.5 An airplane goes into a dive. If at the bottom of the dive the airplane flies in a curve of radius $R = 2.5$ km with the velocity 900 km/h, calculate the force exerted by the seat on the pilot.

3.6 An airplane flying at 1000 km/h must make a circular turn of radius $R = 20$ km. At what angle relative to the plane of motion should the pilot bank the airplane? Assume that the airplane is flying in a horizontal plane.

3.7 A highway makes a bend with a radius of curvature $R = 100$ m. Determine the angle of the bank so as to permit a safe speed of 90 km/h if the coefficient of friction between the tires of a vehicle and the road is 0.2.

3.8 Determine the energy required to place a 1000-kg satellite in a circular orbit about the earth at an altitude of 300 km above the earth's surface. Calculate the period of the orbit.

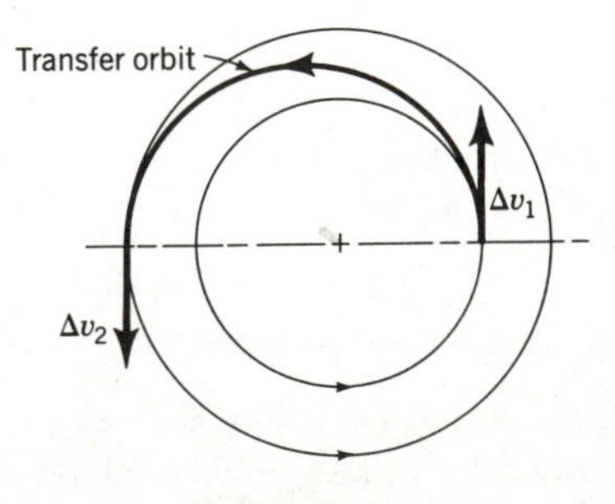

FIGURE 3.16

Pertinent data are as follows:

The universal gravitation constant $G = 6.6685 \times 10^{-11}\ \text{m}^3/\text{kg}\cdot\text{s}^2$.
The mass of the earth $M = 5.9774 \times 10^{24}$ kg.
The radius of the earth $R = 6.3781 \times 10^6$ m.

3.9 A synchronous satellite is one that maintains a fixed position relative to the rotating earth. Find the altitude of an earth synchronous orbit.

3.10 Find the energy required to transfer the satellite of Problem 3.8 from the 300-km circular orbit to a 400-km circular orbit. The transfer is to be performed by means of two impulses, the first placing the satellite in an elliptic transfer orbit and the second placing the satellite in the modified orbit, as shown in Fig. 3.16.

CHAPTER 4
Response of First-Order and Second-Order Systems

4.1 INTRODUCTION

In Chapter 1 we discussed the response of linear systems in a general way by using a generic differential equation to describe the system behavior, without actually deriving the equation. In this chapter, we derive first the differential equations governing the behavior of simple mechanical and electrical systems and then produce their solution by the methods developed in Chapter 1. The discussion is confined to first-order and second-order systems. In deriving the differential equations describing mechanical and electrical systems, it becomes evident that the equations for these two classes of systems are entirely analogous. Hence, solutions obtained for mechanical systems are valid for electrical systems and vice-versa. Although for the most part the various concepts introduced are common to mechanical and electrical systems, applications presented in this chapter tend to favor mechanical systems.

The behavior of first-order systems is markedly different from the behavior of second-order systems. Specifically, the free response of first-order systems tends to have an aperiodic nature, in contrast to the response of second-order systems, which tends to be oscillatory. Exceptions to the latter are mechanical systems with relatively heavy damping and electrical systems with relatively large resistance.

Both first-order and second-order systems are mathematical idealizations of actual physical systems. As far as mechanical systems are concerned, first-order systems are less common. Nevertheless, on many occasions, first-order systems can provide useful information concerning system behavior. Moreover, they are of interest mathematically, as more complex systems can be formulated to resemble first-order systems. Hence, their study is fully warranted. Second-order systems are considerably more common, as they are used as mathematical models for a large variety of systems. Refined models of engineering systems are often of high order. In fact, distributed systems are of infinite order. However, in many instances it is possible to gain substantial insight into the behavior of systems from low-order models. Moreover, as shown later in this text, high-order systems can be decomposed into a set of low-order ones.

Although the behavior of first-order systems is different from the behavior of second-order systems, the mathematical techniques for obtaining their response are the same. For this reason, we choose to treat both first-order and second-order systems in a single chapter and let the nature of the excitation dictate the order of presentation of the material.

4.2 DIFFERENTIAL EQUATIONS OF MOTION FOR MECHANICAL SYSTEMS

In Chapter 3, we presented a very fundamental discussion of particle dynamics, beginning with Newton's laws and ending with the motion of planets and satellites. In the process, we introduced an entire spectrum of basic concepts, such as impulse, momentum, work, and energy. In this section, we expand the discussion by deriving the differential equations of motion for certain low-order mechanical systems of particular interest in vibrations and control. Then, in subsequent sections, we devote a great deal of attention to the solution of these equations of motion.

Before we can derive the system differential equations of motion, it will prove convenient to introduce certain definitions and notations. We wish to distinguish between variables and components, or elements. Variables refer to quantities describing excitation and response, and they are functions of time. Components, or elements, refer to parts of the system and they are identified with the system parameters. Although they can depend on time, only constant parameters will be considered here.

In the case of mechanical systems, the variables can be identified as the force and the displacement. At times, the velocity or the acceleration can play the role of variable. Mechanical components are of three types: two that store energy and one that dissipates energy. In particular, *masses* store kinetic energy, *springs* store potential energy, and *dampers* dissipate energy.

The relation between the excitation and response for the various mechanical components can be derived by means of the free-body diagrams shown in Fig. 4.1. Indeed, these relations are

$$f_m(t) = ma(t) = m\ddot{x}(t) \tag{4.1a}$$

$$f_c(t) = c[v_2(t) - v_1(t)] = c[\dot{x}_2(t) - \dot{x}_1(t)] \tag{4.1b}$$

$$f_k(t) = k[x_2(t) - x_1(t)] \tag{4.1c}$$

where the overdots represent derivatives with respect to time. Equation (4.1a) is merely an expression of Newton's second law of motion, and it states that a force

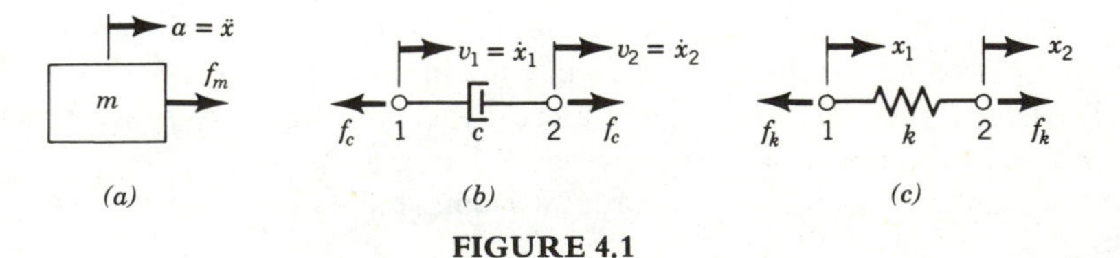

FIGURE 4.1

$f_m(t)$ causes the mass m to move with the acceleration $a(t) = \ddot{x}(t)$, where $x(t)$ is the displacement of the mass. Equation (4.1b) states that a force $f_c(t)$ applied at two terminal points 1 and 2 of a damper with the *coefficient of viscous damping c* will cause the two terminal points to separate with the relative velocity $v_2(t) - v_1(t) = \dot{x}_2(t) - \dot{x}_1(t)$. The equation is an expression of the fact that forces causing smooth shearing in viscous liquid are proportional to the relative velocity between the shearing layers. Viscous dampers are also known as *dashpots*. Finally, Eq. (4.1c) states that a force $f_k(t)$ applied at the two terminal points of a linear spring of *stiffness k* causes an elongation of the spring equal to $x_2(t) - x_1(t)$. The equation reflects the fact that in linear elasticity displacements are proportional to forces. The constant of proportionality k is also known as the *spring constant*. In SI units, the unit of force is the newton (N), the unit of mass is the kilogram (kg), and the unit of displacement is the meter (m). Of course, the unit of time is the second (s). It follows from Eqs. (4.1b) and (4.1c) that the unit of the viscous damping coefficient c is newton-seconds per meter (N·s/m) and that of the spring constant k is newtons per meter (N/m).

In the above discussion, it was assumed implicitly that the excitation–response relation is linear. This assumption is not always valid, and it is perhaps worth examining in detail. Letting $x_1 = 0$ and $x_2 = x$ in Fig. 4.1*c*, a typical force–displacement relation for the spring is as shown in Fig. 4.2. For relatively small spring deflections, the deflections are proportional to the force, i.e., the spring is linear. Beyond a certain deflection $x = x_l$, however, small force increments produce relatively large deflection increments, so that beyond $x = x_l$ the spring becomes *nonlinear*. Such a spring is known as a *softening spring*. Note that a different type of nonlinear spring is the *hardening spring*, for which deflection increments require increasingly large force increments. Clearly, the spring can be regarded as linear provided the deflections satisfy the inequality $|x(t)| < x_l$. The range $-x_l < x(t) < x_l$ is known as the *linear range*.

In the above discussion, we have examined how various mechanical components act separately. We are now in a position to derive the differential equations governing the behavior of the assembled system. We confine ourselves to one of the simplest cases, namely, one in which only one variable is necessary to describe the system behavior. In the case of mechanical systems, this variable is ordinarily the displacement, referred to as a *coordinate*.

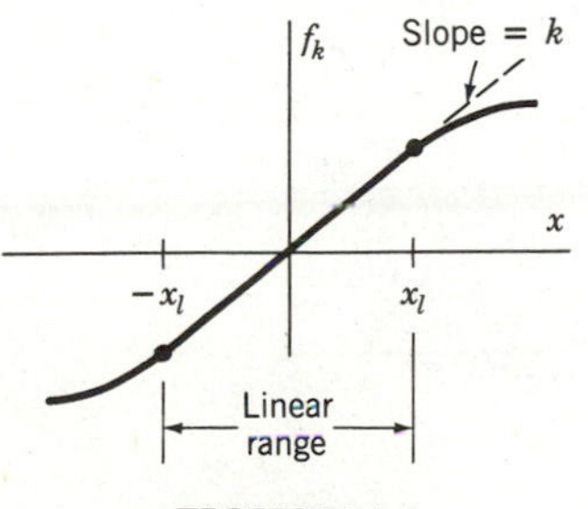

FIGURE 4.2

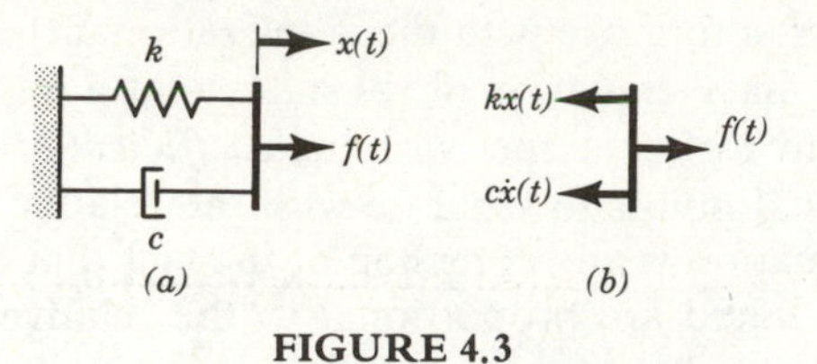

FIGURE 4.3

Let us consider the system shown in Fig. 4.3*a*, which is known as a *damper–spring system.* Figure 4.3*b* shows a free-body diagram of the system. We assume that the bar transmitting the force $f(t)$ to the damper and spring is massless, so that Newton's second law, Eq. (3.4), reduces to

$$\sum F_x = f(t) - c\dot{x}(t) - kx(t) = 0 \tag{4.2}$$

which can be rearranged as follows:

$$c\dot{x}(t) + kx(t) = f(t) \tag{4.3}$$

Equation (4.3) represents an ordinary differential equation of first order, so that the system is called a *first-order system.* The solution of Eq. (4.3) consists of two parts, the first corresponding to $f(t)=0$ and the second corresponding to $f(t)\neq 0$. They are known as the homogeneous solution and the particular solution, respectively. We shall discuss the solution of Eq. (4.3) later in this chapter.

Let us consider now the system of Fig. 4.4*a*. The system is commonly known as a *mass–damper–spring system* and is a simplified physical model representative of a large number of engineering systems, such as a piece of machinery on shock-absorbing mounts, or a buoy in viscous liquid. The corresponding free-body diagram is shown in Fig. 4.4*b*. Denoting the vertical displacement of the mass m from the *unstressed spring position* by $y(t)$, where the displacement is considered as positive in the upward direction, and using Newton's second law, we can write

$$\sum F_y = f(t) - f_c(t) - f_k(t) - mg = m\ddot{y}(t) \tag{4.4}$$

Letting $x_1 = 0$, $x_2 = y$ in Eqs. (4.1b) and (4.1c), introducing the results into Eq. (4.4), and rearranging, we obtain

$$m\ddot{y}(t) + c\dot{y}(t) + ky(t) + mg = f(t) \tag{4.5}$$

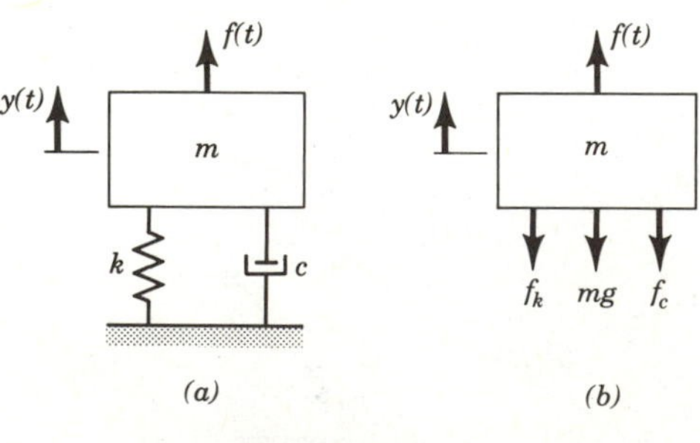

FIGURE 4.4

which represents the system equation of motion. The equation can be simplified by eliminating the effect of the weight mg. Indeed, instead of measuring the displacement of m from the unstressed spring position, we can measure it from the *static equilibrium position*, the latter position being obtained from the former position by letting the mass undergo the static deflection (Fig. 4.5)

$$\delta_{st} = mg/k \tag{4.6}$$

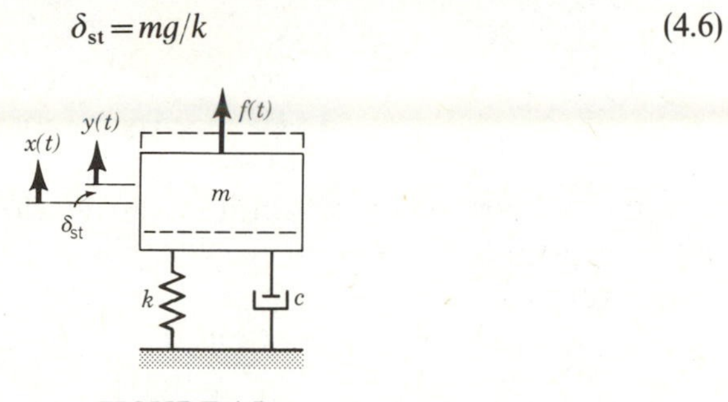

FIGURE 4.5

denoting the displacement from equilibrium by $x(t)$, introducing the coordinate transformation

$$y(t) = x(t) - \delta_{st} \tag{4.7}$$

into Eq. (4.5), and considering Eq. (4.6), we obtain

$$m\ddot{x}(t) + c\dot{x}(t) + kx(t) = f(t) \tag{4.8}$$

so that *by measuring the motion from the static equilibrium position, we can omit the weight mg*. The explanation for this fact is that in the process we omit not only the weight mg but also a prestress in the spring with a resultant force equal to $k\delta_{st}$, because these two forces cancel each other out according to Eq. (4.6).

The time derivative of highest order in Eq. (4.6) is the second derivative. Hence, the mass–damper–spring system is a *second-order system*. It is commonly referred to as a *single-degree-of-freedom system*.

Recalling the definition of linearity introduced in Section 1.3, we conclude that the systems described by Eqs. (4.3) and (4.8) are linear. In Section 1.4 we gave an example of a nonlinear differential equation and made the comment that it represented the equation of motion of a simple pendulum. At this point, it may prove of interest to verify the statement by deriving this equation. To this end, let us consider the simple pendulum shown in Fig. 4.6*a*. Using Newton's second law, in conjunction with the free-body diagram of Fig. 4.6*b*, we can write the equation of motion in the tangential direction

$$\sum F_t = -mg \sin\theta = a_t = mL\ddot{\theta} \tag{4.9}$$

where $a_t = L\ddot{\theta}$ is the acceleration of the mass m in the tangential direction, in which L is the length of the pendulum and $\ddot{\theta}$ is the angular acceleration. Writing the equation of motion in the tangential direction has the advantage that the string tension T, which is in the normal direction, does not appear in the equation.

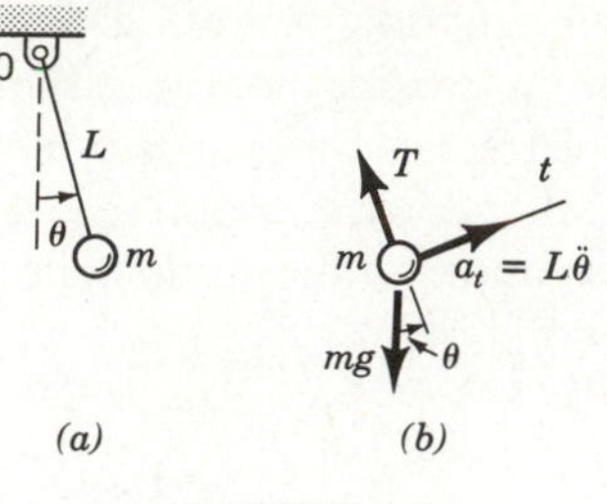

FIGURE 4.6

Note that we could have obtained essentially the same differential equation of motion by writing the moment equation about 0. Equation (4.9) can be rewritten in the form

$$mL\ddot{\theta}+mg\sin\theta=0 \tag{4.10}$$

which is a nonlinear differential equation. Comparing Eqs. (4.8) and (4.10), we conclude that the simple pendulum represents an undamped second-order system with a nonlinear restoring force of magnitude $mg\sin\theta$. Moreover, comparing Eqs. (1.12) and (4.10), we conclude that Eq. (1.12) does indeed describe the motion of a simple pendulum, provided $c(t)=\theta(t)$, $a_0=mL$, and $a_2=mg$.

4.3 DIFFERENTIAL EQUATIONS FOR ELECTRICAL SYSTEMS

Electrical systems are encountered frequently in everyday life. Some of the most common ones are the light bulb, electric heaters, and toasters. They are also some of the simplest. A more complex one is the radio. Although not immediately evident, for the most part the behavior of electrical systems is analogous to the behavior of mechanical systems. In fact, it is possible to simulate mechanical systems by electrical analogs, and vice versa. Electrical systems are ordinarily known as *networks*, or *circuits*, and consist of arrays of electrical components, or elements. Quite often, electrical and mechanical elements are put together into so-called electromechanical devices. Typical examples of systems involving both electrical and mechanical elements are control systems (see Chapter 11). Before we proceed with the derivation of the differential equations describing the behavior of electrical systems, it is advisable to establish relations governing the behavior of the individual elements.

Electrical elements can be divided into three basic types: *inductors*, *resistors*, and *capacitors* (sometimes known as *condensers*). The inductors and capacitors store energy, and the resistors dissipate energy. Note that light bulbs, heaters, and toasters are mere resistors. Clearly, as in the case of mechanical elements, the electrical elements can be identified with the system parameters. As variables, we can identify the *voltage* $v(t)$ and the *current* $i(t)$.

The relations between the voltage and the current for the various electrical elements are as follows:

$$v_L(t) = L\frac{di_L(t)}{dt} \tag{4.11a}$$

$$v_R(t) = Ri_R(t) \tag{4.11b}$$

$$v_C(t) = \frac{1}{C}\int i_C(t)\,dt \tag{4.11c}$$

The elements are shown in Figs. 4.7*a*, 4.7*b*, and 4.7*c* and can be identified as inductor, resistor, and capacitor, respectively. The parameters L, R, and C characterizing them are known as inductance, resistance, and capacitance, respectively. The units of voltage, current, inductance, resistance, and capacitance are volts, amperes (amp), henrys, ohms, and farads, respectively. Note that Eq. (4.11b) is the well-known *Ohm's law*.

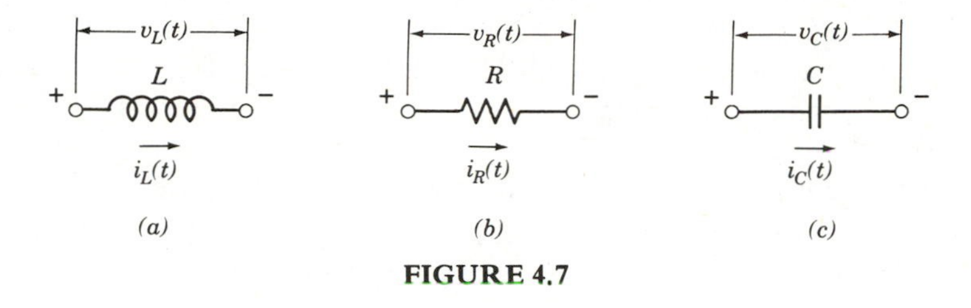

FIGURE 4.7

The analogy between the mechanical components and the electrical elements can be brought out by introducing the *charge* $q(t)$, which is related to the current by

$$i(t) = dq(t)/dt \tag{4.12}$$

where the unit of charge is the coulomb. Introducing Eq. (4.12) into Eqs. (4.11), the analogy becomes self-evident: The inductor is the electrical analog of the mass, the resistor is the analog of the damper, and the reciprocal of the capacitor is the analog of the spring. Moreover, a voltage source plays the role of a driving force. The analogy is made complete by observing that the resistor is the only electrical element dissipating energy.

The behavior of electrical networks is governed by *Kirchhoff's laws.* There are two such laws: the *voltage law* and the *current law.* In this chapter, we consider only the voltage law. The current law is introduced in Chapter 11.

The voltage law can be stated as follows: *The sum of voltage drops in the elements of a loop is equal to the sum of applied voltages.* A loop is an array of elements forming a closed circuit, such as the system shown in Fig. 4.8. This particular loop consists of a resistor R, a capacitor C, and a voltage source $v(t)$. Using Kirchhoff's law, we can write

$$v_R(t) + v_C(t) = v(t) \tag{4.13}$$

Introducing Eqs. (4.11b) and (4.11c) into Eq. (4.13) and recognizing that the current

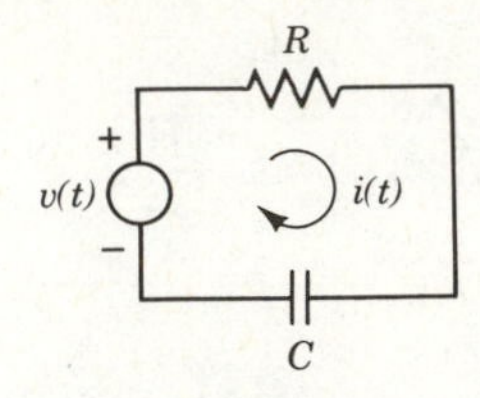

FIGURE 4.8

is the same within a given loop, we obtain

$$Ri(t)+\frac{1}{C}\int i(t)\,dt=v(t) \tag{4.14}$$

Moreover, considering Eq. (4.12), we can rewrite Eq. (4.14) in the form

$$R\dot{q}(t)+\frac{1}{C}\,q(t)=v(t) \tag{4.15}$$

which is entirely analogous to Eq. (4.3). Hence, the system of Fig. 4.8 is the electrical analog of the mechanical system of Fig. 4.3*a*. The system of Fig. 4.8 is known as an *RC* circuit.

Next, let us consider the electrical system shown in Fig. 4.9. It consists of an inductor *L*, a resistor *R*, a capacitor *C*, and a voltage source $v(t)$. We refer to this network as an *LRC* system. Using Kirchhoff's voltage law, we can write

$$v_L(t)+v_R(t)+v_C(t)=v(t) \tag{4.16}$$

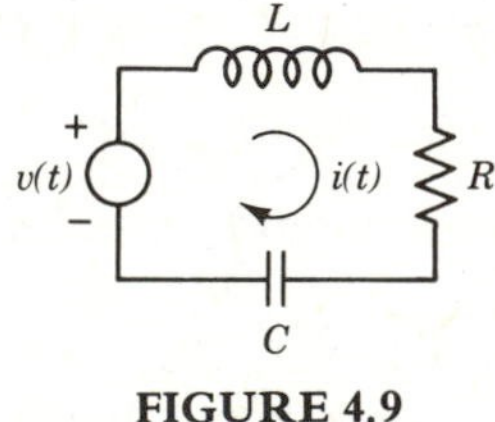

FIGURE 4.9

Introducing Eqs. (4.11) into Eq. (4.16), and recognizing that the current $i(t)$ is the same within a given loop, we obtain

$$L\frac{di(t)}{dt}+Ri(t)+\frac{1}{C}\int i(t)\,dt=v(t) \tag{4.17}$$

Moreover, considering Eq. (4.12), we can rewrite Eq. (4.17) in the form

$$L\ddot{q}(t)+R\dot{q}(t)+\frac{1}{C}\,q(t)=v(t) \tag{4.18}$$

which is entirely analogous to Eq. (4.8) describing a mechanical system.

From the above discussion, we conclude that the behavior of the damper–spring system or of the mass–damper–spring system can be simulated by means of electrical analogs having the form of an RC network or of an LRC network, respectively. The analogous quantities become evident if one compares the coefficients of like derivatives in Eqs. (4.8) and (4.18).

The analogy between mechanical and electrical systems carries the implication that the response of the two types of systems to the same excitation is the same. Hence, in discussing the system response, it is not necessary to distinguish unduly between mechanical and electrical systems.

4.4 FREE RESPONSE OF FIRST-ORDER SYSTEMS

The *free response* is defined as the response of a system in the absence of external excitation. Hence, the free response represents simply the homogeneous part of the solution, which is due entirely to initial conditions. This definition is somewhat artificial, because quite often initial conditions are produced by some form of initial external excitation. Nevertheless, the definition is helpful, as in the case of linear systems it permits the derivation of the homogeneous solution independently of the particular solution.

In Section 4.3, we established an analogy between mechanical and electrical systems that allows us to ignore the distinction between the two types of systems and treat them as if they belonged to a larger single class. In view of this, we propose to classify systems according to the structure of the governing differential equations alone. This permits us to extend the analogy to a large variety of systems, as many mechanical systems are governed by differential equations that are similar in structure to Eq. (4.8), and the only difference lies in the system parameters. To carry out the joint analysis of similar systems, it will prove convenient to introduce certain groupings of parameters, some of them having the same units and some of them being nondimensional.

Let us consider a first-order homogeneous equation having the generic form

$$\dot{x}(t)+ax(t)=0 \tag{4.19}$$

where a is a parameter with the unit of reciprocal of seconds (s^{-1}). Equation (4.19) is subject to the initial condition

$$x(0)=x_0 \tag{4.20}$$

In the case of the mechanical system shown in Fig. 4.3a and described by Eq. (4.3), $x(t)$ is the displacement and

$$a=k/c \tag{4.21}$$

where k is the spring constant and c is the coefficient of viscous damping. In the case of the electrical system shown in Fig. 4.8 and described by Eq. (4.15), $x(t)$ is the charge and

$$a=1/RC \tag{4.22}$$

in which R is the resistance and C is the capacitance.

The classical approach to the solution of Eq. (4.19) is to assume a solution in the exponential form

$$x(t)=Ae^{\lambda t} \tag{4.23}$$

where A and λ are constant scalars. Introducing Eq. (4.23) into Eq. (4.19), we obtain

$$(\lambda+a)Ae^{\lambda t}=0 \tag{4.24}$$

Because $Ae^{\lambda t}$ cannot be zero for a nontrivial solution, Eq. (4.24) implies that

$$\lambda+a=0 \tag{4.25}$$

Equation (4.25) is known as the *characteristic equation* and has the solution

$$\lambda=-a \tag{4.26}$$

so that solution (4.23) becomes

$$x(t)=Ae^{-at} \tag{4.27}$$

where A is a constant of integration that can be determined by invoking the initial condition. Using Eq. (4.20) and introducing the notation

$$\tau=1/a \tag{4.28}$$

where τ is known as the system *time constant*, we can rewrite solution (4.27) in the form

$$x(t)=x_0 e^{-t/\tau}u(t) \tag{4.29}$$

and we note that the solution was multiplied by the unit step function $u(t)$ in recognition of the fact that the response is zero for $t<0$. The solution of Eq. (4.29) is plotted in Fig. 4.10 for several values of τ. We note that the response has an aperiodic nature, as $x(t)$ approaches zero asymptotically for all time constants τ. For small time constants, it approaches zero faster.

The same solution can be obtained by the Laplace transform method (see Appendix). Recalling Eq. (A.3), the transform of Eq. (4.19) can be written as

$$sX(s)-x(0)+aX(s)=0 \tag{4.30}$$

where $X(s)$ is the Laplace transform of $x(t)$. Hence, using Eq. (4.20), we can rewrite Eq. (4.30) in the form

$$X(s)=\frac{1}{s+a}x_0 \tag{4.31}$$

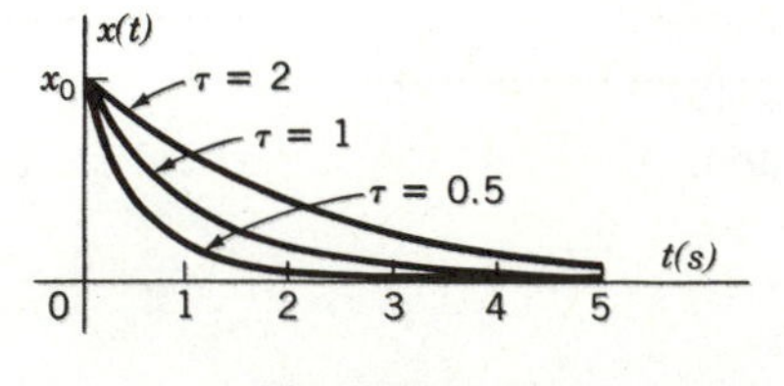

FIGURE 4.10

so that, using the table of Laplace transform pairs in Section A.7, we obtain the inverse transform of $X(s)$ in the form of Eq. (4.29). Note that the value $s=-a$ *is known as a simple pole of* $X(s)$ *and it coincides with the root of the characteristic equation.*

4.5 FREE RESPONSE OF UNDAMPED SECOND-ORDER SYSTEMS. THE HARMONIC OSCILLATOR

Consistent with the approach of Section 4.4, we wish to consider a generic differential equation, applicable to both mechanical and electrical systems. Nevertheless, when the situation demands, we shall favor terminology more common to mechanical engineering than to electrical engineering. Lack of damping implies that the elements associated with energy dissipation, namely, the damper and the resistor, are absent, so that there is no first-order derivative term in Eqs. (4.8) and (4.18). Hence, let us write the differential equation describing the behavior of a second-order undamped system in the form

$$\ddot{x}(t)+\omega_{\mathrm{n}}^2 x(t)=0 \tag{4.32}$$

where ω_{n} is known as the *natural frequency* of the system. The meaning of the term will become evident shortly. In the case of mechanical systems

$$\omega_{\mathrm{n}}=\sqrt{k/m} \tag{4.33}$$

and in the case of electrical systems

$$\omega_{\mathrm{n}}=\sqrt{1/LC} \tag{4.34}$$

The behavior of a large number of diverse physical and engineering systems can be described by Eq. (4.32). A classical example is the simple pendulum of Section 4.2, provided the motion is restricted to small angles, an assumption often referred to as the *small-motions assumption.* Invoking the small-motions assumption, which carries the implication that $\sin\theta\cong\theta$, we can rewrite Eq. (4.10) in the form

$$\ddot{\theta}(t)+\omega_{\mathrm{n}}^2\theta(t)=0 \tag{4.35}$$

where the natural frequency of the pendulum is simply

$$\omega_{\mathrm{n}}=\sqrt{g/L} \tag{4.36}$$

Equation (4.32) is one of the simplest second-order differential equations. Its general solution can be written in the form

$$x(t)=Ae^{\lambda t} \tag{4.37}$$

Introducing Eq. (4.37) into Eq. (4.32) and using the same reasoning as in Section 4.4, we obtain the characteristic equation

$$\lambda^2+\omega_{\mathrm{n}}^2=0 \tag{4.38}$$

which has the roots

$$\left.\begin{matrix}\lambda_1\\ \lambda_2\end{matrix}\right\} = \pm i\omega_n \tag{4.39}$$

Hence, the general solution (4.37) becomes

$$x(t) = A_1 e^{\lambda_1 t} + A_2 e^{\lambda_2 t} = A_1 e^{i\omega_n t} + A_2 e^{-i\omega_n t} \tag{4.40}$$

where A_1 and A_2 are constants of integration. They are complex quantities. Because $x(t)$ must be real, however, A_2 must be the complex conjugate of A_1. It will prove convenient to introduce the notation

$$A_1 = \tfrac{1}{2}Ae^{-i\psi}, \qquad A_2 = \tfrac{1}{2}Ae^{i\psi} \tag{4.41}$$

where A and ψ are real. Inserting Eqs. (4.41) into Eq. (4.40), we obtain

$$x(t) = \tfrac{1}{2}A[e^{i(\omega_n t - \psi)} + e^{-i(\omega_n t - \psi)}] \tag{4.42}$$

so that, recalling formula (1.20) and its complex conjugate, we can reduce Eq. (4.42) to the real form

$$x(t) = A\cos(\omega_n t - \psi) \tag{4.43}$$

where the constants of integration are now A and ψ. It is easy to verify that the solution of Eq. (4.32) can also be expressed as

$$x(t) = B_1 \sin \omega_n t + B_2 \cos \omega_n t \tag{4.44}$$

where B_1 and B_2 are constants of integration. Then, comparing Eqs. (4.43) and (4.44) and recalling that $\cos(\alpha - \beta) = \cos\alpha \cos\beta + \sin\alpha \sin\beta$, we conclude that the two sets of constants of integration, B_1, B_2 and A, ψ, are related by

$$B_1 = A\sin\psi, \qquad B_2 = A\cos\psi \tag{4.45}$$

Equation (4.43) or Eq. (4.44) indicates that the free response of an undamped second-order system consists of simple sinusoidal oscillation. Sine and cosine functions are known as harmonic functions, and, consistent with this, Eq. (4.43) is said to describe *simple harmonic oscillation.* Moreover, systems governed by equations of the type (4.32) or (4.35) are called *harmonic oscillators.* The constants A and ψ are known as the *amplitude* and *phase angle* of the oscillation, respectively.

Solution (4.43) can be conveniently discussed by means of the geometric construction shown in Fig. 4.11. The amplitude A is represented in Fig. 4.11*a* by a vector **A** making an angle $\omega_n t - \psi$ with the vertical axis. Hence, at any time t the projection of the vector **A** on the vertical axis represents the solution $x(t)$, Eq. (4.43). The constants A_1 and A_2 can be interpreted as the Cartesian components of the vector **A**, so that **A** is the diagonal of the rectangle with sides A_1 and A_2, where the angle between A_1 and **A** is constant and can be recognized as the phase angle ψ. As time unfolds, the angle $\omega_n t - \psi$ increases linearly with it, causing the vector **A** to rotate in the plane with the angular velocity ω_n. In the process, the vertical projection of **A** varies harmonically with time. This projection is shown in Fig. 4.11*b*. At $t = 0$ the projection is A_1, and at $t = \psi/\omega_n$ the projection reaches its peak

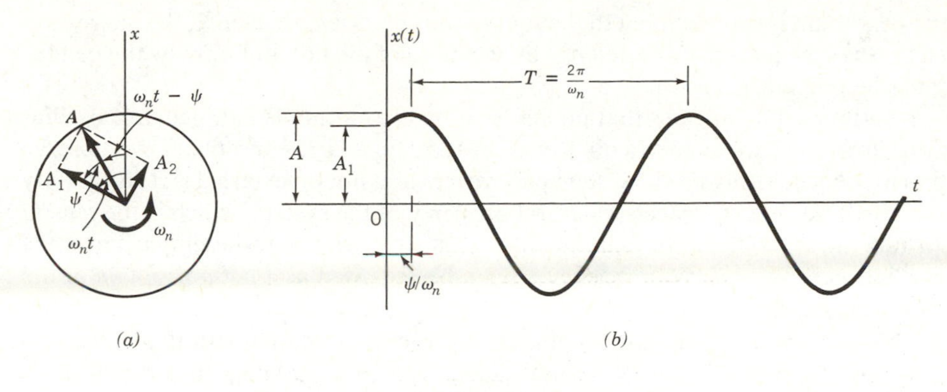

FIGURE 4.11

at a value equal to the amplitude A. Hence, we conclude from Fig. 4.11*b* that the phase angle ψ is a measure of the shift of the cosine function to the right. The solution $x(t)$ reaches a peak value of A once again after the time

$$T = 2\pi/\omega_n \tag{4.46}$$

has elapsed, and at every integer multiple of T thereafter, where T is known as the *period* of oscillation. Hence, the period represents the time between two consecutive peaks, or the time corresponding to one cycle of motion. It is commonly measured in seconds (s). From Eq. (4.46), we conclude that the frequency ω_n is measured in *radians per second* (rad/s). The natural frequency can also be defined as merely the reciprocal of the period, or

$$f_n = 1/T = \omega_n/2\pi \tag{4.47}$$

where f_n has units of *cycles per second* (cps). One cycle per second is commonly known as one *hertz* (Hz). Clearly, high frequencies imply short periods and vice versa.

It will prove of interest to examine the factors determining the period of the mass–spring system and of the simple pendulum. From Eqs. (4.33) and (4.46), we can write

$$T = 2\pi\sqrt{m/k} \tag{4.48}$$

so that the period T varies as the square root of m and is inversely proportional to the square root of k. Hence, the period T can be increased by increasing the mass or by decreasing the spring stiffness, or both. Similarly, using Eqs. (4.36) and (4.46), we obtain the pendulum period

$$T = 2\pi\sqrt{L/g} \tag{4.49}$$

But the quantity g represents the acceleration due to gravity. It is commonly assumed to be constant, although it varies with the altitude as measured from the sea level. For a given location, g can be regarded as constant, so that the period of the pendulum is proportional to the square root of its length. Hence, the interesting fact is that the period is affected not by the mass of the bob but only by the pendulum's length, a fact known to ancient Greeks.

Solution (4.43) indicates that no matter how the motion is initiated, free oscillation always occurs at the frequency ω_n. From Eqs. (4.33) and (4.36), we observe that ω_n depends only on the system parameters and not on external factors (gravity excluded), so that ω_n reflects a natural property of the system, which is the reason for it being called the natural frequency. Consistent with this, the simple harmonic oscillation at the natural frequency ω_n can be regarded as a *natural motion* of the harmonic oscillator.

The concept of harmonic oscillator represents a mathematical idealization more than a physical reality. Indeed, according to Eq. (4.43), once a motion is initiated, it will perpetuate itself ad infinitum. This is in contradiction to observed behavior, which indicates that the motion of a mass–spring system, or of a simple pendulum, will come to rest eventually if allowed to oscillate freely. This behavior can be attributed to the fact that every real system possesses some measure of damping. In the case of the pendulum, factors causing the motion to decay are air resistance and friction at the point of support. Nevertheless, the concept of a harmonic oscillator has its place. For some systems, damping is so small that the behavior is very close to that of a harmonic oscillator. In particular, if the interest lies in motion over a relatively short time interval compared to the period, then small damping may not have any noticeable effect over that interval.

Although the motion of a harmonic oscillator is always sinusiodal and the frequency of oscillation is always the natural frequency ω_n, the amplitude A and the phase angle ψ generally differ from case to case. Hence, the question remains as to what determines A and ψ. As mentioned earlier, A and ψ in Eq. (4.43) represent constants of integration. Mathematically, the determination of two constants of integration requires two conditions to be imposed on the solution. These conditions can be the value of the solution at two distinct times. More commonly, however, the two conditions are chosen as the value of the solution and of its first derivative at a given time, such as $t=0$. In this case, they represent physically the initial displacement and initial velocity. We denote them by

$$x(0)=x_0, \qquad \dot{x}(0)=v_0 \tag{4.50}$$

Introducing Eqs. (4.50) into Eq. (4.43), we obtain

$$x(0)=A\cos\psi=x_0, \qquad \dot{x}(0)=\omega_n A\sin\psi=v_0 \tag{4.51}$$

so that

$$A=\sqrt{x_0^2+(v_0/\omega_n)^2}, \qquad \psi=\tan^{-1}(v_0/\omega_n x_0) \tag{4.52}$$

As a matter of interest, we note that, by inserting Eqs. (4.51) into Eqs. (4.45) and by using Eq. (4.44), we can write the solution directly in terms of the initial conditions in the form

$$x(t)=x_0\cos\omega_n t+\frac{v_0}{\omega_n}\sin\omega_n t \tag{4.53}$$

Before considering the free vibration of damped systems, let us consider a solution of Eq. (4.32) by the Laplace transformation method. Transforming Eq. (4.32), we obtain

$$s^2X(s)-sx(0)-\dot{x}(0)+\omega_n^2X(s)=0 \tag{4.54}$$

which yields the transformed response

$$X(s)=\frac{s}{s^2+\omega_n^2}x(0)+\frac{1}{s^2+\omega_n^2}\dot{x}(0) \tag{4.55}$$

The inverse transformation of $X(s)$ can be obtained by the method of partial fractions described in Section A.3. If we were to expand $X(s)$ in terms of partial fractions, then we would factor out the denominator as follows:

$$s^2+\omega_n^2=(s-s_1)(s-s_2)=(s-i\omega_n)(s+i\omega_n) \tag{4.56}$$

where $s_1=i\omega_n$ and $s_2=-i\omega_n$ are simple poles of $X(s)$ (see Section A.3). Hence, from Eqs. (4.38) and (4.56), we conclude that *the simple poles of $X(s)$ are precisely the roots of the characteristic equation.* Because the functions $s/(s^2+\omega_n^2)$ and $1/s^2+\omega_n^2)$ are listed in the table of Laplace transform pairs given in Section A.7, expansion into partial fractions is actually not necessary. Indeed, using the table of Section A.7, we obtain directly

$$x(t)=x(0)\cos\omega_nt+\frac{\dot{x}(0)}{\omega_n}\sin\omega_nt \tag{4.57}$$

Equation (4.57) represents the response $x(t)$ expressed in terms of the initial displacement $x(0)$ and initial velocity $\dot{x}(0)$, obtained earlier in the form of Eq. (4.53).

4.6 FREE RESPONSE OF DAMPED SECOND-ORDER SYSTEMS

Letting the external excitation be equal to zero, we can write the differential equation governing the free response of a damped second-order system in the form

$$\ddot{x}(t)+2\zeta\omega_n\dot{x}(t)+\omega_n^2x(t)=0 \tag{4.58}$$

where ζ is a nondimensional parameter. Comparing Eqs. (4.8) and (4.58), we conclude that in the case of mechanical systems

$$\zeta=\frac{c}{2m\omega_n} \tag{4.59}$$

in which the natural frequency ω_n is given by Eq. (4.33). The nondimensional parameter ζ is known as the *viscous damping factor.* On the other hand, comparing Eqs. (4.18) and (4.58), it follows that for electrical systems

$$\zeta=\frac{R}{2L\omega_n} \tag{4.60}$$

where the natural frequency ω_n is given by Eq. (4.34). The solution of Eq. (4.58) is subject to the initial conditions (4.50) and can be obtained by the approach of Section 4.5.

Let us assume a solution of Eq. (4.58) in the exponential form

$$x(t) = Ce^{\lambda t} \tag{4.61}$$

where C and λ are constant scalars. Introducing Eq. (4.61) into Eq. (4.58) and using the same argument as for undamped systems, we conclude that the characteristic equation for damped systems is

$$\lambda^2 + 2\zeta\omega_n\lambda + \omega_n^2 = 0 \tag{4.62}$$

Equation (4.62) has the roots

$$\left.\begin{matrix}\lambda_1\\ \lambda_2\end{matrix}\right\} = (-\zeta \pm \sqrt{\zeta^2 - 1})\omega_n \tag{4.63}$$

so that the roots depend on the viscous damping factor ζ. We distinguish the following cases:

i. If $\zeta > 1$, the roots are real, negative, and distinct. They are in the form given by Eq. (4.63).

ii. If $\zeta = 1$, the roots are real, negative, and equal to one another, or

$$\lambda_1 = \lambda_2 = -\omega_n \tag{4.64}$$

iii. If $\zeta < 1$, the roots are complex conjugates with negative real part, or

$$\left.\begin{matrix}\lambda_1\\ \lambda_2\end{matrix}\right\} = (-\zeta \pm i\sqrt{1 - \zeta^2})\omega_n \tag{4.65}$$

The nature of the motion in each case depends on the roots λ_1 and λ_2, and hence on the viscous damping factor ζ. We now discuss the above three cases separately.

For $\zeta > 1$, the solution becomes

$$\begin{aligned} x(t) &= C_1 e^{\lambda_1 t} + C_2 e^{\lambda_2 t} \\ &= C_1 \exp(-\zeta + \sqrt{\zeta^2 - 1})\omega_n t + C_2 \exp(-\zeta - \sqrt{\zeta^2 - 1})\omega_n t \\ &= (C_1 e^{\sqrt{\zeta^2 - 1}\omega_n t} + C_2 e^{-\sqrt{\zeta^2 - 1}\omega_n t})e^{-\zeta\omega_n t} \end{aligned} \tag{4.66}$$

where the constants of integration C_1 and C_2 depend on x_0 and v_0. Because $\zeta > \sqrt{\zeta^2 - 1}$, the response $x(t)$ decays exponentially with time. The motion is *aperiodic*, i.e., it approaches zero without oscillation. When $\zeta > 1$, the system is said to be *overdamped.*

For $\zeta = 1$ the two roots coincide, $\lambda_1 = \lambda_2 = -\omega_n$. In this case, the solution can be verified to have the form

$$x(t) = (C_1 + C_2 t)e^{-\omega_n t} \tag{4.67}$$

where C_1 and C_2 depend on x_0 and v_0. Once again, the motion can be shown to be aperiodic, approaching zero asymptotically. The case $\zeta = 1$ is known as *critical damping*. From Eq. (4.59), we conclude that in the critical damping case the coefficient of viscous damping has the value

$$c_{cr} = 2m\omega_n = 2\sqrt{km} \tag{4.68}$$

For $\zeta < 1$, the solution becomes

$$x(t) = C_1 e^{\lambda_1 t} + C_2 e^{\lambda_2 t} = (C_1 e^{i\omega_d t} + C_2 e^{-i\omega_d t})e^{-\zeta\omega_n t} \tag{4.69}$$

where the notation

$$\omega_d = \sqrt{1-\zeta^2}\,\omega_n \tag{4.70}$$

has been introduced. Because $x(t)$ must be real, C_2 must be the complex conjugate of C_1, $C_2 = \bar{C}_1$, so that Eq. (4.69) reduces to

$$x(t) = 2\,\mathrm{Re}\, C_1 e^{i\omega_d t} e^{-\zeta\omega_n t} \tag{4.71}$$

But $e^{i\omega_d t}$ represents a complex vector of unit magnitude rotating counterclockwise in the complex plane with the angular velocity ω_d, as demonstrated in Section 1.6. Hence, $2\,\mathrm{Re}\, C_1 e^{i\omega_d t}$ represents the projection on the real axis of a rotating complex vector of magnitude $2|C_1|$. Recalling Fig. 4.11*a*, we conclude that $2\,\mathrm{Re}\, C_1 e^{i\omega_d t}$ varies harmonically with time. On the other hand, $e^{-\zeta\omega_n t}$ represents a function decaying exponentially with time, approaching zero asymptotically. Letting

$$2C_1 = Ae^{-i\psi} \tag{4.72}$$

where A and ψ are real quantities, we can reduce Eq. (4.71) to

$$x(t) = Ae^{-\zeta\omega_n t}\cos(\omega_d t - \psi) \tag{4.73}$$

Equation (4.73) permits a simple interpretation of the motion. Indeed, $Ae^{-\zeta\omega_n t}$ can be regarded as a time-dependent amplitude, modulating the harmonic function $\cos(\omega_d t - \psi)$, where ω_d can be interpreted as the frequency *of the damped free vibration*. Moreover, ψ is merely a phase angle. Hence, Eq. (4.73) represents *damped harmonic motion*, with the oscillation being bounded by the envelope $\pm Ae^{-\zeta\omega_n t}$. Because the width of the envelope approaches zero asymptotically as $t \to \infty$, the system comes to rest eventually. The case $\zeta < 1$ is commonly known as the *underdamped* case. Example 4.2 presents a typical response of an underdamped system.

An interesting picture can be obtained by examining how the roots λ_1 and λ_2 change with ζ. Such a picture is shown in the complex λ-plane of Fig. 4.12. In the undamped case, $\zeta = 0$, the roots $\lambda_1 = i\omega_n$ and $\lambda_2 = -i\omega_n$ lie on the imaginary axis. As ζ increases, the roots move along a circle of radius ω_n, until they coalesce on the real axis, when ζ reaches unity. As ζ increases beyond $\zeta = 1$, the two roots split once again, moving along the negative real axis in opposite directions. Because the case $\zeta = 1$ represents merely a point in the λ-plane, critical damping should be regarded as being primarily of academic interest and representing the borderline case between overdamping and underdamping. Figure 4.12 represents a root-locus plot, a subject discussed extensively in Chapter 11.

In all three cases discussed above, the motion is fully determined as soon as the constants of integration are evaluated in terms of the initial conditions. We do

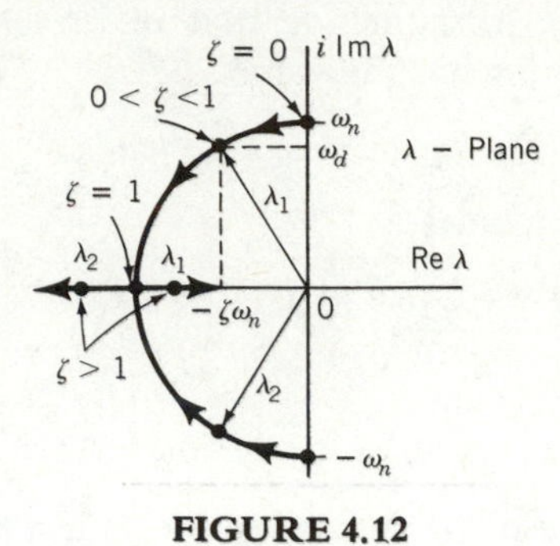

FIGURE 4.12

not pursue this approach here. Instead, we obtain the solution by the Laplace transformation method.

Transforming Eq. (4.58), we can write the transformed response $X(s)$ in the form

$$X(s)=\frac{s+2\zeta\omega_n}{s^2+2\zeta\omega_n s+\omega_n^2}\,x_0+\frac{1}{s^2+2\zeta\omega_n s+\omega_n^2}\,v_0 \tag{4.74}$$

where $x_0=x(0)$ and $v_0=\dot{x}(0)$. The simple poles of $X(s)$ are the roots of the characteristic equation, Eq. (4.62), or

$$s_1=(-\zeta+\sqrt{\zeta^2-1})\omega_n, \qquad s_2=(-\zeta-\sqrt{\zeta^2-1})\omega_n \tag{4.75}$$

and we note once again that the poles s_1 and s_2 coincide with the roots λ_1 and λ_2 of the characteristic equation. But from Section A.3 we can write

$$\begin{aligned}\mathscr{L}^{-1}\frac{s+2\zeta\omega_n}{s^2+2\zeta\omega_n s+\omega_n^2}&=\frac{s_1+2\zeta\omega_n}{s_1-s_2}\,e^{s_1t}+\frac{s_2+2\zeta\omega_n}{s_2-s_1}\,e^{s_2t}\\&=\frac{\zeta+\sqrt{\zeta^2-1}}{2\sqrt{\zeta^2-1}}\,e^{(-\zeta+\sqrt{\zeta^2-1})\omega_n t}-\frac{\zeta-\sqrt{\zeta^2-1}}{2\sqrt{\zeta^2-1}}\,e^{(-\zeta-\sqrt{\zeta^2-1})\omega_n t}\\&=\frac{1}{\sqrt{\zeta^2-1}}(\zeta\sinh\sqrt{\zeta^2-1}\,\omega_n t\\&\quad+\sqrt{\zeta^2-1}\cosh\sqrt{\zeta^2-1}\,\omega_n t)e^{-\zeta\omega_n t}\end{aligned} \tag{4.76}$$

and

$$\begin{aligned}\mathscr{L}^{-1}\frac{1}{s^2+2\zeta\omega_n s+\omega_n^2}&=\frac{1}{s_1-s_2}\,e^{s_1t}+\frac{1}{s_2-s_1}\,e^{s_2t}\\&=\frac{1}{\sqrt{\zeta^2-1}\,\omega_n}\,e^{-\zeta\omega_n t}\sinh\sqrt{\zeta^2-1}\,\omega_n t\end{aligned} \tag{4.77}$$

so that the response to the initial displacement x_0 and initial velocity v_0 is

$$x(t)=\frac{x_0}{\sqrt{\zeta^2-1}}\left(\zeta \sinh \sqrt{\zeta^2-1}\,\omega_n t+\sqrt{\zeta^2-1}\cosh \sqrt{\zeta^2-1}\,\omega_n t\right)e^{-\zeta\omega_n t}$$
$$+\frac{v_0}{\sqrt{\zeta^2-1}\,\omega_n}e^{-\zeta\omega_n t}\sinh \sqrt{\zeta^2-1}\,\omega_n t \tag{4.78}$$

The response in all three cases, $\zeta>1$, $\zeta=1$, and $\zeta<1$, can be derived from Eq. (4.78).

Example 4.1
The damped system described by Eq. (4.58) has the following parameters:

$$\zeta=1.2, \qquad \omega_n=5 \quad \text{rad/s} \tag{a}$$

Plot the response to the initial excitation

$$x_0=0.1 \quad \text{m}, \qquad v_0=0 \tag{b}$$

Introducing Eqs. (a) and (b) into Eq. (4.78), we obtain

$$x(t)=(0.1809 \sinh 3.3166t+0.1 \cosh 3.3166t)e^{-6t} \tag{c}$$

The plot $x(t)$ versus t is shown in Fig. 4.13; it confirms the aperiodic nature of the motion for this overdamped case.

Example 4.2
The damped system described by Eq. (4.58) has the following parameters:

$$\zeta=0.1, \quad \omega_n=5 \quad \text{rad/s} \tag{a}$$

Plot the response to the initial excitation

$$x_0=0, \qquad v_0=0.2 \quad \text{m/s} \tag{b}$$

Because $\zeta<1$, we wish to introduce the notation

$$\sqrt{\zeta^2-1}\,\omega_n=i\sqrt{1-\zeta^2}\,\omega_n=i\omega_d \tag{c}$$

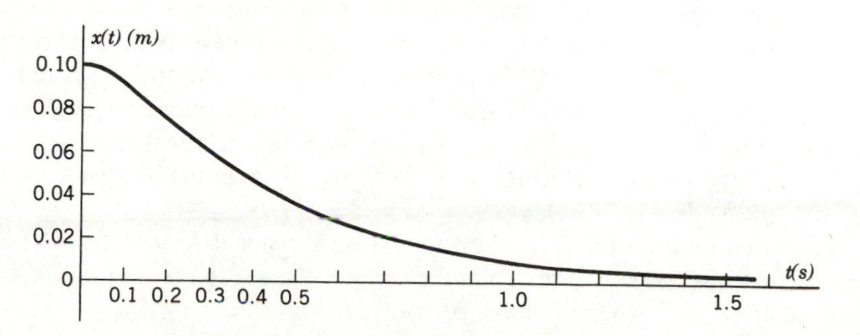

FIGURE 4.13

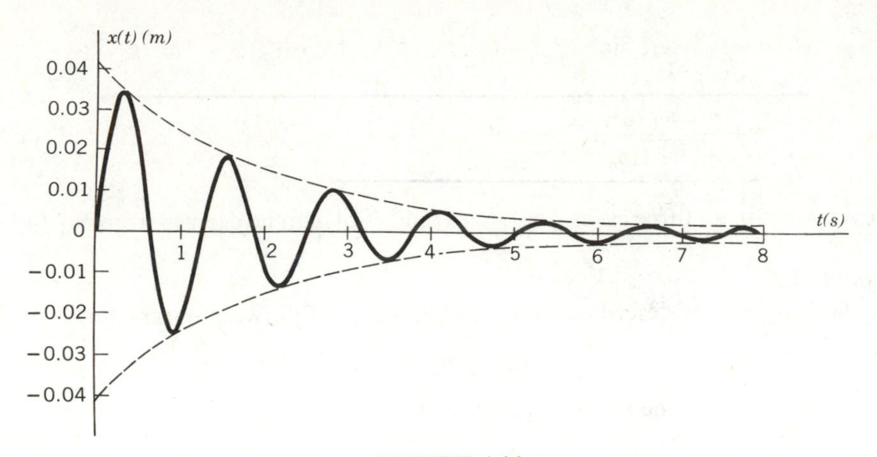

FIGURE 4.14

where ω_d is the frequency of the damped oscillation. Introducing Eq. (c) into Eq. (4.78), we can write

$$x(t)=\frac{v_0}{i\omega_d}\, e^{-\zeta\omega_n t} \sinh i\omega_d t=\frac{v_0}{\omega_d}\, e^{-\zeta\omega_n t} \sin \omega_d t \tag{d}$$

Hence, inserting Eqs. (a) and (b) into Eq. (d), we obtain

$$x(t)=0.0402e^{-0.5t} \sin 4.9749t \tag{e}$$

The plot $x(t)$ versus t is shown in Fig. 4.14, and it represents the damped oscillatory motion characterizing an underdamped system.

4.7 THE LOGARITHMIC DECREMENT

Quite often the amount of damping in a system is not known and must be determined experimentally. This can be done by disturbing the system initially in some fashion and measuring the response, so that the question reduces to how to determine the amount of damping from the observed response. We are interested in a viscously damped system, and in particular in an underdamped system, so that the response has the form of an exponentially decaying oscillation, such as the one shown in Fig. 4.15. Clearly, the rate of decay depends on the amount of damping. Hence, we propose to determine the damping by relating it to an established measure of the decay.

Let us denote the time at which the response reaches a peak by t_1 (Fig. 4.15). Because the motion is periodic, albeit damped, the subsequent peak is reached at the time $t_2=t_1+T$, where T is the period given by

$$T=2\pi/\omega_d \tag{4.79}$$

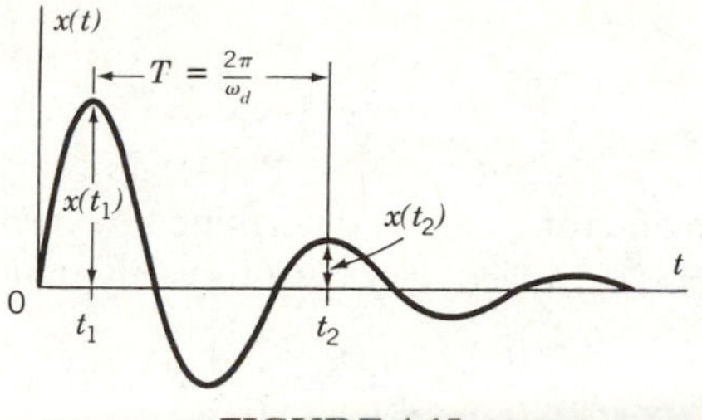

FIGURE 4.15

From Eq. (4.73), the corresponding peak responses have the expressions

$$x(t_1) = Ae^{-\zeta\omega_n t_1}\cos(\omega_d t_1 - \psi) \tag{4.80a}$$

$$x(t_2) = Ae^{-\zeta\omega_n t_2}\cos(\omega_d t_2 - \psi) = Ae^{-\zeta\omega_n(t_1+T)}\cos[\omega_d(t_1+T) - \psi] \tag{4.80b}$$

where T is given by Eq. (4.79). But, recalling Eq. (4.70), we obtain

$$\begin{aligned} e^{-\zeta\omega_n(t_1+T)} &= e^{-\zeta\omega_n t_1}e^{-\zeta\omega_n T} = e^{-\zeta\omega_n t_1}e^{-2\pi\zeta\omega_n/\omega_d} \\ &= e^{-\zeta\omega_n t_1}e^{-2\pi\zeta/(1-\zeta^2)1/2} \end{aligned} \tag{4.81}$$

Moreover,

$$\begin{aligned} \cos[\omega_d(t_1+T) - \psi] &= \cos(\omega_d t_1 - \psi)\cos\omega_d T - \sin(\omega_d t_1 - \psi)\sin\omega_d T \\ &= \cos(\omega_d t_1 - \psi)\cos 2\pi - \sin(\omega_d t_1 - \psi_1)\sin 2\pi \\ &= \cos(\omega_d t_1 - \psi) \end{aligned} \tag{4.82}$$

Using Eqs. (4.80)–(4.82), we can write the ratio between two peak values

$$\frac{x(t_1)}{x(t_2)} = \frac{Ae^{-\zeta\omega_n t_1}\cos(\omega_d t_1 - \psi)}{Ae^{-\zeta\omega_n t_2}\cos(\omega_d t_2 - \psi)} = e^{2\pi\zeta/(1-\zeta^2)1/2} \tag{4.83}$$

In view of the exponential form of the above ratio, it is convenient to introduce the notation

$$\delta = \ln\frac{x(t_1)}{x(t_2)} = \frac{2\pi\zeta}{(1-\zeta^2)^{1/2}} \tag{4.84}$$

where δ is known as the *logarithmic decrement*. Clearly, the logarithmic decrement can be obtained from the response curve by taking the natural logarithm of the ratio of two consecutive peak values, not necessarily the first two. Then, the viscous damping factor can be calculated by solving Eq. (4.84) for ζ. The result can be shown to be

$$\zeta = \frac{\delta}{[(2\pi)^2 + \delta^2]^{1/2}} \tag{4.85}$$

For damping sufficiently small that $\delta << 2\pi$, the viscous damping factor can be approximated by

$$\zeta \cong \frac{\delta}{2\pi} \tag{4.86}$$

The viscous damping factor ζ can be determined by measuring peaks separated by any number of periods. Indeed, it is easy to verify that

$$\frac{x(t_1)}{x(t_2)} = \frac{x(t_2)}{x(t_3)} = \frac{x(t_3)}{x(t_4)} = \cdots \tag{4.87}$$

so that considering peaks at t_1 and at $t_{k+1} = t_1 + kT$, where k is a given integer, we conclude that

$$\frac{x(t_1)}{x(t_{k+1})} = \frac{x(t_1)}{x(t_2)} \frac{x(t_2)}{x(t_3)} \frac{x(t_3)}{x(t_4)} \cdots \frac{x(t_k)}{x(t_{k+1})} = \left[\frac{x(t_1)}{x(t_2)}\right]^k \tag{4.88}$$

Hence, inserting Eq. (4.88) into Eq. (4.84), we obtain the logarithmic decrement in the form

$$\delta = \ln \frac{x(t_1)}{x(t_2)} = \ln \left[\frac{x(t_1)}{x(t_{k+1})}\right]^{1/k} = \frac{1}{k} \ln \frac{x(t_1)}{x(t_{k+1})} \tag{4.89}$$

Then, the viscous damping factor ζ can be determined from Eq. (4.85) or from Eq. (4.86).

It should be pointed out that Eq. (4.83) remains valid even when t_1 and t_2 are any two instants separated by a period T and not necessarily corresponding to peak values for $x(t)$. Measuring peak values, however, is more convenient than measuring arbitrary amplitudes.

Example 4.3

After two complete periods the peak amplitude of a viscously damped second-order system has fallen by 60%. Calculate the viscous damping factor by using both Eqs. (4.85) and (4.86), compare results, and draw conclusions.

Using Eq. (4.89) with $k=2$, we obtain the logarithmic decrement

$$\delta = \frac{1}{2} \ln \frac{x(t_1)}{x(t_3)} = \frac{1}{2} \ln \frac{1}{0.4} = 0.45815 \tag{a}$$

so that, from Eq. (4.85), we can write

$$\zeta = \frac{\delta}{[(2\pi)^2 + \delta^2]^{1/2}} = \frac{0.45815}{[(2\pi)^2 + 0.45815^2]^{1/2}} = 0.0727 \tag{b}$$

and, from Eq. (4.86), we have

$$\zeta = \frac{\delta}{2\pi} = \frac{0.45815}{2\pi} = 0.0729 \tag{c}$$

Comparing Eqs. (b) and (c), we conclude that in the case under consideration Eq. (4.86) yields a value for the damping factor differing by about 0.27% from the value given by the more accurate Eq. (4.85), so that Eq. (4.86) is entirely adequate.

In fact, it is easy to verify that the logarithmic decrement can reach the value $\delta=0.9$ and the error arising from using Eq. (4.86) is still only about 1%.

4.8 RESPONSE OF FIRST-ORDER SYSTEMS TO HARMONIC EXCITATION. FREQUENCY RESPONSE

Let us consider now the case in which the first-order system described by Eq. (4.3) is subjected to the harmonic excitation given by

$$f(t)=f_0 e^{i\omega t}=Ake^{i\omega t} \tag{4.90}$$

where A is a real constant having units of displacement and k is the spring constant. The notation $f_0=Ak$ has the advantage that it permits expressing the response in terms of a nondimensional ratio, as we shall see shortly. Moreover, because $e^{i\omega t}=\cos\omega t+i\sin\omega t$, the notation of Eq. (4.90) enables us to derive the response to $f_0\cos\omega t$ and $f_0\sin\omega t$ simultaneously. Inserting Eq. (4.90) into Eq. (4.3) and dividing through by c, we can write the equation of motion in the form

$$\dot{x}+ax=Aae^{i\omega t} \tag{4.91}$$

where, according to Eq. (4.21), $a=k/c$. The response of a general linear system whose dynamic characteristics are described by a differential operator $D(t)$ to the excitation given by Eq. (4.90) was virtually evaluated in Section 1.7. Indeed, the response was given by Eq. (1.27), so that letting $c(t)=x(t)$ and $r_0=Ak$ in Eq. (1.27) we have

$$x(t)=X(i\omega)e^{i\omega t} \tag{4.92}$$

where

$$X(i\omega)=\frac{Aa}{Z(i\omega)} \tag{4.93}$$

in which $Z(i\omega)$ is the system impedance. Note that the particular solution given by Eq. (4.92) represents a steady-state solution.

For the first-order system at hand, the impedance is

$$Z(i\omega)=a+i\omega \tag{4.94}$$

Dividing the top and bottom of the right side of Eq. (4.93) by a and recalling the definition (4.28) of the time constant, namely, $\tau=1/a=c/k$, we obtain

$$X(i\omega)=\frac{A}{1+i\omega\tau} \tag{4.95}$$

It will prove convenient to introduce the nondimensional ratio

$$G(i\omega)=\frac{X(i\omega)}{A}=\frac{1}{1+i\omega\tau}=\frac{1-i\omega\tau}{1+\omega^2\tau^2} \tag{4.96}$$

where $G(i\omega)$ is recognized as the frequency response (Section 1.6). Inserting Eqs.

(4.95) and (4.96) into Eq. (4.92), we obtain the harmonic response

$$x(t) = AG(i\omega)e^{i\omega t} \tag{4.97}$$

But, the frequency response, as any complex expression, can be written in the form

$$G(i\omega) = |G(i\omega)|e^{i\phi} \tag{4.98}$$

where $|G(i\omega)|$ is the magnitude of $G(i\omega)$ and ϕ is a phase angle.* Introducing Eq. (4.98) into Eq. (4.97), the response becomes

$$x(t) = A|G(i\omega)|e^{i(\omega t + \phi)} \tag{4.99}$$

Note that the nature of the present phase angle ϕ is different from that encountered in the free response.

Comparing Eqs. (4.90) and (4.99), we conclude that the phase angle ϕ represents a measure of the time interval by which the response leads the excitation. As shown later in this section, in the case of the first-order system considered here the phase angle is negative, so that the response lags behind the excitation.

Equation (4.99) contains in essence the response to $Ak \cos \omega t$ and $Ak \sin \omega t$ in a single expression, as anticipated. The two responses can be separated from one another by considering the real and imaginary parts of Eq. (4.99), so that the response to the harmonic excitation $Ak \cos \omega t$ is simply

$$\text{Re } x(t) = A|G(i\omega)| \cos(\omega t + \phi) \tag{4.100a}$$

and the response to the harmonic excitation $Ak \sin \omega t$ is

$$\text{Im } x(t) = A|G(i\omega)| \sin(\omega t + \phi) \tag{4.100b}$$

Later in this chapter we shall present a geometric interpretation of solutions (4.99)–(4.100).

Next, let us examine how the response of the system behaves as the driving frequency ω varies. To this end, we wish to plot the magnitude $|G(i\omega)|$ and the phase angle ϕ of the frequency response $G(i\omega)$ as functions of ω. From complex algebra, we refer to Eq. (4.96) and write

$$|G(i\omega)| = [\text{Re}^2\, G(i\omega) + \text{Im}^2\, G(i\omega)]^{1/2} = \left[\left(\frac{1}{1+\omega^2\tau^2}\right)^2 + \left(\frac{-\omega\tau}{1+\omega^2\tau^2}\right)^2\right]^{1/2}$$

$$= \frac{1}{(1+\omega^2\tau^2)^{1/2}} \tag{4.101}$$

The plot of $|G(i\omega)|$ versus $\omega\tau$ is displayed in Fig. 4.16. Moreover, recognizing that $e^{i\phi} = \cos\phi + i \sin\phi$ and recalling Eqs. (4.96) and (4.98), we obtain the phase angle

$$\phi = \tan^{-1} \frac{\text{Im } G(i\omega)}{\text{Re } G(i\omega)} = \tan^{-1} \frac{-\omega\tau/(1+\omega^2\tau^2)}{1/(1+\omega^2\tau^2)} = \tan^{-1}(-\omega\tau) \tag{4.102}$$

The plot of ϕ versus $\omega\tau$ is shown in Fig. 4.17. It should be pointed out that the plots of Figs. 4.16 and 4.17 are known as *frequency response plots*. They are used extensively in vibrations and in control.

---

*In texts on vibrations, the phase angle is defined as the negative of the one here. The definition given here is consistent with the one given in Chapter 11, and is the definition ordinarily used in control.

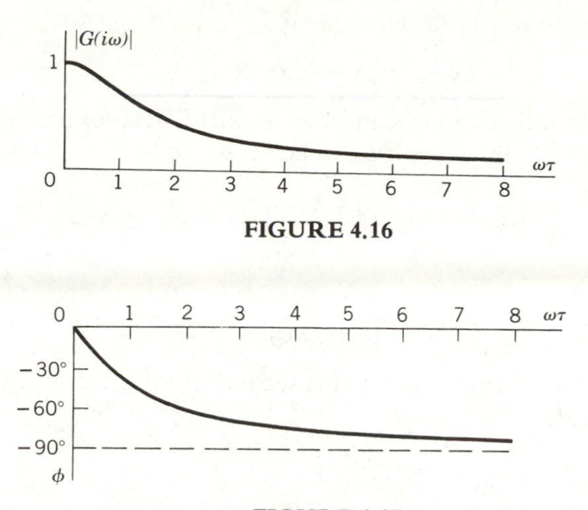

FIGURE 4.16

FIGURE 4.17

The magnitude $|G(i\omega)|$ of the frequency response can be given a physical interpretation. Indeed, using Eq. (4.99), we observe that the magnitude of the force in the spring is

$$|f_s(t)| = k|x(t)| = Ak|G(i\omega)| \tag{4.103}$$

Moreover, the magnitude of the external force is simply $|f(t)| = Ak$. Hence, we can write

$$|G(i\omega)| = \frac{|f_s(t)|}{|f(t)|} \tag{4.104}$$

or, the magnitude of the frequency response represents the nondimensional ratio of the magnitude of the spring force $f_s(t)$ to the magnitude of the external force $f(t)$.

We observe from the plot of $|G(i\omega)|$ versus $\omega\tau$ in Fig. 4.16 that for relatively large values of $\omega\tau$ the response is attenuated greatly. Hence, for a given τ, the system acts like a *filter*, leaving low-frequency inputs largely unaffected and attenuating high-frequency inputs. In many applications, electrical signals are contaminated by undesirable external factors called *noise*. In general, signals have low frequencies and noise has high frequencies. Then, the RC circuit discussed in Section 4.3 can be used as a filter reducing the amplitude of the undesirable noise relative to the amplitude of the signal. In view of this, such an RC circuit is called a *low-pass filter*. Note that in this case the time constant is $\tau = RC$.

4.9 RESPONSE OF SECOND-ORDER SYSTEMS TO HARMONIC EXCITATION

The response of second-order systems to harmonic excitation can be obtained in a way analogous to that used in Section 4.8 for the response of first-order systems. Indeed, the mathematical analogy is complete, and the difference lies in the manner in which the two types of systems respond.

Let us consider the second-order system

$$\ddot{x}+2\zeta\omega_n\dot{x}+\omega_n^2x=A\omega_n^2e^{i\omega t} \tag{4.105}$$

where A is a constant having the same units as $x(t)$. Then, following the procedure outlined in Section 4.8, the response can be shown to be

$$x(t)=X(i\omega)e^{i\omega t} \tag{4.106}$$

where

$$X(i\omega)=\frac{A}{1-(\omega/\omega_n)^2+i2\zeta\omega/\omega_n} \tag{4.107}$$

Defining the frequency response for this second-order system in the form of the nondimensional ratio

$$G(i\omega)=\frac{X(i\omega)}{A}=\frac{1}{1-(\omega/\omega_n)^2+i2\zeta\omega/\omega_n} \tag{4.108}$$

we can write the response once again in the form

$$x(t)=A|G(i\omega)|e^{i(\omega t+\phi)} \tag{4.109}$$

where $|G(i\omega)|$ is the magnitude and ϕ is the phase angle of the frequency response $G(i\omega)$.

To study the nature of the response, let us examine the dependence of $|G(i\omega)|$ and ϕ on the driving frequency ω. To this end, let us write

$$\begin{aligned}|G(i\omega)|&=[\mathrm{Re}^2\, G(i\omega)+\mathrm{Im}^2\, G(i\omega)]^{1/2}\\&=\frac{1}{\{[1-(\omega/\omega_n)^2]^2+(2\zeta\omega/\omega_n)^2\}^{1/2}}\end{aligned} \tag{4.110}$$

Plots of $|G(i\omega)|$ versus ω/ω_n for various values of ζ are shown in Fig. 4.18. For small ζ, the amplitude increases appreciably in the neighborhood of $\omega/\omega_n=1$ and then it falls off as ω/ω_n continues to increase. Note that $|G(i\omega)|$ is called the *magnification factor*, in spite of the fact that for certain values of ω the amplitude of the response is actually reduced instead of being magnified. The curves $|G(i\omega)|$ versus ω/ω_n reach peak values for certain ω/ω_n. To obtain the peak value for any of the curves, we use the standard technique of calculus for the determination of maxima and write

$$\frac{dG(i\omega)}{d(\omega/\omega_n)}=-\frac{1}{2}\,\frac{2[1-(\omega/\omega_n)^2](-2\omega/\omega_n)+8\zeta^2\omega/\omega_n}{\{[1-(\omega/\omega_n)^2]^2+(2\zeta\omega/\omega_n)^2\}^{3/2}}=0 \tag{4.111}$$

which yields

$$\frac{\omega}{\omega_n}=(1-2\zeta^2)^{1/2} \tag{4.112}$$

so that the peaks occur for $\omega/\omega_n<1$. The proximity of the peaks to $\omega/\omega_n=1$ depends on how small ζ is. Moreover, peaks occur only if $1-2\zeta^2$ is positive. Inserting

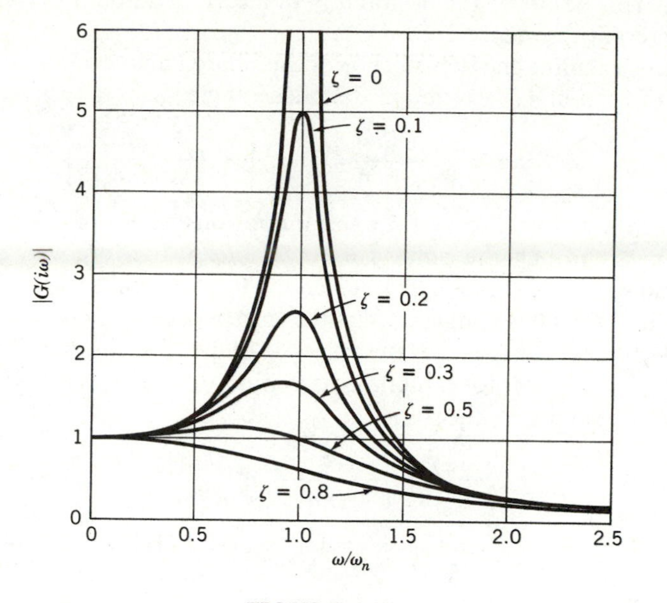

FIGURE 4.18

Eq. (4.112) back into Eq. (4.110), we obtain the peak amplitude

$$Q = |G(i\omega)|_{\max} = \frac{1}{2\zeta(1-\zeta^2)^{1/2}}, \qquad \zeta < \frac{1}{\sqrt{2}} \tag{4.113}$$

which, for small ζ, can be approximated by

$$Q \cong \frac{1}{2\zeta} \tag{4.114}$$

The peak amplitude Q is known as the *quality factor*.

For very small values of ζ, Q becomes very large. In fact, as $\zeta \to 0$, $Q \to \infty$. For $\zeta = 0$, we have no longer a peak but a discontinuity. It is easy to verify from Eq. (4.110) that the discontinuity occurs at $\omega = \omega_n$, at which driving frequency the amplitude becomes infinite. Of course, this is impossible for real physical systems, for which the displacement must remain finite. In fact, for our analysis to remain valid, the displacement must remain sufficiently small to stay within the linear range. Nevertheless, this serves as an indication that undamped systems experience violent vibrations at $\omega = \omega_n$, a phenomenon known as *resonance*. It should be pointed out that solution (4.109) is not valid at resonance, so that a separate solution for the case $\zeta = 0$, $\omega = \omega_n$ must be produced. This is done later in this section. In many engineering systems, the driving frequency is not constant but increases from zero to a given steady operating value, such as when starting a motor driving the system. If the operating value of ω is larger than the natural frequency ω_n, then some high-amplitude vibration can be expected when ω is close to ω_n. This points to the desirability of a certain amount of damping in the system to prevent resonance.

Next, let us examine the dependence of the phase angle ϕ on ω. Following the procedure of Section 4.5, we obtain the phase angle

$$\phi = \tan^{-1}\frac{\text{Im } G(i\omega)}{\text{Re } G(i\omega)} = \tan^{-1}\left[-\frac{2\zeta\omega/\omega_n}{1-(\omega/\omega_n)^2}\right] \tag{4.115}$$

Figure 4.19 shows plots of ϕ versus ω/ω_n for various values of ζ. We observe that all curves pass through the point $\phi = -\pi/2$, $\omega/\omega_n = 1$. Moreover, $\phi > -\pi/2$ for $\omega/\omega_n < 1$ and $\phi < -\pi/2$ for $\omega/\omega_n > 1$. As $\omega/\omega_n \to 0$, $\phi \to 0$, and as $\omega/\omega_n \to \infty$, $\phi \to -\pi$. Hence, because the phase angle is negative, except for $\zeta = 0$, $\omega < \omega_n$, we conclude from Eqs. (4.105) and (4.109) that the response of damped systems lags behind the excitation. For $\zeta = 0$, the plot exhibits a discontinuity at $\omega/\omega_n = 1$. In the undamped case, $\zeta = 0$, the response reduces to

$$x(t) = \frac{A}{|1-(\omega/\omega_n)^2|}\, e^{i(\omega t + \phi)} \tag{4.116}$$

where $\phi = 0$ for $\omega/\omega_n < 1$ and $\phi = -\pi$ for $\omega/\omega_n > 1$. Hence, Eq. (4.116) can be written in the form

$$x(t) = \frac{A}{1-(\omega/\omega_n)^2}\, e^{i\omega t} \tag{4.117}$$

Equation (4.117) states that the displacement is *in phase* with the excitation for $\omega < \omega_n$ and that it is *180° out of phase* with the excitation for $\omega > \omega_n$.

Finally, let us examine the resonance case, which occurs when a harmonic oscillator is driven at the natural frequency. Letting $c = 0$, $\omega = \omega_n$ in Eq. (4.105), considering only the real part of the excitation, and dividing through by m, we obtain

$$\ddot{x}(t) + \omega_n^2 x(t) = \omega_n^2 A \cos \omega_n t \tag{4.118}$$

We shall produce a particular solution of Eq. (4.118) by the Laplace transformation

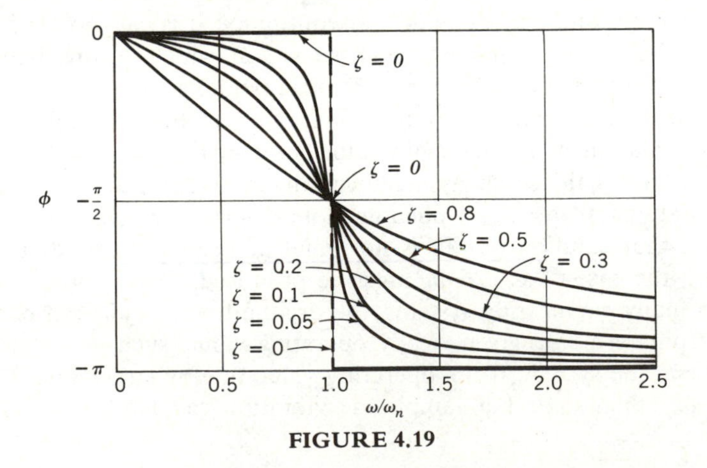

FIGURE 4.19

method. Transforming both sides of Eq. (4.118), letting $x(0)=\dot{x}(0)=0$, and using the Laplace transforms table in the Appendix, we can write

$$(s^2+\omega_n^2)X(s)=\omega_n^2 A\frac{s}{s^2+\omega_n^2} \tag{4.119}$$

or

$$X(s)=\frac{\omega_n^2 As}{(s^2+\omega_n^2)^2} \tag{4.120}$$

Once again using the Laplace transforms table, we can write the inverse transformation

$$x(t)=\mathscr{L}^{-1}X(s)=\frac{A}{2}\omega_n t\sin\omega_n t \tag{4.121}$$

The response lends itself to relatively easy interpretation. The term $(A/2)\omega_n t$ can be regarded as a time-dependent amplitude, modulating the harmonic function $\sin\omega_n t$. Hence, the response will be bounded by the envelope defined by the two straight lines $\pm(A/2)\omega_n t$. As the width of the envelope increases with time, the response is characterized by increasingly large amplitudes (Fig. 4.20). At a certain point, however, the linear range of the spring will be exceeded, at which point either the system breaks down, as in the case of a softening spring, or the motion is contained, as in the case of a stiffening spring. Of course, when the system exceeds the linear range, one must abandon the linear analysis as invalid and consider nonlinear analysis. Because the excitation is a cosine function and the response is a sine function and the two functions are related by the identity $\sin\omega t\equiv\cos(\omega t-\pi/2)$, it follows that the phase angle ϕ has the value $-\pi/2$. Hence the plot ϕ versus ω/ω_n for $\zeta=0$ consists of the straight line $\phi=0$ for $\omega<\omega_n$, the point $\phi=-\pi/2$ for $\omega=\omega_n$, and the straight line $\phi=-\pi$ for $\omega>\omega_n$.

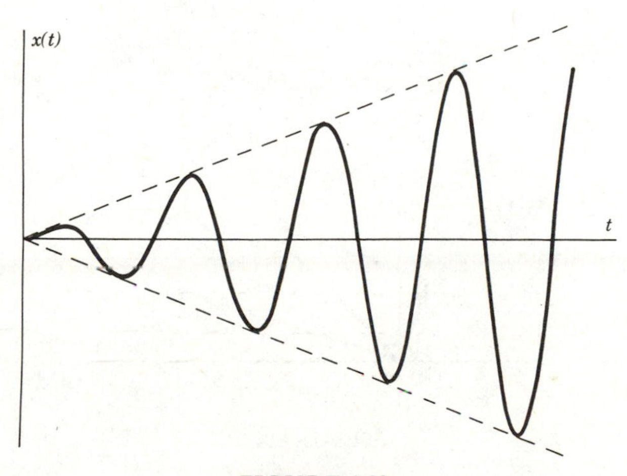

FIGURE 4.20

4.10 GEOMETRIC INTERPRETATION OF THE RESPONSE TO HARMONIC EXCITATION

Equation (4.91), governing the response of a first-order system to harmonic excitation, can be given an interesting geometric interpretation by representing it in the complex plane. Referring to Eq. (4.99), we can write

$$\dot{x}(t) = i\omega A|G(i\omega)|e^{i(\omega t+\phi)} \tag{4.122}$$

so that, considering the identity

$$i = \cos\frac{\pi}{2} + i\sin\frac{\pi}{2} = e^{i\pi/2} \tag{4.123}$$

we can rewrite Eq. (4.122) in the form

$$\dot{x}(t) = \omega A|G(i\omega)|e^{i(\omega t+\phi+\pi/2)} \tag{4.124}$$

Hence, $\dot{x}(t)$ is a vector whose magnitude is equal to the magnitude of $x(t)$ multiplied by ω and whose direction makes an angle $\pi/2$ with the direction of $x(t)$. In view of this, Eq. (4.91) can be satisfied vectorially, as shown in the diagram of Fig. 4.21. Note that, as time unfolds, the entire diagram rotates counterclockwise in the complex plane with the angular velocity ω. The response to the excitation $Aa \cos \omega t$ can be obtained by taking the projection of $x(t)$ on the real axis and the response to $Ak \sin \omega t$ can be obtained by taking the projection of $x(t)$ on the imaginary axis, so that the complex representation of motion yields the two solutions simultaneously.

The geometric interpretation of the response of a second-order system to harmonic excitation can be obtained analogously. From Eq. (4.99), we can write

$$\ddot{x}(t) = \omega^2 A|G(i\omega)|e^{i(\omega t+\phi+\pi)} \tag{4.125}$$

and because

$$-1 = \cos\pi + i\sin\pi = e^{i\pi} \tag{4.126}$$

we have

$$\ddot{x}(t) = \omega^2 A|G(i\omega)|e^{i(\omega t+\phi+\pi)} \tag{4.127}$$

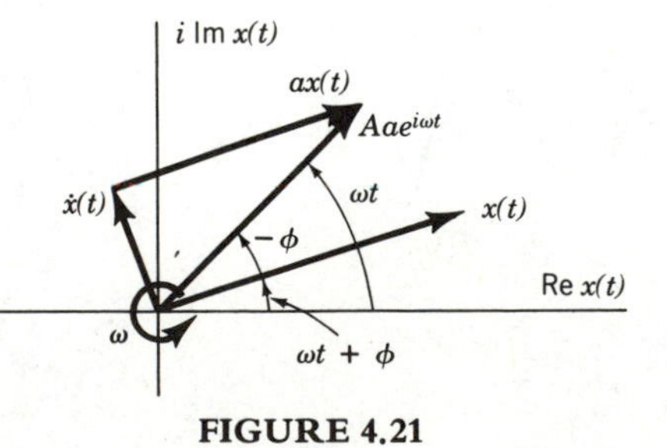

FIGURE 4.21

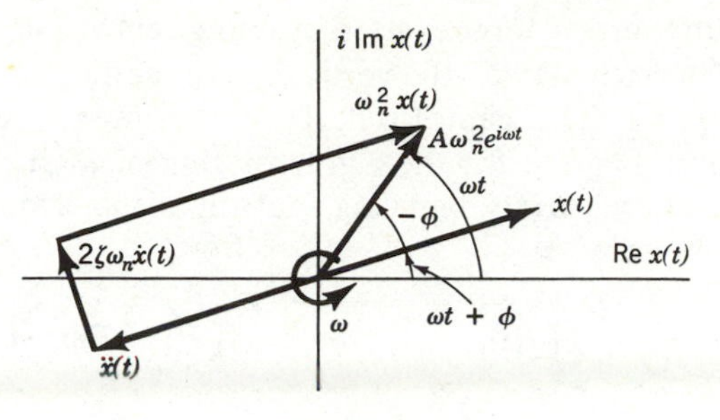

FIGURE 4.22

Hence, $\ddot{x}(t)$ is a vector whose magnitude is equal to the magnitude of $x(t)$ multiplied by ω^2 and whose direction is opposite to that of $x(t)$. The vector diagram describing Eq. (4.105) is shown in Fig. 4.22.

4.11 ROTATING UNBALANCED MASSES

Many engineering systems contain rotating unbalanced masses, sometimes by design but quite often inadvertently. Such masses produce harmonic excitation that can lead to excessive vibration and possible damage.

Let us consider a system consisting of a principal mass $M-m$ supported by two equal springs of combined stiffness k and a damper with coefficient of viscous damping c. Two equal eccentric masses $m/2$ rotate in opposite sense with constant angular velocity ω about symmetrically placed points at distances R from the masses, so that at any time the angle between the horizontal and the rigid links carrying the masses is ωt (Fig. 4.23a). Figures 4.23b and 4.23c show free-body diagrams for the principal mass and for the right eccentric mass, respectively. From Fig. 4.23b, if we measure the displacement $x(t)$ from the static equilibrium position, we can write Newton's second law for the principal mass in the form

$$-2F_x - c\dot{x}(t) - kx(t) = (M-m)\ddot{x}(t) \tag{4.128}$$

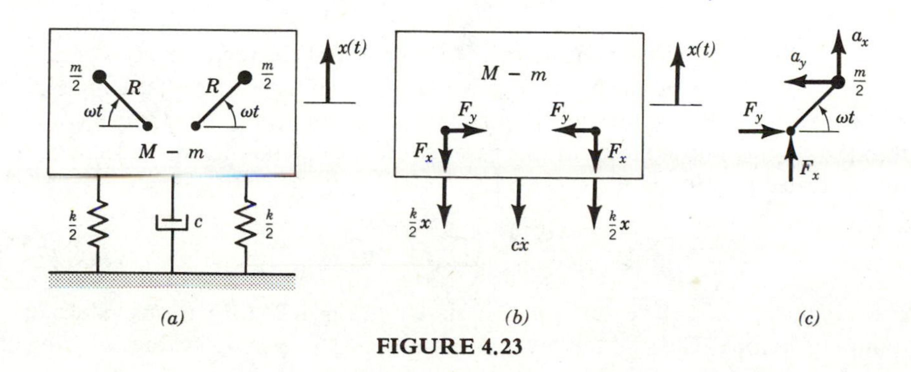

FIGURE 4.23

where F_x represents the vertical force exerted by one eccentric mass on the principal mass. Of course, for the two masses, the vertical forces add up to $2F_x$, whereas the horizontal forces, F_y and $-F_y$, cancel out. To obtain the expression for F_x, we consider the free-body diagram for the right rotating mass shown in Fig. 4.23*c* and write Newton's second law. Observing that the displacement of the mass is $x(t)+R \sin \omega t$, we can write

$$F_x=\frac{m}{2}a_x=\frac{m}{2}\frac{d^2}{dt^2}[x(t)+R \sin \omega t]=\frac{m}{2}[\ddot{x}(t)-R\omega^2 \sin \omega t] \tag{4.129}$$

It must be pointed out that, by measuring $x(t)$ from equilibrium, we were able to cancel out the effect of the weights $(M-m)g$ and $mg/2$ in Eqs. (4.128) and (4.129), respectively. Introducing Eq. (4.129) into Eq. (4.128) and rearranging, we obtain the system equation of motion

$$\ddot{x}(t)+2\zeta\omega_n\dot{x}(t)+\omega_n^2x(t)=\frac{mR\omega^2}{M}\sin \omega t \tag{4.130}$$

where

$$2\zeta\omega_n=\frac{c}{M}, \qquad \omega_n^2=\frac{k}{M} \tag{4.131}$$

Hence, the rotating unbalanced masses produce a harmonic excitation of the system, where the excitation has the frequency ω. Note that, although the system involves three masses, the motion of the reciprocating masses relative to the principal mass is prescribed, so that this is a single-degree-of-freedom system.

The solution of an equation similar to Eq. (4.130) was derived earlier in the form of Eq. (4.109), and to use this solution it is only necessary to recognize that in this case

$$A=\frac{m}{M}\left(\frac{\omega}{\omega_n}\right)^2 R \tag{4.132}$$

Hence, retaining the imaginary part of the solution (4.109), with A as indicated by Eq. (4.132), we obtain

$$x(t)=\frac{mR}{M}\left(\frac{\omega}{\omega_n}\right)^2 |G(i\omega)| \sin(\omega t+\phi) \tag{4.133}$$

where $|G(i\omega)|$ and ϕ are given by Eqs. (4.110) and (4.115), respectively.

Next, let us examine the manner in which the amplitude and phase angle of the response vary with the driving frequency ω. Examining Eq. (4.133), we conclude that the magnification factor in this case requires some modification. Indeed, now the indicated nondimensional ratio is

$$\frac{|x(t)|M}{Rm}=\left(\frac{\omega}{\omega_n}\right)^2 |G(i\omega)| \tag{4.134}$$

Plots of $(\omega/\omega_n)^2|G(i\omega)|$ versus ω/ω_n are shown in Fig. 4.24 for various values of the damping factor ζ. Clearly, the plots ϕ versus ω/ω_n for various values of ζ remain as in Fig. 4.19.

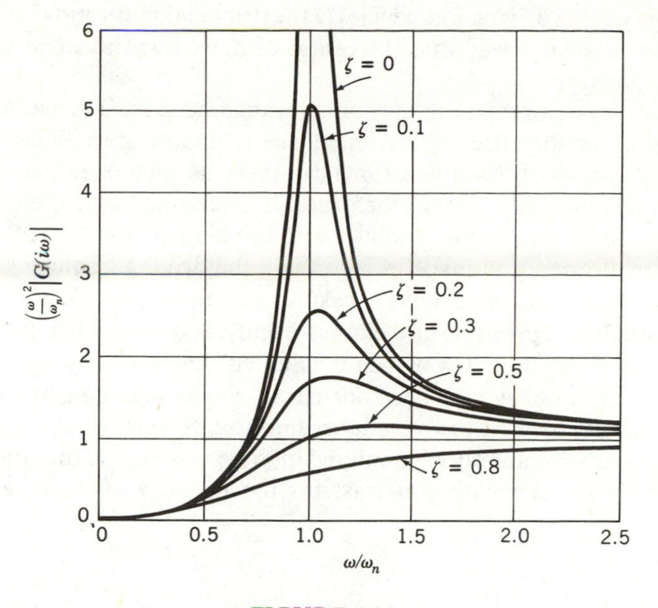

FIGURE 4.24

We note from Fig. 4.24 that the effect of multiplying $|G(i\omega)|$ by $(\omega/\omega_n)^2$ is to shift the peaks from values of ω smaller than ω_n to values larger than ω_n. We also note that $(\omega/\omega_n)^2|G(i\omega)|\to 0$ as $\omega/\omega_n\to 0$ and that $(\omega/\omega_n)^2|G(\omega)|\to 1$ as $\omega/\omega_n\to\infty$. This latter statement leads to an interesting result. From Fig. 4.19, we conclude that $\phi\to-\pi$ as $\omega/\omega_n\to\infty$, which implies that the excitation and response are 180° out of phase in this case. Hence, as $\omega/\omega_n\to\infty$ and $(\omega/\omega_n)^2|G(i\omega)|\to 1$, the displacement of $M-m$ becomes

$$x(t)=\frac{mR}{M}\sin(\omega t-\pi)=-\frac{mR}{M}\sin\omega t \tag{4.135}$$

On the other hand, under the same circumstances, the vertical displacement of the masses $m/2$ becomes

$$x(t)+R\sin\omega t=\frac{M-m}{M}R\sin\omega t \tag{4.136}$$

But in general the position of the mass center of a system of masses is defined as (see Section 5.3)

$$x_C=\frac{\sum_{i=1}^{n} m_i x_i}{\sum_{i=1}^{n} m_i} \tag{4.137}$$

which in this particular case yields

$$x_C=\frac{1}{M}\left[(M-m)\left(-\frac{mR}{M}\sin\omega t\right)+2\frac{m}{2}\frac{M-m}{M}R\sin\omega t\right]=0 \tag{4.138}$$

so that for very large driving frequencies ω the principal mass and the two eccentric masses move in such a way that the center of mass remains at rest in the static equilibrium position.

Examples of systems containing rotating unbalanced masses are very common, and in most cases the effect of the imbalance is undesirable. Washing machines and clothes dryers with rotating drums can serve as such examples if the clothes are not spread uniformly around the drum. It is assumed that the clothes do not move relative to the drum. Automobiles with unbalanced tires are other examples. Because normal operation involves increasing the driving frequency ω from zero to well beyond $\omega=\omega_n$, corrective measures are necessary if vibration is to be eliminated, such as spreading the clothes evenly and balancing the tires. In the discussion just preceding, it was tacitly assumed that washing machines, clothes dryers, and automobiles can be modeled as single-degree-of-freedom systems, which must be regarded only as a crude assumption. Nevertheless, the phenomenon described above is commonly encountered in these systems, so that the assumption has some measure of validity, at least for the purpose of explaining this phenomenon.

4.12 MOTION OF VEHICLES OVER WAVY TERRAIN

Let us consider a vehicle traveling with uniform velocity v over a wavy terrain. We shall model the vehicle as a damped single-degree-of-freedom system and the terrain as the function

$$y(x)=A\sin\frac{2\pi x}{L} \tag{4.139}$$

where L is the wavelength (Fig. 4.25*a*). The forward motion of the vehicle on the wavy terrain $y(x)$ results in a vertical motion $y(t)$ of the wheel. Because uniform forward motion implies the relation $x=vt$, the vertical motion of the wheel is simply

$$y(t)=A\sin\frac{2\pi vt}{L} \tag{4.140}$$

Considering the free-body diagram of Fig. 4.25*b*, we can use Newton's second law to write

$$-c[\dot{z}(t)-\dot{y}(t)]-k[z(t)-y(t)]=m\ddot{z}(t) \tag{4.141}$$

yielding

$$m\ddot{z}(t)+c\dot{z}(t)+kz(t)=c\dot{y}(t)+ky(t) \tag{4.142}$$

Dividing Eq. (4.142) through by m and using Eq. (4.140), we obtain

$$\ddot{z}(t)+2\zeta\omega_n\dot{z}(t)+\omega_n^2 z(t)=\omega_n^2 A[\sin\omega t+2\zeta(\omega/\omega_n)\cos\omega t] \tag{4.143}$$

where ζ and ω_n have the customary meaning and

$$\omega=2\pi v/L \tag{4.144}$$

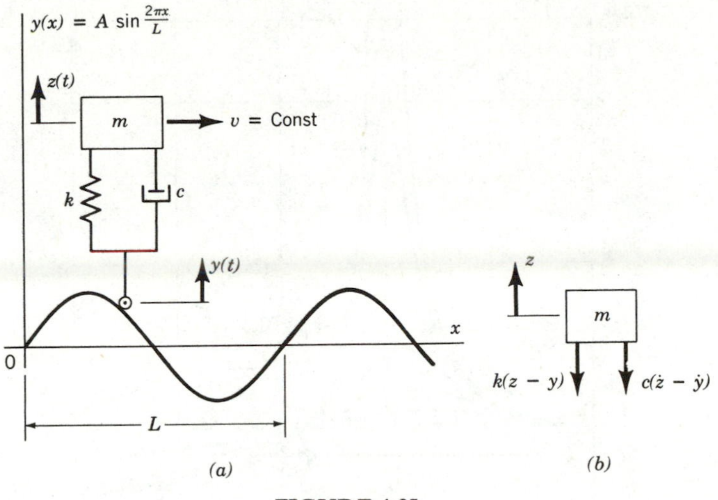

FIGURE 4.25

is the driving frequency. Introducing the notation

$$2\zeta\omega/\omega_n = \tan\alpha \tag{4.145}$$

we can reduce Eq. (4.143) to

$$\begin{aligned}\ddot{z}(t) + 2\zeta\omega_n\dot{z}(t) + \omega_n^2 z(t) &= \omega_n^2 A \frac{1}{\cos\alpha}(\sin\omega t\cos\alpha + \cos\omega t\sin\alpha)\\ &= \omega_n^2 A[1 + (2\zeta\omega/\omega_n)^2]^{1/2}\sin(\omega t + \alpha)\end{aligned} \tag{4.146}$$

so that α can be identified as an excitation phase angle.

The solution to an equation similar to Eq. (4.146), namely, Eq. (4.130), was obtained in Section 4.11 in the form of Eq. (4.133). Hence, we shall produce the solution to Eq. (4.146) by adapting solution (4.133) to the case at hand. Indeed, comparing Eqs. (4.130) and (4.146), we observe that in this case the amplitude of the excitation is multiplied by $[1 + (2\zeta\omega/\omega_n)^2]^{1/2}$ and that the sine function contains the phase angle α. With this in mind, we can modify solution (4.133) and write the solution to Eq. (4.146) directly in the form

$$z(t) = A[1 + (2\zeta\omega/\omega_n)^2]^{1/2}|G(i\omega)|\sin(\omega t + \phi_1) \tag{4.147}$$

where

$$\phi_1 = \phi + \alpha \tag{4.148}$$

is the response phase angle. Note that $|G(i\omega)|$ and ϕ are given by Eqs. (4.110) and (4.115), respectively.

A measure of the magnitude of the response can be obtained from the nondimensional ratio

$$\frac{|z(t)|}{A} = [1 + (2\zeta\omega/\omega_n)^2]^{1/2}|G(i\omega)| \tag{4.149}$$

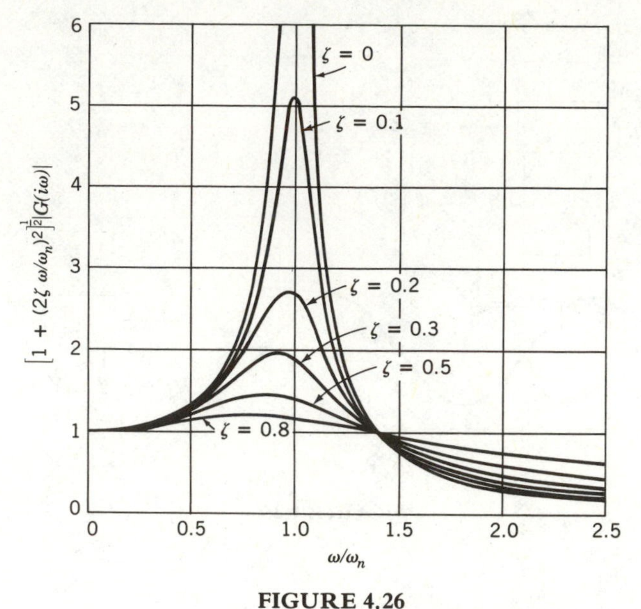

FIGURE 4.26

which is known as *transmissibility*. Figure 4.26 shows plots of $[1+(2\zeta\omega/\omega_n)^2]^{1/2}|G(i\omega)|$ versus ω/ω_n for various values of the damping factor ζ. We note that for small ζ the curves experience peaks in the neighborhood of $\omega=\omega_n$, but for $\omega<\omega_n$. Moreover, the response approaches zero as $\omega/\omega_n\rightarrow\infty$, and it does so faster for smaller damping. The response will be the largest for a given ratio of the velocity v to the wavelength L. Indeed, for small damping, the critical ratio is approximately equal to the natural frequency $f_n=\omega_n/2\pi$, which can be verified by substituting $\omega=\omega_n$ in Eq. (4.144). Shock absorbers in automobiles are really dampers designed to reduce vibration. They generally possess heavy damping. When they are worn out, however, vibration magnification like that discussed above can occur.

To calculate the response phase angle ϕ_1, let us write first

$$\tan\phi_1=\tan(\phi+\alpha)=\frac{\tan\phi+\tan\alpha}{1-\tan\phi\tan\alpha}$$

$$=\frac{-(2\zeta\omega/\omega_n)/[1-(\omega/\omega_n)^2]+2\zeta\omega/\omega_n}{1+\{(2\zeta\omega/\omega_n)/[1-(\omega/\omega_n)^2]\}2\zeta\omega/\omega_n}=-\frac{2\zeta(\omega/\omega_n)^3}{1-(\omega/\omega_n)^2+(2\zeta\omega/\omega_n)^2} \tag{4.150}$$

so that the phase angle ϕ_1 is

$$\phi_1=\tan^{-1}\left[-\frac{2\zeta(\omega/\omega_n)^3}{1-(\omega/\omega_n)^2+(2\zeta\omega/\omega_n)^2}\right] \tag{4.151}$$

Plots of ϕ_1 versus ω/ω_n for various values of ζ are shown in Fig. 4.27.

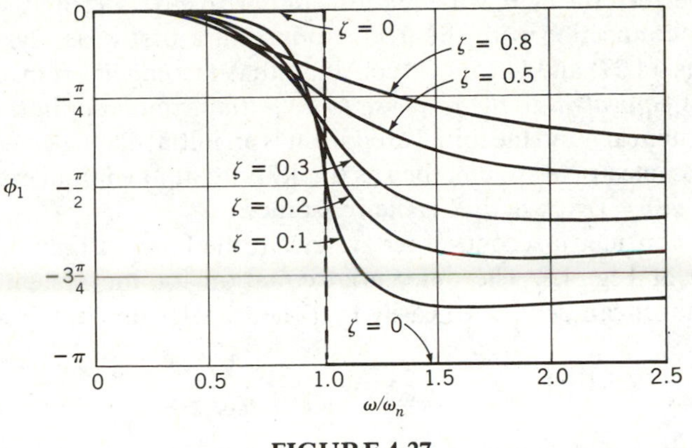

FIGURE 4.27

4.13 IMPULSE RESPONSE

The impulse response, denoted $g(t)$, was defined in Section 1.9 as the response to a unit impulse applied at $t=0$, with the initial conditions being equal to zero. Also in Section 1.9 it was shown that the impulse response is equal to the inverse Laplace transform of the transfer function $G(s)$, or

$$g(t)=\mathscr{L}^{-1}G(s) \tag{4.152}$$

In this section, we propose to derive the impulse response both for a first-order and a second-order system.

The equation of motion for a damper–spring system was shown in Section 4.2 to have the form

$$c\dot{x}(t)+kx(t)=f(t) \tag{4.153}$$

Taking the Laplace transform of both sides of Eq. (4.153), while letting $x(0)$ be equal to zero, we obtain

$$(cs+k)X(s)=F(s) \tag{4.154}$$

so that, using the analogy with Eq. (1.32), we conclude that the transfer function for the first-order system in question has the expression

$$G(s)=\frac{X(s)}{F(s)}=\frac{1}{cs+k}=\frac{1}{c(s+a)} \tag{4.155}$$

where $a=k/c$. Considering Eq. (4.152) and using the table of Laplace transform pairs (Section A.7), we obtain the impulse response

$$g(t)=\mathscr{L}^{-1}G(s)=\mathscr{L}^{-1}\frac{1}{c}\frac{1}{s+a}=\frac{1}{c}e^{-t/\tau}u(t) \tag{4.156}$$

where $\tau=1/a=c/k$ is recognized as the time constant and $u(t)$ is the unit step function. Note that we multiplied the response by $u(t)$ in recognition of the fact that it must be zero for $t<0$. An equation similar to Eq. (4.156) was obtained in Section 4.4 in connection with the free response of a first-order system. Indeed, comparing Eqs. (4.29) and (4.156) we conclude that the *impulse response of a first-order system is equivalent to the response to an initial excitation.* In the case of the mechanical system at hand the initial excitation is an initial displacement, $x(0)=1/c$. In the case of an electrical system, such as the RL circuit, the initial excitation is an initial charge, $q(0)=1/R$, where R is the resistance.

Next, let us consider a second-order system in the form of the mass–damper–spring system of Fig. 4.5. The differential equation for the system is given by Eq. (4.8), so that it can be verified easily that the transfer function has the form

$$G(s)=\frac{1}{ms^2+cs+k}=\frac{1}{m(s^2+2\zeta\omega_n s+\omega_n^2)} \tag{4.157}$$

where ζ is the damping factor and ω_n is the frequency of undamped oscillation. Hence, the impulse response for the mass–damper–spring system is simply

$$g(t)=\mathscr{L}^{-1}G(s)=\mathscr{L}^{-1}\frac{1}{m(s^2+2\zeta\omega_n s+\omega_n^2)} \tag{4.158}$$

To obtain the inverse Laplace transform, it will prove convenient to expand $G(s)$ into partial fractions. It is not difficult to show that

$$G(s)=\frac{1}{m(s_1-s_2)}\left(\frac{1}{s-s_1}-\frac{1}{s-s_2}\right) \tag{4.159}$$

where, assuming that $\zeta<1$,

$$\left.\begin{matrix} s_1 \\ s_2 \end{matrix}\right\}=-\zeta\omega_n\pm i\omega_d \tag{4.160}$$

are the simple poles of $G(s)$, in which $\omega_d=(1-\zeta^2)^{1/2}\omega_n$ is the frequency of damped oscillation. Using the table of Laplace transform pairs (Section A.7), we can write

$$g(t)=\frac{1}{m(s_1-s_2)}(e^{s_1t}+e^{s_2t}) \tag{4.161}$$

Inserting s_1 and s_2 from Eqs. (4.160) into Eq. (4.161), we obtain the impulse response for the mass–damper–system in the form

$$g(t)=\frac{1}{m\omega_d}e^{-\zeta\omega_n t}\sin\omega_d t\, u(t) \tag{4.162}$$

Note that the impulse response, Eq. (4.162), could have been obtained directly from Eq. (4.158) and the table of Laplace transform pairs in Section A.7.

An equation similar to Eq. (4.162) was obtained in Example 4.2. Comparing Eq. (d) of Example 4.2 with Eq. (4.162), we conclude that *the impulse response of the mass–damper–spring system is equal to the response to the initial velocity* $\dot{x}(0)=v_0=1/m$. Similarly, it is not difficult to see that for the LRC circuit of Fig. 4.9 the impulse response is equal to the response to the initial current $i(0)=1/L$.

4.14 STEP RESPONSE

The step response, denoted by $\mathscr{s}(t)$, is defined as the response to a unit step function applied at $t=0$, with the initial conditions being equal to zero. It was shown in Section 1.10 that the step response of a general linear system has the expression

$$\mathscr{s}(t)=\mathscr{L}^{-1}\frac{G(s)}{s} \tag{4.163}$$

where $G(s)$ is the transfer function.

Recalling Eq. (4.155), we can write the step response of the damper–spring system described by Eq. (4.153) in the form

$$\mathscr{s}(t)=\mathscr{L}^{-1}\frac{1}{cs(s+a)} \tag{4.164}$$

It is not difficult to show that the partial fractions expansion of the function on the right side of Eq. (4.164) is

$$\frac{1}{cs(s+a)}=\frac{1}{k}\left(\frac{1}{s}-\frac{1}{s+a}\right) \tag{4.165}$$

so that, using the table of Laplace transform pairs (Section A.7), we can write

$$\mathscr{s}(t)=\frac{1}{k}(1-e^{-t/\tau})\mathscr{u}(t) \tag{4.166}$$

where $\tau=1/a=c/k$ is the time constant and $\mathscr{u}(t)$ is the unit step function. The step response is plotted in Fig. 4.28.

Next, let us determine the step response of an undamped second-order system. The transfer function of such a system is obtained by simply letting $c=0$ in Eq. (4.157), so that the step response is

$$\mathscr{s}(t)=\mathscr{L}^{-1}\frac{1}{ms(s^2+\omega_n^2)} \tag{4.167}$$

But

$$\frac{1}{ms(s^2+\omega_n^2)}=\frac{1}{k}\left[1-\frac{1}{2(s-i\omega_n)}-\frac{1}{2(s+i\omega_n)}\right] \tag{4.168}$$

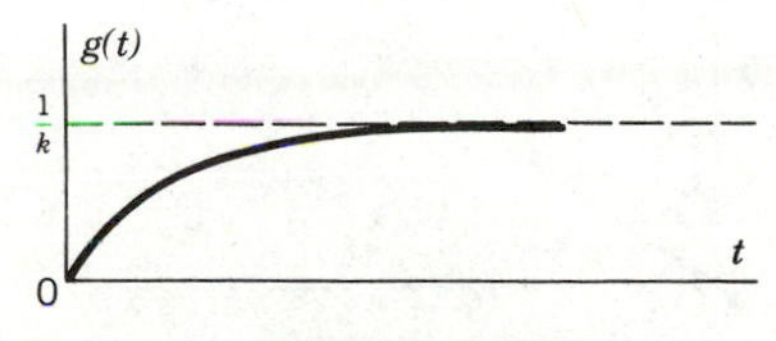

FIGURE 4.28

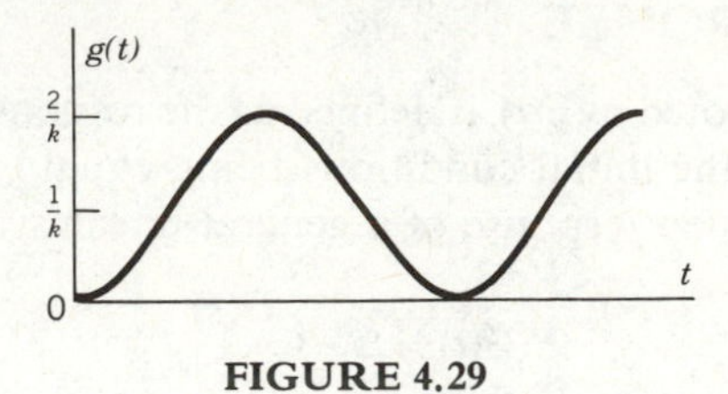

FIGURE 4.29

so that, using the table of Laplace transform pairs once again, we obtain the step response

$$\mathscr{s}(t)=\frac{1}{k}\left(1-\frac{1}{2}e^{i\omega_n t}-\frac{1}{2}e^{-i\omega_n t}\right)\mathscr{u}(t)=\frac{1}{k}(1-\cos\omega_n t)\,\mathscr{u}(t) \tag{4.169}$$

It should be pointed out that the resolution into partial fractions was not really necessary here, because the function on the right side of Eq. (4.167) can be found in the table of Laplace transform pairs in Section A.7. The response is plotted in Fig. 4.29.

4.15 RESPONSE TO ARBITRARY EXCITATION

The response to any arbitrary excitation can be obtained by means of the convolution integral derived in Section 1.11. Because we have used the symbol τ for the time constant, we rewrite Eq. (1.59) in the form

$$x(t)=\int_0^t g(t-\sigma)f(\sigma)\,d\sigma=\int_0^t g(\sigma)f(t-\sigma)\,d\sigma \tag{4.170}$$

where σ is a dummy variable of integration.

As an application of the convolution integral, let us consider the response of the damper–spring system to a force in the form of the ramp function shown in Fig. 4.30. The force can be expressed in the form

$$f(t)=\frac{f_0}{T}\,t\mathscr{u}(t) \tag{4.171}$$

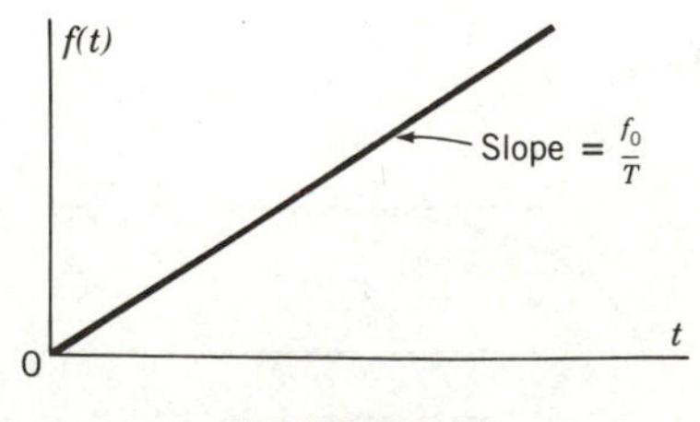

FIGURE 4.30

In Section 4.13, we showed that the impulse response of the damper–spring system is

$$g(t)=\frac{1}{c}\,e^{-t/\tau}\,u(t) \tag{4.172}$$

Inserting Eqs. (4.171) and (4.172) into the first form of the convolution integral, we can write

$$\begin{aligned}x(t)&=\frac{f_0}{Tc}\int_0^t \sigma e^{-(t-\sigma)/\tau}\,d\sigma=\frac{f_0}{Tc}\,e^{-t/\tau}\int_0^t \sigma e^{\sigma/\tau}\,d\sigma\\&=\frac{f_0}{Tc}\,e^{-t/\tau}\left[\frac{e^{\sigma/\tau}}{(1/\tau)^2}\left(\frac{\sigma}{\tau}-1\right)\right]\Bigg|_0^t u(t)=\frac{f_0}{Tk}\,[t-\tau(1-e^{-t/\tau})]u(t)\end{aligned} \tag{4.173}$$

The response is plotted in Fig. 4.31. Note that, compared with an equivalent displacement input equal to $f(t)/k$, the output $x(t)$ exhibits a steady-state error equal to f_0c/Tk^2.

As a second illustration, let us consider the response of a mass–spring system to the rectangular pulse shown in Fig. 4.32. The force can be expressed in the form

$$f(t)=\begin{cases} f_0, & 0<t<T\\ 0, & \text{everywhere else}\end{cases} \tag{4.174}$$

Moreover, letting $\zeta=0$ in Eq. (4.162), we obtain the impulse response

$$g(t)=\frac{1}{m\omega_n}\sin\omega_n t\;u(t) \tag{4.175}$$

Inserting Eqs. (4.174) and (4.175) into the first form of the convolution integral,

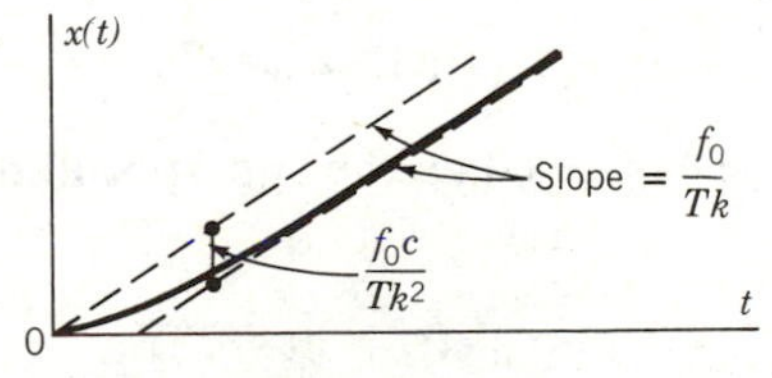

FIGURE 4.31

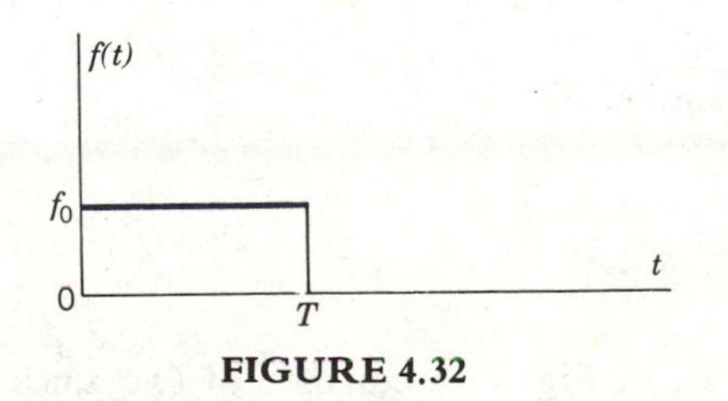

FIGURE 4.32

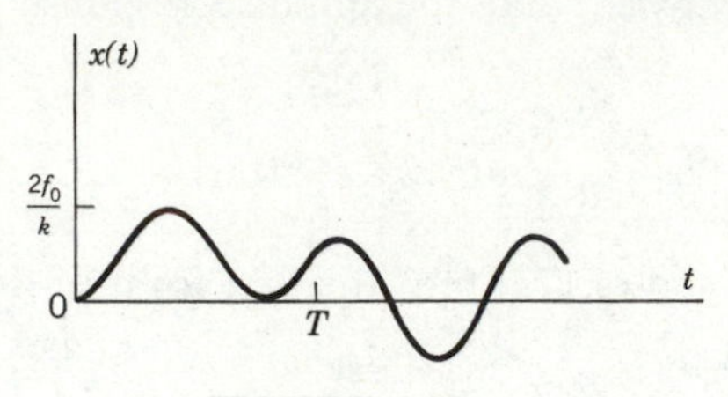

FIGURE 4.33

Eq. (4.170), we obtain

$$x(t)=\frac{f_0}{m\omega_n}\int_0^t \sin \omega_n(t-\sigma)\,d\sigma=\frac{f_0}{m\omega_n}\,\mathrm{Im}\int_0^t e^{i\omega_n(t-\sigma)}\,d\sigma,\qquad 0<t<T \tag{4.176a}$$

$$x(t)=\frac{f_0}{m\omega_n}\int_0^T \sin \omega_n(t-\sigma)\,d\sigma=\frac{f_0}{m\omega_n}\,\mathrm{Im}\int_0^T e^{i\omega_n(t-\sigma)}\,d\sigma,\qquad t>T \tag{4.176b}$$

Evaluation of the intergrals yields

$$x(t)=\begin{cases}\dfrac{f_0}{k}(1-\cos \omega_n t), & 0<t<T\\[2ex] \dfrac{f_0}{k}[\cos \omega_n(t-T)-\cos \omega_n t], & t>T\end{cases} \tag{4.177}$$

A typical plot is shown in Fig. 4.33.

The above result can be obtained, perhaps in a more direct fashion, by regarding the rectangular pulse as a superposition of two step functions. Indeed, recalling developments from Section 1.8, we can write the input in the form

$$f(t)=f_0[\mathscr{u}(t)-\mathscr{u}(t-T)] \tag{4.178}$$

so that the response can be expressed as the superposition of two step responses as follows:

$$x(t)=f_0[\mathscr{s}(t)-\mathscr{s}(t-T)] \tag{4.179}$$

Hence, using Eq. (4.169), we obtain

$$x(t)=\frac{f_0}{k}\{(1-\cos \omega_n t)\mathscr{u}(t)-[1-\cos \omega_n(t-T)]\mathscr{u}(t-T)\} \tag{4.180}$$

which is identical to Eqs. (4.177).

PROBLEMS

4.1 The system shown in Fig. 4.34 consists of two linear springs arranged in series. Determine the equivalent spring constant, defined as $k_{eq}=f/\delta$.

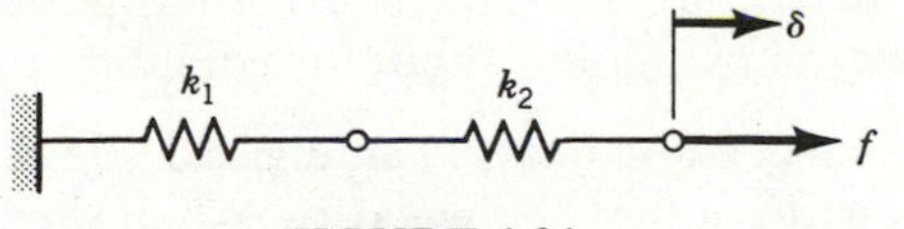

FIGURE 4.34

4.2 Figure 4.35 depicts a mass–damper system. Show that the system can be described by a first-order differential equation, and give an expression for the time constant. Then, indicate the electrical analog, the corresponding differential equation, and the time constant.

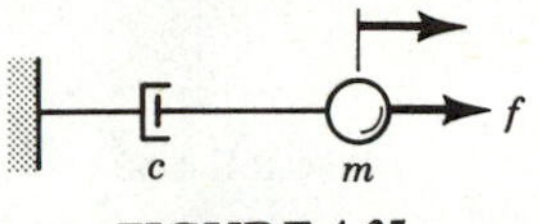

FIGURE 4.35

4.3 A massless rigid bar hinged at point 0 is supported by two linear springs, as shown in Fig. 4.36. Derive the differential equation for the angular motion θ under the assumption that the angle θ is sufficiently small that $\sin\theta \cong \theta$ and $\cos\theta \cong 1$. Determine the static equilibrium position θ_{st}, show how the effect of the weight Mg can be eliminated from the equation of motion, and calculate the natural frequency of oscillation about the equilibrium position.

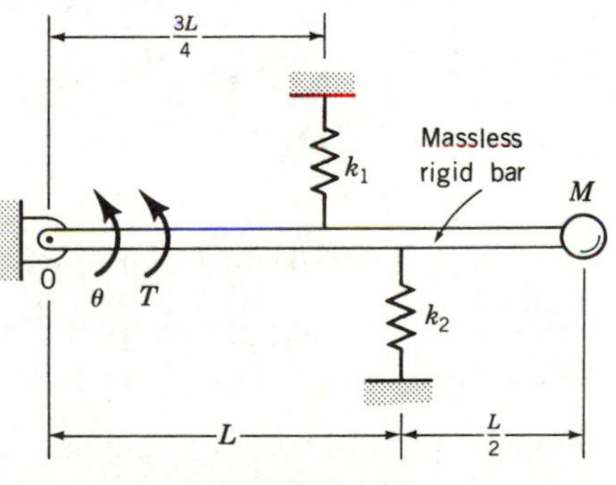

FIGURE 4.36

4.4 A mass m is suspended through a pulley-and-spring mechanism, as shown in Fig. 4.37. Let the spring be linear, and derive the differential equation of

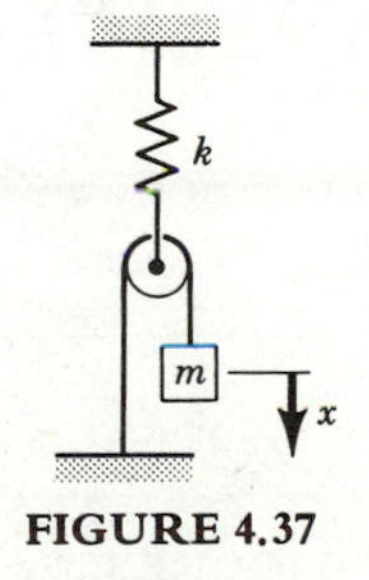

FIGURE 4.37

motion. Then, determine the static equilibrium position, and calculate the natural frequency of oscillation about the equilibrium.

4.5 A cylindrical piece of wood floats in an inviscid liquid, as shown in Fig. 4.38. Denote the mass density of the wood by γ_W and that of the liquid by γ_L, where $\gamma_W < \gamma_L$, and derive the differential equation for the bobbing motion of the wood. Then, determine the static equilibrium position, and calculate the natural frequency of oscillation about the equilibrium.

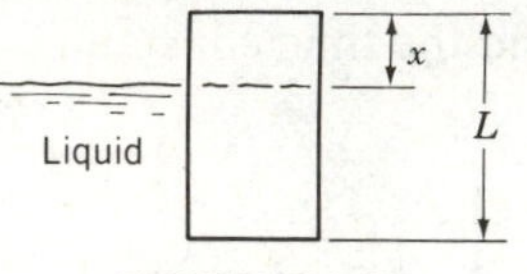

FIGURE 4.38

4.6 Use Eq. (4.66) and obtain the response for an overdamped mass–damper–spring system to the initial conditions

$$x(0)=x_0, \qquad \dot{x}(0)=0$$

Check your results by using the data of Example 4.1.

4.7 Show that solution (4.66) can also be written in the form

$$x(t)=(A_1 \cosh \sqrt{\zeta^2-1}\omega_n t + A_2 \sinh \sqrt{\zeta^2-1}\omega_n t)e^{-\zeta\omega_n t}$$

Then, set $A_1=C_1$ and $A_2\sqrt{\zeta^2-1}\omega_n=C_2$, let $\zeta \to 1$, and prove Eq. (4.67).

4.8 Determine the constants A and ψ in Eq. (4.73) for the initial conditions

$$x(0)=0, \qquad \dot{x}(0)=v_0$$

Verify your result by comparing with the result of Example 4.2.

4.9 Devise a construction similar to that of Fig. 4.11 to represent the solution (4.73). *Hint:* Note that $Ae^{-\zeta\omega_n t}$ plays the role of a decaying amplitude.

4.10 The system shown in Fig. 4.39 has the following parameters:

$$M=100 \quad \text{kg}, \qquad c=50 \quad \text{N}\cdot\text{s/m}, \qquad k=1200 \quad \text{N/m}$$

Derive and plot the response of the system to the initial excitation

$$\theta(0)=0.15 \quad \text{rad}, \qquad \dot{\theta}(0)=0$$

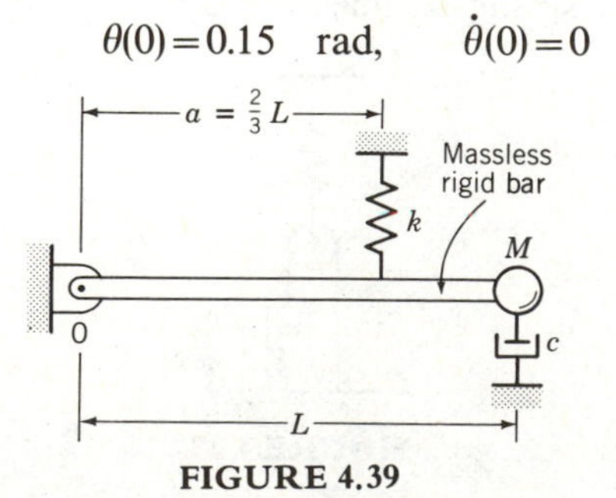

FIGURE 4.39

4.11 A cantilevered steel blade has a mass M weighing 98.1 N at the right end (Fig. 4.40). When in equilibrium, the mass has been observed to have the deflection $\delta = 0.1$ cm. A spring of stiffness $k = 20{,}000$ N/m has one end attached to the mass and the other end experiencing the displacement $y(t) = A \cos \omega t$. Determine the following: (1) the equivalent spring constant of the blade (defined as the ratio of the force to the displacement at the end), (2) the width of the blade if the ratio of the thickness to the length of the blade is $h/L = 1/20$, and (3) the resonance frequency.

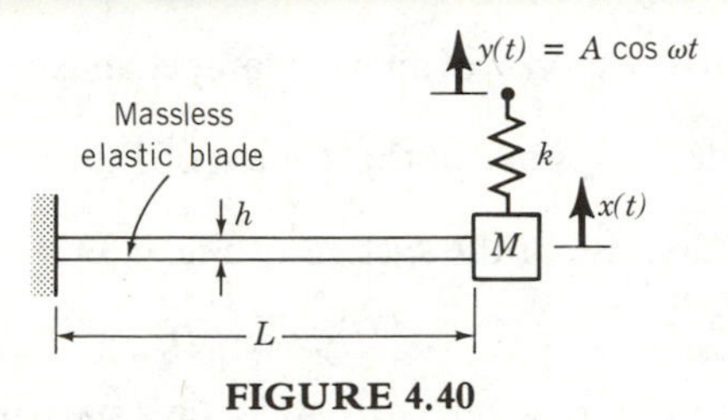

FIGURE 4.40

4.12 Derive the response of the system of Problem 4.2 to the excitation

$$f(t) = f_0 \cos \omega t$$

4.13 Derive the differential equation of motion for the system shown in Fig. 4.41, in which $y(t)$ represents a displacement excitation. The spring constant is $k = 12{,}000$ N/m, but two of the system parameters, m and c, are not known and must be determined experimentally. To this end, the system is excited harmonically according to

$$y(t) = 0.02 \sin \omega t \quad \text{(m)}$$

Determine m and c if the oscillation reaches a peak amplitude of 0.1 m for the driving frequency of $\omega = 39.6$ rad/s.

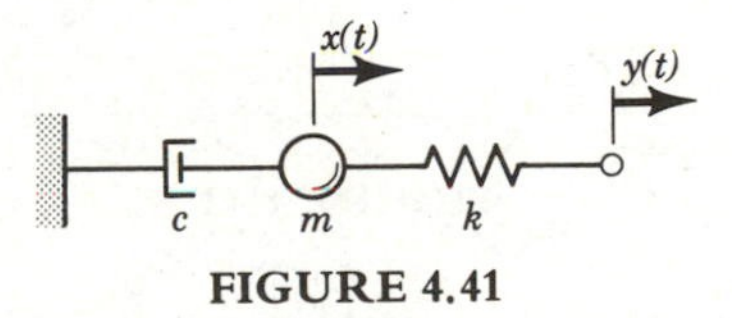

FIGURE 4.41

4.14 Consider the system of Fig. 4.23*a*, and determine the value of the ratio ω/ω_n for which the vibration reaches a peak. Then, determine the peak value.

4.15 A clothes dryer has a mass of 80 kg, and it can handle clothes with a mass of 7.5 kg. The drum has a radius of 0.18 m. Design a suspension system such that the static displacement will not exceed 0.001 m and the peak vibration amplitude will not exceed 0.018 m. Assume that the clothes do not move relative to the drum.

4.16 A vehicle travels over rippled pavement at 50 km/h. The pavement resembles a sine function of amplitude equal to 0.02 m and wavelength equal to 2 m.

If the vehicle has a mass of 1500 kg and a frequency of undamped oscillation of 18 rad/s, and if its shock absorbers have a viscous damping factor equal to 0.5, determine the force experienced by the vehicle.

4.17 The system of Fig. 4.35 is subjected to the impulsive force

$$f(t)=\hat{f}\,\delta(t)$$

where $\hat{f}$ is the magnitude of the impulse. Plot the velocity and displacement of mass m as a function of time.

4.18 Obtain the impulse response of an overdamped mass–damper–spring system.

4.19 Determine the step response of the system of Fig. 4.35.

4.20 Determine the response of the system of Fig. 4.39 if a torque

$$T(t)=T_0\,u(t)$$

is applied about point 0, where $u(t)$ is the unit step function applied at $t=0$. Assume that the system is underdamped.

4.21 Repeat Problem 4.20 for the case in which the system is overdamped.

4.22 The vehicle of Fig. 4.25*a* runs over a curb, where the curb can be represented by a step function of amplitude A. Determine the response of the vehicle for the case in which the damping is zero.

4.23 Repeat Problem 4.22 for the case in which the curb has the shape shown in Fig. 4.42.

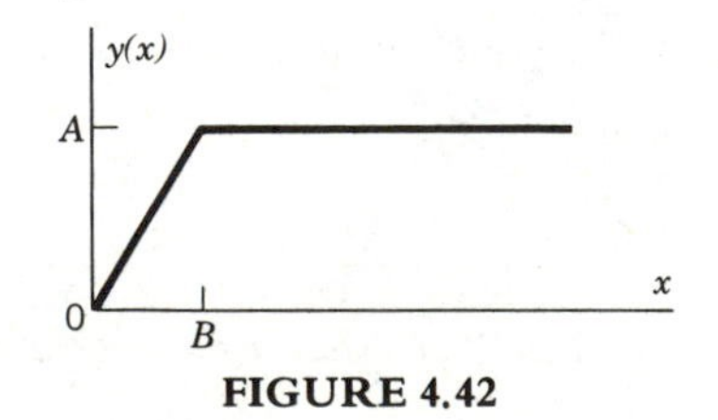

FIGURE 4.42

4.24 Determine the response of an underdamped mass–damper–spring system to the rectangular pulse given by Eq. (4.178).

4.25 Repeat Problem 4.24 for the case in which the excitation has the form

$$f(t)=f_0\,r(t)$$

where $r(t)$ is the unit ramp function (Section 1.8).

4.26 Repeat Problem 4.24 for the excitation

$$f(t)=f_0[r(t)-r(t-T)]$$

4.27 Determine the response of the system of Fig. 4.35 to the sinusoidal pulse

$$f(t)=\begin{cases} f_0 \sin \omega t, & 0<t<\pi/\omega \\ 0, & \text{everywhere else} \end{cases}$$

by the convolution integral.

4.28 Solve Problem 4.25 by the convolution integral.

4.29 Determine the response of a mass–spring system to the sinusoidal pulse of Problem 4.27. Use the convolution integral.

4.30 Determine the response of an underdamped mass–damper–spring system to the sinusoidal pulse of Problem 4.27. Use the convolution integral.

CHAPTER 5
Dynamics of Systems of Particles

5.1 INTRODUCTION

Newton's second law was postulated for a single particle. Many physical phenomena can be explained by means of a model consisting of a single particle, but many more phenomena require a more elaborate model. We recall that, as a first approximation, planets and satellites can be treated as single particles moving around a fixed center of force. A more accurate theory reveals that the motion is really that of two particles moving around their common center of mass. In fact, an even more accurate theory would take into account the perturbing effect of more distant bodies, such as the moon, the other planets, or the sun, as the case may be. Many other dynamical systems can be regarded as systems of particles, including such seemingly disparate ones as rigid bodies and variable-mass systems.

In this chapter, we propose to extend many of the concepts and principles derived in Chapter 3 for a single particle to systems of particles. As applications, we consider the two-body problem and variable-mass systems. The first is a very important problem in dynamics, since two-body systems are encountered frequently. Examples of two-body systems are the earth and the sun, the earth and the moon, and the earth and a man-made satellite. Variable-mass systems are also very important. Indeed, the rocket engine is a prime example of a variable-mass system. The motion of rigid bodies is treated separately in the following chapter.

5.2 THE EQUATION OF MOTION FOR A SYSTEM OF PARTICLES

Let us consider the system of particles shown in Fig. 5.1 and denote the mass of a typical particle by $m_i\,(i=1, 2, \ldots, n)$. For convenience, we distinguish between forces external and internal to the system and denote them by $\mathbf{F}_i$ and $\mathbf{f}_i$, respectively. The internal force is the resultant of the interaction forces $\mathbf{f}_{ij}$ exerted by the particles $m_j\,(j=1, 2, \ldots, n, j\neq i)$ on the particle m_i (Fig. 5.2), or

$$\mathbf{f}_i=\sum_{j=1}^{n} \mathbf{f}_{ij} \tag{5.1}$$

Hence, according to Newton's second law, the equation of motion for particle m_i is

$$\mathbf{F}_i+\sum_{j=1}^{n} \mathbf{f}_{ij}=m_i\ddot{\mathbf{r}}_i=m_i\mathbf{a}_i \tag{5.2}$$

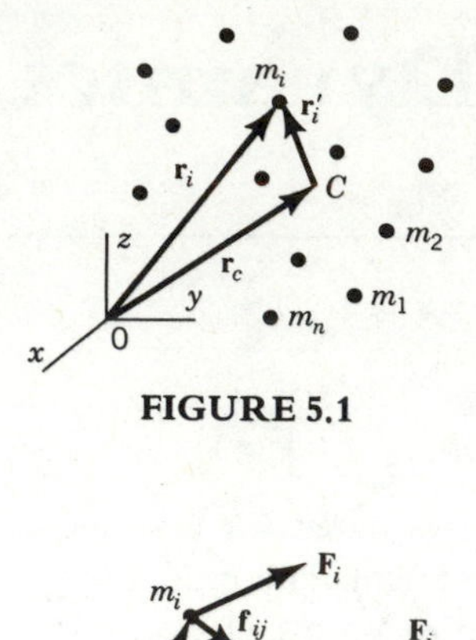

FIGURE 5.1

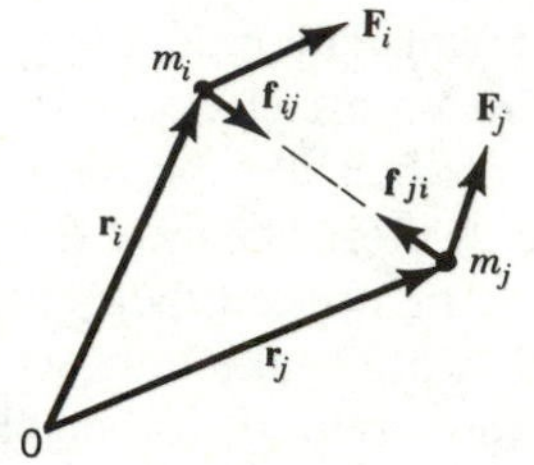

FIGURE 5.2

where $\ddot{\mathbf{r}}_i = \mathbf{a}_i$ is the acceleration of particle m_i relative to the inertial space xyz. Moreover, it is understood that the summation in Eq. (5.2) excludes the term for which $j = i$, as there is no corresponding force.

The equation of motion for the system of particles is obtained by extending Eq. (5.2) over the entire system of particles and then summing the corresponding equations. The result is

$$\sum_{i=1}^{n} \mathbf{F}_i + \sum_{i=1}^{n} \sum_{j=1}^{n} \mathbf{f}_{ij} = \sum_{i=1}^{n} m_i \ddot{\mathbf{r}}_i = \sum_{i=1}^{n} m_i \mathbf{a}_i \tag{5.3}$$

But, by Newton's third law,

$$\mathbf{f}_{ij} = -\mathbf{f}_{ji} \tag{5.4}$$

so that

$$\sum_{i=1}^{n} \sum_{j=1}^{n} \mathbf{f}_{ij} = \mathbf{0} \tag{5.5}$$

because the internal forces cancel out in pairs. Note that Eq. (5.4) carries the implication that the forces $\mathbf{f}_{ij}$ pass through the particles m_i and m_j, which implies further that there are no internal torques. Introducing the resultant of the external force

$$\mathbf{F} = \sum_{i=1}^{n} \mathbf{F}_i \tag{5.6}$$

and considering Eq. (5.5), we obtain the equation of motion for the system of particles in the form

$$\mathbf{F} = \sum_{i=1}^{n} m_i \ddot{\mathbf{r}}_i = \sum_{i=1}^{n} m_i \mathbf{a}_i \tag{5.7}$$

Equation (5.7) is not very convenient for studying the motion of the system of particles, and in the next section we seek a more suitable form.

5.3 EQUATION OF MOTION IN TERMS OF THE MASS CENTER

Equation (5.7) represents a relation between the resultant force acting on the system of particles and the motion of the individual particles in the system. In many instances, however, the interest lies not in the motion of the individual particles but in an average motion of the system. In this section, we propose to derive an equation capable of describing such an average motion.

Let us consider the system of particles shown in Fig. 5.1 and denote by m the total mass of the system, so that

$$m = \sum_{i=1}^{n} m_i \tag{5.8}$$

The absolute position of the particle m_i is given by the radius vector $\mathbf{r}_i$ from the origin 0 of the inertial frame xyz to the particle m_i. The *center of mass C* of the system under consideration is defined as a point in space representing a *weighted average position of the system*, where the weighting factor for each particle is the mass of the particle. Denoting the radius vector from 0 to C by $\mathbf{r}_C$, the mathematical definition of the center of mass is given by

$$\mathbf{r}_C = \frac{1}{m} \sum_{i=1}^{n} m_i \mathbf{r}_i \tag{5.9}$$

Note that the point C is not necessarily a material point. Quite often, it is convenient to measure the motion not from the fixed point 0 but from the mass center C. Denoting the position of m_i relative to C by $\mathbf{r}'_i$ (Fig. 5.1), we can write

$$\mathbf{r}_i = \mathbf{r}_C + \mathbf{r}'_i \tag{5.10}$$

Introducing Eq. (5.10) into Eq. (5.9), we obtain

$$\mathbf{r}_C = \frac{1}{m} \sum_{i=1}^{n} m_i (\mathbf{r}_C + \mathbf{r}'_i) = \mathbf{r}_C + \frac{1}{m} \sum_{i=1}^{n} m_i \mathbf{r}'_i \tag{5.11}$$

which yields

$$\sum_{i=1}^{n} m_i \mathbf{r}'_i = \mathbf{0} \tag{5.12}$$

Hence, an equivalent definition of the mass center C is a point in space such that if the position of each particle is measured relative to that point, then the weighted average position is zero.

From Eq. (5.9), we obtain immediately the absolute velocity of the mass center

$$\mathbf{v}_C = \dot{\mathbf{r}}_C = \frac{1}{m} \sum_{i=1}^{n} m_i \dot{\mathbf{r}}_i = \frac{1}{m} \sum_{i=1}^{n} m_i \mathbf{v}_i \tag{5.13}$$

and the absolute acceleration of the mass center

$$\mathbf{a}_C = \ddot{\mathbf{r}}_C = \frac{1}{m} \sum_{i=1}^{n} m_i \ddot{\mathbf{r}}_i = \frac{1}{m} \sum_{i=1}^{n} m_i \mathbf{a}_i \tag{5.14}$$

where $\mathbf{v}_i$ and $\mathbf{a}_i$ are the absolute velocity and acceleration of particle m_i, respectively. As a matter of interest, we note from Eq. (5.12) that

$$\sum_{i=1}^{n} m_i \dot{\mathbf{r}}_i' = \sum_{i=1}^{n} m_i \mathbf{v}_i' = \mathbf{0} \tag{5.15}$$

and

$$\sum_{i=1}^{n} m_i \ddot{\mathbf{r}}_i' = \sum_{i=1}^{n} m_i \mathbf{a}_i' = \mathbf{0} \tag{5.16}$$

where $\mathbf{v}_i'$ and $\mathbf{a}_i'$ are the velocity and acceleration of m_i relative to C, respectively.

Equation (5.14) permits us to derive an equation for the average motion of the system, which is taken as the motion of the mass center. Indeed, introducing Eq. (5.14) into Eq. (5.7), we obtain the desired equation in the form

$$\mathbf{F} = m\mathbf{a}_C \tag{5.17}$$

Equation (5.17) implies that the motion of the mass center of the system of particles is the same as the motion of a fictitious body equal in mass to the total mass m of the system, concentrated at the mass center, and being acted on by the resultant of the external forces.

5.4 LINEAR MOMENTUM

Equation (5.17) can be written in a different form. To this end, let us introduce the linear momentum of m_i

$$\mathbf{p}_i = m_i \mathbf{v}_i \tag{5.18}$$

Then, if we consider Eq. (5.13), we can write the linear momentum for the system of particles in the form

$$\mathbf{p} = \sum_{i=1}^{n} \mathbf{p}_i = \sum_{i=1}^{n} m_i \mathbf{v}_i = m\mathbf{v}_C \tag{5.19}$$

The rate of change of the linear momentum is

$$\dot{\mathbf{p}} = m\dot{\mathbf{v}}_C = m\mathbf{a}_C \tag{5.20}$$

so that, comparing Eqs. (5.17) and (5.20), we conclude that

$$\mathbf{F}=\dot{\mathbf{p}} \tag{5.21}$$

or the resultant of the external forces acting on the system of particles is equal to the time rate of change of the system linear momentum.

Multiplying Eq. (5.21) by dt, integrating, and using a procedure analogous to that for a single particle (Section 3.4), we can write

$$\hat{\mathbf{F}}=\Delta\mathbf{p} \tag{5.22}$$

where $\hat{\mathbf{F}}=\int_{t_1}^{t_2}\mathbf{F}\,dt$ is the resultant of the external impulses on all particles and $\Delta\mathbf{p}=\mathbf{p}(t_2)-\mathbf{p}(t_1)$ is the change in the system linear momentum between the instants t_1 and t_2.

If $\mathbf{F}=\mathbf{0}$, Eq. (5.21) yields

$$\mathbf{p}=\text{const} \tag{5.23}$$

which states that in the absence of external forces the linear momentum of the system of particles remains constant. This statement represents the principle of *conservation of linear momentum* for a system of particles.

Example 5.1

A man is at rest at the rear end of a barge approaching the shore with the velocity v_0 relative to the shore, when he decides to walk toward the front end. If his velocity relative to the barge is uniform and equal to v_1, determine the velocity of the barge and plot it as a function of time. The mass of the man is m, that of the barge is M, and the length of the barge is L.

Ignoring the water friction, we conclude that the linear momentum is conserved. Denoting by v_2 the velocity of the barge while the man is walking (Fig. 5.3), we have

$$(m+M)v_0=m(v_1+v_2)+Mv_2 \tag{a}$$

which yields

$$v_2=\frac{(m+M)v_0-mv_1}{m+M} \tag{b}$$

Of course, when the man reaches the front of the barge, the velocity of the barge becomes once again v_0.

The velocity of the barge as a function of time is shown in Fig. 5.4, where t_0 denotes the time when the man beings his walk. Note that the shaded area in Fig. 5.4 is simply equal to the distance the barge is set back by the man's change in position relative to the barge. (Prove this statement.)

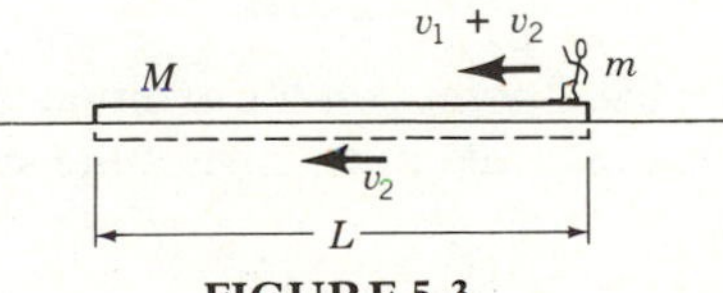

FIGURE 5.3

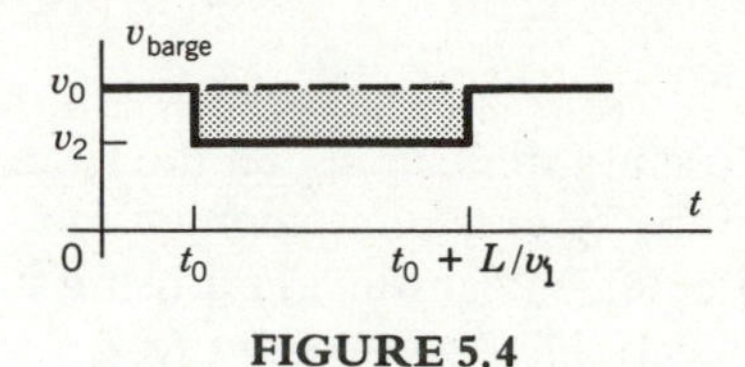

FIGURE 5.4

5.5 ANGULAR MOMENTUM

Next, let us define the angular momentum of the particle m_i about the fixed point 0 as

$$\mathbf{H}_{0i} = \mathbf{r}_i \times \mathbf{p}_i = \mathbf{r}_i \times m_i \mathbf{v}_i \tag{5.24}$$

so that the angular momentum about 0 of the system of particles is

$$\mathbf{H}_0 = \sum_{i=1}^{n} \mathbf{H}_{0i} = \sum_{i=1}^{n} \mathbf{r}_i \times m_i \mathbf{v}_i \tag{5.25}$$

Considering Eq. (5.2), we can write

$$\begin{aligned} \dot{\mathbf{H}}_0 &= \sum_{i=1}^{n} \dot{\mathbf{r}}_i \times m_i \mathbf{v}_i + \sum_{i=1}^{n} \mathbf{r}_i \times m_i \dot{\mathbf{v}}_i = \sum_{i=1}^{n} \mathbf{r}_i \times m_i \mathbf{a}_i \\ &= \sum_{i=1}^{n} \mathbf{r}_i \times \left(\mathbf{F}_i + \sum_{j=1}^{n} \mathbf{f}_{ij}\right) = \sum_{i=1}^{n} \mathbf{r}_i \times \mathbf{F}_i \end{aligned} \tag{5.26}$$

where we took into account that $\sum \dot{\mathbf{r}}_i \times m_i \mathbf{v}_i = \sum m_i \mathbf{v}_i \times \mathbf{v}_i = \mathbf{0}$ by the definition of the vector product and that

$$\sum_{i=1}^{n} \sum_{j=1}^{n} \mathbf{r}_i \times \mathbf{f}_{ij} = \mathbf{0} \tag{5.27}$$

Equation (5.27) is true because $\mathbf{f}_{ij} = -\mathbf{f}_{ji}$ and, in addition, the two vectors are collinear (see Fig. 5.2), so that $\mathbf{r}_i \times \mathbf{f}_{ij} = -\mathbf{r}_j \times \mathbf{f}_{ji}$. Recognizing that

$$\mathbf{M}_0 = \sum_{i=1}^{n} \mathbf{r}_i \times \mathbf{F}_i \tag{5.28}$$

is the moment about 0 of the external forces acting on the system of particles, we can rewrite Eq. (5.26) in the form

$$\mathbf{M}_0 = \dot{\mathbf{H}}_0 \tag{5.29}$$

or, the moment about the fixed point 0 of the external forces acting on the system of particles is equal to the time rate of change of the system angular momentum about the same fixed point.

By analogy with Eq. (5.22) for the linear momentum, we can use Eq. (5.29) to obtain

$$\hat{\mathbf{M}}_0 = \Delta \mathbf{H}_0 \tag{5.30}$$

where $\hat{\mathbf{M}}_0 = \sum \mathbf{r}_i \times \hat{\mathbf{F}}_i$ is the resultant of all the external angular impulses about the fixed point 0 and $\Delta \mathbf{H}_0$ is the change in the system angular momentum about 0.

If $\mathbf{M}_0 = \mathbf{0}$, Eq. (5.29) yields

$$\mathbf{H}_0 = \text{const} \tag{5.31}$$

which states that in the absence of external torques about the fixed point 0 the angular momentum of the system of particles about 0 remains constant. This statement is known as the *conservation of angular momentum about a fixed point.*

The above moment-angular momentum relations were derived for the fixed point 0. The question arises whether similar relations exist also for a moving point. The answer is affirmative, provided the moving point is the mass center of the system of particles. To show this, let us consider the angular momentum of the particle m_i about C. Because the radius vector from C to m_i is $\mathbf{r}_i'$, we have

$$\mathbf{H}_{Ci} = \mathbf{r}_i' \times \mathbf{p}_i = \mathbf{r}_i' \times m_i \mathbf{v}_i \tag{5.32}$$

so that the angular momentum about C of the system of particles is

$$\mathbf{H}_C = \sum_{i=1}^{n} \mathbf{H}_{Ci} = \sum_{i=1}^{n} \mathbf{r}_i' \times m_i \mathbf{v}_i \tag{5.33}$$

But, taking the time derivative of Eq. (5.10), we can write

$$\mathbf{v}_i = \mathbf{v}_C + \mathbf{v}_i' \tag{5.34}$$

Inserting Eq. (5.34) into Eq. (5.33) and considering Eq. (5.12), we obtain

$$\begin{aligned} \mathbf{H}_C &= \sum_{i=1}^{n} \mathbf{r}_i' \times m_i(\mathbf{v}_C + \mathbf{v}_i') = \left(\sum_{i=1}^{n} m_i \mathbf{r}_i' \right) \times \mathbf{v}_C + \sum_{i=1}^{n} \mathbf{r}_i' \times m_i \mathbf{v}_i' \\ &= \sum_{i=1}^{n} \mathbf{r}_i' \times m_i \mathbf{v}_i' \end{aligned} \tag{5.35}$$

Introducing the definition

$$\mathbf{H}_C' = \sum_{i=1}^{n} \mathbf{r}_i' \times m_i \mathbf{v}_i' \tag{5.36}$$

where $\mathbf{H}_C'$ is known as an *apparent angular momentum* and represents the angular momentum of the system of particles as seen by an observer whose position coincides with that of C at all times, we have

$$\mathbf{H}_C = \mathbf{H}_C' \tag{5.37}$$

or the apparent angular momentum about C of a system of particles is equal to the actual angular momentum about C. This property is characteristic of the mass center and does not extend to any other moving point.

Next, let us consider Eqs. (5.33) and (5.34) and write

$$\begin{aligned}\dot{\mathbf{H}}_C &= \sum_{i=1}^{n} \dot{\mathbf{r}}_i' \times m_i \mathbf{v}_i + \sum_{i=1}^{n} \mathbf{r}_i' \times m_i \dot{\mathbf{v}}_i \\ &= \sum_{i=1}^{n} \mathbf{v}_i' \times m_i(\mathbf{v}_C + \mathbf{v}_i') + \sum_{i=1}^{n} \mathbf{r}_i' \times m_i \mathbf{a}_i \\ &= \left(\sum_{i=1}^{n} m_i \mathbf{v}_i'\right) \times \mathbf{v}_C + \sum_{i=1}^{n} m_i \mathbf{v}_i' \times \mathbf{v}_i' + \sum_{i=1}^{n} \mathbf{r}_i' \times m_i \mathbf{a}_i = \sum_{i=1}^{n} \mathbf{r}_i' \times m_i \mathbf{a}_i \end{aligned} \tag{5.38}$$

where we recognize that $\sum m_i \mathbf{v}_i' = \mathbf{0}$ by virtue of Eq. (5.15). Inserting Eq. (5.2) into Eq. (5.38), we obtain

$$\dot{\mathbf{H}}_C = \sum_{i=1}^{n} \mathbf{r}_i' \times \left(\mathbf{F}_i + \sum_{j=1}^{n} \mathbf{f}_{ij}\right) = \sum_{i=1}^{n} \mathbf{r}_i' \times \mathbf{F}_i \tag{5.39}$$

where, consistent with Eq. (5.27), $\sum\sum \mathbf{r}_i' \times \mathbf{f}_{ij} = \mathbf{0}$. Recognizing that the right side of Eq. (5.39) is the moment of the external forces about C, or

$$\mathbf{M}_C = \sum_{i=1}^{n} \mathbf{r}_i' \times \mathbf{F}_i \tag{5.40}$$

we can rewrite Eq. (5.39) in the form

$$\mathbf{M}_C = \dot{\mathbf{H}}_C \tag{5.41}$$

or the moment of the external forces about C is equal to the time rate of change of the system angular momentum about C. Note that the same result is obtained if one uses the apparent angular momentum $\mathbf{H}_C'$ instead of the actual angular momentum $\mathbf{H}_C$. Moreover, using the analogy with Eq. (5.30), we have

$$\hat{\mathbf{M}}_C = \Delta \mathbf{H}_C \tag{5.42}$$

where the notation is obvious.

If $\mathbf{M}_C = \mathbf{0}$, then Eq. (5.41) yields

$$\mathbf{H}_C = \text{const} \tag{5.43}$$

or in the absence of external torques about C the system angular momentum about C remains constant, which is the statement of *conservation of angular momentum about the mass center.*

It should be stressed again that the simple relations between the moment and angular momentum, Eqs. (5.29) and (5.41), hold true only if the reference point is a fixed point or the center of mass. If the reference point for the moment and angular momentum is an arbitrary moving point, then an extra term generally appears in the relation.

Example 5.2

The system shown in Fig. 5.5 consists of three mass particles, $m_1 = 2m$, $m_2 = m_3 = 3m$, connected by rigid bars welded at point 0. Particle m_1 is struck by a force generating

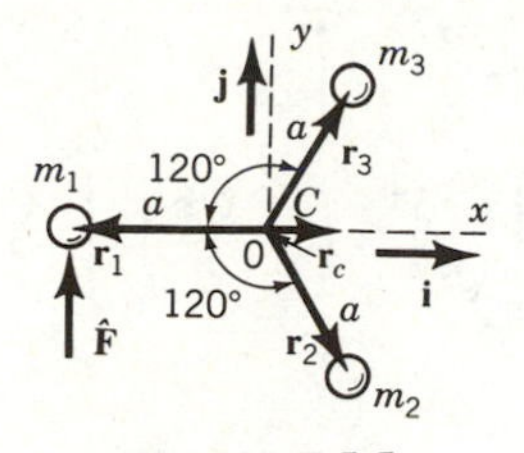

FIGURE 5.5

the impulse $\hat{\mathbf{F}}$, as shown. Calculate the velocity of each particle at the termination of the impulse. The system is initially at rest.

We solve the problem by working with the center of mass. First, we calculate the velocity of the mass center and then the velocity of each particle relative to the mass center. Because the initial linear momentum is zero, we can write from Eq. (5.22)

$$\hat{\mathbf{F}}=\hat{F}\mathbf{j}=\Delta\mathbf{p}=\mathbf{p} \tag{a}$$

where $\mathbf{p}$ is the linear momentum at the termination of the impulse, or

$$\mathbf{p}=\left(\sum_{i=1}^{3} m_i\right)\mathbf{v}_C=8m\mathbf{v}_C \tag{b}$$

Comparing Eqs. (a) and (b), we obtain the velocity of the mass center

$$\mathbf{v}_C=\frac{\hat{F}}{8m}\mathbf{j} \tag{c}$$

Next, we wish to calculate the velocities of the particles relative to C. From Eq. (5.42), we can write

$$\hat{\mathbf{M}}_C=(\mathbf{r}_1-\mathbf{r}_C)\times\hat{\mathbf{F}}=\Delta\mathbf{H}_C=\mathbf{H}_C \tag{d}$$

where $\mathbf{H}_C$ is the angular momentum about C. To calculate $\hat{\mathbf{M}}_C$ and $\mathbf{H}_C$, we need the position of the mass center. Assuming that point 0 coincides initially with the origin of the inertial system xy, we have

$$\mathbf{r}_C=\frac{\sum_{i=1}^{3} m_i\mathbf{r}_i}{\sum_{i=1}^{3} m_i}=\frac{1}{8m}\left[2m(-a\mathbf{i})+3m\left(\frac{a}{2}\mathbf{i}-\frac{\sqrt{3}}{2}a\mathbf{j}\right)+3m\left(\frac{a}{2}\mathbf{i}+\frac{\sqrt{3}}{2}a\mathbf{j}\right)\right]=\frac{a}{8}\mathbf{i} \tag{e}$$

Denoting by $\boldsymbol{\omega}=\omega\mathbf{k}$ the angular velocity of the system following the impulse, the angular momentum about C is

$$\begin{aligned}\mathbf{H}_C=\mathbf{H}'_C&=\sum_{i=1}^{3}\mathbf{r}'_i\times m_i\mathbf{v}'_i=\sum_{i=1}^{3}(\mathbf{r}_i-\mathbf{r}_C)\times m_i\mathbf{v}'_i=\sum_{i=1}^{3}(\mathbf{r}_i-\mathbf{r}_C)\times[m_i\boldsymbol{\omega}\times(\mathbf{r}_i-\mathbf{r}_C)]\\&=2m\left(-a\mathbf{i}-\frac{a}{8}\mathbf{i}\right)\times\left[\omega\mathbf{k}\times\left(-a\mathbf{i}-\frac{a}{8}\mathbf{i}\right)\right]\\&\quad+3m\left(\frac{a}{2}\mathbf{i}-\frac{\sqrt{3}a}{2}\mathbf{j}-\frac{a}{8}\mathbf{i}\right)\times\left[\omega\mathbf{k}\times\left(\frac{a}{2}\mathbf{i}-\frac{\sqrt{3}a}{2}\mathbf{j}-\frac{a}{8}\mathbf{i}\right)\right]\end{aligned}$$

$$+3m\left(\frac{a}{2}\mathbf{i}+\frac{\sqrt{3}a}{2}\mathbf{j}-\frac{a}{8}\mathbf{i}\right)\times\left[\omega\mathbf{k}\times\left(\frac{a}{2}\mathbf{i}+\frac{\sqrt{3}a}{2}\mathbf{j}-\frac{a}{8}\mathbf{i}\right)\right]$$

$$=ma^2\omega\left\{2\left(\frac{9}{8}\right)^2+3\left[\left(\frac{3}{8}\right)^2+\left(-\frac{\sqrt{3}}{2}\right)^2\right]+3\left[\left(\frac{3}{8}\right)^2+\left(\frac{\sqrt{3}}{2}\right)^2\right]\right\}\mathbf{k}$$

$$=\frac{63}{8}ma^2\omega\mathbf{k} \tag{f}$$

so that Eq. (d) yields

$$\left(-a\mathbf{i}-\frac{a}{8}\mathbf{i}\right)\times\hat{F}\mathbf{j}=-\frac{9}{8}a\hat{F}\mathbf{k}=\frac{63}{8}ma^2\omega\mathbf{k} \tag{g}$$

from which we obtain

$$\omega=-\frac{1}{7}\frac{\hat{F}}{ma} \tag{h}$$

The velocities of the particles relative to C are

$$\begin{aligned}
\mathbf{v}_1'&=\boldsymbol{\omega}\times(\mathbf{r}_1-\mathbf{r}_C)=-\frac{1}{7}\frac{\hat{F}}{ma}\mathbf{k}\times\left(-a\mathbf{i}-\frac{a}{8}\mathbf{i}\right)=\frac{9}{56}\frac{\hat{F}}{m}\mathbf{j}\\
\mathbf{v}_2'&=\boldsymbol{\omega}\times(\mathbf{r}_2-\mathbf{r}_C)=-\frac{1}{7}\frac{\hat{F}}{ma}\mathbf{k}\times\left(\frac{a}{2}\mathbf{i}-\frac{\sqrt{3}a}{2}\mathbf{j}-\frac{a}{8}\mathbf{i}\right)=-\frac{1}{56}(4\sqrt{3}\mathbf{i}+3\mathbf{j})\frac{\hat{F}}{m}\\
\mathbf{v}_3'&=\boldsymbol{\omega}\times(\mathbf{r}_3-\mathbf{r}_C)=-\frac{1}{7}\frac{\hat{F}}{ma}\mathbf{k}\times\left(\frac{a}{2}\mathbf{i}+\frac{\sqrt{3}a}{2}\mathbf{j}-\frac{a}{8}\mathbf{i}\right)=\frac{1}{56}(4\sqrt{3}\mathbf{i}-3\mathbf{j})\frac{\hat{F}}{m}
\end{aligned} \tag{i}$$

Combining Eqs. (c) and (i), we obtain the absolute velocities

$$\begin{aligned}
\mathbf{v}_1&=\mathbf{v}_C+\mathbf{v}_1'=\frac{\hat{F}}{8m}\mathbf{j}+\frac{9}{56}\frac{\hat{F}}{m}\mathbf{j}=\frac{2}{7}\frac{\hat{F}}{m}\mathbf{j}\\
\mathbf{v}_2&=\mathbf{v}_C+\mathbf{v}_2'=\frac{\hat{F}}{8m}\mathbf{j}-\frac{1}{56}(4\sqrt{3}\mathbf{i}+3\mathbf{j})\frac{\hat{F}}{m}=-\frac{1}{14}(\sqrt{3}\mathbf{i}-\mathbf{j})\frac{\hat{F}}{m}\\
\mathbf{v}_3&=\mathbf{v}_C+\mathbf{v}_3'=\frac{\hat{F}}{8m}\mathbf{j}+\frac{1}{56}(4\sqrt{3}i-3\mathbf{j})\frac{\hat{F}}{m}=\frac{1}{14}(\sqrt{3}\mathbf{i}+\mathbf{j})\frac{\hat{F}}{m}
\end{aligned} \tag{j}$$

5.6 KINETIC ENERGY

Let us now derive an expression for the kinetic energy of a system of particles. The kinetic energy for particle m_i is simply

$$T_i=\tfrac{1}{2}m_i\dot{\mathbf{r}}_i\cdot\dot{\mathbf{r}}_i \tag{5.44}$$

so that the kinetic energy of the system of particles is

$$T=\sum_{i=1}^{n}T_i=\frac{1}{2}\sum_{i=1}^{n}m_i\dot{\mathbf{r}}_i\cdot\dot{\mathbf{r}}_i \tag{5.45}$$

Using Eq. (5.10) and recalling Eq. (5.15), we can write

$$T = \frac{1}{2} \sum_{i=1}^{n} m_i(\dot{\mathbf{r}}_C + \dot{\mathbf{r}}_i') \cdot (\dot{\mathbf{r}}_C + \dot{\mathbf{r}}_i')$$

$$= \frac{1}{2} \dot{\mathbf{r}}_C \cdot \dot{\mathbf{r}}_C \sum_{i=1}^{n} m_i + \dot{\mathbf{r}}_C \cdot \sum_{i=1}^{n} m_i \dot{\mathbf{r}}_i' + \frac{1}{2} \sum_{i=1}^{n} m_i \dot{\mathbf{r}}_i' \cdot \dot{\mathbf{r}}_i'$$

$$= \frac{1}{2} m \dot{\mathbf{r}}_C \cdot \dot{\mathbf{r}}_C + \frac{1}{2} \sum_{i=1}^{n} m_i \dot{\mathbf{r}}_i' \cdot \dot{\mathbf{r}}_i' \tag{5.46}$$

Introducing the notation

$$T_{\text{tr}} = \frac{1}{2} m \dot{\mathbf{r}}_C \cdot \dot{\mathbf{r}}_C \tag{5.47}$$

$$T_{\text{rel}} = \frac{1}{2} \sum_{i=1}^{n} m_i \dot{\mathbf{r}}_i' \cdot \dot{\mathbf{r}}_i' \tag{5.48}$$

we obtain

$$T = T_{\text{tr}} + T_{\text{rel}} \tag{5.49}$$

so that the kinetic energy of a system of particles can be separated into two parts, the first representing the kinetic energy that would be obtained if all the particles were translating with the velocity of the mass center and the second representing the kinetic energy due to motion relative to the mass center. This separation is possible only if the moving reference point is the mass center.

Example 5.3

Calculate the kinetic energy for the system of Example 5.2.

The kinetic energy is given by Eq. (5.49), where from Eq. (5.47),

$$T_{\text{tr}} = \frac{1}{2} \left(\sum_{i=1}^{3} m_i \right) v_C^2 = \frac{1}{2} 8m \left(\frac{\hat{F}}{8m} \right)^2 = \frac{1}{16} \frac{\hat{F}^2}{m} \tag{a}$$

and, from Eq. (5.48),

$$T_{\text{rel}} = \frac{1}{2} \sum_{i=1}^{3} m_i (v_i')^2 = \frac{1}{2} \left\{ 2m \left(\frac{9}{56} \frac{\hat{F}}{m} \right)^2 + 3m \left(-\frac{1}{56} \right)^2 [(4\sqrt{3})^2 + 3^2] \left(\frac{\hat{F}}{m} \right)^2 \right.$$
$$\left. + 3m \left(\frac{1}{56} \right)^2 [(4\sqrt{3})^2 + (-3)^2] \left(\frac{\hat{F}}{m} \right)^2 \right\} = \frac{151}{784} \frac{\hat{F}^2}{m} \tag{b}$$

5.7 THE TWO-BODY PROBLEM

Let us consider a system consisting of two particles moving under mutual attraction forces, with no external forces present. We propose to show that the system can be treated as a single particle under a central force.

Denoting by C the position of the mass center and by $\mathbf{r}_C$ the radius vector from

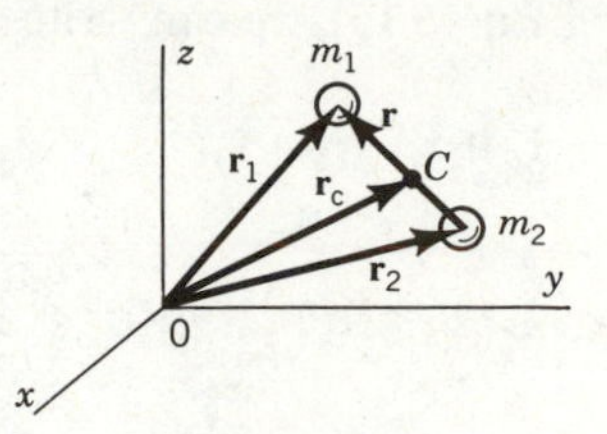

FIGURE 5.6

the fixed point 0 to C (Fig. 5.6), we can write

$$\mathbf{r}_C = \frac{m_1\mathbf{r}_1 + m_2\mathbf{r}_2}{m_1 + m_2} \tag{5.50}$$

Then, denoting the radius vector from m_2 to m_1 by

$$\mathbf{r} = \mathbf{r}_1 - \mathbf{r}_2 \tag{5.51}$$

and eliminating $\mathbf{r}_2$ and $\mathbf{r}_1$ from Eq. (5.50) in sequence, we obtain

$$\mathbf{r}_1 = \mathbf{r}_C + \frac{m_2}{m_1 + m_2}\mathbf{r}, \qquad \mathbf{r}_2 = \mathbf{r}_C - \frac{m_1}{m_1 + m_2}\mathbf{r} \tag{5.52}$$

Letting $\mathbf{f}_{12}$ be the force exerted by m_2 on m_1 and $\mathbf{f}_{21}$ the force exerted by m_1 on m_2, we can write Newton's second law for the two particles in the form

$$\mathbf{f}_{12} = m_1\ddot{\mathbf{r}}_1 = m_1\ddot{\mathbf{r}}_C + \frac{m_1 m_2}{m_1 + m_2}\ddot{\mathbf{r}} \tag{5.53a}$$

$$\mathbf{f}_{21} = m_2\ddot{\mathbf{r}}_2 = m_2\ddot{\mathbf{r}}_C - \frac{m_1 m_2}{m_1 + m_2}\ddot{\mathbf{r}} \tag{5.53b}$$

Adding Eqs. (5.53) and recalling that $\mathbf{f}_{21} = -\mathbf{f}_{12}$, we conclude that

$$\ddot{\mathbf{r}}_C = \mathbf{0} \tag{5.54}$$

or, the mass center C moves with uniform velocity. This is consistent with the fact that in the absence of external forces the system linear momentum is conserved.

In view of Eq. (5.54), Eq. (5.53a) reduces to

$$\mathbf{f}_{12} = \frac{m_1 m_2}{m_1 + m_2}\ddot{\mathbf{r}} \tag{5.55}$$

which can be interpreted as describing the motion of a single body of equivalent mass $m_1 m_2/(m_1 + m_2)$. Because C is unaccelerated, it can be regarded as the origin of an inertial system. Moreover, because $\mathbf{f}_{12}$ passes through C at all times, it is a central force, or

$$\mathbf{f}_{12} = -f\mathbf{u}_r \tag{5.56}$$

where f is the magnitude of $\mathbf{f}_{12}$ and $\mathbf{u}_r$ is a unit vector along $\mathbf{r}$.

As a matter of interest, let us introduce the notation $m_1 = m$ and $m_2 = M$. Then, inserting Eq. (5.56) into Eq. (5.55), we obtain

$$\frac{mM}{m+M}\ddot{\mathbf{r}} = -f\mathbf{u}_r \tag{5.57}$$

If $M >> m$, Eq. (5.57) can be approximated by

$$m\ddot{\mathbf{r}} = -f\mathbf{u}_r \tag{5.58}$$

and, under the same assumption, it also follows from the second of Eqs. (5.52) that

$$\mathbf{r}_2 = \mathbf{r}_C \tag{5.59}$$

so that C coincides with the center of the massive particle M. This is essentially the justification for the assumptions made in the beginning of Section 3.8.

There are many two-body systems in nature. The sun and any one of the planets in the solar system form a two-body system. Of course, this implies that the effect of other planets is negligible, which is indeed the case. Another two-body system consists of the earth and the moon. More modern examples are the earth and any man-made satellite orbiting the earth.

Finally, we wish to reconsider Kepler's third law. To this end, we use Newton's inverse square law to write the radial component of Eq. (5.57) in the form

$$\frac{mM}{m+M}(\ddot{r} - r\dot{\theta}^2) = -\frac{GMm}{r^2} \tag{5.60}$$

Dividing Eq. (5.60) through by $mM/(m+M)$, we obtain

$$\ddot{r} - r\dot{\theta}^2 = -\frac{\mu}{r^2} \tag{5.61}$$

where

$$\mu = G(m+M) \tag{5.62}$$

Introducing Eq. (5.62) into Eq. (3.98), we conclude that the period for elliptic orbits is

$$T = 2\pi\frac{a^{3/2}}{[G(m+M)]^{1/2}} \tag{5.63}$$

so that the orbital period of a planet depends not only on the semimajor axis a, as stated by Kepler's third law, but also on the mass m, which varies from planet to planet. Because the mass of the sun is considerably larger than the mass of any planet in the solar system, the correction to the period given by Kepler's third law is relatively small.

Example 5.4

Let the earth and the sun constitute a two-body system and calculate the orbital period of the earth in two ways, first by using Eq. (3.98) and then by using Eq. (5.63).

Compare the results, and verify the statement made at the end of this section. Pertinent data are as follows:

Semimajor axis of the earth's orbit	$a=1.49527\times 10^{11}$ m
Universal gravitational constant	$G=6.668462\times 10^{-11}$ m^3/(kg$\cdot$s^2)
Mass of the sun	$M=1.987323\times 10^{30}$ kg
Mass of the earth	$m=5.977414\times 10^{24}$ kg

Using Eq. (3.98), we obtain

$$
\begin{aligned}
T &= 2\pi \frac{a^{3/2}}{(GM)^{1/2}} = 2\pi \frac{(1.49527\times 10^{11})^{3/2}}{(6.668462\times 1.987323\times 10^{10})^{1/2}} \\
&= 3.15582\times 10^{7} \quad \text{s} \\
&= 365.256949 \quad \text{days} \qquad \text{(a)}
\end{aligned}
$$

On the other hand, using Eq. (5.63), we can write

$$
\begin{aligned}
T &= 2\pi \frac{a^{3/2}}{G^{1/2}(M+m)^{1/2}} = 2\pi \frac{a^{3/2}}{(GM)^{1/2}}\left(1+\frac{m}{M}\right)^{-1/2} \\
&= 2\pi \frac{a^{3/2}}{(GM)^{1/2}}\left[1-\frac{1}{2}\frac{m}{M}+\frac{3}{8}\left(\frac{m}{M}\right)^{2}+\cdots\right] \\
&= 365.256949(1-1.50388249\times 10^{-6}) = 365.2564 \quad \text{days} \qquad \text{(b)}
\end{aligned}
$$

From Eqs. (a) and (b), we conclude that the difference in the two results is very small, thus corroborating the statement made at the end of this section. The difference is not insignificant, however. Indeed, the measured value of the orbital period of the earth is

$$
T = 365.25636048 \quad \text{days} \qquad \text{(c)}
$$

so that the value given by Eq. (b) is appreciably closer to the measured value. The difference in the values given in (b) and (c) can be attributed to perturbation from various sources, such as the other planets in the solar system.

5.8 VARIABLE-MASS SYSTEMS. ROCKET MOTION

In Example 5.1, we presented an application of the conservation of linear momentum for a system consisting of two bodies subjected to internal forces alone. As an introduction to variable-mass systems, let us examine a similar system. Figure 5.7*a* shows a cart full of rocks traveling with velocity v at time t_1. Someone on the cart throws away one of the rocks in a direction opposite to that of the motion. As a result, at time t_2 a lighter cart is traveling with the velocity $v+\Delta v$, and the thrown rock is traveling with the velocity $v+\Delta v-v_{\text{rel}}$, as shown in Fig. 5.7*b*, where v_{rel} is the velocity of the rock relative to the cart. It is assumed that there are no forces between the ground and the cart. Because the force between the cart and the rock

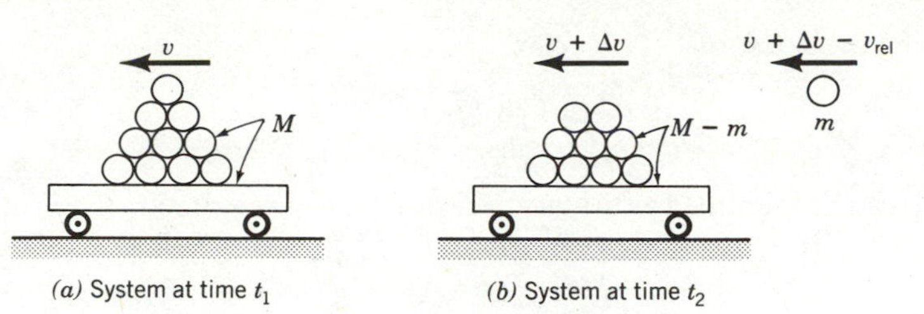

(a) System at time t_1 (b) System at time t_2

FIGURE 5.7

is internal, the momentum is conserved, which permits us to write

$$Mv=(M-m)(v+\Delta v)+m(v+\Delta v-v_{\text{rel}}) \tag{5.64}$$

where M is the mass of the cart and the rocks at time t_1 and m is the mass of the thrown rock. Canceling terms in Eq. (5.64) and rearranging, we obtain

$$\Delta v=\frac{mv_{\text{rel}}}{M} \tag{5.65}$$

which is the velocity with which the cart is propelled forward by the force expelling the mass m. If rocks are being thrown off the cart repeatedly, then the cart will accelerate accordingly. This is essentially the principle of jet propulsion, the main difference being that in jet propulsion there is a continuous stream of particles being expelled at very high relative velocity.

Let us now consider a simple model simulating a rocket engine, as shown in Fig. 5.8. Assuming that there is an external force $F(t)$ acting on the system and letting Δt be a small time increment, the impulse–momentum principle yields

$$F(t)\,\Delta t=[m(t)+\Delta m(t)][v(t)+\Delta v(t)]-\Delta m[v(t)+\Delta v(t)-v_{\text{ex}}]-m(t)v(t) \tag{5.66}$$

where v_{ex} is the exhaust velocity, which is the velocity of the hot gases flowing relative to the engine. The exhaust velocity v_{ex} is a constant depending on the type of fuel. Canceling terms and rearranging, we conclude that Eq. (5.66) yields

$$F(t)=m(t)\frac{\Delta v(t)}{\Delta t}+\frac{\Delta m(t)}{\Delta t}v_{\text{ex}} \tag{5.67}$$

Letting $\Delta t\rightarrow 0$, we obtain the differential equation governing the rocket motion

$$F(t)=m(t)\frac{dv(t)}{dt}+\frac{dm(t)}{dt}v_{\text{ex}} \tag{5.68}$$

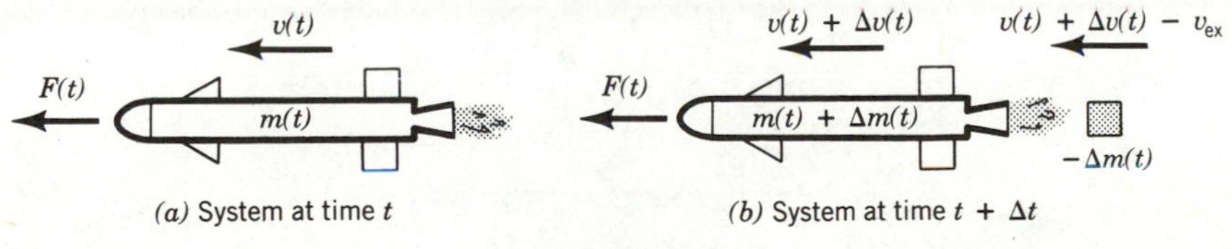

(a) System at time t (b) System at time $t+\Delta t$

FIGURE 5.8

Introducing the notation

$$-\frac{dm(t)}{dt}v_{ex}=T(t) \tag{5.69}$$

where $T(t)$ can be identified as the *engine thrust*, we can rewrite Eq. (5.68) in the form

$$T(t)+F(t)=m(t)\frac{dv(t)}{dt} \tag{5.70}$$

Note that in the above derivation we assumed that the rocket gains mass when in fact it loses mass. This fact is taken into account ordinarily by letting dm/dt be negative. Moreover, the force $F(t)$ is in general due to drag and gravity, so that $F(t)$ is usually negative and hence acting in a direction opposite to that shown in Fig. 5.8.

As an illustration, let us consider the vertical flight of a sounding rocket (Fig. 5.9) and calculate the altitude it will achieve at burnout (i.e., when the entire fuel is exhausted). Neglecting air resistance, we can assume that the force $F(t)$ is simply due to gravity, where g is assumed to be constant. Hence, the differential equation of motion is

$$-m(t)g=m(t)\frac{dv(t)}{dt}+\frac{dm(t)}{dt}v_{ex} \tag{5.71}$$

Rearranging Eq. (5.71) yields

$$dv=-g\,dt-v_{ex}\frac{dm}{m} \tag{5.72}$$

Letting $v(0)=v_0$, $m(0)=m_0$ and integrating Eq. (5.72), we obtain

$$v(t)=v_0-gt-v_{ex}\ln\frac{m}{m_0} \tag{5.73}$$

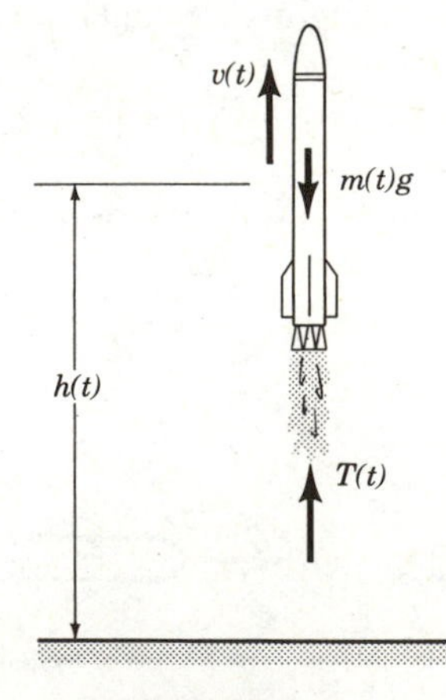

FIGURE 5.9

Denoting the burnout time by t_{bo} and the mass at burnout by m_{bo}, and assuming that mass is expelled at constant rate, we can write

$$m(t)=m_0-\frac{m_0-m_{bo}}{t_{bo}}t \tag{5.74}$$

so that Eq. (5.73) becomes

$$v(t)=v_0-gt-v_{ex}\ln\left(1-\frac{m_0-m_{bo}}{m_0t_{bo}}t\right) \tag{5.75}$$

But the velocity is the rate of change in altitude, $v(t)=dh(t)/dt$, so that integrating once again, we obtain the altitude at burnout

$$h(t_{bo})=v_0t_{bo}-\frac{1}{2}gt_{bo}^2-v_{ex}\int_0^{t_{bo}}\ln\left(1-\frac{m_0-m_{bo}}{m_0t_{bo}}t\right)dt \tag{5.76}$$

To evaluate the integral, let us introduce the change of variable

$$1-\frac{m_0-m_{bo}}{m_0t_{bo}}t=\tau, \qquad dt=-\frac{m_0t_{bo}}{m_0-m_{bo}}d\tau \tag{5.77}$$

which calls for the following change in the limits of integration

$$t=0\rightarrow\tau=1, \qquad t=t_{bo}\rightarrow\tau=m_{bo}/m_0 \tag{5.78}$$

Hence, the integral can be evaluated as follows:

$$\begin{aligned}\int_0^{t_{bo}}\ln\left(1-\frac{m_0t_{bo}}{m_0-m_{bo}}t\right)dt &= -\frac{m_0t_{bo}}{m_0-m_{bo}}\int_1^{m_{bo}/m_0}\ln\tau\,d\tau\\ &= -\frac{m_0t_{bo}}{m_0-m_{bo}}\tau(\ln\tau-1)\Big|_1^{m_{bo}/m_0}\\ &= -\frac{m_0t_{bo}}{m_0-m_{bo}}\left[\frac{m_{bo}}{m_0}\left(\ln\frac{m_{bo}}{m_0}-1\right)+1\right]\\ &= -t_{bo}+\frac{m_{bo}t_{bo}}{m_0-m_{bo}}\ln\frac{m_0}{m_{bo}}\end{aligned} \tag{5.79}$$

so that the altitude at burnout is

$$h(t_{bo})=v_0t_{bo}-\frac{1}{2}gt_{bo}^2+v_{ex}t_{bo}\left(1-\frac{m_{bo}}{m_0-m_{bo}}\ln\frac{m_0}{m_{bo}}\right) \tag{5.80}$$

PROBLEMS

5.1 A dumbell consisting of two particles of mass m_1 and m_2 and connected by a rigid link of length L is falling with the translational velocity v_0 and with zero rotation when m_2 strikes a ledge, as shown in Fig. 5.10. Assume that m_2 rebounds with the velocity v_2, and calculate the following: (1) the velocity v_1

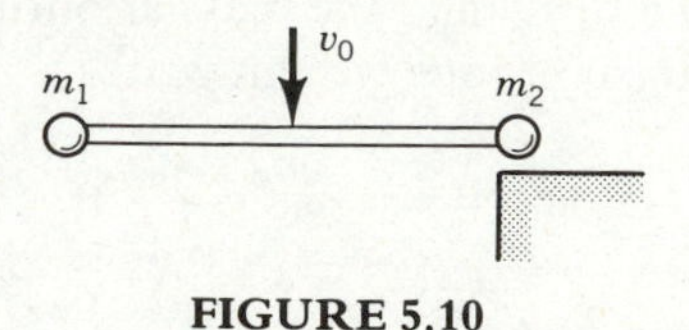

FIGURE 5.10

of m_1 immediately after impact and (2) the impulse $\hat{F}$ imparted to the dumbbell by the ledge.

5.2 A bullet of mass m is traveling with the velocity $\mathbf{v}_0 = 50\mathbf{i}$ m/s when it hits a rock of mass $2m$ lying at rest (Fig. 5.11*a*). As a result of the impact, the rock breaks into two fragments, one of mass $m/2$ traveling with a velocity of 30 m/s and in a direction making an angle of 30° with the vector $\mathbf{v}_0$ and the second of mass $3m/2$ traveling with a velocity of 20 m/s and in a direction making an angle of $-15°$ with $\mathbf{v}_0$, as shown in Fig. 5.11*b*. Find the velocity $\mathbf{v}_1$ of the bullet after impact by magnitude and direction, and calculate the energy lost during impact.

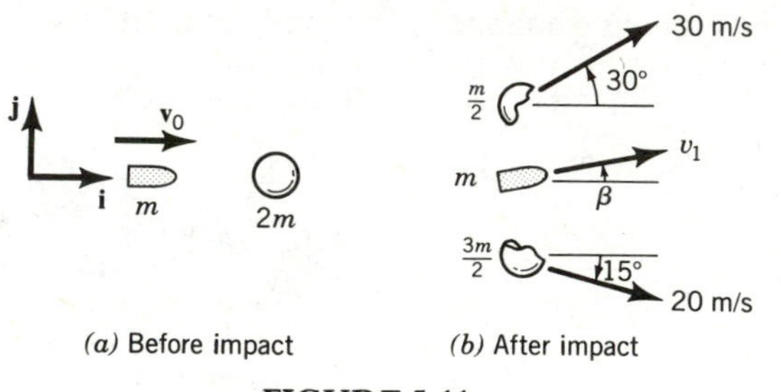

FIGURE 5.11

5.3 A bullet of mass m is fired with a velocity v_0 into a block of wood of mass M (Fig. 5.12). If the bullet becomes embedded in the block, calculate the maximum height reached by the block and the bullet.

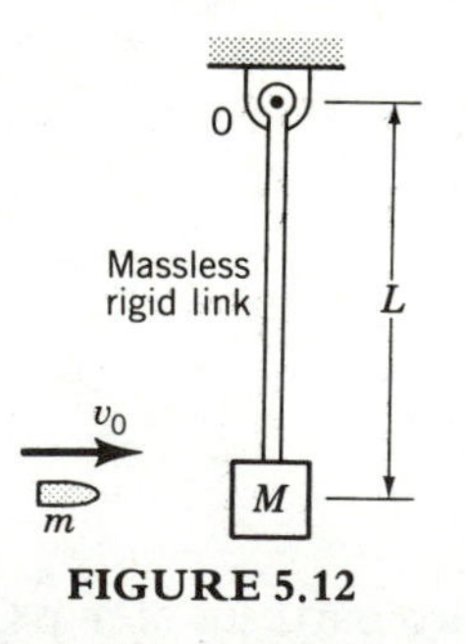

FIGURE 5.12

5.4 Two particles of mass m each are connected by an inextensible string of length L passing through a small hole at point 0, as shown in Fig. 5.13. Initially the string is held fixed at 0 while one of the particles moves on the frictionless

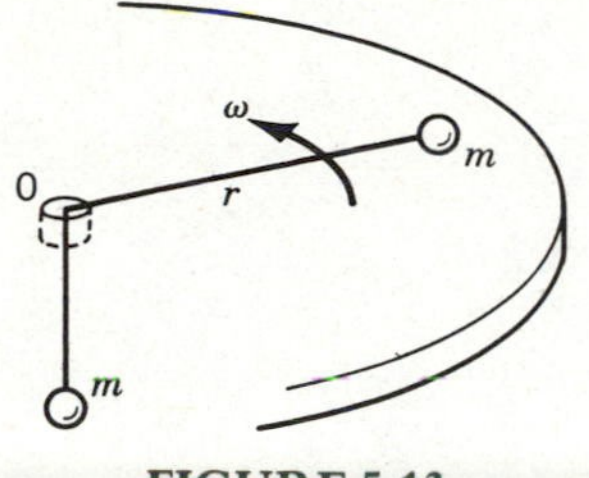

FIGURE 5.13

table in a circle of radius r_0 with the constant angular velocity ω_0. Upon release of the string, the hanging mass begins to fall. Find the minimum value of r and the maximum tension in the string.

5.5 A sounding rocket is launched vertically from the earth's surface. Assuming that the exhaust velocity is $v_{ex} = 2500$ m/s, the burnout time is $t_{bo} = 10$ s and the mass at burnout is 10% of the initial mass, calculate the maximum velocity and the maximum altitude reached by the rocket. The acceleration due to gravity can be assumed to be constant.

CHAPTER 6
Dynamics of Rigid Bodies

6.1 INTRODUCTION

Rigid bodies can be regarded as systems of particles of a special type, namely, one in which the distance between any two particles is constant. Hence, all the developments of Chapter 5 for systems of particles are applicable to rigid bodies, provided the velocity of a point in the rigid body relative to another is due only to the angular velocity of the rigid body. This procedure guarantees automatically that the distance between any two points in the body remains constant. In this regard, we must observe that the three-particle system of Example 5.2 was really a rigid body and was treated as such. In general, however, one thinks of a rigid body more in terms of a continuous body rather than a collection of particles, so that many of the definitions of Chapter 5 must be modified by letting the mass of a typical particle m_i approach a differential element of mass dm and by replacing the process of summation over the system of particles by integration over the rigid body.

6.2 LINEAR AND ANGULAR MOMENTUM

First, let us consider the angular momentum of a rigid body rotating with the angular velocity $\boldsymbol{\omega}$ about the fixed point 0. We let axes XYZ be an inertial system and envision a set of axes xyz embedded in the body and rotating together with it with the same angular velocity $\boldsymbol{\omega}$, as shown in Fig. 6.1. Axes xyz are known as *body axes*. By analogy with Eq. (5.25), the angular momentum about 0 is defined as

$$\mathbf{H}_0=\int_m \mathbf{r}\times\mathbf{v}\,dm \tag{6.1}$$

Because the velocity $\mathbf{v}$ of any point in the body is due entirely to rotation about 0, we can write (see Section 2.4)

$$\mathbf{v}=\boldsymbol{\omega}\times\mathbf{r} \tag{6.2}$$

so that Eq. (6.1) becomes

$$\mathbf{H}_0=\int_m \mathbf{r}\times(\boldsymbol{\omega}\times\mathbf{r})\,dm \tag{6.3}$$

From vector analysis, however, we have for any three vectors

$$\mathbf{A}\times(\mathbf{B}\times\mathbf{C})=(\mathbf{A}\cdot\mathbf{C})\mathbf{B}-(\mathbf{A}\cdot\mathbf{B})\mathbf{C} \tag{6.4}$$

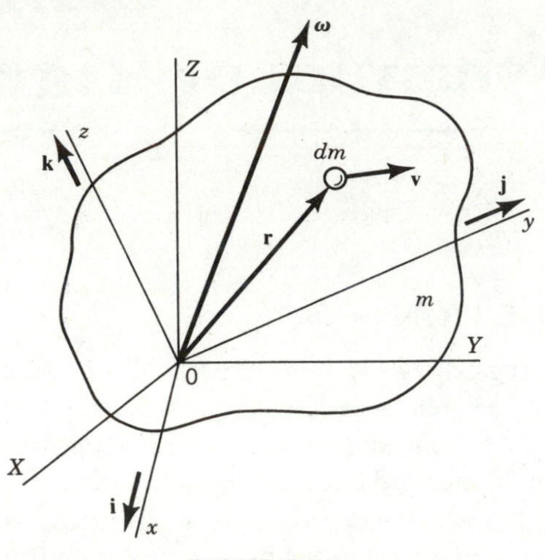

FIGURE 6.1

so that, letting $\mathbf{A}=\mathbf{C}=\mathbf{r}$, $\mathbf{B}=\boldsymbol{\omega}$, we can write Eq. (6.3) in the form

$$\mathbf{H}_0=\int_m [(\mathbf{r}\cdot\mathbf{r})\boldsymbol{\omega}-(\mathbf{r}\cdot\boldsymbol{\omega})\mathbf{r}]\,dm \tag{6.5}$$

To obtain a more detailed expression for $\mathbf{H}_0$, namely, one in terms of the x, y, z components of $\mathbf{r}$ and $\boldsymbol{\omega}$, we let

$$\mathbf{r}=x\mathbf{i}+y\mathbf{j}+z\mathbf{k} \tag{6.6}$$

and

$$\boldsymbol{\omega}=\omega_x\mathbf{i}+\omega_y\mathbf{j}+\omega_z\mathbf{k} \tag{6.7}$$

so that

$$\begin{aligned}(\mathbf{r}\cdot\mathbf{r})\boldsymbol{\omega}-(\mathbf{r}\cdot\boldsymbol{\omega})\mathbf{r}=&[(r^2-x^2)\omega_x-xy\omega_y-xz\omega_z]\mathbf{i}\\&+[-xy\omega_x+(r^2-y^2)\omega_y-yz\omega_z]\mathbf{j}\\&+[-xz\omega_x-yz\omega_y+(r^2-z^2)\omega_z]\mathbf{k}\end{aligned} \tag{6.8}$$

where r is the magnitude of $\mathbf{r}$. Introducing Eq. (6.8) into Eq. (6.5), we obtain

$$\begin{aligned}\mathbf{H}_0=&(I_{xx}\omega_x-I_{xy}\omega_y-I_{xz}\omega_z)\mathbf{i}+(-I_{xy}\omega_x+I_{yy}\omega_y-I_{yz}\omega_z)\mathbf{j}\\&+(-I_{xz}\omega_x-I_{yz}\omega_y+I_{zz}\omega_z)\mathbf{k}\end{aligned} \tag{6.9}$$

where

$$I_{xx}=\int_m (r^2-x^2)\,dm,\quad I_{yy}=\int_m (r^2-y^2)\,dm,\quad I_{zz}=\int_m (r^2-z^2)\,dm \tag{6.10}$$

are mass moments of inertia about the body axes xyz and

$$I_{xy}=I_{yx}=\int_m xy\,dm, \qquad I_{xz}=I_{zx}=\int_m xz\,dm, \qquad I_{yz}=I_{zy}=\int_m yz\,dm \tag{6.11}$$

are mass products of inertia about the same axes.

Next, let us consider a rigid body translating and rotating relative to the inertial space, as shown in Fig. 6.2. The angular momentum about 0 retains the form (6.1), but this time the absolute velocity $\mathbf{v}$ has the expression

$$\mathbf{v}=\mathbf{v}_C+\boldsymbol{\omega}\times\boldsymbol{\rho} \tag{6.12}$$

where $\boldsymbol{\rho}$ is the radius vector from C to the differential element of mass dm, so that

$$\mathbf{r}=\mathbf{r}_C+\boldsymbol{\rho} \tag{6.13}$$

where, by analogy with Eq. (5.9),

$$\mathbf{r}_C=\frac{1}{m}\int_m \mathbf{r}\,dm \tag{6.14}$$

defines the position of the *mass center* C relative to 0. Substituting Eq. (6.13) into Eq. (6.14), we conclude that the alternative definition of the mass center is

$$\int_m \boldsymbol{\rho}\,dm=\mathbf{0} \tag{6.15}$$

Introducing Eqs. (6.12) and (6.13) into Eq. (6.1), we obtain

$$\begin{aligned}\mathbf{H}_0&=\int_m (\mathbf{r}_C+\boldsymbol{\rho})\times(\mathbf{v}_C+\boldsymbol{\omega}\times\boldsymbol{\rho})\,dm\\&=\mathbf{r}_C\times\mathbf{v}_C\int_m dm+\left(\int_m \boldsymbol{\rho}\,dm\right)\times\mathbf{v}_C+\mathbf{r}_C\times\left(\boldsymbol{\omega}\times\int_m \boldsymbol{\rho}\,dm\right)\\&\quad+\int_m \boldsymbol{\rho}\times(\boldsymbol{\omega}\times\boldsymbol{\rho})\,dm\end{aligned} \tag{6.16}$$

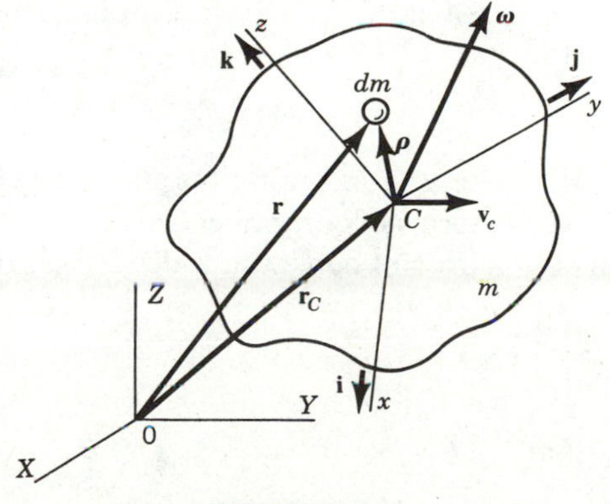

FIGURE 6.2

Moreover, we recognize that

$$\mathbf{v}_C \int_m dm = m\mathbf{v}_C = \mathbf{p} \tag{6.17}$$

is the linear momentum of the body and that

$$\int_m \boldsymbol{\rho} \times (\boldsymbol{\omega} \times \boldsymbol{\rho})\, dm = \mathbf{H}_C = \mathbf{H}'_C \tag{6.18}$$

is the angular momentum of the body about the mass center C, where the actual angular momentum $\mathbf{H}_C$ is equal to the apparent angular momentum $\mathbf{H}'_C$, as for any system of particles. Considering Eqs. (6.17) and (6.18), we can reduce Eq. (6.16) to

$$\mathbf{H}_0 = \mathbf{r}_C \times \mathbf{p} + \mathbf{H}_C \tag{6.19}$$

which states that the angular momentum of the body about 0 is equal to the moment about 0 of the linear momentum obtained if the entire mass were concentrated at the mass center C plus the angular momentum of the body about C. This statement represents a *translation theorem for the angular momentum.* Note that the angular momentum about C has a form similar to that given by (Eq. 6.9), except that here the moments and products of inertia are about body axes through C.

6.3 THE EQUATIONS OF MOTION

Next, we wish to derive the equations of motion for rigid bodies. The equation for the translational motion retains the form (5.21), as for any system of particles, or

$$\mathbf{F} = \dot{\mathbf{p}} = m\mathbf{a}_C \tag{6.20}$$

where $\mathbf{a}_C$ is the acceleration of the mass center. Moreover, for $\mathbf{F} = \mathbf{0}$, the linear momentum of the rigid body is conserved, as indicated by Eq. (5.23).

The equation for the rotational motion of a rigid body retains the same general form (5.29), as for any system of particles. Hence, for pure rotation about the fixed point 0, we have

$$\mathbf{M}_0 = \dot{\mathbf{H}}_0 \tag{6.21}$$

where $\mathbf{H}_0$ is given by Eq. (6.9). But, in taking the time derivative of $\mathbf{H}_0$, we must recall that $\mathbf{H}_0$ is expressed in terms of components about rotating axes, so that the associated unit vectors $\mathbf{i}$, $\mathbf{j}$, $\mathbf{k}$ are not constant. Indeed, their time rate of change is

$$\frac{d\mathbf{i}}{dt} = \boldsymbol{\omega} \times \mathbf{i}, \qquad \frac{d\mathbf{j}}{dt} = \boldsymbol{\omega} \times \mathbf{j}, \qquad \frac{d\mathbf{k}}{dt} = \boldsymbol{\omega} \times \mathbf{k} \tag{6.22}$$

Introducing the notation

$$\mathbf{H}_0 = H_{0x}\mathbf{i} + H_{0y}\mathbf{j} + H_{0z}\mathbf{k} \tag{6.23}$$

we can write

$$\begin{aligned}\dot{\mathbf{H}}_0 &= \frac{d}{dt}(H_{0x}\mathbf{i}+H_{0y}\mathbf{j}+H_{0z}\mathbf{k}) \\ &= \dot{H}_{0x}\mathbf{i}+H_{0x}\frac{d\mathbf{i}}{dt}+\dot{H}_{0y}\mathbf{j}+H_{0y}\frac{d\mathbf{j}}{dt}+\dot{H}_{0z}\mathbf{k}+H_{0z}\frac{d\mathbf{k}}{dt} \\ &= \dot{H}_{0x}\mathbf{i}+\dot{H}_{0y}\mathbf{j}+\dot{H}_{0z}\mathbf{k}+H_{0x}\boldsymbol{\omega}\times\mathbf{i}+H_{0y}\boldsymbol{\omega}\times\mathbf{j}+H_{0z}\boldsymbol{\omega}\times\mathbf{k} \\ &= \dot{H}_{0x}\mathbf{i}+\dot{H}_{0y}\mathbf{j}+\dot{H}_{0z}\mathbf{k}+\boldsymbol{\omega}\times(H_{0x}\mathbf{i}+H_{0y}\mathbf{j}+H_{0z}\mathbf{k}) \\ &= \dot{\mathbf{H}}'_0+\boldsymbol{\omega}\times\mathbf{H}_0 \qquad (6.24)\end{aligned}$$

where $\dot{\mathbf{H}}'_0$ is the rate of change of $\mathbf{H}_0$ regarding the reference frame as fixed. Incidentally, formula (6.24) is valid for the derivative of any vector expressed in terms of components about rotating axes, as demonstrated in Section 2.6. Inserting Eq. (6.24) into Eq. (6.21), we obtain the moment equations of motion in the vector form

$$\mathbf{M}_0=\dot{\mathbf{H}}'_0+\boldsymbol{\omega}\times\mathbf{H}_0 \qquad (6.25)$$

Considering Eqs. (6.7) and (6.9) and introducing the notation

$$\mathbf{M}_0=M_{0x}\mathbf{i}+M_{0y}\mathbf{j}+M_{0z}\mathbf{k} \qquad (6.26)$$

we can write Eq. (6.25) by components as follows:

$$\begin{aligned}M_{0x} &= I_{xx}\dot{\omega}_x-I_{xy}\dot{\omega}_y-I_{xz}\dot{\omega}_z-I_{xz}\omega_x\omega_y+I_{xy}\omega_x\omega_z \\ &\quad +(I_{zz}-I_{yy})\omega_y\omega_z-I_{yz}(\omega_y^2-\omega_z^2) \\ M_{0y} &= -I_{xy}\dot{\omega}_x+I_{yy}\dot{\omega}_y-I_{yz}\dot{\omega}_z+I_{yz}\omega_x\omega_y+(I_{xx}-I_{zz})\omega_x\omega_z \\ &\quad -I_{xy}\omega_y\omega_z-I_{xz}(\omega_z^2-\omega_x^2) \\ M_{0z} &= -I_{xz}\dot{\omega}_x-I_{yz}\dot{\omega}_y+I_{zz}\dot{\omega}_z+(I_{yy}-I_{xx})\omega_x\omega_y-I_{yz}\omega_x\omega_z \\ &\quad +I_{xz}\omega_y\omega_z-I_{xy}(\omega_x^2-\omega_y^2)\end{aligned} \qquad (6.27)$$

Note that the reason for working with body axes is that the moments of inertia about these axes are constant.

The moment equations about the mass center C have an entirely analogous form, or

$$\mathbf{M}_C=\dot{\mathbf{H}}_C=\dot{\mathbf{H}}'_C+\boldsymbol{\omega}\times\mathbf{H}_C \qquad (6.28)$$

where the components of $\mathbf{H}_C$ have the same expressions as in Eqs. (6.9), *the only difference being that the moments and products of inertia are about body axes through C.*

Equations (6.27) are very complicated, and they are seldom used. Fortunately, there exists a much simpler and more useful form. Indeed, letting axes x, y, and z be the principal axes 1, 2, and 3, respectively, we can write

$$\begin{aligned}&I_{xx}=I_1, \qquad I_{yy}=I_2, \qquad I_{zz}=I_3 \\ &I_{xy}=I_{xz}=I_{yz}=0\end{aligned} \qquad (6.29)$$

where I_1, I_2, and I_3 are the principal mass moments of inertia. Introducing Eqs. (6.29) into Eqs. (6.27) and replacing x, y, and z by 1, 2, and 3, respectively, we obtain

$$\begin{aligned} M_1 &= I_1\dot{\omega}_1 + (I_3 - I_2)\omega_2\omega_3 \\ M_2 &= I_2\dot{\omega}_2 + (I_1 - I_3)\omega_1\omega_3 \\ M_3 &= I_3\dot{\omega}_3 + (I_2 - I_1)\omega_1\omega_2 \end{aligned} \tag{6.30}$$

which are known as *Euler's moment equations.* They are valid for two important cases: (1) when the moment and the mass moments of inertia are about principal axes with the origin at the fixed point 0, and (2) when they are about principal axes with the origin at the mass center C. We note, in passing, that Eqs. (6.30) are nonlinear in the angular velocities ω_1, ω_2, and ω_3.

6.4 KINETIC ENERGY

Using the analogy with the kinetic energy for a system of particles, Eq. (5.45), the kinetic energy for a rigid body can be written in the form

$$T = \frac{1}{2}\int_m \mathbf{v}\cdot\mathbf{v}\, dm \tag{6.31}$$

where $\mathbf{v}$ is the velocity of the differential element dm relative to an inertial space. But, according to Eq. (6.12), the velocity $\mathbf{v}$ can be expressed as the sum of the translational velocity $\mathbf{v}_C$ of the mass center C and the velocity $\boldsymbol{\omega}\times\boldsymbol{\rho}$ of dm relative to C, where the latter is due entirely to the rotation of the body about C. Hence, introducing Eq. (6.12) into Eq. (6.31), we obtain

$$\begin{aligned} T &= \frac{1}{2}\int_m (\mathbf{v}_C + \boldsymbol{\omega}\times\boldsymbol{\rho})\cdot(\mathbf{v}_C + \boldsymbol{\omega}\times\boldsymbol{\rho})\, dm \\ &= \frac{1}{2}\mathbf{v}_C\cdot\mathbf{v}_C\int_m dm + \mathbf{v}_C\cdot\boldsymbol{\omega}\times\int_m \boldsymbol{\rho}\, dm + \frac{1}{2}\int_m (\boldsymbol{\omega}\times\boldsymbol{\rho})\cdot(\boldsymbol{\omega}\times\boldsymbol{\rho})\, dm \end{aligned} \tag{6.32}$$

But, according to Eq. (6.15), $\int_m \boldsymbol{\rho}\, dm = \mathbf{0}$. It follows that Eq. (6.32) reduces to

$$T = T_{\mathrm{tr}} + T_{\mathrm{rot}} \tag{6.33}$$

where

$$T_{\mathrm{tr}} = \tfrac{1}{2}m\mathbf{v}_C\cdot\mathbf{v}_C \tag{6.34a}$$

is the kinetic energy that would be obtained if the rigid body were undergoing pure translation with the velocity $\mathbf{v}_C$ and

$$T_{\mathrm{rot}} = \frac{1}{2}\int_m (\boldsymbol{\omega}\times\boldsymbol{\rho})\cdot(\boldsymbol{\omega}\times\boldsymbol{\rho})\, dm \tag{6.34b}$$

is the kinetic energy of rotation of the rigid body about the mass center C. Denoting the Cartesian components of $\boldsymbol{\rho}$ by x, y, and z and recalling Eq. (6.7), we can write

$$\boldsymbol{\omega} \times \boldsymbol{\rho} = (\omega_y z - \omega_z y)\mathbf{i} + (\omega_z x - \omega_x z)\mathbf{j} + (\omega_x y - \omega_y x)\mathbf{k} \tag{6.35}$$

so that, introducing Eq. (6.35) into Eq. (6.34b), we obtain

$$\begin{aligned} T_{\text{rot}} &= \frac{1}{2}\int_m [(\omega_y z - \omega_z y)^2 + (\omega_z x - \omega_x z)^2 + (\omega_x y - \omega_y x)^2]\, dm \\ &= \frac{1}{2}(I_{xx}\omega_x^2 + I_{yy}\omega_y^2 + I_{zz}\omega_z^2 - 2I_{xy}\omega_x\omega_y - 2I_{xz}\omega_x\omega_z - 2I_{yz}\omega_y\omega_z) \end{aligned} \tag{6.36}$$

In the special case in which x, y, and z are the principal axes 1, 2, and 3, the rotational kinetic energy reduces to

$$T_{\text{rot}} = \tfrac{1}{2}(I_1\omega_1^2 + I_2\omega_2^2 + I_3\omega_3^2) \tag{6.37}$$

It should be reiterated that the separation of the kinetic energy into two terms, one due to translation and the other due to rotation, is possible only if the mass center C is used as the origin of the body axes.

6.5 PLANAR MOTION OF A RIGID BODY

A case of particular interest is the one in which the motion takes place in a given plane. Assuming that the motion takes place in the xy plane, we have

$$v_z = a_z = 0 \tag{6.38a}$$

$$\omega_x = \dot{\omega}_x = \omega_y = \dot{\omega}_y = 0, \qquad \omega_z = \omega \tag{6.38b}$$

Moreover, we shall assume that one of the dimensions of the body is infinitesimally small and that this small dimension is in the z-direction. Hence, we have

$$I_{xz} = I_{yz} = 0 \tag{6.39}$$

The various quantities derived in Sections 6.2–6.4, such as momentum and kinetic energy, as well as the equations of motion can be given a simpler form by considering Eqs. (6.38) and (6.39). Certain special cases of motion involve further simplifications. Hence, before proceeding to the more general case, we propose to study these special cases.

i. Pure Translation

In this case, the rotation of the body is zero

$$\omega = \dot{\omega} = 0 \tag{6.40}$$

so that every point of the body undergoes the same translational motion. Using Eq. (6.20) and considering Eqs. (6.38a), we can write the force equations in the form

$$F_x = ma_{Cx} \tag{6.41a}$$

$$F_y = ma_{Cy} \tag{6.41b}$$

Moreover, the only moment equation is about the z-axis. Considering Eqs. (6.38b) and (6.39), the moment equation about the mass center C is simply

$$M_{Cz} = 0 \tag{6.42}$$

The kinetic energy is due to translation alone, and from Eq. (6.33) it has the explicit form

$$T = \tfrac{1}{2} m \mathbf{v}_C \cdot \mathbf{v}_C = \tfrac{1}{2} m (v_{Cx}^2 + v_{Cy}^2) \tag{6.43}$$

where v_{Cx} and v_{Cy} are the Cartesian components of the velocity vector of the mass center.

Example 6.1

Derive the equations of motion for the horizontal translation of the body shown in Fig. 6.3*a*. The horizontal reactions at the points A and B are proportional to the vertical reactions at the same points, where the proportionality constant is the coefficient of friction μ. Then, determine the magnitude of the force $\mathbf{F}$ when the body is on the verge of tipping over.

Figure 6.3*b* shows the corresponding free-body diagram. Using Eqs. (6.41) and (6.42) we obtain simply

$$F_x = F \cos \beta - \mu(N_A + N_B) = ma_{Cx} \tag{a}$$

$$F_y = F \sin \beta + N_A + N_B - mg = 0 \tag{b}$$

$$M_{Cz} = -F \cos \beta \frac{H}{2} + (N_B - N_A - F \sin \beta) \frac{L}{2} - \mu(N_A + N_B)\left(D + \frac{H}{2}\right) = 0 \tag{c}$$

Just before tipping over, the reaction N_A reduces to zero. Then, from Eqs. (b) and (c), we obtain

$$F = \frac{mg(L - 2\mu D - \mu H)}{H \cos \beta + (2L - 2\mu D - \mu H) \sin \beta} \tag{d}$$

ii. Rotation about a Fixed Point

Let us consider the case of planar motion about a fixed point 0, insert Eqs. (6.38b) and (6.39) into Eq. (6.9), recognize that there is only one angular momentum

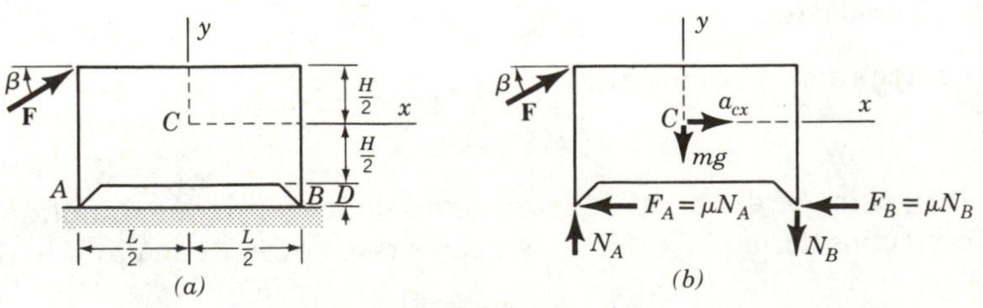

FIGURE 6.3

component, and write

$$H_0 = I_0\omega \tag{6.44}$$

where we introduced the notation $I_{zz} = I_0$ for the mass moment of inertia of the body about 0. Similarly, using Eq. (6.21), we can write

$$M_0 = I_0\alpha \tag{6.45}$$

where $\alpha = \dot{\omega}$ is the angular acceleration. If the interest lies in the rotational motion alone, then Eq. (6.45) is the only one needed. On the other hand, if the reactions at point 0 are also of interest, then Eqs. (6.41) must also be considered.

The kinetic energy can be attributed entirely to rotation about 0, and, from Eq. (6.36), it has the form

$$T = \tfrac{1}{2} I_0 \omega^2 \tag{6.46}$$

Example 6.2

A uniform bar of total mass m is hinged at point 0, as shown in Fig. 6.4*a*. If the bar is released from rest in the horizontal position, determine: (i) the angular acceleration of the bar immediately after release, (ii) the reaction at point 0 at the same time, and (iii) the angular velocity of the bar when it passes through the vertical position.

Considering counterclockwise moments and angular motions as positive, Eq. (6.45) in conjunction with the free-body diagram of Fig. 6.4*b* yields

$$M_0 = -\tfrac{1}{6} Lmg = I_0\alpha \tag{a}$$

where the mass moment of inertia of the bar about point 0 can be obtained through integration as follows:

$$I_0 = \int_m x^2\, dm = \frac{m}{L}\int_{-L/3}^{2L/3} x^2\, dx = \frac{m}{L}\,\tfrac{1}{3}x^3\Big|_{-L/3}^{2L/3} = \tfrac{1}{9} mL^2 \tag{b}$$

Hence, from Eqs. (a) and (b), the angular acceleration immediately after release is

$$\alpha = -\tfrac{3}{2}\frac{g}{L} \tag{c}$$

To obtain the reaction at point 0, we use Eq. (6.41b) in conjunction with the

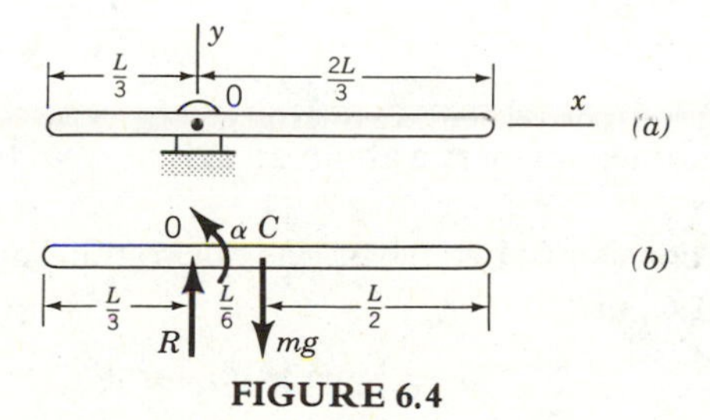

FIGURE 6.4

free-body diagram of Fig. 6.4*b*, or

$$F_y = R - mg = ma_{Cy} = m\tfrac{1}{6}L\alpha = m\tfrac{1}{6}L\left(-\tfrac{3}{2}\frac{g}{L}\right) = -\tfrac{1}{4}mg \tag{d}$$

from which we obtain

$$R = \tfrac{3}{4}mg \tag{e}$$

Finally, to determine the angular velocity of the bar when it passes through the vertical position, we invoke the conservation of energy

$$E = T + V = \text{const} \tag{f}$$

where E is the total energy, T is the kinetic energy, and V is the potential energy. Denoting the horizontal configuration by the subscript 1 and the vertical one by the subscript 2, letting the potential energy in the configuration 1 be equal to zero, and recalling that the angular velocity is zero when the bar is horizontal, we can write

$$E = T_1 + V_1 = 0 \tag{g}$$

In the configuration 2, the potential energy is due to the lowering of the mass center C, or

$$V_2 = -\tfrac{1}{6}Lmg \tag{h}$$

and the kinetic energy is

$$T_2 = \tfrac{1}{2}I_O\omega_2^2 = \tfrac{1}{2}\tfrac{1}{9}mL^2\omega_2^2 \tag{i}$$

Because $T_2 + V_2 = 0$, we obtain

$$\omega_2 = -\sqrt{\frac{3g}{L}} \tag{j}$$

where the negative sign was chosen in recognition of the fact that the angular velocity is in the clockwise sense.

iii. General Case of Planar Motion

In this case it is often convenient to express the equations of motion in terms of the mass center C. The force equations remain in the form (6.41) and the moment equation has the scalar form

$$M_C = I_C\alpha \tag{6.47}$$

where I_C is the mass moment of inertia about an axis normal to the plane of motion and passing through C.

The kinetic energy consists of two parts, one due to the translation of C and one due to rotation about C, or

$$T = \tfrac{1}{2}m(v_{Cx}^2 + v_{Cy}^2) + \tfrac{1}{2}I_C\omega^2 \tag{6.48}$$

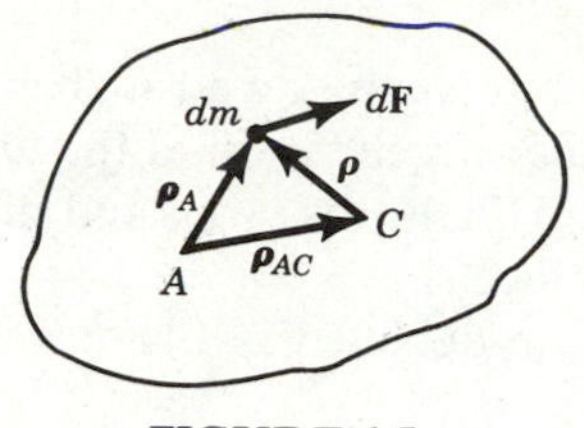

FIGURE 6.5

On occasions it is more convenient to express the moment equation about a point other than the mass center C. Let us consider the system of Fig. 6.5 and write the moment about the arbitrary point A in the form

$$\mathbf{M}_A = \int \boldsymbol{\rho}_A \times d\mathbf{F} = \int (\boldsymbol{\rho}_{AC} + \boldsymbol{\rho}) \times d\mathbf{F} = \boldsymbol{\rho}_{AC} \times \mathbf{F} + \mathbf{M}_C \tag{6.49}$$

In view of Eq. (6.20), however, Eq. (6.49) becomes

$$\mathbf{M}_A = \boldsymbol{\rho}_{AC} \times m\mathbf{a}_C + \mathbf{M}_C \tag{6.50}$$

where we note that for planar motion

$$\mathbf{M}_C = I_C \boldsymbol{\alpha} \tag{6.51}$$

in which the vectors $\mathbf{M}_C$ and $\boldsymbol{\alpha}$ are both normal to the plane of motion.

Next, let us consider the kinematical equation

$$\mathbf{a}_C = \mathbf{a}_A + \mathbf{a}_{C/A} \tag{6.52}$$

which for planar motion reduces to (see Section 2.5)

$$\mathbf{a}_C = \mathbf{a}_A - \omega^2 \boldsymbol{\rho}_{AC} + \boldsymbol{\alpha} \times \boldsymbol{\rho}_{AC} \tag{6.53}$$

Introducing Eq. (6.53) into Eq. (6.50) and considering Eq. (6.51), we obtain

$$\mathbf{M}_A = \boldsymbol{\rho}_{AC} \times m\mathbf{a}_A + I_A \boldsymbol{\alpha} \tag{6.54}$$

where

$$I_A = I_C + m\rho_{AC}^2 \tag{6.55}$$

is recognized as the mass moment of inertia of the body about point A.

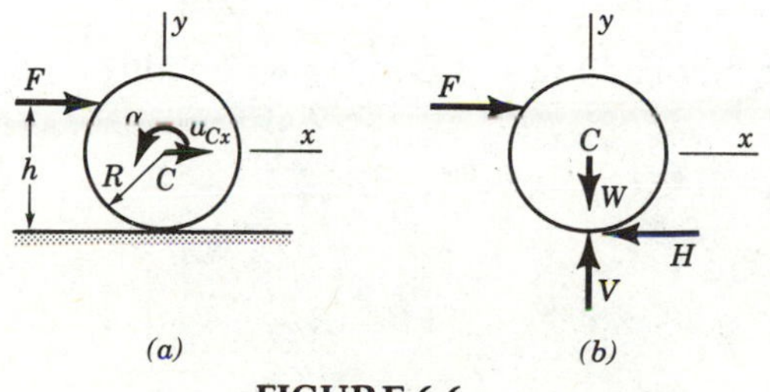

FIGURE 6.6

Example 6.3

The disk shown in Fig. 6.6*a* is traveling on a rough surface while acted upon by a force F. Determine the angular acceleration of the disk for the two cases: (1) the disk rolls without slipping and (2) the disk rolls and slips. The coefficient of friction between the disk and the surface is μ.

From the free-body diagram depicted in Fig. 6.6*b*, we can write the equations of motion

$$\begin{aligned} F_x &= F - H = ma_{Cx} \\ F_y &= V - W = 0 \\ M_C &= -F(h-R) - HR = I_C\alpha \end{aligned} \tag{a}$$

In the case of roll without slip the friction force must satisfy the inequality $|H| < \mu V = \mu W$. On the other hand, the translational and rotational accelerations are related by

$$a_{Cx} = -\alpha R \tag{b}$$

Eliminating H from the first and third of Eqs. (a), using Eq. (b), and recalling that for a disk $I_C = mR^2/2$, we obtain

$$a_{Cx} = \frac{FRh}{I_C + mR^2} = \frac{2Fh}{3mR} \tag{c}$$

from which it follows that

$$\alpha = -\frac{2Fh}{3mR^2} \tag{d}$$

Note that, from the first of Eqs. (a), we can write

$$H = F - ma_{Cx} = F\left(1 - \frac{2h}{3R}\right) \tag{e}$$

so that, for no slip, F must satisfy the inequality

$$F < \frac{\mu W}{1 - 2h/3R} \tag{f}$$

In the case of rolling and slipping, the kinematical relation (b) is no longer valid. On the other hand, the friction force is

$$H = \mu W \tag{g}$$

Hence, solving the first and third of Eqs. (a), we obtain

$$a_{Cx} = \frac{F - \mu W}{m} \tag{h}$$

and

$$\alpha = -\frac{2[F(h-R) + \mu WR]}{mR^2} \tag{i}$$

respectively.

Example 6.4
Solve the problem of Example 3.6 but, instead of a particle as shown in Fig. 3.11, the body consists of a disk of radius R. Assume that the disk rolls without slipping.

Because no energy is lost in roll without slip, the energy is conserved. Hence, the only difference between the solution of Example 3.6 and the one here is that here we must include the kinetic energy of rotation. It follows that the kinetic energy in position 2 is

$$T_2=\tfrac{1}{2}mv_2^2+\tfrac{1}{2}I_C\omega_2^2 \tag{a}$$

Recalling that $I_C=mR^2/2$ and recognizing that $v_2=-R\omega_2$, we obtain

$$T_2=\tfrac{3}{4}mv_2^2 \tag{b}$$

The conservation of energy yields

$$E=T_2+V_2=-mgH+\tfrac{3}{4}mv_2^2=0 \tag{c}$$

so that the velocity in position 2 is

$$v_2=\sqrt{4gH/3} \tag{d}$$

which is lower than the value obtained in Example 3.6. This is to be expected, because in the case of this example part of the kinetic energy is due to rotation.

6.6 ROTATION OF A RIGID BODY ABOUT A FIXED AXIS

Let us consider the body of Fig. 6.1 and assume that the motion takes place about a fixed axis. Hence, the fixed origin 0 must lie on this axis. For convenience, we take the axis of rotation as the z-axis. Rotation about the fixed axis z is characterized by

$$\omega_x=\omega_y=0, \qquad \omega_z=\omega \tag{6.56}$$

Introducing Eqs. (6.56) into Eqs. (6.27), we obtain the moment equations

$$\begin{aligned} M_{0x}&=-I_{xz}\dot{\omega}+I_{yz}\omega^2 \\ M_{0y}&=-I_{yz}\dot{\omega}-I_{xz}\omega^2 \\ M_{0z}&=I_{zz}\dot{\omega} \end{aligned} \tag{6.57}$$

The force equations remain in the form (6.41).

Various pieces of machinery involve rotors in the form of rigid masses of relatively large moments of inertia mounted on a shaft and spinning rapidly about the axis coinciding with the shaft. The shaft is supported at both ends by bearings. Generally the rotor is symmetric. In this case the products of inertia I_{xz} and I_{yz} are zero, and the first two of Eqs. (6.57) reduce to zero identically, so that there are no torques M_{0x} and M_{0y}. Moreover, the mass center C lies on the z-axis, and hence a_{Cx} and a_{Cy} are zero. It follows that for a symmetric rotor, the only forces present are the bearing reactions, which are simply static forces due to the weight of the rotor. On the other hand, if there is some asymmetry in the rotor, so that not all products of inertia are zero, then this asymmetry will generate some dynamic bearing reactions.

As an illustration, let us consider the disk shown in Fig. 6.7*a*, where the normal to the disk makes an angle β with respect to the shaft. The disk rotates uniformly about z, so that $\omega=\dot{\theta}=\text{const}$, $\dot{\omega}=0$. Figure 6.7*b* depicts the free-body diagram. Axes XYZ are inertial and axes xyz are body axes, where z is along the shaft, x is embedded in the disk and normal to z, and y is normal to both. Note that in the position shown, x makes an angle θ with respect to X. The object is to determine the reactions in that position. The products of inertia entering into Eqs. (6.57) are I_{xz} and I_{yz}. But, xyz are not principal axes, which is likely to cause some difficulty in determining I_{xz} and I_{yz} by direct integration. Indeed, it is easier to determine I_{xz} and I_{yz} by means of a coordinate transformation from axes xyz to the principal axes $x'y'z'$ (Fig. 6.7*c*). From Fig. 6.7*c*, we can write the relation between the two sets of axes as follows:

$$\begin{aligned} x &= x' \\ y &= y' \cos\beta + z' \sin\beta \\ z &= -y' \sin\beta + z' \cos\beta \end{aligned} \tag{6.58}$$

Inserting Eqs. (6.58) into the second and third of Eqs. (6.11), we obtain

$$\begin{aligned} I_{xz} &= \int_m xz\, dm = \int_m x'(-y' \sin\beta + z' \cos\beta)\, dm \\ &= -\sin\beta \int_m x'y'\, dm + \cos\beta \int_m x'z'\, dm \\ &= -\sin\beta\, I_{x'y'} + \cos\beta\, I_{x'z'} \end{aligned} \tag{6.59a}$$

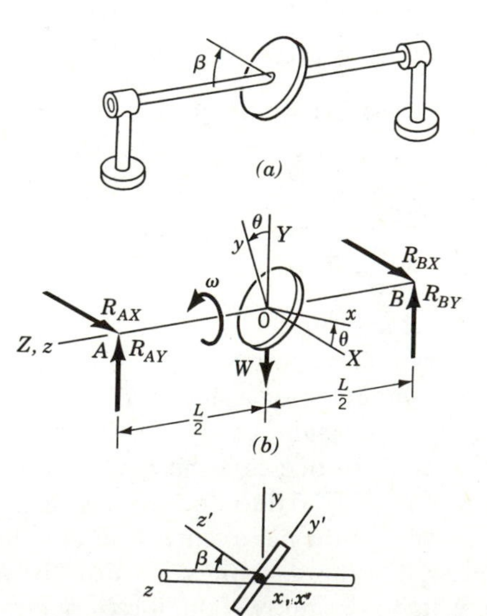

FIGURE 6.7

$$I_{yz}=\int_m yz\,dm=\int_m (y'\cos\beta+z'\sin\beta)(-y'\sin\beta+z'\cos\beta)\,dm$$

$$=\sin\beta\cos\beta\int_m (z'^2-y'^2)\,dm+(\cos^2\beta-\sin^2\beta)\int_m y'z'\,dm$$

$$=\sin\beta\cos\beta\int_m [(r'^2-y'^2)-(r'^2-z'^2)]\,dm$$

$$+(\cos^2\beta-\sin^2\beta)\int_m y'z'\,dm$$

$$=\sin\beta\cos\beta(I_{y'y'}-I_{z'z'})+(\cos^2\beta-\sin^2\beta)I_{y'z'} \tag{6.59b}$$

But, because $x'y'z'$ are principal axes, the products of inertia are zero. Moreover, the moments of inertia of a thin disk are

$$I_{x'x'}=I_{y'y'}=\tfrac{1}{4}mR^2, \qquad I_{z'z'}=\tfrac{1}{2}mR^2 \tag{6.60}$$

where m is the total mass and R is the radius of the disk. Hence, Eqs. (6.59) yield

$$I_{xz}=0, \qquad I_{yz}=-\tfrac{1}{4}mR^2\sin\beta\cos\beta \tag{6.61}$$

Letting $\dot{\omega}=0$ in Eqs. (6.57) and using Eqs. (6.61), we obtain

$$\begin{aligned} M_{0x}&=I_{yz}\omega^2=-\tfrac{1}{4}mR^2\omega^2\sin\beta\cos\beta \\ M_{0y}&=0 \\ M_{0z}&=0 \end{aligned} \tag{6.62}$$

For convenience, we resolve M_{0x} in terms of components along X and Y as follows:

$$M_{0X}=M_{0x}\cos\theta,\; M_{0Y}=M_{0x}\sin\theta \tag{6.63}$$

Because the acceleration of the mass center is zero, the force equations yield

$$\begin{aligned} F_X&=R_{AX}+R_{BX}=0 \\ F_Y&=R_{AY}+R_{BY}-W=0 \end{aligned} \tag{6.64}$$

On the other hand, the moment equations about 0 are

$$\begin{aligned} M_{0X}&=(R_{BY}-R_{AY})\frac{L}{2}=-\frac{1}{4}mR^2\omega^2\sin\beta\cos\beta\cos\theta \\ M_{0Y}&=(R_{AX}-R_{BX})\frac{L}{2}=-\frac{1}{4}mR^2\omega^2\sin\beta\cos\beta\sin\theta \end{aligned} \tag{6.65}$$

Because $\theta=\omega t$, Eqs. (6.64) and (6.65) yield

$$R_{BX} = -R_{AX} = \frac{1}{4}\frac{mR^2\omega^2}{L}\sin\beta\cos\beta\sin\omega t$$

$$R_{AY} = \frac{1}{2}W - \frac{1}{4}\frac{mR^2\omega^2}{L}\sin\beta\cos\beta\cos\omega t \tag{6.66}$$

$$R_{BY} = \frac{1}{2}W + \frac{1}{4}\frac{mR^2\omega^2}{L}\sin\beta\cos\beta\cos\omega t$$

Hence, in addition to the static reactions equal to half the weight, there are dynamic reactions which vary harmonically with a frequency equal to the spin frequency ω. These dynamic reactions tend to wear out the bearings.

6.7 SYSTEMS OF RIGID BODIES

At times, one is faced with the problem not of a single rigid body but of a number of rigid bodies. The bodies are connected at given points through hinges, thus ensuring that the set of rigid bodies acts as a system. This, of course, gives rise to reaction forces at the hinges, so that at a given point connecting two of the bodies there are forces equal in magnitude and opposite in directions acting on the two bodies. One such example was encountered earlier in connection with the rotating unbalanced masses depicted in Fig. 4.23.

Consider a system of N rigid bodies. Then, using Eqs. (6.20) and (6.28), we can write the force and moment equations for each of the bodies in the form

$$\mathbf{F}_i = m_i\mathbf{a}_{Ci}, \qquad i = 1, 2, \ldots, N \tag{6.67a}$$

$$\mathbf{M}_{Ci} = \dot{\mathbf{H}}_{Ci}, \qquad i = 1, 2, \ldots, N \tag{6.67b}$$

where $\mathbf{F}_i$ is the resultant of the force vectors, m_i is the mass, $\mathbf{a}_{Ci}$ is the acceleration of the mass center C_i, $\mathbf{M}_{Ci}$ is the resultant of the moment vectors about the mass center, and $\mathbf{H}_{Ci}$ is the angular momentum vector about the mass center, all quantities pertaining to a typical body $i\,(i = 1, 2, \ldots, N)$. Equations (6.67) represent a system of simultaneous equations of motion. In writing Eqs. (6.67), we must draw a free-body diagram for each of the bodies and be sure to include in the forces the reaction forces between any two bodies, as shown in Figs. 4.23*b* and 4.23*c*. In this regard we observe that, although the reaction forces are internal and they cancel out in pairs if the system of bodies is considered as a whole, when the bodies are considered separately the reaction forces become external and must be treated as any other external force acting on a given body. Of course, the same forces act in opposite directions on the adjacent body. Note that the reaction forces are not known in advance and must be treated as unknowns, together with the motions of the bodies. Quite often they can be eliminated by combining several equations.

In deriving explicit expressions for Eqs. (6.67), we must choose a consistent notation for the motions of the various bodies in the system, as the motions of the individual bodies are not free but constrained kinematically. To arrive at a kinematically consistent set of accelerations for the mass centers of the bodies in the system, we are likely to use equations similar to Eq. (2.57).

In the case of planar motion, the vector Eq. (6.67a) reduces to two scalar equations, and Eq. (6.67b) reduces to a single scalar equation. Moreover, because the angular velocity vectors are all normal to the plane of motion, Eq. (6.67b) can be reduced to

$$M_{Ci} = I_{Ci}\ddot{\theta}_i, \qquad i = 1, 2, \ldots, N \tag{6.68}$$

where I_{Ci} is the mass moment of inertia of the body i about an axis normal to the plane of motion and passing through the mass center and $\ddot{\theta}_i$ is the angular acceleration of the body.

Example 6.5

Derive the equations for the planar motion of the system of two bodies shown in Fig. 6.8*a*. Note that one of the bodies can be treated as a particle and the other as a thin, slender rigid bar.

Figures 6.8*b* and 6.8*c* show the free-body diagrams for the two bodies. Because the body of Fig. 6.8*b* can be treated as a particle, there are only two scalar force equations and no moment equation. On the other hand, for the bar of Fig. 6.8*c* there are two force equations and one moment equation. From Figs. 6.8*b* and 6.8*c*, the equations can be written as follows:

$$F_x - kx = Ma_{Ox} \tag{a}$$

$$f + F_y - Mg = Ma_{Oy} \tag{b}$$

$$F - F_x = ma_{Cx} \tag{c}$$

$$-F_y - mg = ma_{Cy} \tag{d}$$

$$(F + F_x)a\cos\theta + F_y a\sin\theta = I_C\ddot{\theta} \tag{e}$$

where $I_C = m(2a)^2/12 = ma^2/3$ is the mass moment of inertia of the bar about the mass center C. Equations (a) and (c) on the one hand and Eqs. (c), (d), and (e) on the

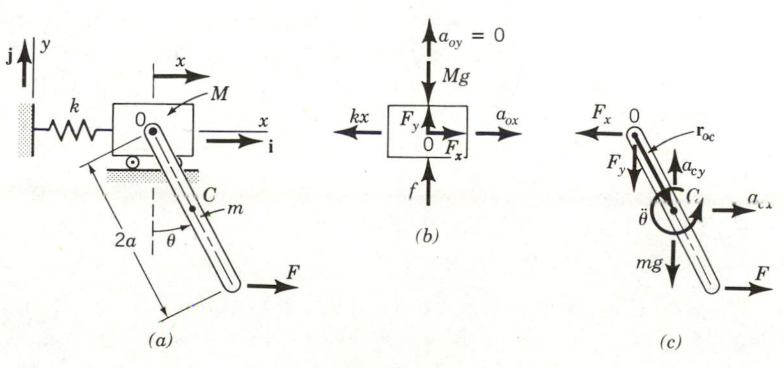

FIGURE 6.8

other hand can be combined into

$$
\begin{aligned}
&M a_{Ox} + m a_{Cx} + kx = F \\
&I_C \ddot{\theta} + ma(a_{Cx} \cos\theta + a_{Cy} \sin\theta) + mga \sin\theta = 2Fa \cos\theta
\end{aligned}
\tag{f}
$$

which are entirely free of F_x and F_y.

The system possesses only two degrees of freedom, and the generalized coordinates x and θ determine the motion of the system fully. Indeed, the accelerations a_{Cx} and a_{Cy} can all be described in terms of these coordinates. To this end, we observe that

$$\mathbf{a}_O = a_{Ox}\mathbf{i} + a_{Oy}\mathbf{j} = \ddot{x}\mathbf{i} \tag{g}$$

so that

$$a_{Ox} = \ddot{x}, \qquad a_{Oy} = 0 \tag{h}$$

Moreover, Eq. (2.57) can be written in the form

$$\mathbf{a}_C = \mathbf{a}_{Cx}\mathbf{i} + \mathbf{a}_{Cy}\mathbf{j} = \mathbf{a}_O + \boldsymbol{\omega} \times (\boldsymbol{\omega} \times \mathbf{r}_{OC}) + \dot{\boldsymbol{\omega}} \times \mathbf{r}_{OC} \tag{i}$$

where

$$\mathbf{r}_{OC} = a(\sin\theta\,\mathbf{i} - \cos\theta\,\mathbf{j}), \qquad \boldsymbol{\omega} = \dot{\theta}\mathbf{k}, \qquad \dot{\boldsymbol{\omega}} = \ddot{\theta}\mathbf{k} \tag{j}$$

in which $\mathbf{k}$ is a unit vector normal to the plane of motion. Inserting Eqs. (g), (h), and (j) into Eq. (i), we obtain simply

$$\mathbf{a}_C = [\ddot{x} + a(\ddot{\theta}\cos\theta - \dot{\theta}^2 \sin\theta)]\mathbf{i} + a(\ddot{\theta}\sin\theta + \dot{\theta}^2\cos\theta)\mathbf{j} \tag{k}$$

so that

$$a_{Cx} = \ddot{x} + a(\ddot{\theta}\cos\theta - \dot{\theta}^2\sin\theta), \qquad a_{Cy} = a(\ddot{\theta}\sin\theta + \dot{\theta}^2\cos\theta) \tag{l}$$

Introducing Eqs. (h) and (l) into Eqs. (f), we obtain the equations of motion

$$
\begin{aligned}
&(M+m)\dot{x} + ma(\ddot{\theta}\cos\theta - \dot{\theta}^2\sin\theta) + kx = F \\
&\tfrac{4}{3}ma^2\ddot{\theta} + ma\ddot{x}\cos\theta + mga\sin\theta = 2Fa\cos\theta
\end{aligned}
\tag{m}
$$

which can be identified as two nonlinear ordinary differential equations in x and θ.

Note that for any $x = x(t)$ and $\theta = \theta(t)$, one can use Eqs. (a) and (d) in conjunction with Eqs. (h) and (l) to determine the hinge reactions F_x and F_y. Moreover, introducing the latter into Eq. (b) and recalling that $a_{Oy} = 0$, we can determine the ground reaction f.

6.8 MOTION OF A TORQUE-FREE SYMMETRIC BODY

Let us consider a torque-free symmetric rigid body rotating about the mass center C. Letting $M_1 = M_2 = M_3 = 0$ and $I_2 = I_1$ in Eqs. (6.30), we obtain

$$\begin{aligned} I_1\dot{\omega}_1+(I_3-I_1)\omega_2\omega_3&=0 \\ I_1\dot{\omega}_2+(I_1-I_3)\omega_1\omega_3&=0 \\ I_3\dot{\omega}_3&=0 \end{aligned} \tag{6.69}$$

The third of Eqs. (6.69) can be integrated immediately to yield

$$\omega_3=\text{const} \tag{6.70}$$

Introducing the notation

$$(I_3-I_1)\omega_3/I_1=\Omega=\text{const} \tag{6.71}$$

the first two of Eqs. (6.69) become

$$\begin{aligned} \dot{\omega}_1+\Omega\omega_2&=0 \\ \dot{\omega}_2-\Omega\omega_1&=0 \end{aligned} \tag{6.72}$$

Multiplying the first of Eqs. (6.72) by ω_1 and the second by ω_2 and adding the results, we obtain

$$\omega_1\dot{\omega}_1+\omega_2\dot{\omega}_2=0 \tag{6.73}$$

which, upon integration, yields

$$\omega_1^2+\omega_2^2=\omega_{12}^2=\text{const} \tag{6.74}$$

where ω_{12} is the projection of the angular velocity $\boldsymbol{\omega}$ on the 1, 2-plane. Combining Eqs. (6.70) and (6.74), we conclude that

$$|\boldsymbol{\omega}|=(\omega_1^2+\omega_2^2+\omega_3^2)^{1/2}=\text{const} \tag{6.75}$$

or, the magnitude of the angular velocity vector $\boldsymbol{\omega}$ is constant.

Because $\mathbf{M}_C=\mathbf{0}$, the angular momentum about C is conserved

$$\mathbf{H}_C=\text{const} \tag{6.76}$$

which means that both the magnitude and the direction of $\mathbf{H}_C$ are constant. The projection of $\mathbf{H}_C$ on the 1, 2 plane is

$$H_{12}=(I_1^2\omega_1^2+I_1^2\omega_2^2)^{1/2}=I_1(\omega_1^2+\omega_2^2)^{1/2}=I_1\omega_{12} \tag{6.77}$$

so that the projections of $\boldsymbol{\omega}$ and $\mathbf{H}_C$ on the 1, 2-plane are along the same line. Hence, axes 3, $\mathbf{H}_C$, and $\boldsymbol{\omega}$ are in the same plane, so that the motion can be interpreted geometrically as the rotation of the plane defined by 3, $\mathbf{H}_C$, and $\boldsymbol{\omega}$ about $\mathbf{H}_C$, as shown in Fig. 6.9. Letting α be the angle between 3 and $\mathbf{H}_C$, we have

$$\tan\alpha=\frac{H_{12}}{H_3}=\frac{I_1\omega_{12}}{I_3\omega_3}=\text{const} \tag{6.78}$$

Moreover, letting β be the angle between 3 and $\boldsymbol{\omega}$, we can write

$$\tan\beta=\frac{\omega_{12}}{\omega_3}=\text{const} \tag{6.79}$$

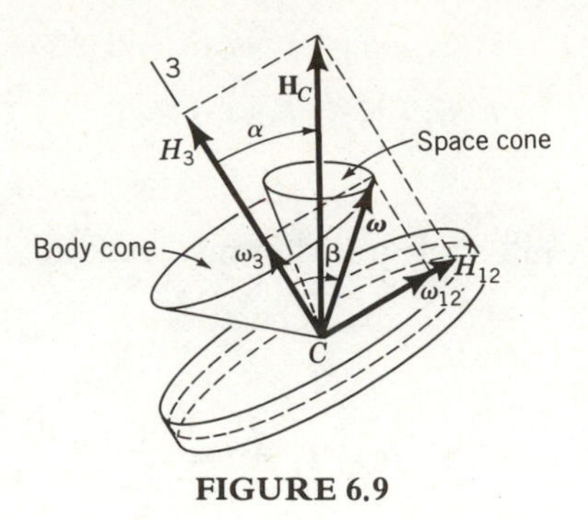

FIGURE 6.9

so that Eq. (6.78) can be rewritten in the form

$$\tan\alpha = \frac{I_1}{I_3}\tan\beta \tag{6.80}$$

There are two cases:

i. $I_3 > I_1$, which implies that $\alpha < \beta$.
ii. $I_3 < I_1$, which implies that $\alpha > \beta$.

The first case represents a flat body, such as a disk, and the second one represents a slender body, such as a rod. Figure 6.9 shows the relative position of the vectors $\boldsymbol{\omega}$ and $\mathbf{H}_C$ for the first case.

Because the angle $\beta - \alpha$ is constant and $\mathbf{H}_C$ is fixed in space, the angular velocity vector $\boldsymbol{\omega}$ rotates about $\mathbf{H}_C$. In the process, it generates a cone fixed in space called the *space cone.* The component ω_{12} of the velocity vector moves relative to axes 1, 2. To study this motion, let us introduce the complex notation

$$\omega_{12} = \omega_1 + i\omega_2 \tag{6.81}$$

Then, multiplying the second of Eqs. (6.72) by i and adding the result to the first, we obtain

$$\dot{\omega}_{12} - i\Omega\omega_{12} = 0 \tag{6.82}$$

which has the solution

$$\omega_{12}(t) = \omega_{12}(0)e^{i\Omega t} \tag{6.83}$$

where $\omega_{12}(0)$ is some initial angular velocity, so that the vector ω_{12} rotates in the plane 1, 2 with the angular velocity Ω. This is equivalent to saying that the vector $\boldsymbol{\omega}$ rotates relative to the body with the same angular velocity Ω. In the process, the vector $\boldsymbol{\omega}$ generates a cone in the body frame known as the *body cone.* Because the vector $\boldsymbol{\omega}$ is a generatrix for both the space cone and the body cone, the two cones are tangent to one another. Hence, the motion of the body can be visualized as the rolling of the body cone on the space cone. For $I_3 > I_1$, $\alpha < \beta$, and the space cone lies inside the body cone. For $I_3 < I_1$, $\alpha > \beta$, and the space cone lies outside the body cone. To an observer in an inertial space, the motion of the body appears as a wobbling, spinning motion, such as in imperfect throwing of a discus or a football. As a crude approximation, the earth can be regarded as an example of a symmetric torque-free body. It is common to assume that the earth is an oblate spheroid with the polar axis coinciding with the symmetry axis for which $(I_3-I_1)/I_1 \cong 0.0033$. Moreover, assuming that $\omega_3 \cong |\boldsymbol{\omega}| = 1$ rotation per day, we obtain from Eq. (6.71)

$$\Omega = (I_3 - I_1)\omega_3/I_1 \cong 0.0033 \quad \text{rotation per day} \tag{6.84}$$

so that the vector $\boldsymbol{\omega}$ completes one sweep around the body cone approximately every 300 days, and, in the process, it traces a circle on the earth around the North Pole. A phenomenon resembling this wobbling motion has actually been observed and is known as the *variation in latitude*. The radius of the circle is about 3.1 m and the period is approximately 433 days instead of 300 days. The discrepancy is attributed to factors unaccounted for, such as the earth not being perfectly rigid.

6.9 ROTATION ABOUT A FIXED POINT. GYROSCOPES

An important class of dynamical systems is characterized by symmetric rigid bodies spinning about the symmetry axis. This is the class of gyroscopic systems, and it includes the rotating earth discussed in Section 6.8 and the spinning top. It also includes a number of devices in which a symmetric rotor spins rapidly relative to the shaft. Such systems are known as *gyroscopes*, and they are used as devices measuring angular motions or as torquing devices. They are essential components in inertial navigation, which is a self-contained system of navigation based only on Newton's laws. In this section we present some basic features of gyroscopic motion.

Quite often, the treatment of gyroscopic systems is facilitated by referring the motion to a set of axes other than the body axes. We recall that the advantage of referring the motion to body axes is that the moments of inertia about these axes are constant. But, in the case of bodies exhibiting axial symmetry, the body axes are not the only axes about which the moments of inertia are constant. Considering an axisymmetric body and letting $\xi\eta\zeta$ be a set of principal axes, with ζ coinciding with the symmetry axis and ξ and η coinciding with any two orthogonal transverse axes, it is clear that the moments of inertia about ξ and η are equal to one another, $I_\xi = I_\eta$. Moreover, the moments of inertia I_ξ, I_η, and I_ζ about axes ξ, η, and ζ, respectively, remain the same even when the body rotates relative to axes $\xi\eta\zeta$, provided the only component of relative rotation is about the symmetry axis, which is assumed to be the case here. Note that, for the moments of inertia about $\xi\eta\zeta$ to remain unchanged, it is not necessary that the body be cylindrical but only that it possess the inertial symmetry defined by $I_\xi = I_\eta$. It is easy to verify that this inertial symmetry is possessed by a parallelepiped with a square cross-sectional area. In this case, if we let axes ξ and η lie in the plane of the square cross section, then $I_\xi = I_\eta$ regardless of the directions of ξ and η. In the following, we wish to produce a set of moment equations of motion in terms of axes $\xi\eta\zeta$.

The derivation of the moment equations of motion in terms of axes $\xi\eta\zeta$ is based on Eq. (6.25). In carrying out this derivation, we distinguish between the angular velocity $\boldsymbol{\omega}$ of axes $\xi\eta\zeta$ and the angular velocity $\boldsymbol{\Omega}$ of the body about axes $\xi\eta\zeta$, where $\boldsymbol{\Omega}$ differs from $\boldsymbol{\omega}$ only in the ζ-component. Moreover, it is important to recognize that the symbol $\boldsymbol{\omega}$ appearing in Eq. (6.25) represents the angular velocity of the reference axes and that the angular velocity entering into the angular momentum $\mathbf{H}_0$ is the angular velocity $\boldsymbol{\Omega}$ of the body. Because the only difference between $\boldsymbol{\omega}$ and $\boldsymbol{\Omega}$ is in the ζ-component, we can write

$$\boldsymbol{\omega} = \omega_\xi \mathbf{i} + \omega_\eta \mathbf{j} + \omega_\zeta \mathbf{k} \tag{6.85a}$$

$$\boldsymbol{\Omega} = \Omega_\xi \mathbf{i} + \Omega_\eta \mathbf{j} + \Omega_\zeta \mathbf{k} = \omega_\xi \mathbf{i} + \omega_\eta \mathbf{j} + \Omega_\zeta \mathbf{k} \tag{6.85b}$$

The angular momentum vector has the general expression

$$\mathbf{H}_0 = H_\xi \mathbf{i} + H_\eta \mathbf{j} + H_\zeta \mathbf{k} \tag{6.86}$$

Similarly, the moment about 0 can be expressed in the form

$$\mathbf{M}_0 = M_\xi \mathbf{i} + M_\eta \mathbf{j} + M_\zeta \mathbf{k} \tag{6.87}$$

Hence, inserting Eqs. (6.85a), (6.86), and (6.87) into Eq. (6.25), we obtain the equations

$$\begin{aligned} M_\xi &= \dot{H}_\xi + \omega_\eta H_\zeta - \omega_\zeta H_\eta \\ M_\eta &= \dot{H}_\eta + \omega_\zeta H_\xi - \omega_\xi H_\zeta \\ M_\zeta &= \dot{H}_\zeta + \omega_\xi H_\eta - \omega_\eta H_\xi \end{aligned} \tag{6.88}$$

which are known as the *modified Euler equations.* But, recalling the symmetry of the rotor, we can write the angular momentum components about axes ξ, η, and ζ

$$H_\xi = I_\xi \Omega_\xi = I_\xi \omega_\xi, \qquad H_\eta = I_\eta \Omega_\eta = I_\xi \omega_\eta, \qquad H_\zeta = I_\zeta \Omega_\zeta \tag{6.89}$$

where I_ξ, $I_\eta = I_\xi$, and I_ζ are the associated principal moments of inertia. It follows that Eqs. (6.88) can be written in the more explicit form

$$\begin{aligned} M_\xi &= I_\xi \dot{\omega}_\xi + \omega_\eta (I_\zeta \Omega_\zeta - I_\xi \omega_\zeta) \\ M_\eta &= I_\xi \dot{\omega}_\eta + \omega_\xi (I_\xi \omega_\zeta - I_\zeta \Omega_\zeta) \\ M_\zeta &= I_\zeta \dot{\Omega}_\zeta \end{aligned} \tag{6.90}$$

Next, we propose to use Eqs. (6.90) to develop the theory explaining the phenomenon of gyroscopic motion. To this end, we consider the mathematical model shown in Fig. 6.10. The model consists of a shaft pivoted at the fixed point and a wheel spinning rapidly relative to the shaft. The wheel, also known as a rotor, is at a distance L from point 0. Before we can derive the equations of motion, it is necessary to choose a set of angles so as to be able to describe the motion of the rotor. A set which enjoys wide acceptance consists of *Euler's angles.* These are

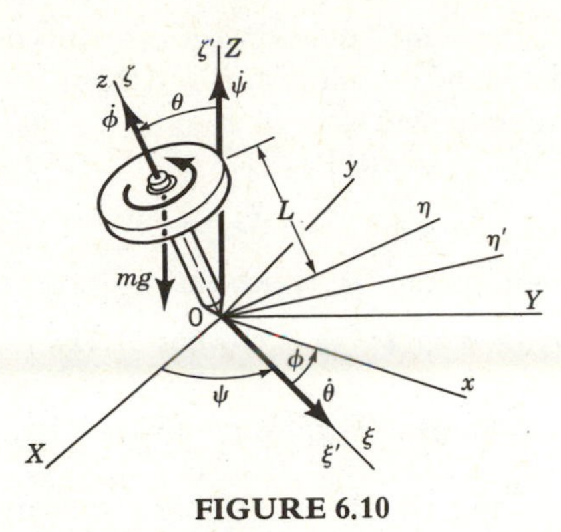

FIGURE 6.10

three successive angular displacements, ψ, θ, and ϕ, defining the angular position of the rotor relative to an inertial space. To demonstrate how Euler's angles provide a description of the body orientation, we envision a body-axes triad, initially coincident with a set of inertial axes XYZ. A rotation through an angle ψ about axis Z brings the triad into coincidence with the set of axes $\xi'\eta'\zeta'$. A further rotation θ about axis ξ' brings the triad to the set of axes $\xi\eta\zeta$, where ξ, sometimes referred to as the *nodal axis*, remains in the horizontal plane at all times. In fact, ξ is the intersection of the horizontal plane defined by axes X and Y and the plane defined by axes ξ and η, where the angle between the two planes is θ. Finally, a rotation ϕ about ζ causes the triad to coincide with the body axes xyz. The sequence of rotations is shown in Fig. 6.10. It is easy to see from Fig. 6.10 that axes $\xi\eta\zeta$ undergo the angular velocities $\dot{\psi}$ about $Z=\zeta'$ and $\dot{\theta}$ about $\xi'=\xi$. Hence, *the angular velocity components of the reference frame $\xi\eta\zeta$ along ξ, η, and ζ* are

$$\omega_\xi=\dot{\theta}, \qquad \omega_\eta=\dot{\psi}\sin\theta, \qquad \omega_\zeta=\dot{\psi}\cos\theta \tag{6.91}$$

But the wheel rotates relative to axes $\xi\eta\zeta$ with the angular velocity $\dot{\phi}$ about $\zeta=z$. Hence, *the angular velocity components of the body along ξ, η, and ζ* are

$$\Omega_\xi=\omega_\xi=\dot{\theta}, \qquad \Omega_\eta=\omega_\eta=\dot{\psi}\sin\theta, \qquad \Omega_\zeta=\dot{\phi}+\omega_\zeta=\dot{\phi}+\dot{\psi}\cos\theta \tag{6.92}$$

Moreover, the components of the moment along ξ, η, and ζ are

$$M_\xi=mgL\sin\theta, \qquad M_\eta=0, \qquad M_\zeta=0 \tag{6.93}$$

Introducing Eqs. (6.91)–(6.93) into Eqs. (6.90), we obtain

$$\begin{gathered}
I_\xi\ddot{\theta}+\dot{\psi}\sin\theta[I_\zeta(\dot{\phi}+\dot{\psi}\cos\theta)-I_\xi\dot{\psi}\cos\theta]=mgL\sin\theta \\
I_\xi\frac{d}{dt}(\dot{\psi}\sin\theta)+\dot{\theta}[I_\xi\dot{\psi}\cos\theta-I_\zeta(\dot{\phi}+\dot{\psi}\cos\theta)]=0 \\
I_\zeta\frac{d}{dt}(\dot{\phi}+\dot{\psi}\cos\theta)=0
\end{gathered} \tag{6.94}$$

which represent the differential equations of motion describing the motion of a gyroscope. Incidentally, the same equations govern the behavior of a spinning top, a child's toy, where L is the distance from 0 to the mass center of the top.

Equations (6.94) can be simplified somewhat by observing that the third equation can be integrated immediately to yield

$$\dot{\phi}+\dot{\psi}\cos\theta=\Omega_\zeta=\text{const} \tag{6.95}$$

so that the first two equations can be reduced to

$$\begin{aligned} &I_\xi\ddot{\theta}+\dot{\psi}\sin\theta(I_\zeta\Omega_\zeta-I_\xi\dot{\psi}\cos\theta)=mgL\sin\theta \\ &I_\xi\frac{d}{dt}(\dot{\psi}\sin\theta)+\dot{\theta}(I_\xi\dot{\psi}\cos\theta-I_\zeta\Omega_\zeta)=0 \end{aligned} \tag{6.96}$$

Equations (6.96) represent two simultaneous differential equations for the angular motions $\theta(t)$ and $\psi(t)$, where the motions are known as *nutation* and *precession*, respectively. The equations are highly nonlinear and admit no closed-form solution. Hence, we wish to explore a special solution known as *steady precession* of the spin axis and defined by

$$\theta=\theta_0=\text{const},\qquad \dot{\psi}=\dot{\psi}_0=\text{const} \tag{6.97}$$

It is easy to verify that the second of Eqs. (6.96) is satisfied identically. On the other hand, the first of Eqs. (6.96) yields the quadratic equation

$$\dot{\psi}_0^2-\frac{I_\zeta\Omega_\zeta}{I_\xi\cos\theta_0}\dot{\psi}_0+\frac{mgL}{I_\xi\cos\theta}=0 \tag{6.98}$$

which has the solutions

$$\left.\begin{matrix}\dot{\psi}_{01}\\ \dot{\psi}_{02}\end{matrix}\right\}=\frac{I_\zeta\Omega_\zeta}{2I_\xi\cos\theta_0}\pm\frac{I_\zeta\Omega_\zeta}{2I_\xi\cos\theta_0}\left(1-\frac{4mgLI_\xi\cos\theta_0}{I_\zeta^2\Omega_\zeta^2}\right)^{1/2} \tag{6.99}$$

where the roots $\dot{\psi}_{01}$ and $\dot{\psi}_{02}$ are called *fast precession* and *slow precession*, respectively. Because the roots must be real, steady precession is possible only if the spin is sufficiently high that the discriminant in Eq. (6.99) is positive. This leads to the condition

$$\Omega_\zeta^2>\frac{4mgLI_\xi\cos\theta_0}{I_\zeta^2} \tag{6.100}$$

For relatively large values of Ω_ζ, we can retain only the first two terms in a binomial expansion of the type $(1-\varepsilon)^{1/2}=1-\frac{1}{2}\varepsilon-\frac{1}{2}\cdot\frac{1}{4}\varepsilon^2-\ldots$, where ε is a small quantity, and write the approximation

$$\left(1-\frac{4mgLI_\xi\cos\theta_0}{I_\zeta^2\Omega_\zeta^2}\right)^{1/2}\cong 1-\frac{2mgLI_\xi\cos\theta_0}{I_\zeta^2\Omega_\zeta^2} \tag{6.101}$$

Hence, inserting Eq. (6.101) into Eq. (6.99), we obtain

$$\dot{\psi}_{01}\cong\frac{I_\zeta\Omega_\zeta}{I_\xi\cos\theta_0},\qquad \dot{\psi}_{02}\cong\frac{mgL}{I_\zeta\Omega_\zeta} \tag{6.102}$$

Because of the high-energy requirement, the fast precession is generally unattainable, so that in referring to steady precession we really mean $\dot{\psi}_{02}=\dot{\psi}_0=mgL/I_\zeta\Omega_\zeta$.

Note that the same value for the steady precession can be obtained from the first of Eqs. (6.96) by assuming that the term $I_\xi \dot{\psi}_0 \cos\theta_0$ is small relative to the term $I_\zeta \Omega_\zeta$.

The steady precession can be demonstrated in a simpler fashion by referring directly to Eq. (6.25). In the case of steady precession, we have

$$\mathbf{H}_0 = I_\zeta \Omega_\zeta \mathbf{k}, \qquad \dot{\mathbf{H}}'_0 = \mathbf{0} \tag{6.103}$$

Moreover, the torque has the vector form

$$\mathbf{M}_0 = mgL \sin\theta \, \mathbf{i} \tag{6.104}$$

Introducing Eqs. (6.103) and (6.104) into Eq. (6.25), we obtain

$$\begin{aligned} \mathbf{M}_0 &= \boldsymbol{\omega} \times \mathbf{H}_0 = (\omega_\xi \mathbf{i} + \omega_\eta \mathbf{j} + \omega_\zeta \mathbf{k}) \times I_\zeta \Omega_\zeta \mathbf{k} \\ &= -\omega_\xi I_\zeta \Omega_\zeta \mathbf{j} + \omega_\eta I_\zeta \Omega_\zeta \mathbf{i} = mgL \sin\theta \, \mathbf{i} \end{aligned} \tag{6.105}$$

from which we conclude that $\omega_\xi = 0$ and, recalling the second of Eqs. (6.91), that

$$\dot{\psi} = \frac{mgL}{I_\zeta \Omega_\zeta} \tag{6.106}$$

where $\Omega_\zeta = \text{const}$ by virtue of the third component of $\dot{\mathbf{H}}'_0 = \mathbf{0}$.

A physical interpretation of the above result should prove most rewarding. Assume first that the shaft is held fixed in the position $\theta = \theta_0$ while the rotor is at rest. Then, if the support is removed, the torque $mgL \sin\theta$ about the ξ-axis will cause the shaft to drop in the $\eta\zeta$-plane, thus rotating with the velocity $\dot{\theta}$ and acceleration $\ddot{\theta}$ about the ξ-axis. On the other hand, if the rotor is spinning with the angular velocity Ω_ζ when the support is removed, where Ω_ζ satisfied inequality (6.100), then the shaft begins to precess uniformly about the Z-axis. This behavior is at odds with intuition, according to which one might expect the shaft to fall. This behavior has prompted the statement that a gyroscope defies gravity, which is not quite correct, since it is the moment due to gravity that causes the precession. We note that the steady precession is proportional to the product of the mass m and the distance L and inversely proportional to the magnitude of the angular momentum. We also note that if the rotor is spinning originally in the opposite sense, then the precession is also in the opposite sense.

6.10 SMALL OSCILLATIONS ABOUT STEADY PRECESSION

Next, let us examine how the uniformly precessing gyroscope behaves if the spin axis is imparted some small initial angular velocities $\omega_\xi(0)$ and $\omega_\eta(0)$, where $\omega_\eta(0)$ is in addition to the steady precession. To this end, we assume that the solution of Eqs. (6.96) can be written in the form

$$\theta(t) = \theta_0 + \theta_1(t), \qquad \dot{\psi}(t) = \dot{\psi}_0 + \dot{\psi}_1(t) \tag{6.107}$$

where θ_0 and $\dot{\psi}_0$ define the steady precession and $\theta_1(t)$ and $\dot{\psi}_1(t)$ are small perturbations from the steady precession. Introducing Eqs. (6.107) into Eqs. (6.96), making the approximation

$$\begin{aligned}\sin\theta &= \sin(\theta_0+\theta_1) \cong \sin\theta_0 + \theta_1\cos\theta_0\\ \cos\theta &= \cos(\theta_0+\theta_1) \cong \cos\theta_0 - \theta_1\sin\theta_0\end{aligned} \tag{6.108}$$

and ignoring nonlinear terms in θ_1 and $\dot{\psi}_1$, we obtain

$$\begin{aligned}&I_\xi\ddot{\theta}_1 + [\dot{\psi}_0(I_\zeta\Omega_\zeta - I_\xi\dot{\psi}_0\cos\theta_0) - mgL]\sin\theta_0\\ &\quad + \dot{\psi}_1(I_\zeta\Omega_\zeta - I_\xi\dot{\psi}_0\cos\theta_0)\sin\theta_0 - \dot{\psi}_0 I_\xi(\dot{\psi}_1\cos\theta_0 - \theta_1\dot{\psi}_0\sin\theta_0)\sin\theta_0\\ &\quad + [\dot{\psi}_0(I_\zeta\Omega_\zeta - I_\xi\dot{\psi}_0\cos\theta_0) - mgL]\theta_1\cos\theta_0 = 0\\ &I_\xi(\ddot{\psi}_1\sin\theta_0 + \dot{\theta}_1\dot{\psi}_0\cos\theta_0) + \dot{\theta}_1(I_\xi\dot{\psi}_0\cos\theta_0 - I_\zeta\Omega_\zeta) = 0\end{aligned} \tag{6.109}$$

The terms in Eqs. (6.109) not containing the perturbations are called *zero-order terms* and those linear in the perturbations are called *first-order terms.* Because these terms are of different orders of magnitude, according to the perturbation theory,* the zero-order terms and the first-order terms must be set equal to zero separately. There is only one equation for the zero-order terms, namely,

$$[\dot{\psi}_0(I_\zeta\Omega_\zeta - I_\xi\dot{\psi}_0\cos\theta_0) - mgL]\sin\theta_0 = 0 \tag{6.110}$$

Equation (6.110) admits two solutions: (1) the trivial one given by $\sin\theta_0$, which implies $\theta_0=0$, and (2) the nontrivial one. In the trivial case, the rotor spins in the upright position, appearing motionless. Note that in the case of a spinning top, $\theta_0=0$ corresponds to the so-called *sleeping top.* In the nontrivial case, Eq. (6.110) yields

$$\dot{\psi}_0(I_\zeta\Omega_\zeta - I_\xi\dot{\psi}_0\cos\theta_0) - mgL = 0 \tag{6.111}$$

which can be recognized as being identical to the equation for the steady precession, Eq. (6.98). Then, considering Eq. (6.111), we can reduce the equations for the first-order terms to

$$\begin{aligned}I_\xi\ddot{\theta}_1 + (\dot{\psi}_0\sin\theta_0)^2\theta_1 + (I_\zeta\Omega_\zeta - 2I_\xi\dot{\psi}_0\cos\theta_0)\sin\theta_0\,\dot{\psi}_1 &= 0\\ -(I_\zeta\Omega_\zeta - 2I_\xi\dot{\psi}_0\cos\theta_0)\dot{\theta}_1 + I_\xi\sin\theta_0\ddot{\psi}_1 &= 0\end{aligned} \tag{6.112}$$

Equations (6.112) are known as *perturbation equations*, and they represent the linearized equations of motion about the steady precession. Introducing the notation

$$\frac{I_\zeta\Omega_\zeta - 2I_\xi\dot{\psi}_0\cos\theta_0}{I_\xi} = \sigma, \qquad \dot{\psi}_0\sin\theta_0 = \omega_{\eta 0} \tag{6.113}$$

we can rewrite Eqs. (6.112) in the simple form

$$\begin{aligned}\ddot{\theta}_1 + \omega_{\eta 0}^2\theta_1 + \sigma\sin\theta_0\,\dot{\psi}_1 &= 0\\ -\sigma\dot{\theta}_1 + \sin\theta_0\,\ddot{\psi}_1 &= 0\end{aligned} \tag{6.114}$$

*See L. Meirovitch, *Methods of Analytical Dynamics*, McGraw-Hill, New York, 1970, Chapter 8.

The solution of Eqs. (6.114) can be obtained conveniently by the Laplace transformation method. Transforming Eqs. (6.114), we obtain

$$
\begin{aligned}
s^2\Theta_1 - s\theta_1(0) - \dot{\theta}_1(0) + \omega_{\eta 0}^2\Theta_1 + \sigma \sin\theta_0[s\Psi_1 - \psi_1(0)] &= 0 \\
-\sigma[s\Theta_1 - \theta_1(0)] + \sin\theta_0[s^2\Psi_1 - s\psi_1(0) - \dot{\psi}_1(0)] &= 0
\end{aligned}
\tag{6.115}
$$

where Θ_1 and Ψ_1 are the Laplace transforms of θ_1 and ψ_1, respectively. Moreover, the initial conditions are

$$
\theta_1(0) = \psi_1(0) = 0, \qquad \dot{\theta}_1(0) = \omega_\xi(0), \qquad \sin\theta_0\,\dot{\psi}_1 = \omega_\eta(0) \tag{6.116}
$$

so that Eqs. (6.115) become

$$
\begin{aligned}
(s^2 + \omega_{\eta 0}^2)\Theta_1 + \sigma \sin\theta_0\, s\Psi_1 &= \omega_\xi(0) \\
-\sigma s\Theta_1 + \sin\theta_0\, s^2\Psi_1 &= \omega_\eta(0)
\end{aligned}
\tag{6.117}
$$

Equations (6.117) have the solution

$$
\begin{aligned}
\Theta_1 &= \frac{\begin{vmatrix} \omega_\xi(0) & s\sigma\sin\theta_0 \\ \omega_\eta(0) & s^2\sin\theta_0 \end{vmatrix}}{\begin{vmatrix} s^2+\omega_{\eta 0}^2 & s\sigma\sin\theta_0 \\ -s\sigma & s^2\sin\theta_0 \end{vmatrix}} = \frac{s[s\omega_\xi(0) - \omega_\eta(0)]\sin\theta_0}{s^2(s^2+\omega_{\eta 0}^2+\sigma^2)\sin\theta_0} = \frac{\omega_\xi(0)}{s^2+\Omega^2} - \frac{\sigma\omega_\eta(0)}{s(s^2+\Omega^2)} \\
\Psi_1 &= \frac{\begin{vmatrix} s^2+\omega_{\eta 0}^2 & \omega_\xi(0) \\ -s\sigma & \omega_\eta(0) \end{vmatrix}}{\begin{vmatrix} s^2+\omega_{\eta 0}^2 & s\sigma\sin\theta_0 \\ -s\sigma & s^2\sin\theta_0 \end{vmatrix}} = \frac{\sigma s\omega_\xi(0) + (s^2+\omega_{\eta 0}^2)\omega_\eta(0)}{s^2(s^2+\omega_{\eta 0}^2+\sigma^2)\sin\theta_0} \\
&= \frac{\sigma\omega_\xi(0)}{s(s^2+\Omega^2)\sin\theta_0} + \left[\frac{1}{s^2} - \frac{\sigma^2}{s^2(s^2+\Omega^2)}\,\frac{\omega_\eta(0)}{\sin\theta_0}\right]
\end{aligned}
\tag{6.118}
$$

where

$$
\Omega^2 = \omega_{\eta 0}^2 + \sigma^2 \tag{6.119}
$$

Hence, using the table of Laplace transforms in Section A.7, we can write the inverse transformations

$$
\begin{aligned}
\theta_1(t) &= \frac{\omega_\xi(0)}{\Omega}\sin\Omega t - \frac{\sigma\omega_\eta(0)}{\Omega^2}(1 - \cos\Omega t) \\
\psi_1(t) &= \frac{\sigma\omega_\xi(0)}{\Omega^2\sin\theta_0}(1 - \cos\Omega t) + \frac{\omega_\eta(0)}{\sin\theta_0}\left[t - \frac{\sigma^2}{\Omega^2}(\Omega t - \sin\Omega t)\right]
\end{aligned}
\tag{6.120}
$$

But, in general the spin velocity is much larger than the precessional velocity, so that

$$
\Omega \cong \sigma \cong \frac{I_\zeta\Omega_\zeta}{I_\xi} \tag{6.121}
$$

in which case Eqs. (6.120) reduce to

$$\theta_1(t)=\frac{\omega_\xi(0)}{\Omega}\sin\Omega t-\frac{\omega_\eta(0)}{\Omega}(1-\cos\Omega t)$$
$$\psi_1(t)=\frac{\omega_\xi(0)}{\Omega\sin\theta_0}(1-\cos\Omega t)+\frac{\omega_\eta(0)}{\Omega\sin\theta_0}\sin\Omega t \tag{6.122}$$

Finally, considering Eqs. (6.107), we obtain

$$\theta(t)=\theta_0+\frac{\omega_\xi(0)}{\Omega}\sin\Omega t-\frac{\omega_\eta(0)}{\Omega}(1-\cos\Omega t)$$
$$\psi(t)=\dot{\psi}_0 t+\frac{\omega_\xi(0)}{\Omega\sin\theta_0}(1-\cos\Omega t)+\frac{\omega_{\xi\eta}(0)}{\Omega\sin\theta_0}\sin\Omega t \tag{6.123}$$

where $\dot{\psi}_0$ is the steady precession given by Eq. (6.106).

To interpret the motion geometrically, let us introduce the notation

$$\omega_\xi(0)=\omega_{\xi\eta}(0)\cos\gamma,\qquad \omega_\eta(0)=\omega_{\xi\eta}(0)\sin\gamma \tag{6.124}$$

Then, Eqs. (6.123) can be rewritten as

$$\theta(t)=\theta_0-\frac{\omega_\eta(0)}{\Omega}+\frac{\omega_{\xi\eta}(0)}{\Omega}\sin(\Omega t+\gamma)$$
$$\psi(t)=\dot{\psi}_0 t+\frac{\omega_\xi(0)}{\Omega\sin\theta_0}-\frac{\omega_{\xi\eta}(0)}{\Omega}\cos(\Omega t+\gamma) \tag{6.125}$$

It is convenient to envision the spin axis as tracing a path on an imaginary unit sphere centered at 0, so that the angles $\theta(t)$ and $\psi(t)$ can be regarded as being equivalent to arcs on the unit sphere. Ignoring the steady precession for the moment, we conclude from Eqs. (6.125) that the spin axis traces a circle centered at $\theta=\theta_0-\omega_\eta(0)/\Omega$, $\psi=\omega_\xi(0)/\Omega\sin\theta_0$ and of radius $\omega_{\xi\eta}(0)/\Omega$, as shown in Fig. 6.11, where the circular velocity is Ω. The steady precession causes the circle to advance horizontally at the constant rate $\dot{\psi}_0$. Of course, because of this precession, the

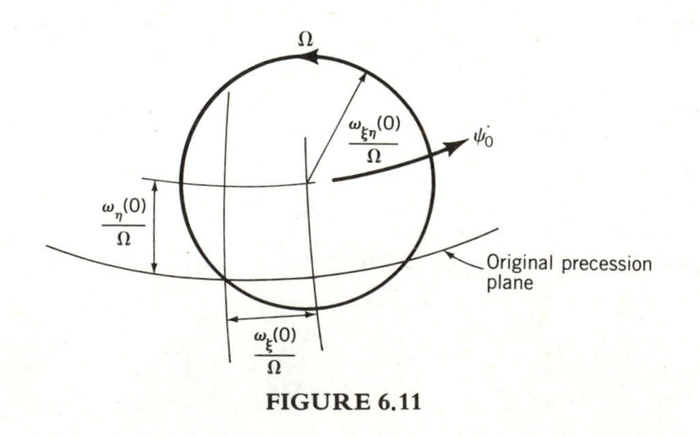

FIGURE 6.11

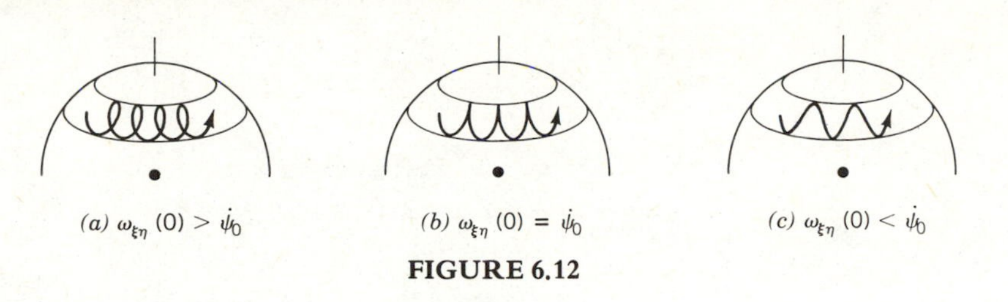

FIGURE 6.12

circle does not close, and the spin axis follows one of the trajectories shown in Fig. 6.12, where the type of trajectory depends on the ratio $\omega_{\xi\eta}(0)/\dot{\psi}_0$, as indicated in the figure.

PROBLEMS

6.1 A truck carries a board leaning against the cab at an angle β, as shown in Fig. 6.13. If the coefficient of friction between the truck floor and the board is $\mu = 0.3$, determine the safe acceleration of the truck to prevent the board from sliding and the angle β to prevent the board from tipping.

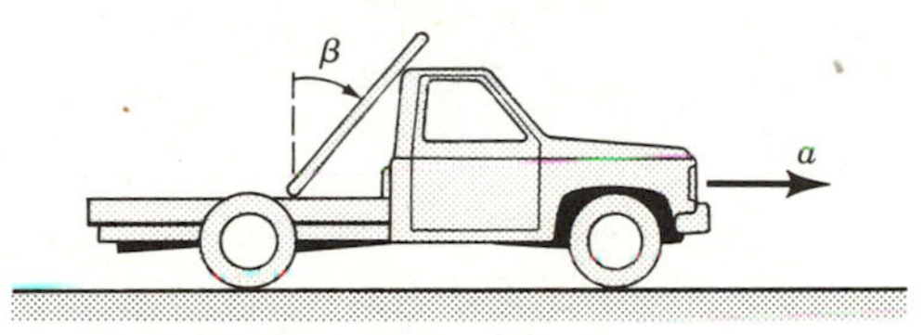

FIGURE 6.13

6.2 Upon applying the brakes, the automobile shown in Fig. 6.14 was observed to decelerate uniformly at 6 m/s^2. Assume that all four wheels are locked, and determine the reactions on the wheels, as well as the coefficient of friction between the wheels and the road.

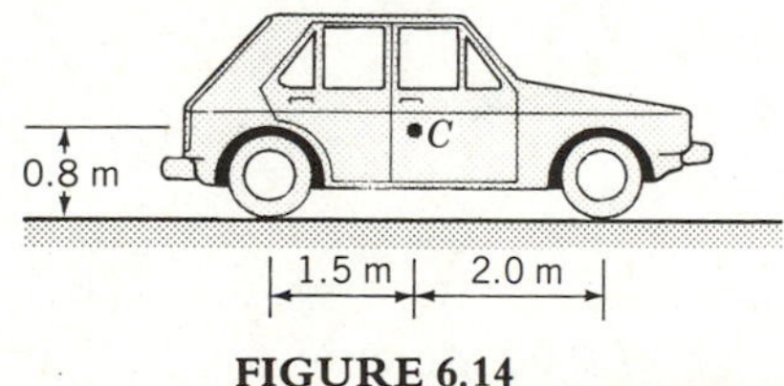

FIGURE 6.14

6.3 A monorail system consists of a car suspended on a horizontal cable (Fig. 6.15). The car is driven by one of the wheels. If the coefficient of friction between the driving wheel and the cable is $\mu = 0.5$, determine the following: (1) which wheel should be the driving wheel and (2) the maximum acceleration of the car.

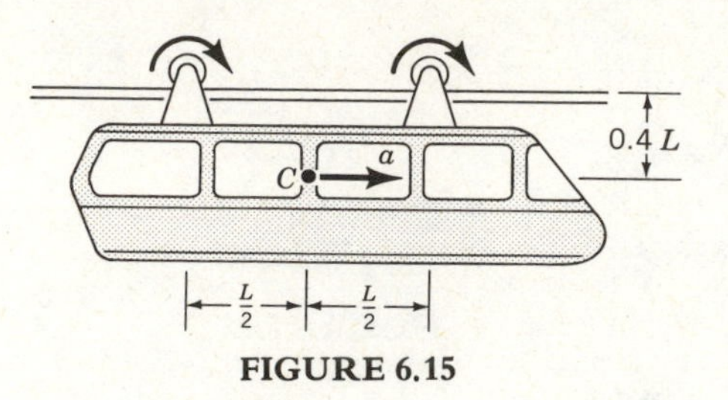

FIGURE 6.15

6.4 A uniform bar hinged at point 0 is at rest in the position shown (Fig. 6.16). A force F strikes horizontally at a point P a distance h from 0. Determine h so that the horizontal reaction at 0 is zero. Note that such a point is called the center of percussion.

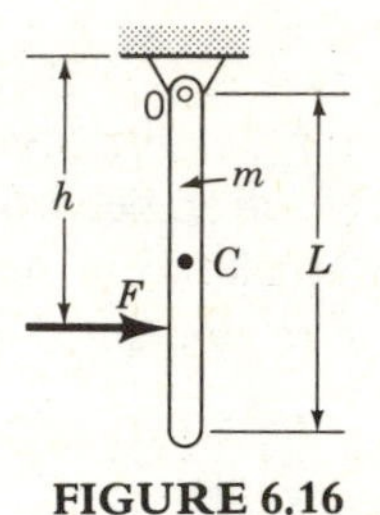

FIGURE 6.16

6.5 A uniform rectangular door hangs at an angle α with respect to the vertical (Fig. 6.17). If disturbed from equilibrium, the door will oscillate. Derive the equation of motion and then assume small motions and calculate the natural frequency of oscillation.

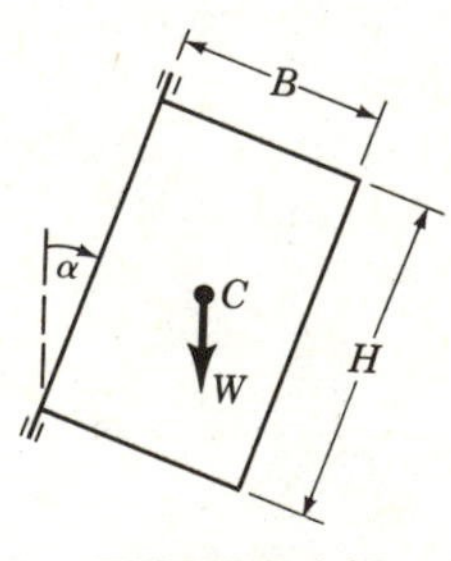

FIGURE 6.17

6.6 A pendulum consists of a massless rigid rod of length L and a thin disk of mass m and radius R (Fig. 6.18). Assume that the pendulum is released from rest in the position $\theta = \pi/2$ and calculate the reactions at 0 when the pendulum reaches an arbitrary angle θ. Derive the differential equation of motion, linearize the equation by assuming small oscillations about $\theta = 0$, and calculate the natural frequency.

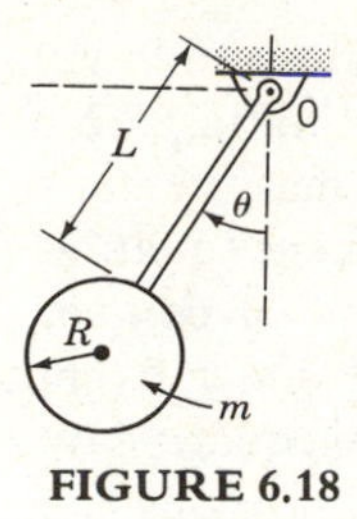

FIGURE 6.18

6.7 A uniform bar of mass m and length $\sqrt{2}R$ slides inside a smooth cylindrical surface (Fig. 6.19). Derive the equation of motion when the bar is in the position shown, and determine the forces exerted by the surface on the bar.

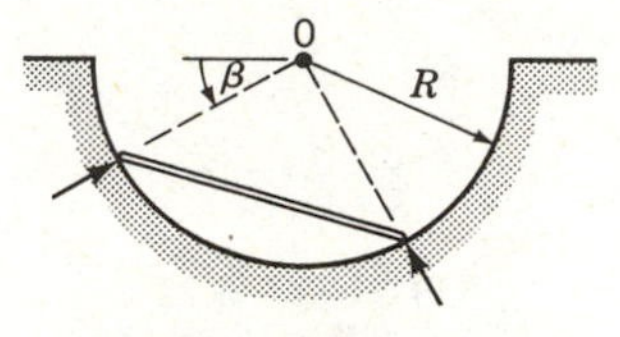

FIGURE 6.19

6.8 A thin disk of mass m and radius R is at rest at the end of a ledge (Fig. 6.20) when nudged slightly to the left. Assume that the friction is such that the disk rotates initially about 0 and then leaves the ledge. Find the angle θ, the angular velocity ω, and the angular acceleration α when the disk leaves the ledge.

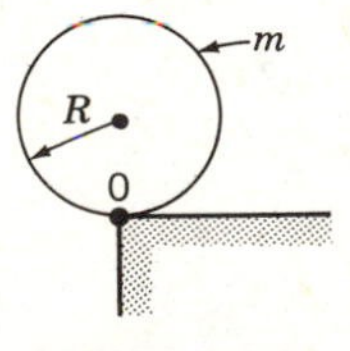

FIGURE 6.20

6.9 A thin uniform bar of mass m and length L is suspended by a string at one end, as shown in Fig. 6.21, when the string is cut. Initially the bar rotates about 0, and then it begins to slip. If the coefficient of friction between the corner of the ledge and the bar is μ, determine the angle between the bar and the horizontal when slip begins.

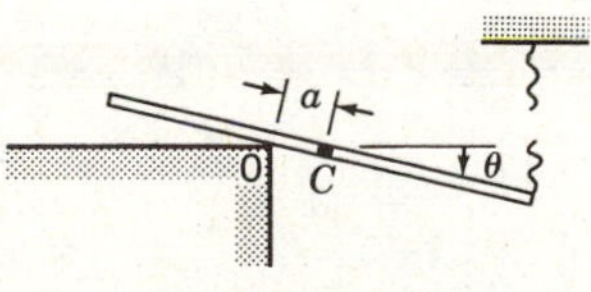

FIGURE 6.21

6.10 A thin disk of mass m and radius r rolls without slip inside a rough cylindrical surface of radius R, as shown in Fig. 6.22. Derive the differential equation for the angular motion θ of the mass center C in two ways: (1) by writing two force equations and one moment equation about C and then eliminating the constraint forces at the point of contact 0 and (2) by writing a moment equation about the point of contact. Compare results and draw conclusions. Then, linearize the differential equation by assuming small oscillations about $\theta=0$, and calculate the natural frequency.

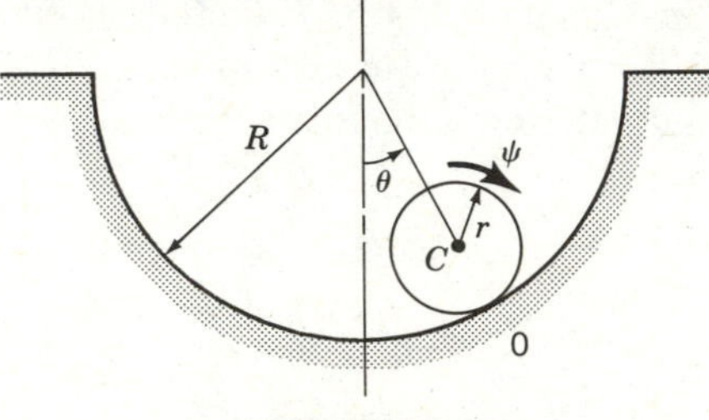

FIGURE 6.22

6.11 The uniform thin bar shown in Fig. 6.23 is initially at rest in the vertical position on a rough horizontal surface when it begins to fall under gravity. When the angle between the bar and the surface reaches 60°, the bar begins to slip. Determine the coefficient of friction between the bar and the surface.

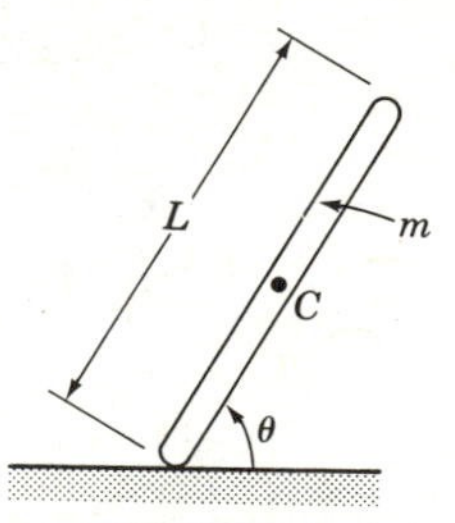

FIGURE 6.23

6.12 A dumbbell is attached to a shaft of total length L, as shown in Fig. 6.24.

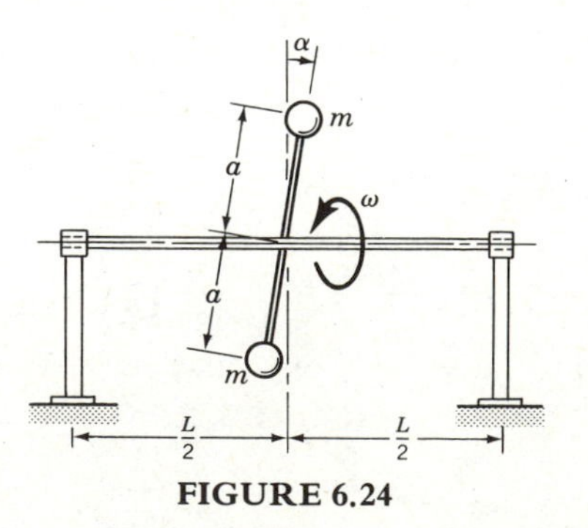

FIGURE 6.24

Determine the bearing reactions if the system rotates with the uniform angular velocity ω about an horizontal axis passing through the bearings.

6.13 A cylinder of total mass m and radius R is attached to a shaft of total length L, as shown in Fig. 6.25. Determine the bearing reactions if the system rotates with the angular velocity $\omega(t)$ about an horizontal axis passing through the bearings.

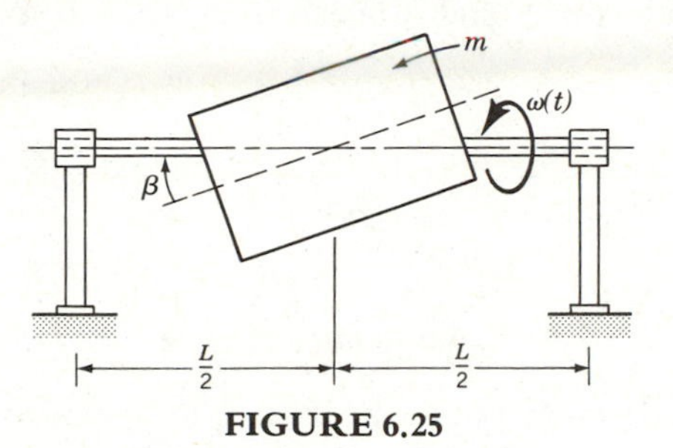

FIGURE 6.25

6.14 Figure 6.26 shows a simplified model of an airfoil section regarded as a rigid body of mass m and mass moment of inertia I_E about point E (representing the elastic axis of the aircraft wing). Point C denotes the center of mass of the section, and the distance between E and C is denoted by s. The elasticity of the structure is simulated by a spring k_x restraining the vertical motion x (known as plunge) and by a torsional spring k_θ restraining the rotational motion θ (known as pitch). Derive the equations of motion under the assumption that the angle θ is small.

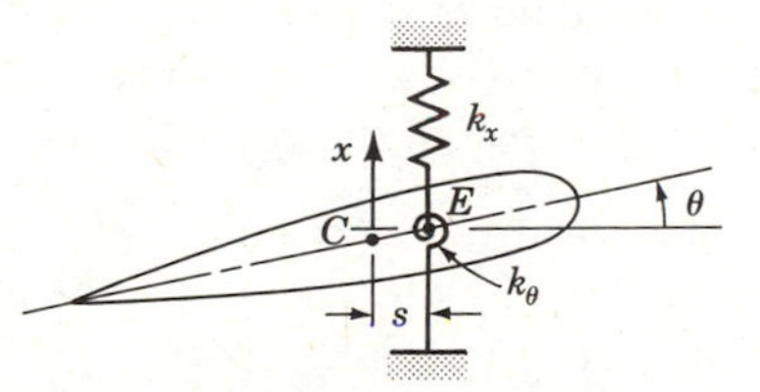

FIGURE 6.26

6.15 The system depicted in Fig. 6.27 consists of two uniform rigid links of masses m_1 and m_2 and lengths L_1 and L_2. Use Eqs. (6.67) and derive the system

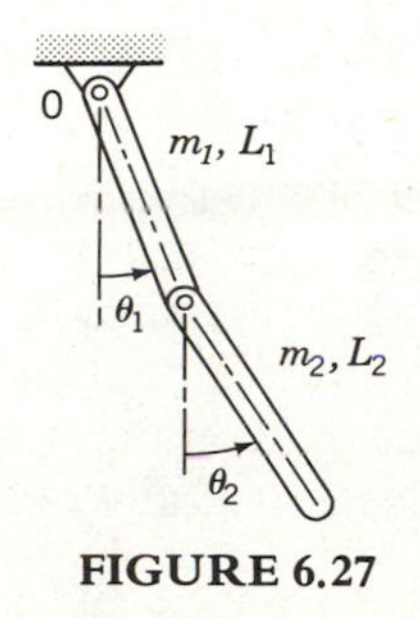

FIGURE 6.27

equations of motion. Then, eliminate the constraint forces, and reduce the equations to two equations in terms of θ_1 and θ_2.

6.16 The system shown in Fig. 6.28 consists of two identical uniform rigid links of mass m and length L, a mass M, and two springs k_1 and k_2. The pulley at point B is massless. Use Eqs. (6.67) and derive the system equations of motion. Then, eliminate the constraint forces at A, and reduce the equations to two equations in terms of x and θ. Assume that $x=\theta=0$ when the springs are unstretched.

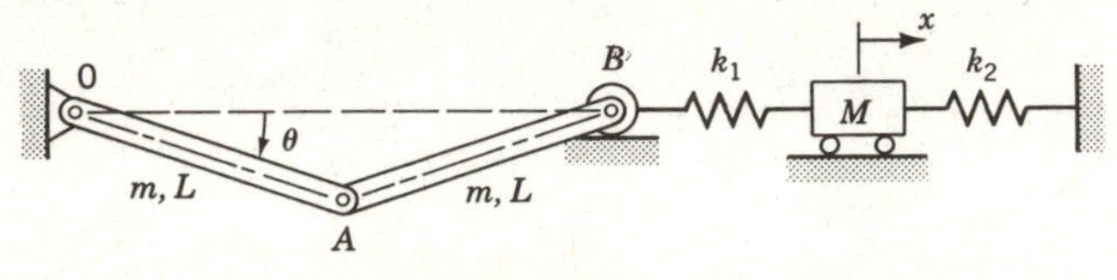

FIGURE 6.28

6.17 A symmetric communications satellite is designed to spin about the symmetry axis at 5π rad/s. The satellite has the form of a drum of radius 2 m, and its moments of inertia about the symmetry axis and a transverse axis are 2000 kg·m^2 and 1600 kg·m^2, respectively. During deployment, the satellite is hit accidentally by an impulsive force of magnitude 300 N·s, where the force is parallel to the symmetry axis and at a distance of 0.32 m from the rim. Determine the subsequent rotational motion of the satellite, and give it a geometric interpretation. *Hints:* (1) The impulsive force imparts an initial angular velocity to the satellite, and (2) at the termination of the impulse, the satellite is torque-free once again.

6.18 A motorcycle is rounding a curve of radius R with the velocity v. Calculate the torque necessary to overcome the gyroscopic effect of the wheels, thus permitting the motorcycle to round the curve. How is this torque produced?

6.19 A single-engine turboprop aircraft has a propeller of moment of inertia I_p rotating clockwise (as seen by the pilot) with the angular velocity ω_p and an engine rotor of moment of inertia I_e rotating counterclockwise with the angular velocity ω_e. The pilot has the elevator and rudder set to make a right turn of radius R at a speed v. If the pilot did not take into consideration the gyroscopic effect in attempting to make the turn, determine whether the nose of the aircraft tends to rise or fall.

6.20 Projectiles are stabilized in flight by means of a system of spiral grooves cut in the gun barrel in a process known as rifling. As a result, the projectile leaves the gun barrel with a given spin velocity Ω_ζ. If the spin axis deviates from the velocity vector **v** by a given angle α, then the drag force **D** acting at the center of pressure P, which lies ahead of the mass center C at a distance h, will cause the projectile to wobble about the vector **v** at the constant angular rate ω (Fig. 6.29). Determine the minimum spin Ω_ζ necessary to prevent the projectile from tumbling. Neglect gravity effects and assume that the pro-

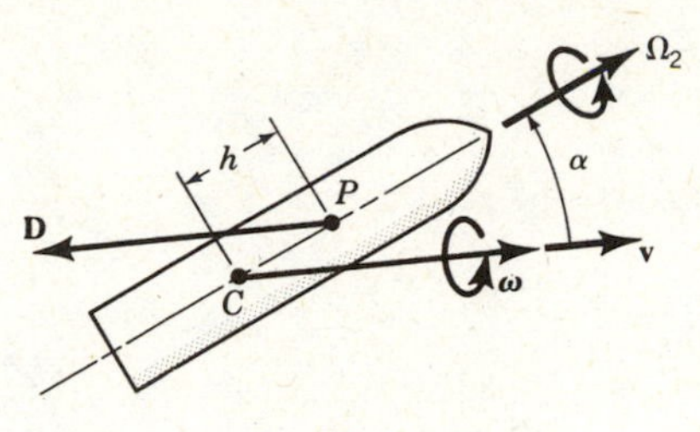

FIGURE 6.29

jectile is a rigid cylinder with centroidal mass moments of inertia I_S and I_T about the symmetry and a transverse axes, respectively.

6.21 The shaft of the gyroscope shown in Fig. 6.10 is precessing steadily when struck by the impulsive force $\mathbf{F}(t)=\hat{F}\,\delta(t)\mathbf{i}$ at a point a distance $0.8L$ from the pivot 0. Determine the subsequent motion.

CHAPTER 7
Elements of Analytical Dynamics

7.1 INTRODUCTION

Newton's laws were formulated for a single particle and can be extended to systems of particles and rigid bodies, as shown in Chapters 5 and 6. A basic tool in the application of Newton's second law is the free-body diagram, which is a diagram of a given mass and all the forces acting upon it. For a system consisting of several bodies, we must use one free-body diagram for each of the bodies. In drawing these free-body diagrams, interacting forces between the bodies, which result from kinematical constraints and are internal to the system, must be regarded as external to the individual bodies and must be included in the free-body diagrams. Quite often, the object is to determine the motion of a system only, and the interacting forces present no particular interest. In describing the motion of a system, Newtonian mechanics uses physical coordinates and forces, which are generally vector quantities. For this reason, Newtonian mechanics is often referred to as *vectorial mechanics*. The inclusion of interacting forces between bodies in free-body diagrams and the use of physical coordinates and forces, both vector quantities, can be regarded as two disadvantages of Newtonian mechanics when compared with other approaches.

D'Alembert's principle is a variational principle permitting the derivation of the equations of motion without considering explicitly the interacting forces. It is still basically a vectorial approach, so that it goes only part way toward removing the objections to Newtonian mechanics.

Another variational approach to mechanics is known as Lagrangian mechanics, or analytical mechanics, and it removes both objections to Newtonian mechanics. Lagrangian mechanics permits the derivation of the equations of motion from two scalar functions, the kinetic energy and the potential energy, as well as from a differential expression known as virtual work in the case of nonconservative systems. The equations of motion can be derived without the need of free-body diagrams. The kinetic energy and the potential energy can be expressed in terms of so-called generalized displacements and generalized velocities, and the virtual work can be expressed in terms of generalized virtual displacements and generalized forces, all scalar quantities not necessarily representing physical quantities. The basic tool for deriving the equations of motion consists of a set of general differential equations known as Lagrange's equations. Explicit equations of motions for a given system can be produced as soon as the kinetic energy, the potential energy, and the virtual work are given for the system. The advantage of Lagrange's equations becomes more and more evident as the number of degrees of freedom of the system increases.

In this chapter we develop the framework of the variational approach to mechanics, including the principle of virtual work, d'Alembert's principle, and Lagrange's equations.

7.2 GENERALIZED COORDINATES

In Chapter 5, we considered the dynamics of a system of N particles, each particle having the mass m_i. The position of a typical particle is given by the radius vector $\mathbf{r}_i(t)(i=1, 2, \ldots, N)$, where $\mathbf{r}_i$ can be written in terms of the cartesian components x_i, y_i, z_i, as follows:

$$\mathbf{r}_i = x_i\mathbf{i} + y_i\mathbf{j} + z_i\mathbf{k}, \qquad i=1, 2, \ldots, N \tag{7.1}$$

The motion of the system is defined completely if the rectangular coordinates of all the particles are known functions of time,

$$x_i = x_i(t), \qquad y_i = y_i(t), \qquad z_i = z_i(t), \qquad i=1, 2, \ldots, N \tag{7.2}$$

Quite often it is more convenient to express the motion not in terms of the rectangular coordinates x_i, y_i, z_i but in terms of a different set of coordinates, say, $q_1, q_2, \ldots, q_n$, where $n=3N$. The relation between the rectangular coordinates x_i, y_i, z_i $(i=1, 2, \ldots, N)$ and the new coordinates q_k $(k=1, 2, \ldots, n)$ can be written in the general form

$$\begin{aligned}
x_1 &= x_1(q_1, q_2, \ldots, q_n) \\
y_1 &= y_1(q_1, q_2, \ldots, q_n) \\
z_1 &= z_1(q_1, q_2, \ldots, q_n) \\
x_2 &= x_2(q_1, q_2, \ldots, q_n) \\
&\vdots \\
z_N &= z_N(q_1, q_2, \ldots, q_n)
\end{aligned} \tag{7.3}$$

Equations (7.3) represent a *coordinate transformation* whose purpose is to facilitate the treatment of dynamical problems. For example, in Section 3.7 we found it convenient to describe the planar motion of a particle in terms of the polar coordinates r and θ, related to the cartesian coordinates x and y by

$$x = r\cos\theta, \qquad y = r\sin\theta \tag{7.4}$$

Hence, letting $r=q_1$, $\theta=q_2$, we can write

$$x = x(q_1, q_2) = q_1\cos q_2, \qquad y = y(q_1, q_2) = q_1\sin q_2 \tag{7.5}$$

Clearly, Eqs. (7.5) represent a special case of the general coordinate transformation (7.3)

Equations (7.3) represent a transformation between two sets of $n=3N$ coordinates. Implied in this is the assumption that the N particles are free to move unrestricted in a three-dimensional space. In many physical systems, however, the particles are not free but subject to constraints restricting their freedom of motion. The particle in planar motion mentioned above is an example of constrained motion. Indeed, because the particle is confined to planar motion, Eqs. (7.5) should be completed by writing

$$x=q_1 \cos q_2, \qquad y=q_1 \sin q_2, \qquad z=0 \tag{7.6}$$

where $z=0$ is to be interpreted as a *constraint equation.* Another simple example of a constrained system is a dumbbell, consisting of two mass particles connected by a rigid rod. Denoting the cartesian coordinates of the two particles by x_1, y_1, z_1 and x_2, y_2, z_2 and the length of the rod by L, the fact that the length of the rod does not change can be expressed in the form

$$(x_2-x_1)^2+(y_2-y_1)^2+(z_2-z_1)^2=L^2=\text{const} \tag{7.7}$$

Equation (7.7) represents a constraint equation implying that not all six coordinates $x_1, y_1, z_1, x_2, y_2, z_2$ are independent. Indeed, the motion of the dumbbell is fully determined by only five of these coordinates, as Eq. (7.7) can be used to determine the sixth.

In general, if a system of N particles moving in a three-dimensional space is subject to c kinematical constraints, such as that expressed by Eq. (7.7), the motion of the system can be described completely by n coordinates $q_1, q_2, \ldots, q_n$, where

$$n=3N-c \tag{7.8}$$

is the *number of degrees of freedom* of the system. Hence, the number of degrees of freedom of a system can be defined as the minimum number of coordinates required to describe the motion of a system completely. The n coordinates $q_1, q_2, \ldots, q_n$ describing the motion of the system are known as *generalized coordinates.*

The generalized coordinates $q_1, q_2, \ldots, q_n$ may not always have physical meaning, nor are they necessarily unique. For example, any set of five of the six cartesian coordinates used to describe the motion of the dumbbell discussed above can serve as generalized coordinates, although they are not the most convenient ones. In all likelihood, the most convenient set of generalized coordinates consists of three coordinates describing the translation of the mass center of the dumbbell relative to an inertial space and two angular coordinates describing the orientation of the rigid rod relative to the same inertial space.

7.3 THE PRINCIPLE OF VIRTUAL WORK

The principle of virtual work is a first variational principle of mechanics, and it represents a statement of the static equilibrium of a mechanical system. Although we shall use this principle to illustrate how to calculate the position of static equilibrium of a system, our real interest lies in its ability to facilitate the transition from Newtonian to Lagrangian mechanics. Before the principle can be discussed, it is necessary to introduce a new class of displacements, known as *virtual displacements.*

Let us consider once again the system of N particles of Section 7.2 and define the virtual displacements $\delta x_1, \delta y_1, \delta z_1, \ldots, \delta z_N$ as infinitesimal changes in the coordinates $x_1, y_1, z_1, x_2, \ldots, z_N$. The virtual displacements must be consistent with the constraints of the system, but they are otherwise arbitrary. They are not true displacements but small variations in the system coordinates resulting from imagining the system in a slightly displaced position, a process taking place without any change in time, so that the forces and constraints do not change during this process. This is in direct contrast with actual displacements, which require a certain amount of time to evolve, during which time the forces and constraints may change. The virtual displacements obey the rules of differential calculus and in the event the system is subject to a constraint of the form

$$f(x_1, y_1, z_1, x_2, \ldots, z_N, t) = C \tag{7.9}$$

they must be such that the equation

$$f(x_1+\delta x_1, y_1+\delta y_1, z_1+\delta z_1, x_2+\delta x_2, \ldots, z_N+\delta z_N, t) = C \tag{7.10}$$

is also satisfied, and note that the time t has not been varied in Eq. (7.10). Expanding Eq. (7.10) in a Taylor series about the position $x_1, y_1, z_1, x_2, \ldots, z_N$, we obtain

$$f(x_1, y_1, z_1, x_2, \ldots, z_N, t) + \sum_{i=1}^{N}\left(\frac{\partial f}{\partial x_i}\delta x_i + \frac{\partial f}{\partial y_i}\delta y_i + \frac{\partial f}{\partial z_i}\delta z_i\right) + 0(\delta^2) = C \tag{7.11}$$

where $0(\delta^2)$ denotes nonlinear terms in the virtual displacements. Considering Eq. (7.10) and ignoring the nonlinear terms as being of higher order in magnitude, we conclude that the virtual displacements must satisfy the relation

$$\sum_{i=1}^{N}\left(\frac{\partial f}{\partial x_i}\delta x_i + \frac{\partial f}{\partial y_i}\delta y_i + \frac{\partial f}{\partial z_i}\delta z_i\right) = 0 \tag{7.12}$$

so that only $3N-1$ of the virtual displacements are arbitrary. As an example, in the case of the dumbbell, which is subject to the constraint equation (7.7), the virtual displacements must satisfy

$$(x_2-x_1)(\delta x_2-\delta x_1)+(y_2-y_1)(\delta y_2-\delta y_1)+(z_2-z_1)(\delta z_2-\delta z_1)=0 \tag{7.13}$$

so that, if five of the virtual displacements $\delta x_1, \delta y_1, \delta z_1, \delta x_2, \delta y_2, \delta z_2$ are given, the sixth can be determined by means of Eq. (7.13). In general, for a system of N particles subject to c constraints, i.e., for a system possessing $n=3N-c$ degrees of freedom, only n virtual displacements are arbitrary, so that there are as many arbitrary virtual displacements as the number of independent coordinates.

Next, let us assume that each of the particles in the system is acted upon by a set of forces with resultant $\mathbf{R}_i$ $(i=1, 2, \ldots, N)$. For a system in equilibrium the resultant force is zero, $\mathbf{R}_i=\mathbf{0}$, so that

$$\overline{\delta W}_i = \mathbf{R}_i \cdot \delta \mathbf{r}_i = 0, \qquad i=1, 2, \ldots, N \tag{7.14}$$

where $\overline{\delta W_i}$ is the *virtual work* performed by the resultant force on the ith particle over the virtual displacement vector $\delta \mathbf{r}_i$. Note that the overbar in $\overline{\delta W_i}$ indicates that $\overline{\delta W_i}$ is in general a mere infinitesimal expression and not the variation of a function W_i, because a work function W_i exists only when the system is conservative. Summing over the system, we have simply

$$\overline{\delta W} = \sum_{i=1}^{N} \overline{\delta W_i} = \sum_{i=1}^{N} \mathbf{R}_i \cdot \delta \mathbf{r}_i = 0 \tag{7.15}$$

where $\overline{\delta W}$ is the virtual work for the entire system.

The above result appears quite trivial, as nothing new has been gained. When the system is subject to constraints, however, Eq. (7.15) leads to some interesting conclusions. Hence, let us assume that the resultant forces $\mathbf{R}_i$ consist of *impressed*, or *applied*, *forces* $\mathbf{F}_i$ and *constraint forces* $\mathbf{f}_i$, or

$$\mathbf{R}_i = \mathbf{F}_i + \mathbf{f}_i, \qquad i = 1, 2, \ldots, N \tag{7.16}$$

Introducing Eqs. (7.16) into Eq. (7.15), we can write

$$\overline{\delta W} = \sum_{i=1}^{N} \mathbf{F}_i \cdot \delta \mathbf{r}_i + \sum_{i=1}^{N} \mathbf{f}_i \cdot \delta \mathbf{r}_i = 0 \tag{7.17}$$

We shall confine ourselves to systems for which the virtual work performed by the constraint forces is zero, or

$$\sum_{i=1}^{N} \mathbf{f}_i \cdot \delta \mathbf{r}_i = 0 \tag{7.18}$$

As an example, consider a particle on a smooth surface. The constraint force in this case is normal to the surface and any virtual displacement consistent with this constraint must be parallel to the surface. Clearly, the constraint force vector and the virtual displacement vector are normal to one another, so that their scalar product is zero. It follows that the work performed by the constraint force over a virtual displacement consistent with the constraint is zero. Hence, Eq. (7.17) reduces to

$$\sum_{i=1}^{N} \mathbf{F}_i \cdot \delta \mathbf{r}_i = 0 \tag{7.19}$$

Equation (7.19) represents the mathematical expression of the *principle of virtual work*, which can be stated as follows: *The work performed by the applied forces through virtual displacements compatible with the system constraints is zero.*

In the case of a conservative system (Section 3.6), we can write

$$\sum_{i=1}^{N} \mathbf{F}_i \cdot \delta \mathbf{r}_i = -\delta V = 0 \tag{7.20}$$

where δV is the variation in the potential energy, which has the form

$$\delta V = \sum_{i=1}^{N} \left(\frac{\partial V}{\partial x_i} \delta x_i + \frac{\partial V}{\partial y_i} \delta y_i + \frac{\partial V}{\partial z_i} \delta z_i \right) \tag{7.21}$$

so that Eq. (7.20) can be rewritten

$$\sum_{i=1}^{N}(F_{xi}\delta x_i+F_{yi}\delta y_i+F_{zi}\delta z_i)=-\sum_{i=1}^{N}\left(\frac{\partial V}{\partial x_i}\delta x_i+\frac{\partial V}{\partial y_i}\delta y_i+\frac{\partial V}{\partial z_i}\delta z_i\right)=0 \quad (7.22)$$

where F_{xi}, F_{yi}, F_{zi} are the cartesian components of $\mathbf{F}_i$. For no constraints acting on the system, the virtual displacements δx_i, δy_i, δz_i $(i=1, 2, \ldots, N)$ are all independent. Moreover, because the virtual displacements are arbitrary and can be assigned values at will, the only way Eq. (7.22) can remain valid for all possible values of the virtual displacements is for the quantities multiplying the virtual displacements to be zero independently. Hence we must have

$$F_{xi}=-\frac{\partial V}{\partial x_i}=0, \qquad F_{yi}=-\frac{\partial V}{\partial y_i}=0, \qquad F_{zi}=-\frac{\partial V}{\partial z_i}=0, \qquad i=1,2,\ldots,N \quad (7.23)$$

Equations (7.23) represent the equilibrium conditions for a system of N free particles, and they state that the components of the applied force acting on every particle must be zero. In terms of the potential energy function V, these are the conditions for the function V to have a *stationary value*. Hence, *the potential energy has a stationary value when the system is in static equilibrium.* Equations (7.23) are written in terms of cartesian components, but they are not so restricted. Indeed, the equilibrium conditions can be expressed in terms of any other set of coordinates, including curvilinear coordinates.

Example 7.1

Consider a mass on a smooth inclined plane attached to a spring of stiffness k. Use the principle of virtual work to determine the equilibrium position.

Figure 7.1 shows the mass and the forces acting upon it. We denote the displacement of the spring in the equilibrium position by the vector $\mathbf{r}$ as measured parallel to the smooth surface from the unstretched position. There are two applied forces: the first due to the weight of the mass and equal to $\mathbf{W}=-mg\,\mathbf{j}$ and the second the restoring force $-k\mathbf{r}$ due to the elasticity of the spring. In addition,

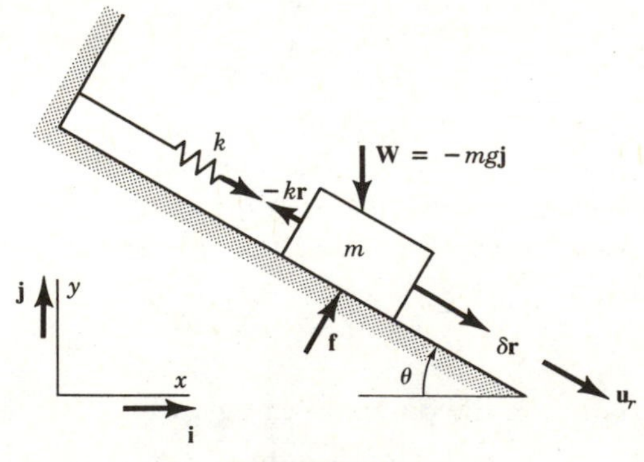

FIGURE 7.1

there is the constraint force $\mathbf{f}$ acting in a direction normal to the surface. Letting $\delta\mathbf{r}$ be the virtual displacement vector, as shown in Fig. 7.1, we can write the virtual work expression

$$(-k\mathbf{r}-mg\mathbf{j})\cdot\delta\mathbf{r}=0 \tag{a}$$

The vectors $\mathbf{r}$ and $\delta\mathbf{r}$ can be written in the form

$$\mathbf{r}=r\mathbf{u}_r, \qquad \delta\mathbf{r}=\delta r\mathbf{u}_r \tag{b}$$

where r and δr are the corresponding magnitudes and $\mathbf{u}_r$ is a unit vector in the direction of the vector $\mathbf{r}$, as shown in Fig. 7.1. Moreover, we have

$$\mathbf{j}\cdot\delta\mathbf{r}=-\delta r\sin\theta \tag{c}$$

where θ is the angle between the plane and the horizontal, so that inserting Eqs. (b) and (c) into Eq. (a), we obtain

$$(-kr+mg\sin\theta)\,\delta r=0 \tag{d}$$

Due to the arbitrariness of the virtual displacement δr, Eq. (d) can be satisfied only if the coefficient of δr is zero. Letting the coefficient of δr in Eq. (d) be equal to zero, we obtain the equilibrium condition

$$-kr+mg\sin\theta=0 \tag{e}$$

which can be solved for the equilibrium position, with the result

$$r=\frac{mg\sin\theta}{k} \tag{f}$$

Example 7.2

Determine the equilibrium position for the system of Example 7.1 by differentiating the potential energy expression.

The potential energy consists of two parts: the strain energy due to the elongation of the spring and the gravitational potential energy. Hence, using the unstretched spring position as a reference position, we can write

$$V=\tfrac{1}{2}kr^2-mgr\sin\theta=0 \tag{a}$$

so that

$$F_r=-\frac{\partial V}{\partial r}=-kr+mg\sin\theta=0 \tag{b}$$

where F_r is the force in the direction of the vector $\mathbf{r}$. Equation (b) is identical to Eq. (e) of Example 7.1. Hence, the coefficient of δr in Eq. (d) of Example 7.1 can be identified as F_r, so that Eq. (d) of Example 7.1 represents the virtual work expression

$$F_r\,\delta r=0 \tag{c}$$

7.4 D'ALEMBERT'S PRINCIPLE

As pointed out in Section 7.3, the principle of virtual work represents the statement of static equilibrium of a system. However, our interest lies not in problems of static equilibrium but in problems of dynamics. As shown in Section 5.2, the equations of motion for a system of particles can be written by invoking Newton's second law. In writing Newton's equations of motion, one must consider any constraint forces present. But, constraint forces are seldom known explicitly. Indeed, more often than not constraints appear only in the form of kinematical relations, such as Eq. (7.7). As a result, the constraint forces must be carried along as unknowns, which requires that the equations of motion be supplemented by the constraint equations. This increases the complexity of the problem, so that an approach permitting the elimination of the constraint forces from the problem formulation is highly desirable. One such approach is known as d'Alembert's principle.

Newton's equations of motion for a system of particles $m_i\ (i=1,2,\ldots,N)$ are

$$\mathbf{F}_i+\mathbf{f}_i=m_i\ddot{\mathbf{r}}_i, \qquad i=1,2,\ldots,N \tag{7.24}$$

where $\mathbf{F}_i$ are applied forces, $\mathbf{f}_i$ are constraint forces, and $\ddot{\mathbf{r}}_i$ is the acceleration of the mass m_i. Equations (7.24) can be rewritten in the form

$$\mathbf{F}_i+\mathbf{f}_i-m_i\ddot{\mathbf{r}}_i=\mathbf{0} \qquad i=1,2,\ldots,N \tag{7.25}$$

where $-m_i\ddot{\mathbf{r}}_i$ is referred to as an *inertia force*. Equations (7.25) can be regarded as representing the *dynamic equilibrium* of the system of particles. If we add the inertia force to the forces acting on m_i, we can use the analogy with the virtual work for the static case to write the virtual work for the individual particles

$$(\mathbf{F}_i+\mathbf{f}_i-m_i\ddot{\mathbf{r}}_i)\cdot\delta\mathbf{r}_i=0, \qquad i=1,2,\ldots,N \tag{7.26}$$

Summing over the entire system, we obtain

$$\sum_{i=1}^{N}(\mathbf{F}_i+\mathbf{f}_i-m_i\ddot{\mathbf{r}}_i)\cdot\delta\mathbf{r}_i=0 \tag{7.27}$$

But, as indicated by Eq. (7.18), the virtual work performed by the constraint forces over virtual displacements $\delta\mathbf{r}_i$ compatible with the system constraints is zero. Hence, Eq. (7.27) reduces to

$$\sum_{i=1}^{N}(\mathbf{F}_i-m_i\ddot{\mathbf{r}}_i)\cdot\delta\mathbf{r}_i=0 \tag{7.28}$$

Equation (7.28) represents the mathematical statement of *d'Alembert's principle*. The sum of the applied force $\mathbf{F}_i$ and the inertia force $-m_i\ddot{\mathbf{r}}_i$ is known as the *effective force*. Hence, d'Alembert's principle can be enunciated as follows: *The virtual work performed by the effective forces through virtual displacements compatible with the system constraints is zero.*

Equation (7.28) is suitable for a system of particles. On occasions, one is faced with a system of rigid bodies instead of a system of particles. If the motion is planar, then use of Eqs. (6.20) and (6.47) permits us to write d'Alembert's principle for a system of rigid bodies in the form

$$\sum_{i=1}^{N} [(\mathbf{F}_i - m_i \ddot{\mathbf{r}}_{Ci}) \cdot \delta \mathbf{r}_{Ci} + (M_{Ci} - I_{Ci}\ddot{\theta}_i)\, \delta\theta_i] = 0 \tag{7.29}$$

where the subscript i identifies now a rigid body in the system. Otherwise, the notation is consistent with that in Eqs. (6.20) and (6.47). Once again, the virtual displacements $\delta \mathbf{r}_{Ci}$ and $\delta\theta_i$ $(i = 1, 2, \ldots, N)$ must be consistent with the system constraints.

Example 7.3

Derive the equations of motion for the system of Fig. 6.8*a* by means of d'Alembert's principle. Use x and θ as independent coordinates.

The system of Fig. 6.8*a* consists of two rigid bodies, where the first behaves like a particle, so that it is subjected to no moments and possesses no moment of inertia. Hence, Eq. (7.29) reduces to

$$(\mathbf{F}_1 - m_1 \ddot{\mathbf{r}}_{C1}) \cdot \delta \mathbf{r}_{C1} + (\mathbf{F}_2 - m_2 \ddot{\mathbf{r}}_{C2}) \cdot \delta \mathbf{r}_{C2} + (M_{C2} - I_{C2}\ddot{\theta}_2)\, \delta\theta_2 = 0 \tag{a}$$

In our case,

$$m_1 = M, \qquad m_2 = m, \qquad I_{C2} = \tfrac{1}{12} m (2a)^2 = \tfrac{1}{3} m a^2 \tag{b}$$

Moreover,

$$\mathbf{F}_1 = -kx\mathbf{i} - Mg\mathbf{j}, \qquad \mathbf{F}_2 = F\mathbf{i} - mg\mathbf{j}, \qquad M_{C2} = Fa\cos\theta \tag{c}$$

and

$$\begin{aligned} \mathbf{r}_{C1} &= \mathbf{r}_O = (L + x)\mathbf{i} \\ \mathbf{r}_{C2} &= \mathbf{r}_C = \mathbf{r}_O + \mathbf{r}_{OC} = (L + x + a\sin\theta)\mathbf{i} - a\cos\theta\,\mathbf{j} \end{aligned} \tag{d}$$

where $\theta = \theta_2$ and L is the unstretched length of the spring. From Eqs. (d), we can write the virtual displacements

$$\delta \mathbf{r}_{C1} = \delta x\,\mathbf{i}, \qquad \delta \mathbf{r}_{C2} = (\delta x + a\,\delta\theta\cos\theta)\mathbf{i} + a\,\delta\theta\sin\theta\,\mathbf{j} \tag{e}$$

and, of course, $\delta\theta_2 = \delta\theta$. Also, differentiating Eqs. (d) twice with respect to time, we obtain the accelerations

$$\begin{aligned} \ddot{\mathbf{r}}_{C1} &= \ddot{x}\mathbf{i} \\ \ddot{\mathbf{r}}_{C2} &= [\ddot{x} + a(\ddot{\theta}\cos\theta - \dot{\theta}^2\sin\theta)]\mathbf{i} + a(\ddot{\theta}\sin\theta + \dot{\theta}^2\cos\theta)\mathbf{j} \end{aligned} \tag{f}$$

Inserting Eqs. (b)–(f) into Eq. (a), we can write

$$\begin{aligned} &(-kx\mathbf{i} - Mg\mathbf{j} - M\ddot{x}\mathbf{i}) \cdot \delta x\mathbf{i} + \{F\mathbf{i} - mg\mathbf{j} - m[\ddot{x} + a(\ddot{\theta}\cos\theta - \dot{\theta}^2\sin\theta)]\mathbf{i} \\ &\quad - ma(\ddot{\theta}\sin\theta + \dot{\theta}^2\cos\theta)\mathbf{j}\} \cdot [(\delta x + a\,\delta\theta\cos\theta)\mathbf{i} + a\,\delta\theta\sin\theta\mathbf{j}] \\ &\quad + (Fa\cos\theta - \tfrac{1}{3}ma^2\ddot{\theta})\,\delta\theta = \{-kx - M\ddot{x} + F - m[\ddot{x} + a(\ddot{\theta}\cos\theta \\ &\quad - \dot{\theta}^2\sin\theta)]\}\,\delta x + \langle\{F - m[\ddot{x} + a(\ddot{\theta}\cos\theta - \dot{\theta}^2\sin\theta)]\}a\cos\theta \\ &\quad - [mg + ma(\ddot{\theta}\sin\theta + \dot{\theta}^2\cos\theta)]a\sin\theta + Fa\cos\theta - \tfrac{1}{3}ma^2\ddot{\theta} \rangle\,\delta\theta = 0 \end{aligned} \tag{g}$$

Because the generalized displacements x and θ are independent, the generalized virtual displacements δx and $\delta\theta$ are arbitrary. Hence, Eq. (g) can be satisfied only if the coefficients of δx and $\delta\theta$ are each equal to zero. Setting these coefficients equal to zero and collecting terms, we obtain the equations of motion

$$\begin{gathered}(M+m)\ddot{x}+ma(\ddot{\theta}\cos\theta-\dot{\theta}^2\sin\theta)+kx=F\\ ma\ddot{x}\cos\theta+\tfrac{4}{3}ma^2\ddot{\theta}+mga\sin\theta=2Fa\cos\theta\end{gathered} \tag{h}$$

Equations (h) are identical to Eqs. (m) obtained in Example 6.5 by means of Newton's second law (as extended to systems of rigid bodies).

7.5 LAGRANGE'S EQUATIONS OF MOTION

In deriving the equations of motion, d'Alembert's principle has an advantage over Newton's second law in that the constraint forces do not come into play. However, d'Alembert's principle is still basically a vectorial approach involving accelerations. As a result, the derivation of the equations of motion remains a relatively cumbersome process. Hence, a simpler approach to the derivation of the equations of motion is highly desirable. Such an approach is provided by the Lagrangian mechanics, and it permits the derivation of the equations of motion from two scalar functions, the kinetic energy and the potential energy, as well as the virtual work due to nonconservative forces. It is here that d'Alembert's principle proves its worth since it can be used to facilitate the transition from the Newtonian mechanics to Lagrangian mechanics. Lagrangian mechanics works with the generalized coordinates $q_k\,(k=1,2,\ldots,n)$ instead of the physical coordinates $\mathbf{r}_i\,(i=1,2,\ldots,N)$, and the equations of motion in terms of the generalized coordinates are known as Lagrange's equations.

Considering an n-degree-of-freedom system and assuming that the coordinates $\mathbf{r}_i$ do not depend explicitly on time, we can write the coordinate transformation in the general form

$$\mathbf{r}_i=\mathbf{r}_i(q_1,q_2,\ldots,q_n),\qquad i=1,2,\ldots,N \tag{7.30}$$

The velocities $\dot{\mathbf{r}}_i$ are obtained by simply taking the total time derivative of Eqs. (7.30), yielding

$$\dot{\mathbf{r}}_i=\frac{\partial\mathbf{r}_i}{\partial q_1}\dot{q}_1+\frac{\partial\mathbf{r}_i}{\partial q_2}\dot{q}_2+\cdots+\frac{\partial\mathbf{r}_i}{\partial q_n}\dot{q}_n=\sum_{k=1}^{n}\frac{\partial\mathbf{r}_i}{\partial q_k}\dot{q}_k,\qquad i=1,2,\ldots,N \tag{7.31}$$

Because the quantities $\partial\mathbf{r}_i/\partial q_k$ do not depend explicitly on the generalized velocities $\dot{q}_k$, Eqs. (7.31) yield

$$\frac{\partial\dot{\mathbf{r}}_i}{\partial\dot{q}_k}=\frac{\partial\mathbf{r}_i}{\partial q_k},\qquad i=1,2,\ldots,N;\qquad k=1,2,\ldots,n \tag{7.32}$$

Moreover, by analogy with Eqs. (7.31), we can write

$$\delta\mathbf{r}_i=\frac{\partial\mathbf{r}_i}{\partial q_1}\delta q_1+\frac{\partial\mathbf{r}_i}{\partial q_2}\delta q_2+\cdots+\frac{\partial\mathbf{r}_i}{\partial q_n}\delta q_n=\sum_{i=1}^{n}\frac{\partial\mathbf{r}_i}{\partial q_k}\delta q_k,\quad k=1,2,\ldots,n \tag{7.33}$$

In view of Eqs. (7.33), the second term in Eq. (7.28) becomes

$$\sum_{i=1}^{N} m_i \ddot{\mathbf{r}}_i \cdot \delta \mathbf{r}_i = \sum_{i=1}^{N} m_i \ddot{\mathbf{r}}_i \cdot \sum_{k=1}^{n} \frac{\partial \mathbf{r}_i}{\partial q_k} \delta q_k$$
$$= \sum_{k=1}^{n} \sum_{i=1}^{N} \left(m_i \ddot{\mathbf{r}}_i \cdot \frac{\partial \mathbf{r}_i}{\partial q_k} \right) \delta q_k \tag{7.34}$$

Concentrating on a typical term on the right side of (7.34), we observe that

$$m_i \ddot{\mathbf{r}}_i \cdot \frac{\partial \mathbf{r}_i}{\partial q_k} = \frac{d}{dt}\left(m_i \dot{\mathbf{r}}_i \cdot \frac{\partial \mathbf{r}_i}{\partial q_k} \right) - m_i \dot{\mathbf{r}}_i \cdot \frac{d}{dt}\left(\frac{\partial \mathbf{r}_i}{\partial q_k} \right) \tag{7.35}$$

Considering Eqs. (7.32) and assuming that the order of the total derivatives with respect to time and partial derivatives with respect to q_k is interchangeable, we can write Eq. (7.35) in the form

$$m_i \ddot{\mathbf{r}}_i \cdot \frac{\partial \mathbf{r}_i}{\partial q_k} = \frac{d}{dt}\left(m_i \dot{\mathbf{r}}_i \cdot \frac{\partial \dot{\mathbf{r}}_i}{\partial \dot{q}_k} \right) - m_i \dot{\mathbf{r}}_i \cdot \frac{\partial \dot{\mathbf{r}}_i}{\partial q_k}$$
$$= \left[\frac{d}{dt}\left(\frac{\partial}{\partial \dot{q}_k} \right) - \frac{\partial}{\partial q_k} \right] \left(\tfrac{1}{2} m_i \dot{\mathbf{r}}_i \cdot \dot{\mathbf{r}}_i \right) \tag{7.36}$$

But the second term in parentheses on the right side of Eq. (7.36) is recognized as the kinetic energy of particle i (see Section 5.6). Hence, insertion of Eq. (7.36) into Eq. (7.34) yields

$$\sum_{i=1}^{N} m_i \ddot{\mathbf{r}}_i \cdot \delta \mathbf{r}_i = \sum_{k=1}^{n} \left\{ \left[\frac{d}{dt}\left(\frac{\partial}{\partial \dot{q}_k} \right) - \frac{\partial}{\partial q_k} \right] \left(\sum_{i=1}^{N} \frac{1}{2} m_i \dot{\mathbf{r}}_i \cdot \dot{\mathbf{r}}_i \right) \right\} \delta q_k$$
$$= \sum_{k=1}^{n} \left[\frac{d}{dt}\left(\frac{\partial T}{\partial \dot{q}_k} \right) - \frac{\partial T}{\partial q_k} \right] \delta q_k \tag{7.37}$$

where, in view of transformation (7.31),

$$T = \frac{1}{2} \sum_{i=1}^{N} m_i \dot{\mathbf{r}}_i \cdot \dot{\mathbf{r}}_i = T(q_1, q_2, \ldots, q_n, \dot{q}_1, \dot{q}_2, \ldots, \dot{q}_n) \tag{7.38}$$

is the kinetic energy of the entire system expressed as a function of the generalized coordinates and velocities.

It remains to write the forces $\mathbf{F}_i(\mathbf{r}_1, \mathbf{r}_2, \ldots, \mathbf{r}_N, \dot{\mathbf{r}}_1, \dot{\mathbf{r}}_2, \ldots, \dot{\mathbf{r}}_N, t)$ in terms of the generalized coordinates q_k $(k = 1, 2, \ldots, n)$. This is done by using the virtual work expression, in conjunction with transformation (7.33), as follows:

$$\overline{\delta W} = \sum_{i=1}^{N} \mathbf{F}_i \cdot \delta \mathbf{r}_i = \sum_{i=1}^{N} \mathbf{F}_i \cdot \sum_{k=1}^{n} \frac{\partial \mathbf{r}_i}{\partial q_k} \delta q_k$$
$$= \sum_{k=1}^{n} \left(\sum_{i=1}^{N} \mathbf{F}_i \cdot \frac{\partial \mathbf{r}_i}{\partial q_k} \right) \delta q_k \tag{7.39}$$

The virtual work, however, can be regarded as the product of n *generalized forces* Q_k acting over the virtual displacements δq_k,

$$\overline{\delta W}=\sum_{k=1}^{n} Q_k\,\delta q_k \tag{7.40}$$

so that, comparing Eqs. (7.39) and (7.40), we conclude that the generalized forces have the form

$$Q_k=\sum_{i=1}^{N} \mathbf{F}_i\cdot\frac{\partial \mathbf{r}_i}{\partial q_k}, \qquad k=1,2,\ldots,n \tag{7.41}$$

In actual situations, the generalized forces are derived by identifying physically a set of generalized coordinates and writing the virtual work directly in the form (7.40), rather than by using formula (7.41) (see Example 7.4). We note that the generalized forces are not necessarily forces. They can be moments or any other quantities such that the product $Q_k\delta q_k$ has units of work.

If the forces acting on the system can be divided into conservative forces, which are derivable from the potential energy $V=V(q_1, q_2, \ldots, q_n)$, and nonconservative forces, which are not, then the first term in Eq. (7.28) becomes

$$\begin{aligned}\sum_{i=1}^{N} \mathbf{F}_i\cdot\delta\mathbf{r}_i=\overline{\delta W}=\delta W_{\mathrm{c}}+\overline{\delta W}_{\mathrm{nc}}&=-\delta V+\sum_{k=1}^{n} Q_{k\mathrm{nc}}\,\delta q_k\\ &=-\left(\frac{\partial V}{\partial q_1}\,\delta q_1+\frac{\partial V}{\partial q_2}\,\delta q_2+\cdots+\frac{\partial V}{\partial q_n}\,\delta q_n\right)+\sum_{k=1}^{n} Q_{k\mathrm{nc}}\,\delta q_k\\ &=-\sum_{k=1}^{n}\left(\frac{\partial V}{\partial q_k}-Q_{k\mathrm{nc}}\right)\delta q_k\end{aligned} \tag{7.42}$$

where $Q_{k\mathrm{nc}}$ $(k=1,2,\ldots,n)$ are nonconservative generalized forces. Note that the overbar was omitted from δW_{c}, because δW_{c} represents the variation of the work function W_{c} due to conservative forces, where W_{c} is the negative of the potential energy V. Introducing Eqs. (7.37) and (7.42) into Eq. (7.28), we obtain

$$-\sum_{k=1}^{n}\left[\frac{d}{dt}\left(\frac{\partial T}{\partial \dot{q}_k}\right)-\frac{\partial T}{\partial q_k}+\frac{\partial V}{\partial q_k}-Q_{k\mathrm{nc}}\right]\delta q_k=0 \tag{7.43}$$

However, by definition, the generalized virtual displacements δq_k are both arbitrary and independent. Hence, letting $\delta q_k=0$, $k\neq j$, and $\delta q_j\neq 0$, we conclude that Eq. (7.43) can be satisfied if and only if the coefficient of δq_j is zero. The procedure can be repeated n times for $j=1,2,\ldots,n$. Moreover, with the understanding that Q_j represents nonconservative forces, we can drop the subscript nc in $Q_{j\mathrm{nc}}$ and arrive at the set of equations

$$\frac{d}{dt}\left(\frac{\partial T}{\partial \dot{q}_j}\right)-\frac{\partial T}{\partial q_j}+\frac{\partial V}{\partial q_j}=Q_j, \qquad j=1,2,\ldots,n \tag{7.44}$$

which are the famous *Lagrange's equations of motion*. In general, the potential energy does not depend on the generalized velocities $\dot{q}_j$ $(j=1, 2, \ldots, n)$. In view of this, we can introduce the *Lagrangian* defined by

$$L=T-V \tag{7.45}$$

and reduce Eqs. (7.44) to the more compact form

$$\frac{d}{dt}\left(\frac{\partial L}{\partial \dot{q}_j}\right)-\frac{\partial L}{\partial q_j}=Q_j, \qquad j=1, 2, \ldots, n \tag{7.46}$$

In many problems there are no nonconservative forces involved, in which cases $Q_j=0$ $(j=1, 2, \ldots, n)$. Hence, *Lagrange's equations for conservative systems* are simply

$$\frac{d}{dt}\left(\frac{\partial L}{\partial \dot{q}_j}\right)-\frac{\partial L}{\partial q_j}=0, \qquad j=1, 2, \ldots, n \tag{7.47}$$

The Lagrangian approach is very efficient for deriving the system equations of motion, especially when the number of degrees of freedom of the system is large. All the differential equations of motion are derived from two scalar functions, namely, the kinetic energy T and the potential energy V, and the virtual work δW_{nc} associated with the nonconservative forces. The equations apply to linear as well as nonlinear systems. Although it appears that the identification of the generalized coordinates and generalized forces is a major stumbling block in using this approach, this is actually not the case. Indeed, in most physical systems this aspect presents no particular difficulty, as illustrated in Example 7.4.

Example 7.4

Derive Lagrange's equations of motion for the system of Example 7.3.

As in Example 7.3, we propose to use x and θ as generalized coordinates, so that $q_1=x$, $q_2=\theta$. Accordingly, the nonconservative generalized forces are denoted by $Q_1=X$, $Q_2=\Theta$. To calculate the kinetic energy, we must express the velocities of the two bodies in terms of these coordinates. From Eqs. (d) of Example 7.3, we can write

$$\mathbf{v}_O=\dot{\mathbf{r}}_O=\dot{x}\mathbf{i}, \qquad \mathbf{v}_C=\dot{\mathbf{r}}_C=(\dot{x}+a\dot{\theta}\cos\theta)\mathbf{i}+a\dot{\theta}\sin\theta\,\mathbf{j} \tag{a}$$

so that the kinetic energy can be written in the form

$$\begin{aligned} T&=\tfrac{1}{2}M\mathbf{v}_O\cdot\mathbf{v}_O+\tfrac{1}{2}m\mathbf{v}_C\cdot\mathbf{v}_C+\tfrac{1}{2}I_C\dot{\theta}^2 \\ &=\tfrac{1}{2}M\dot{x}^2+\tfrac{1}{2}m[(\dot{x}+a\dot{\theta}\cos\theta)^2+(a\dot{\theta}\sin\theta)^2]+\tfrac{1}{2}\tfrac{1}{12}m(2a)^2\dot{\theta}^2 \\ &=\tfrac{1}{2}M\dot{x}^2+ma\dot{x}\dot{\theta}\cos\theta+\tfrac{2}{3}ma^2\dot{\theta}^2 \end{aligned} \tag{b}$$

The potential energy is due to the deformation of the spring and the rise of the mass center of the bar. Its expression is

$$V=\tfrac{1}{2}kx^2+mga(1-\cos\theta) \tag{c}$$

Hence, the Lagrangian of the system is

$$L=T-V=\tfrac{1}{2}(M+m)\dot{x}^2+ma\dot{x}\dot{\theta}\cos\theta+\tfrac{2}{3}ma^2\dot{\theta}^2$$
$$-\tfrac{1}{2}kx^2-mga(1-\cos\theta) \tag{d}$$

It remains to calculate the virtual work associated with the nonconservative force F. Denoting the point of application of the force by B, we can write the position vector of B in terms of cartesian components as follows:

$$\mathbf{r}_B=(L+x+2a\sin\theta)\mathbf{i}-2a\cos\theta\,\mathbf{j} \tag{e}$$

so that the virtual displacement of B is

$$\delta\mathbf{r}_B=(\delta x+2a\,\delta\theta\cos\theta)\mathbf{i}+2a\,\delta\theta\sin\theta\,\mathbf{j} \tag{f}$$

Moreover, the force F can be written in the vector form

$$\mathbf{F}=F\mathbf{i} \tag{g}$$

Hence, the nonconservative virtual work is

$$\overline{\delta W}_{\text{nc}}=\mathbf{F}\cdot\delta\mathbf{r}_B=F\mathbf{i}\cdot[(\delta x+2a\,\delta\theta\cos\theta)\mathbf{i}+2a\,\delta\theta\sin\theta\,\mathbf{j}]$$
$$=F(\delta x+2a\,\delta\theta\cos\theta) \tag{h}$$

Equation (h) is in the form (7.40), so that the coefficients of δx and $\delta\theta$ are the nonconservative generalized forces, or

$$X=F,\qquad \Theta=2Fa\cos\theta \tag{i}$$

The derivatives entering into Lagrange's equations are as follows:

$$\begin{aligned}
\frac{\partial L}{\partial\dot{x}}&=(M+m)\dot{x}+ma\dot{\theta}\cos\theta\\
\frac{d}{dt}\left(\frac{\partial L}{\partial\dot{x}}\right)&=(M+m)\ddot{x}+ma(\ddot{\theta}\cos\theta-\dot{\theta}^2\sin\theta)\\
\frac{\partial L}{\partial x}&=-kx\\
\frac{\partial L}{\partial\dot{\theta}}&=ma\dot{x}\cos\theta+\tfrac{4}{3}ma^2\dot{\theta}\\
\frac{d}{dt}\left(\frac{\partial L}{\partial\dot{\theta}}\right)&=ma(\ddot{x}\cos\theta-\dot{x}\dot{\theta}\sin\theta)+\tfrac{4}{3}ma^2\ddot{\theta}\\
\frac{\partial L}{\partial\theta}&=-ma\dot{x}\dot{\theta}\sin\theta-mga\sin\theta
\end{aligned} \tag{j}$$

Introducing Eqs. (i) and (j) into Eqs. (7.46) with $q_1=x$, $q_2=\theta$, $Q_1=X$ and $Q_2=\Theta$, we obtain Lagrange's equations of motion for the system in the explicit form

$$\begin{aligned}
(M+m)\ddot{x}+ma(\ddot{\theta}\cos\theta-\dot{\theta}^2\sin\theta)+kx&=F\\
ma\ddot{x}\cos\theta+\tfrac{4}{3}ma^2\ddot{\theta}+mga\sin\theta&=2Fa\cos\theta
\end{aligned} \tag{k}$$

Equations (k) are identical to Eqs. (h) obtained in Example 7.3 by means of d'Alembert's principle and Eqs. (m) obtained in Example 6.5 by means of the Newtonian approach.

Comparing the amount of work involved in Examples 6.5, 7.3, and 7.4, it appears that the Lagrangian approach permits the derivation of the equations of motion in the most expeditious manner, at least in this example. Actually this is true in most cases and not merely in this particular example. In particular, we observe that equations of motion having the same form as Eqs. (k) were obtained in Example 6.5 only after the elimination of the internal forces at the hinge connecting the rigid bar to the lumped mass. In other cases the equations derived by means of Newton's second law do not appear to correspond exactly to those derived by means of Lagrange's equations. In such cases, to show that the equations derived by the two approaches are indeed identical, it is necessary to combine linearly several equations in one set or the other.

PROBLEMS

7.1 Determine the position of static equilibrium of the system of Fig. 6.28 by means of the virtual work principle.

7.2 Determine the position of static equilibrium of the airfoil section of Fig. 6.26 by the virtual work principle.

7.3 Derive the equations of motion for the system of Fig. 6.26 by means of d'Alembert's principle.

7.4 Derive the equations of motion for the system of Fig. 6.28 by means of d'Alembert's principle.

7.5 Derive Lagrange's equation of motion for the system of Fig. 6.18.

7.6 Derive Lagrange's equation of motion for the system of Fig. 6.19.

7.7 Derive Lagrange's equation of motion for the system of Fig. 6.22.

7.8 Derive Lagrange's equations of motion for the system of Fig. 6.26.

7.9 Derive Lagrange's equations of motion for the system of Fig. 6.27.

7.10 Derive Lagrange's equations of motion for the system of Fig. 6.28.

7.11 Derive Lagrange's equations of motion for the gyroscope shown in Fig. 6.10.

CHAPTER 8
Vibration of Linear Multi-Degree-of-Freedom Systems

8.1 INTRODUCTION

In Chapter 7, we learned efficient methods for the derivation of the equations of motion for multi-degree-of-freedom systems. In particular, we learned how to use Lagrange's equations to derive the equations of motion from two scalar functions, the kinetic energy and the potential energy, and from an infinitesimal expression, the virtual work. Lagrange's equations are valid for linear and nonlinear systems. The problem of determining the response of multi-degree-of-freedom systems remains.

There is a significant difference between linear and nonlinear systems, as far as producing the system response is concerned. This is particularly true of multi-degree-of-freedom systems. Indeed, whereas a theory permitting the derivation of the response of linear multi-degree-of-freedom systems exists, no such theory exists for general nonlinear systems. More often than not, nonlinear systems are linearized by restricting the motion to the neighborhood of special points in the state space known as equilibrium points.

In this chapter, we first use Lagrange's equations to derive the equations of motion for linear multi-degree-of-freedom systems. Then, concentrating on undamped systems, we introduce the concepts of normal-mode vibration and orthogonality. The orthogonality of the modes of vibration forms the basis for a procedure for the derivation of the system response known as modal analysis, whereby an n-degree-of-freedom system is reduced to n independent single-degree-of-freedom systems. Under certain circumstances, the same approach can be used also for damped systems, but this is not true in general. The special cases in which the modal analysis developed for undamped systems can be used to treat damped systems are discussed.

8.2 LAGRANGE'S EQUATIONS OF MOTION FOR LINEAR SYSTEMS

In Section 7.5 we derived general Lagrange's equations for a multi-degree-of-freedom system. In this section we propose to apply Lagrange's equations to the derivation of the equations of motion for a linear n-degree-of-freedom system.

In linear systems the dependent variables appear in the equations of motion to the first power only. Because the various terms in the equations of motion result from differentiations of the Lagrangian with respect to the generalized displacements and generalized velocities, it follows that terms of powers higher than 2 in the Lagrangian lead to nonlinear terms in the equations of motion. If the displacements and velocities are sufficiently small that nonlinear terms can be ignored, then the resulting system is said to be *linearized*. The assumption under which the nonlinear terms can be ignored is known as the *small-motions assumption*. In this chapter we refer to a linearized system as simply linear.

As a preliminary to the derivation of Lagrange's equations of motion for a linear system, we must first derive expressions for the kinetic and potential energy containing at most quadratic terms in the generalized coordinates and velocities. Introducing Eq. (7.31) into Eq. (7.38), we can write the system kinetic energy in the form

$$T=\frac{1}{2}\sum_{i=1}^{N} m_i\dot{\mathbf{r}}_i\cdot\dot{\mathbf{r}}_i=\frac{1}{2}\sum_{i=1}^{N} m_i\left(\sum_{r=1}^{n}\frac{\partial\mathbf{r}_i}{\partial q_r}\dot{q}_r\right)\cdot\left(\sum_{s=1}^{n}\frac{\partial\mathbf{r}_i}{\partial q_s}\dot{q}_s\right)$$
$$=\frac{1}{2}\sum_{r=1}^{n}\sum_{s=1}^{n}\left(\sum_{i=1}^{N} m_i\frac{\partial\mathbf{r}_i}{\partial q_r}\cdot\frac{\partial\mathbf{r}_i}{\partial q_s}\right)\dot{q}_r\dot{q}_s=\frac{1}{2}\sum_{r=1}^{n}\sum_{s=1}^{n} m_{rs}\dot{q}_r\dot{q}_s \tag{8.1}$$

where the terms

$$m_{rs}=\sum_{i=1}^{N} m_i\frac{\partial\mathbf{r}_i}{\partial q_r}\cdot\frac{\partial\mathbf{r}_i}{\partial q_s}=m_{sr} \tag{8.2}$$

are symmetric in r and s. The reason for replacing the dummy index in the second sum in Eq. (8.1) is to ensure that cross products are not missed inadvertently. Although for a general nonlinear system m_{rs} can contain terms in the generalized displacements, in the case of a linear system they cannot, as the terms $\dot{q}_r\dot{q}_s$ $(r, s=1, 2, \ldots, n)$ are already of degree 2. Hence, we consider the case in which m_{rs} are constant, in which case m_{rs} $(r, s=1, 2, \ldots, n)$ are known as *mass coefficients*, or *inertia coefficients*. As pointed out earlier, the coefficients are symmetric.

Next, let us turn our attention to the potential energy. In general, the potential energy is a nonlinear function of the generalized displacements $q_1, q_2, \ldots, q_n$. Assuming without loss of generality that $q_1=q_2=\cdots=q_n=0$, represents an equilibrium point and expanding the system potential energy in a Taylor series about that point, we can write

$$V(q_1, q_2, \ldots, q_n)=V(0, 0\ldots, 0)+\sum_{r=1}^{n}\left.\frac{\partial V}{\partial q_r}\right|_{q_r=0} q_r$$
$$+\frac{1}{2}\sum_{r=1}^{n}\sum_{s=1}^{n}\left.\frac{\partial^2 V}{\partial q_r\partial q_s}\right|_{q_r=q_s=0} q_r q_s+O(q^3) \tag{8.3}$$

where $O(q^3)$ denotes higher-order terms in the generalized coordinates. But, because the term $V(0, 0\ldots 0)$ is a mere constant and Lagrange's equations involve derivatives of V, it can be ignored in series (8.3). Moreover, consistent with Eqs.

(7.23), the equations

$$\frac{\partial V}{\partial q_r}=0, \qquad r=1,2,\ldots,n \tag{8.4}$$

can be identified as the conditions of static equilibrium, so that the linear terms in series (8.3) are zero. Hence, ignoring the higher-order terms and introducing the notation

$$k_{rs}=\left.\frac{\partial^2 V}{\partial q_r \partial q_s}\right|_{q_r=q_s=0}=\left.\frac{\partial^2 V}{\partial q_s \partial q_r}\right|_{q_r=q_s=0}=k_{sr} \tag{8.5}$$

Eq. (8.3) can be rewritten in the form

$$V=\frac{1}{2}\sum_{r=1}^{n}\sum_{s=1}^{n} k_{rs}q_r q_s \tag{8.6}$$

where the constant symmetric coefficients k_{rs} $(r,s=1,2,\ldots,n)$ are called *stiffness coefficients.*

The kinetic energy, Eq. (8.1), and the potential energy, Eq. (8.6), represent quadratic forms, the first in the generalized velocities and the second in the generalized coordinates. Hence, a linear system is consistent with quadratic forms for the kinetic energy and potential energy.

At this point, we can proceed with the derivation of the equations of motion. To this end, we use Lagrange's equations in the general form (7.44). Differentiating Eq. (8.1) with respect to $\dot{q}_j$, we obtain

$$\frac{\partial T}{\partial \dot{q}_j}=\frac{1}{2}\sum_{r=1}^{n}\sum_{s=1}^{n} m_{rs}\left(\frac{\partial \dot{q}_r}{\partial \dot{q}_j}\dot{q}_s+\dot{q}_r\frac{\partial \dot{q}_s}{\partial \dot{q}_j}\right)=\frac{1}{2}\sum_{r=1}^{n}\sum_{s=1}^{n} m_{rs}(\dot{q}_s\,\delta_{rj}+\dot{q}_r\,\delta_{sj}) \tag{8.7}$$

where δ_{rj} is the Kronecker delta, defined as being equal to one for $r=j$ and equal to zero for $r\neq j$. Of course, δ_{sj} is also a Kronecker delta and is defined in a similar manner. The effect of the Kronecker delta δ_{rj} is to reduce the series over r on the right side of Eq. (8.7) to one term only, namely, the term corresponding to $r=j$. Similarly, δ_{sj} reduces the series over s to the term for which $s=j$. Hence, Eq. (8.7) reduces to

$$\frac{\partial T}{\partial \dot{q}_j}=\frac{1}{2}\sum_{s=1}^{n} m_{js}\dot{q}_s+\frac{1}{2}\sum_{r=1}^{n} m_{rj}\dot{q}_r \tag{8.8}$$

But, r and s are mere dummy indices and can be replaced by any other ones without affecting the result. Moreover, we recall that the mass coefficients are symmetric. Hence, the two series on the right side of Eq. (8.8) are equal to one another, so that Eq. (8.8) can be rewritten in the form

$$\frac{\partial T}{\partial \dot{q}_j}=\sum_{s=1}^{n} m_{js}\dot{q}_s \tag{8.9}$$

from which it follows immediately that

$$\frac{d}{dt}\left(\frac{\partial T}{\partial \dot{q}_j}\right)=\sum_{s=1}^{n} m_{js}\ddot{q}_s \tag{8.10}$$

Similarly, differentiating Eq. (8.6) with respect to q_j, we can write directly

$$\frac{\partial V}{\partial q_j}=\sum_{s=1}^{n} k_{js}q_s \tag{8.11}$$

Moreover, because the mass coefficients are constant, the kinetic energy contains no generalized coordinates, so that

$$\frac{\partial T}{\partial q_j}=0 \tag{8.12}$$

Inserting Eqs. (8.10)–(8.12) into Eq. (7.44), we obtain Lagrange's equations of motion for a linear system in the form

$$\sum_{s=1}^{n} [m_{js}\ddot{q}_s(t)+k_{js}q_s(t)]=Q_j(t), \qquad j=1, 2, \ldots, n \tag{8.13}$$

where Q_j $(j=1, 2, \ldots, n)$ are the generalized nonconservative forces. The nonconservative forces Q_j can be derived from the virtual work expression, as shown in Section 7.5. Equations (8.13) represent the equations of motion for a *linear undamped system.*

There is one class of forces that can be derived more conveniently by means other than the virtual work. These are the damping forces of viscous type, which can be accounted for explicitly in the Lagrange's equations of motion by means of *Rayleigh's dissipation function*, defined as

$$\mathscr{F}=\frac{1}{2}\sum_{r=1}^{n}\sum_{s=1}^{n} c_{rs}\dot{q}_r\dot{q}_s \tag{8.14}$$

where c_{rs} are known as *viscous damping coefficients.* The coefficients are symmetric, $c_{rs}=c_{sr}$ $(r, s=1, 2, \ldots, n)$. It should be pointed out here that the derivation of Rayleigh's dissipation function for a given system follows the same pattern as the derivation of the potential energy (see Example 8.1). Then, to account for viscous damping in the equations of motion explicitly, Lagrange's equations (7.44) must be rewritten as follows:

$$\frac{d}{dt}\left(\frac{\partial T}{\partial \dot{q}_j}\right)-\frac{\partial T}{\partial q_j}+\frac{\partial \mathscr{F}}{\partial \dot{q}_j}+\frac{\partial V}{\partial q_j}=Q_j, \qquad j=1, 2, \ldots, n \tag{8.15}$$

where now Q_j include only external forces. By analogy with Eq. (8.9), we can write

$$\frac{\partial \mathscr{F}}{\partial \dot{q}_j}=\sum_{s=1}^{n} c_{js}\dot{q}_s \tag{8.16}$$

so that Eqs. (8.13) become

$$\sum_{s=1}^{n} [m_{js}\ddot{q}_s(t)+c_{js}\dot{q}_s(t)+k_{js}q_s(t)]=Q_j(t), \qquad j=1, 2, \ldots, n \tag{8.17}$$

Equations (8.17) represent the equations of motion of a *viscously damped linear system.*

Equations (8.17) can be cast in matrix form. To this end, we introduce the $n \times n$ matrices

$$M=[m_{rs}], \qquad C=[c_{rs}], \qquad K=[k_{rs}] \tag{8.18}$$

where M is known as the *mass matrix*, C as the *damping matrix*, and K as the *stiffness matrix*. All three matrices are symmetric, a fact expressed by

$$M^{\mathrm{T}}=M, \qquad C^{\mathrm{T}}=C, \qquad K^{\mathrm{T}}=K \tag{8.19}$$

where the superscript T denotes the *transpose* of the matrix. Moreover, defining the n-dimensional *generalized displacement vector* and *generalized force vector*

$$\mathbf{q}(t)=[q_1(t) \quad q_2(t) \quad \dots \quad q_n(t)]^{\mathrm{T}}, \qquad \mathbf{Q}(t)=[Q_1(t) \quad Q_2(t) \quad \dots \quad Q_n(t)]^{\mathrm{T}} \tag{8.20}$$

we can rewrite Eqs. (8.17) in the compact matrix form

$$M\ddot{\mathbf{q}}(t)+C\dot{\mathbf{q}}(t)+K\mathbf{q}(t)=\mathbf{Q}(t) \tag{8.21}$$

The equations of motion for a linear system can be derived in a more direct manner by observing that the kinetic energy T, Rayleigh's dissipation function $\mathscr{F}$, and the potential energy V can be expressed in terms of the mass matrix M, the damping matrix C, and the stiffness matrix K, respectively, as follows:

$$T=\tfrac{1}{2}\dot{\mathbf{q}}^{\mathrm{T}}(t)M\dot{\mathbf{q}}(t) \tag{8.22a}$$

$$\mathscr{F}=\tfrac{1}{2}\dot{\mathbf{q}}^{\mathrm{T}}(t)C\dot{\mathbf{q}}(t) \tag{8.22b}$$

$$V=\tfrac{1}{2}\mathbf{q}^{\mathrm{T}}(t)K\mathbf{q}(t) \tag{8.22c}$$

In addition, the virtual work can be written in the form

$$\overline{\delta W}=\mathbf{Q}^{\mathrm{T}}(t)\,\delta\mathbf{q}(t) \tag{8.23}$$

where $\delta\mathbf{q}(t)$ is the generalized virtual displacement vector. Hence, the equations of motion can be considered as being derived as soon as the kinetic energy, Rayleigh's dissipation function, the potential energy, and the virtual work expressions have been written in the general form (8.22) and (8.23)

Example 8.1

Derive the equations of motion for the three-degree-of-freedom system of Fig. 8.1.

We propose to derive the equations of motion by simply deriving the kinetic energy, Rayleigh's dissipation function, and the potential energy and expressing

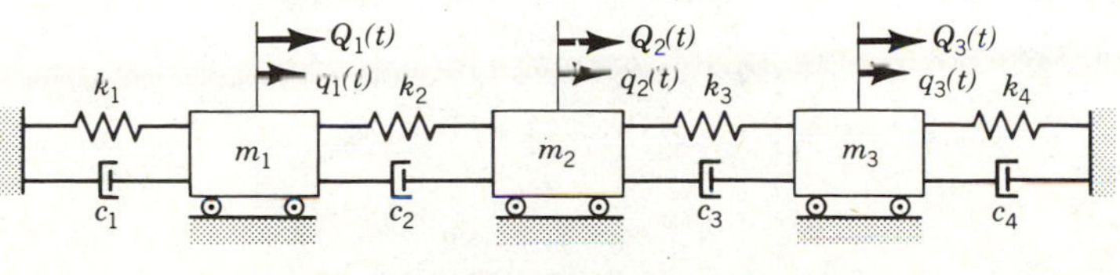

FIGURE 8.1

them in the matrix from (8.22). The kinetic energy is simply

$$T=\tfrac{1}{2}(m_1\dot{q}_1^2+m_2\dot{q}_2^2+m_3\dot{q}_3^2)=\tfrac{1}{2}\begin{bmatrix}\dot{q}_1\\ \dot{q}_2\\ \dot{q}_3\end{bmatrix}^{\mathrm{T}}\begin{bmatrix}m_1 & 0 & 0\\ 0 & m_2 & 0\\ 0 & 0 & m_3\end{bmatrix}\begin{bmatrix}\dot{q}_1\\ \dot{q}_2\\ \dot{q}_3\end{bmatrix} \tag{a}$$

As pointed out earlier, the Rayleigh dissipation function is derived in a manner similar to the potential energy. Hence, each of the dashpots contributes a term proportional to the coefficient of viscous damping and to the square of the velocity of separation of the terminal points. In the case of the system of Fig. 8.1, we have

$$\begin{aligned}\mathscr{F}&=\tfrac{1}{2}[c_1\dot{q}_1^2+c_2(\dot{q}_2-\dot{q}_1)^2+c_3(\dot{q}_3-\dot{q}_2)^2+c_4\dot{q}_3^2]\\ &=\tfrac{1}{2}\begin{bmatrix}\dot{q}_1\\ \dot{q}_2\\ \dot{q}_3\end{bmatrix}^{\mathrm{T}}\begin{bmatrix}c_1+c_2 & -c_2 & 0\\ -c_2 & c_2+c_3 & -c_3\\ 0 & -c_3 & c_3+c_4\end{bmatrix}\begin{bmatrix}\dot{q}_1\\ \dot{q}_2\\ \dot{q}_3\end{bmatrix}\end{aligned} \tag{b}$$

Similarly, the potential energy has the form

$$\begin{aligned}V&=\tfrac{1}{2}[k_1q_1^2+k_2(q_2-q_1)^2+k_3(q_3-q_2)^2+k_4q_3^2]\\ &=\tfrac{1}{2}\begin{bmatrix}q_1\\ q_2\\ q_3\end{bmatrix}^{\mathrm{T}}\begin{bmatrix}k_1+k_2 & -k_2 & 0\\ -k_2 & k_2+k_3 & -k_3\\ 0 & -k_3 & k_3+k_4\end{bmatrix}\begin{bmatrix}q_1\\ q_2\\ q_3\end{bmatrix}\end{aligned} \tag{c}$$

It follows that the equations of motion have the general form (8.21), in which the mass matrix is

$$M=\begin{bmatrix}m_1 & 0 & 0\\ 0 & m_2 & 0\\ 0 & 0 & m_3\end{bmatrix} \tag{d}$$

the damping matrix is

$$C=\begin{bmatrix}c_1+c_2 & -c_2 & 0\\ -c_2 & c_2+c_3 & -c_3\\ 0 & -c_3 & c_3+c_4\end{bmatrix} \tag{e}$$

and the stiffness matrix is

$$K=\begin{bmatrix}k_1+k_2 & -k_2 & 0\\ -k_2 & k_2+k_3 & -k_3\\ 0 & -k_3 & k_3+k_4\end{bmatrix} \tag{f}$$

Moreover, $\mathbf{q}(t)$ and $\mathbf{Q}(t)$ have the general form (8.20), in which $n=3$.

8.3 NORMAL-MODE VIBRATION. THE EIGENVALUE PROBLEM

Equation (8.21) represents a set of n simultaneous differential equations of motion. To obtain the system response, we must solve these equations of motion. The methods of solution depend on the type of system under consideration. In particular, the problem is considerably simpler when the system is undamped, $C=0$. We propose to examine the undamped case first and the damped case later.

Letting $C=0$ in Eq. (8.21), we obtain the differential equation of motion of an undamped system in the matrix form

$$M\ddot{\mathbf{q}}(t)+K\mathbf{q}(t)=\mathbf{Q}(t) \tag{8.24}$$

Before we discuss the general forced-vibration case, it will be more convenient to examine the free-vibration case, described by the homogeneous equation

$$M\ddot{\mathbf{q}}(t)+K\mathbf{q}(t)=\mathbf{0} \tag{8.25}$$

Following the pattern of Chapter 4, we consider a solution of Eq. (8.25) in the exponential form

$$\mathbf{q}(t)=e^{\lambda t}\mathbf{u} \tag{8.26}$$

where λ is a constant scalar and $\mathbf{u}$ is a constant n-dimensional vector. Introducing Eq. (8.26) into Eq. (8.25), we obtain

$$[\lambda^2 M+K]\mathbf{u}=\mathbf{0} \tag{8.27}$$

which represents a set of n homogeneous algebraic equations with λ^2 playing the role of a parameter and the components of $\mathbf{u}$ playing the role of unknowns. The problem of determining the values of the parameter λ^2 for which Eq. (8.27) possesses nontrivial solutions is known as the *eigenvalue problem.*

From linear algebra, a set of homogeneous algebraic equations has a nontrivial solution if and only if the determinant of the coefficients is equal to zero, or

$$\Delta(\lambda^2)=\det[\lambda^2 M+K]=|\lambda^2 M+K|=0 \tag{8.28}$$

Equation (8.28) is known as the *characteristic equation* and $\Delta(\lambda^2)$ is called the *characteristic determinant*, or *characteristic polynomial*. The characteristic polynomial is a polynomial in λ^2 of degree n. Hence, it has n roots, $\lambda_1^2, \lambda_2^2, \ldots, \lambda_n^2$, which are known as *characteristic values*, or *eigenvalues*. If each of the eigenvalues is inserted into Eq. (8.27), we obtain

$$[\lambda_r^2 M+K]\mathbf{u}_r=\mathbf{0}, \qquad r=1, 2, \ldots, n \tag{8.29}$$

where $\mathbf{u}_r$ is the solution of the eigenvalue problem corresponding to $\lambda^2=\lambda_r^2$ $(r=1, 2, \ldots, n)$. The vector $\mathbf{u}_r$ is called the *eigenvector* belonging to the eigenvalue λ_r^2. Of course, there are n such eigenvectors.

In the case of elastic systems, in which the matrix K is due entirely to restoring forces, all the eigenvalues are negative, or

$$\lambda_r^2=-\omega_r^2, \qquad r=1, 2, \ldots, n \tag{8.30}$$

so that

$$\lambda_r = \pm i\omega_r, \qquad r=1, 2, \ldots, n \tag{8.31}$$

where $i=\sqrt{-1}$. Hence, solution (8.26) can be rewritten

$$\mathbf{q}_r(t) = (a_r e^{i\omega_r t} + b_r e^{-i\omega_r t})\mathbf{u}_r, \qquad r=1, 2, \ldots, n \tag{8.32}$$

But, $\mathbf{q}_r(t)$ is a real quantity, and so is $\mathbf{u}_r$. It follows that the expression in parentheses on the right side of Eq. (8.32) must also be real, so that $b_r = \bar{a}_r$, where $\bar{a}_r$ is the complex conjugate a_r. Hence, Eq. (8.32) reduces to

$$\begin{aligned}\mathbf{q}_r(t) &= (a_r e^{i\omega_r t} + \bar{a}_r e^{-i\omega_r t})\mathbf{u}_r \\ &= 2\,\mathrm{Re}(a_r e^{i\omega_r t})\mathbf{u}_r, \qquad r=1, 2, \ldots, n\end{aligned} \tag{8.33}$$

Moreover, introducing the notation

$$\mathrm{Re}\, a_r = \tfrac{1}{2}c_r \cos\phi_r, \qquad \mathrm{Im}\, a_r = -\tfrac{1}{2}c_r \sin\phi_r \tag{8.34}$$

so that

$$a_r = \tfrac{1}{2}c_r e^{-i\phi_r}, \qquad r=1, 2, \ldots, n \tag{8.35}$$

we can rewrite Eq. (8.33) in the form

$$\begin{aligned}q_r(t) &= 2\,\mathrm{Re}(\tfrac{1}{2}c_r e^{-i\phi_r} e^{i\omega_r t})\mathbf{u}_r \\ &= c_r \cos(\omega_r t - \phi_r)\mathbf{u}_r, \qquad r=1, 2, \ldots, n\end{aligned} \tag{8.36}$$

Equations (8.36) indicate that the free vibration of the system can take place in n distinct ways. In each of these ways, the motion is harmonic with amplitude c_r, frequency ω_r, and phase angle ϕ_r. Moreover, the displacement of each mass is proportional to the corresponding component of the eigenvector $\mathbf{u}_r$. The frequencies ω_r and the eigenvectors $\mathbf{u}_r$ ($r=1, 2, \ldots, n$) are characteristics of the system, determined by the matrices M and K. This implies that, for a given system, the motions $\mathbf{q}_r(t)$ ($r=1, 2, \ldots, n$) are unique except for the amplitude and the phase angle. For this reason, $\mathbf{q}_r(t)$ are called *natural motions*. Consistent with this, the frequencies ω_r are called *natural frequencies*, and the eigenvectors $\mathbf{u}_r$ are known as *natural modes*. They are also referred to as *modal vectors*. Because the modal vectors are solutions of a set of homogeneous algebraic equations, the magnitude of a model vector $\mathbf{u}_r$ is not unique, as $\alpha_r\mathbf{u}_r$ represents the same modal vector, where α_r is an arbitrary constant scalar. To render a modal vector unique, we can assign a certain value to its magnitude, a procedure known as *vector normalization*. If a given natural mode has been normalized, then it represents a *normal mode*. One normalization scheme used frequently is to assign the value unity to a given component of the vector. Another normalization process is to choose the magnitude of $\mathbf{u}_r$ so as to satisfy

$$\mathbf{u}_r^T M \mathbf{u}_r = 1, \qquad r=1, 2, \ldots, n \tag{8.37}$$

Then, premultiplying Eq. (8.29) by $\mathbf{u}_r^T$ and considering Eq. (8.30), we conclude that if $\mathbf{u}_r$ satisfies Eq. (8.37) it also satisfies

$$\mathbf{u}_r^T K \mathbf{u}_r = \omega_r^2, \qquad r = 1, 2, \ldots, n \tag{8.38}$$

The natural modes possess a very important property known as orthogonality. But, a set of orthogonal vectors represents by definition a set of independent vectors. As we shall demonstrate later in this chapter, the implication of the orthogonality of the modal vectors $\mathbf{u}_r$ $(r = 1, 2, \ldots, n)$ is that the natural motions $\mathbf{q}_r(t)$ can be excited independently of one another. In general, however, the free vibration of an n-degree-of-freedom system can be expressed as a linear combination of the natural motions, or

$$\mathbf{q}(t) = \sum_{r=1}^{n} \mathbf{q}_r(t) = \sum_{r=1}^{n} c_r \cos(\omega_r t - \phi_r)\mathbf{u}_r \tag{8.39}$$

where the amplitudes c_r and the phase angles ϕ_r $(r = 1, 2, \ldots, n)$ play the role of constants of integration. They can be determined by imposing certain conditions on the solution. Ordinarily, the vector $\mathbf{q}(t)$ is required to satisfy

$$\mathbf{q}(0) = \mathbf{q}_0, \qquad \dot{\mathbf{q}}(0) = \dot{\mathbf{q}}_0 \tag{8.40}$$

where $\mathbf{q}_0$ and $\dot{\mathbf{q}}_0$ represent the vectors of initial displacements and velocities, respectively. Equations (8.40) are known as *initial conditions.* Inserting Eqs. (8.40) into Eq. (8.39) and using the orthogonality property of the normal modes, we can solve for the amplitudes c_r and phase angles ϕ_r $(r = 1, 2, \ldots, n)$.

Next, let us consider the solution of the eigenvalue problem. In view of Eq. (8.30) the characteristic equation can be written in the form

$$\Delta(\omega^2) = \det[K - \omega^2 M] = |K - \omega^2 M| = 0 \tag{8.41}$$

where $\Delta(\omega^2)$ is the characteristic determinant and represents a polynomial of degree n in ω^2. Hence, the roots of $\Delta(\omega^2)$ are the squares of the natural frequencies, $\omega_1^2, \omega_2^2, \ldots, \omega_n^2$. Then, upon solving the characteristic equation, we can obtain the modal vectors $\mathbf{u}_r$ by solving the equations

$$[K - \omega_r^2 M]\mathbf{u}_r = \mathbf{0}, \qquad r = 1, 2, \ldots, n \tag{8.42}$$

which implies that one must solve a set of n homogeneous algebraic equations for every ω_r^2 $(r = 1, 2, \ldots, n)$. But, because the matrix $[K - \omega_r^2 M]$ of the coefficients is singular, not all equations in a given set are independent. In fact, ordinarily only $n - 1$ equations are independent, so that it is possible to omit one of the equations in the set and solve the remaining $n - 1$ equations for the components of $\mathbf{u}_r$. Observing that there are $n - 1$ equations and n components of $\mathbf{u}_r$ and recalling that the magnitude of $\mathbf{u}_r$ is arbitrary, we can assign an arbitrary value to one of the components of $\mathbf{u}_r$ and solve $n - 1$ nonhomogeneous equations for the remaining $n - 1$ components. By assigning a value to one of the components of $\mathbf{u}_r$, the vector $\mathbf{u}_r$ has been normalized. This does not prevent us from normalizing the same vector again, perhaps by insisting that the vector $\mathbf{u}_r$ satisfy Eq. (8.37).

The procedure for the solution of the eigenvalue problem based on the characteristic determinant discussed above is of practical value only when the system possesses a small number of degrees of freedom. In fact, a direct solution can be obtained only when $n=2$, in which case the characteristic equation reduces to a quadratic equation, yielding ω_1^2 and ω_2^2. Then, the associated two-dimensional modal vectors $\mathbf{u}_1$ and $\mathbf{u}_2$ are obtained by solving a single algebraic equation in each case. For $n \geqslant 3$, no direct solution is possible, and we must solve for the roots of the characteristic polynomial in an iterative manner. Before solving for the roots, however, it is necessary to derive the characteristic polynomial by expanding the characteristic determinant $|K-\omega^2 M|$. For large n, this approach becomes prohibitive, so that other approaches are recommended.

The solution of the eigenvalue problem for $n \geqslant 3$ is essentially a computational problem. There exists a large variety of computational algorithms for the solution of the eigenvalue problem, all of them representing iteration techniques of one type or another. Among these, we mention the matrix iteration based on the power method, the Jacobi method, the method based on Sturm's theorem, and the *QR* method. The first two methods yield both the natural frequencies and the modal vectors, whereas the last two yield only the natural frequencies. The modal vectors must be obtained by solving nonhomogeneous algebraic equations, as discussed above. In this regard, one may wish to consider the inverse iteration method, which involves Gaussian elimination in conjunction with back substitution. Discussion of the computational algorithms mentioned here lies beyond the scope of this text, and the interested reader can find details of all these algorithms in the book by L. Meirovitch entitled *Computational Methods in Structural Dynamics.** It should also be pointed out that computer codes for all these algorithms are available in virtually all the libraries of scientific computing centers.

Example 8.2

Derive the equations of motion for the two-degree-of-freedom undamped system shown in Fig. 8.2, derive and solve the associated eigenvalue problem, and obtain the response to the initial conditions

$$\mathbf{q}_0 = \begin{bmatrix} 0 \\ 0 \end{bmatrix}, \qquad \dot{\mathbf{q}}_0 = \begin{bmatrix} 0 \\ 1 \end{bmatrix} \tag{a}$$

The kinetic energy is simply

$$T = \tfrac{1}{2}\left(m\dot{q}_1^2 + 2m\dot{q}_2^2\right) = \frac{1}{2}\begin{bmatrix} \dot{q}_1 \\ \dot{q}_2 \end{bmatrix}^{\mathrm{T}} \begin{bmatrix} m & 0 \\ 0 & 2m \end{bmatrix} \begin{bmatrix} \dot{q}_1 \\ \dot{q}_2 \end{bmatrix} \tag{b}$$

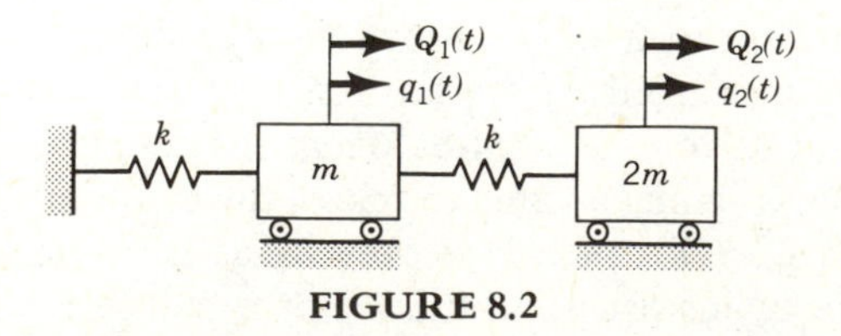

FIGURE 8.2

*Sijthoff-Noordhoff International Publishers, Inc., The Netherlands, 1980.

so that the mass matrix is

$$M = m\begin{bmatrix} 1 & 0 \\ 0 & 2 \end{bmatrix} \tag{c}$$

Moreover, the potential energy has the expression

$$\begin{aligned} V &= \tfrac{1}{2}[kq_1^2 + k(q_2 - q_1)^2] = \tfrac{1}{2}(2kq_1^2 - 2kq_1q_2 + kq_2^2) \\ &= \tfrac{1}{2}\begin{bmatrix} q_1 \\ q_2 \end{bmatrix}^{\mathrm{T}}\begin{bmatrix} 2k & -k \\ -k & k \end{bmatrix}\begin{bmatrix} q_1 \\ q_2 \end{bmatrix} \end{aligned} \tag{d}$$

so that the stiffness matrix is

$$K = k\begin{bmatrix} 2 & -1 \\ -1 & 1 \end{bmatrix} \tag{e}$$

Hence, the equations of motion have the matrix form

$$\begin{bmatrix} m & 0 \\ 0 & 2m \end{bmatrix}\begin{bmatrix} \ddot{q}_1(t) \\ \ddot{q}_2(t) \end{bmatrix} + \begin{bmatrix} 2k & -k \\ -k & k \end{bmatrix}\begin{bmatrix} q_1(t) \\ q_2(t) \end{bmatrix} = \begin{bmatrix} Q_1(t) \\ Q_2(t) \end{bmatrix} \tag{f}$$

To derive the eigenvalue problem, we set $Q_1(t) = Q_2(t) = 0$ in Eq. (f). Then, consistent with Eq. (8.32), if we let

$$q_1(t) = e^{i\omega t}u_1, \qquad q_2(t) = e^{i\omega t}u_2 \tag{g}$$

and introduce Eqs. (g) into Eq. (f), we obtain the eigenvalue problem

$$\begin{bmatrix} 2k - \omega^2 m & -k \\ -k & k - 2\omega^2 m \end{bmatrix}\begin{bmatrix} u_1 \\ u_2 \end{bmatrix} = \begin{bmatrix} 0 \\ 0 \end{bmatrix} \tag{h}$$

so that the characteristic equation has the form

$$\Delta(\omega^2) = \begin{vmatrix} 2k - \omega^2 m & -k \\ -k & k - 2\omega^2 m \end{vmatrix} = 2(m\omega^2)^2 - 5m\omega^2 k + k^2 = 0 \tag{i}$$

which represents a quadratic equation in ω^2 having the roots

$$\left.\begin{matrix} \omega_1^2 \\ \omega_1^2 \end{matrix}\right\} = \left[\frac{5}{4} \mp \sqrt{\left(\frac{5}{4}\right)^2 - \frac{1}{2}}\right]\frac{k}{m} = \begin{cases} 0.2192k/m \\ 2.2808k/m \end{cases} \tag{j}$$

Introducing the notation

$$\mathbf{u}_1 = \begin{bmatrix} u_{11} \\ u_{21} \end{bmatrix}, \qquad \mathbf{u}_2 = \begin{bmatrix} u_{12} \\ u_{22} \end{bmatrix} \tag{k}$$

and inserting Eqs. (j) into Eq. (h), we obtain

$$\begin{bmatrix} 2 - 0.2192 & -1 \\ -1 & 1 - 2 \times 0.2192 \end{bmatrix}\begin{bmatrix} u_{11} \\ u_{21} \end{bmatrix} = \begin{bmatrix} 0 \\ 0 \end{bmatrix} \tag{l}$$

$$\begin{bmatrix} 2 - 2.2808 & -1 \\ -1 & 1 - 2 \times 2.2808 \end{bmatrix}\begin{bmatrix} u_{12} \\ u_{22} \end{bmatrix} = \begin{bmatrix} 0 \\ 0 \end{bmatrix} \tag{m}$$

Letting arbitrarily $u_{21}=u_{22}=1$, we see that Eqs. (l) and (m) yield the modal vectors

$$\mathbf{u}_1=\begin{bmatrix}0.5616\\1\end{bmatrix}, \quad \mathbf{u}_2=\begin{bmatrix}-3.5613\\1\end{bmatrix} \tag{n}$$

Note that, with u_{21} known, Eq. (l) represents two algebraic equations in the single unknown u_{11} and it does not matter which of the two equations we solve for u_{11}, as both equations yield the same result. A similar statement can be made concerning u_{12} and Eq. (m).

Although, to produce the response to the initial conditions (a), it is not necessary to normalize the modal vectors $\mathbf{u}_1$ and $\mathbf{u}_2$ according to the scheme given by Eq. (8.37), we shall normalize the modal vectors to illustrate the procedure. Moreover, the normalized modal vectors will be used in a later example. Hence, let

$$\mathbf{u}_1=\alpha_1\begin{bmatrix}0.5616\\1\end{bmatrix} \tag{o}$$

Introducing Eq. (o) into Eq. (8.37) with $r=1$, we obtain

$$\alpha_1^2\begin{bmatrix}0.5616\\1\end{bmatrix}^{\mathrm{T}}\begin{bmatrix}m & 0\\0 & 2m\end{bmatrix}\begin{bmatrix}0.5616\\1\end{bmatrix}=m\alpha_1^2(0.5616^2+2\times 1^2)=1 \tag{p}$$

Solving Eq. (p) for α_1, we have

$$\alpha_1=\frac{1}{\sqrt{(0.5616^2+2)m}}=0.6572m^{-1/2} \tag{q}$$

so that the normalized first modal vector is

$$\mathbf{u}_1=0.6572m^{-1/2}\begin{bmatrix}0.5616\\1\end{bmatrix}=m^{-1/2}\begin{bmatrix}0.3691\\0.6572\end{bmatrix} \tag{r}$$

Following the same procedure, the normalized second modal vector can be shown to be

$$\mathbf{u}_2=m^{-1/2}\begin{bmatrix}-0.9294\\0.2610\end{bmatrix} \tag{s}$$

To obtain the response, we first use Eq. (8.39) and write

$$\begin{bmatrix}q_1(t)\\q_2(t)\end{bmatrix}=c_1\cos(\omega_1 t-\phi_1)\begin{bmatrix}0.5616\\1\end{bmatrix}+c_2\cos(\omega_2 t-\phi_2)\begin{bmatrix}-3.5613\\1\end{bmatrix} \tag{t}$$

where we used the modal vectors as given by Eqs. (n). Then, using the initial conditions (a), we can write

$$\begin{aligned}\mathbf{q}(0)&=c_1\cos\phi_1\begin{bmatrix}0.5616\\1\end{bmatrix}+c_2\cos\phi_2\begin{bmatrix}-3.5613\\1\end{bmatrix}=\begin{bmatrix}0\\0\end{bmatrix}\\ \dot{\mathbf{q}}(0)&=c_1\omega_1\sin\phi_1\begin{bmatrix}0.5616\\1\end{bmatrix}+c_2\omega_2\sin\phi_2\begin{bmatrix}-3.5613\\1\end{bmatrix}=\begin{bmatrix}0\\1\end{bmatrix}\end{aligned} \tag{u}$$

where

$$\omega_1=\sqrt{0.2192k/m}=0.4682\sqrt{k/m}$$
$$\omega_2=\sqrt{2.2808k/m}=1.5102\sqrt{k/m} \tag{v}$$

Equations (u) represent four equations in the unknowns c_1, c_2, ϕ_1, and ϕ_2. Their solution is

$$c_1=1.8449\sqrt{m/k}, \qquad c_2=0.0902\sqrt{m/k}, \qquad \phi_1=\phi_2=\pi/2 \tag{w}$$

so that the response is

$$\begin{bmatrix} q_1(t) \\ q_2(t) \end{bmatrix}=1.8449\sqrt{\frac{m}{k}}\sin 0.4682\sqrt{\frac{k}{m}}\,t\begin{bmatrix} 0.5616 \\ 1 \end{bmatrix}$$
$$+0.0902\sqrt{\frac{m}{k}}\sin 1.5102\sqrt{\frac{k}{m}}\,t\begin{bmatrix} -3.5613 \\ 1 \end{bmatrix} \tag{x}$$

It should be pointed out that the same response would have been obtained had we used the normalized modal vectors given by Eqs. (r) and (s) instead of those given by Eqs. (n). In this case, the constants c_1 and c_2 would have acquired different values, but the final result would have been the same. Normalization is a convenience having no effect on the response.

In Section 8.5, we shall derive the system response by a different approach known as modal analysis.

8.4 ORTHOGONALITY OF THE MODAL VECTORS

As pointed out in Section 8.3, the modal vectors possess the orthogonality property, a property that is not only interesting but also very useful, as it facilitates the derivation of the system response. This is not ordinary orthogonality, but orthogonality in a certain sense, as we shall demonstrate in this section.

Let us consider two distinct solutions ω_r, $\mathbf{u}_r$ and ω_s, $\mathbf{u}_s$ of the eigenvalue problem (8.42) and write Eq. (8.42) associated with these solutions in the form

$$K\mathbf{u}_r=\omega_r^2 M\mathbf{u}_r \tag{8.43a}$$

$$K\mathbf{u}_s=\omega_s^2 M\mathbf{u}_s \tag{8.43b}$$

Premultiplying Eq. (8.43a) by $\mathbf{u}_s^{\mathrm{T}}$ and Eq. (8.43b) by $\mathbf{u}_r^{\mathrm{T}}$, we obtain

$$\mathbf{u}_s^{\mathrm{T}}K\mathbf{u}_r=\omega_r^2\mathbf{u}_s^{\mathrm{T}}M\mathbf{u}_r \tag{8.44a}$$

$$\mathbf{u}_r^{\mathrm{T}}K\mathbf{u}_s=\omega_s^2\mathbf{u}_r^{\mathrm{T}}M\mathbf{u}_s \tag{8.44b}$$

But, because the stiffness matrix K and the mass matrix M are symmetric, $K^{\mathrm{T}}=K$, $M^{\mathrm{T}}=M$, we can write

$$(\mathbf{u}_r^{\mathrm{T}}K\mathbf{u}_s)^{\mathrm{T}}=\mathbf{u}_s^{\mathrm{T}}K^{\mathrm{T}}\mathbf{u}_r=\mathbf{u}_s^{\mathrm{T}}K\mathbf{u}_r \tag{8.45a}$$

$$(\mathbf{u}_r^{\mathrm{T}}M\mathbf{u}_s)^{\mathrm{T}}=\mathbf{u}_s^{\mathrm{T}}M^{\mathrm{T}}\mathbf{u}_r=\mathbf{u}_s^{\mathrm{T}}M\mathbf{u}_r \tag{8.45b}$$

where we used the fact that the transpose of a product of matrices is equal to the product of the transposed matrices in reversed order. In this regard, we note that a vector can be regarded as a column matrix. Transposing Eq. (8.44b), subtracting the result from Eq. (8.44a), and considering Eqs. (8.45), we obtain

$$0=(\omega_r^2-\omega_s^2)\mathbf{u}_s^{\mathrm{T}}M\,\mathbf{u}_r \tag{8.46}$$

so that, assuming that $\omega_r^2 \neq \omega_s^2$, we conclude that Eq. (8.46) is satisfied provided

$$\mathbf{u}_s^{\mathrm{T}}M\,\mathbf{u}_r=0, \qquad \omega_r^2 \neq \omega_s^2 \tag{8.47}$$

which expresses the *orthogonality of the modal vectors.* This is not ordinary orthogonality, however, but *orthogonality with respect to the mass matrix.* Moreover, inserting Eq. (8.47) into Eq. (8.44a), we obtain

$$\mathbf{u}_s^{\mathrm{T}}K\,\mathbf{u}_r=0, \qquad \omega_r^2 \neq \omega_s^2 \tag{8.48}$$

so that *the modal vectors are also orthogonal with respect to the stiffness matrix.*

Orthogonality of a set of vectors implies independence of the set of vectors, by definition. As it turns out, because of the symmetry of the eigenvalue problem, the modal vectors $\mathbf{u}_r$ and $\mathbf{u}_s$ are independent whether ω_r^2 is different from ω_s^2 or is equal to ω_s^2 (see text by Meirovitch*). Hence, if a square of a natural frequency is a multiple root of the characteristic polynomial, the modal vectors belonging to the repeated root are all independent. Moreover, a linear combination of the modal vectors belonging to a repeated root is also a modal vector. Hence, the modal vectors belonging to the repeated root can be chosen in such linear combinations so as to render them orthogonal. Of course, they are orthogonal to the modal vectors belonging to the other roots. It follows that *the modal vectors are orthogonal with respect to the mass matrix* (*and with respect to the stiffness matrix*) *whether the system has repeated natural frequencies or not.*

When $s=r$, the product $\mathbf{u}_s^{\mathrm{T}}M\,\mathbf{u}_r$ is not zero but a real and positive quantity. In fact, we used a process called normalization in Section 8.3 to adjust the magnitude of $\mathbf{u}_r$ by imposing the condition $\mathbf{u}_r^{\mathrm{T}}M\,\mathbf{u}_r=1$, as indicated by Eq. (8.37). If all the modal vectors are normalized according to Eq. (8.37), then the set of modal vectors is said to be *orthonormal.* The orthonormality of the eigenvectors can be expressed in the form

$$\mathbf{u}_s^{\mathrm{T}}M\,\mathbf{u}_r=\delta_{rs}, \qquad r,s=1,2,\ldots,n \tag{8.49}$$

where δ_{rs} is the Kronecker delta, defined in Section 8.2 as equal to zero if $s \neq r$ and equal to unity if $s=r$. Moreover, from Eq. (8.44a), we conclude that

$$\mathbf{u}_s^{\mathrm{T}}K\,\mathbf{u}_r=\omega_r^2\,\delta_{rs}, \qquad r,s=1,2,\ldots,n \tag{8.50}$$

The orthogonality of the modal vectors has important implications in vibrations. In particular, it permits the motion of a multi-degree-of-freedom dynamical system to be expressed in terms of the modal vectors multiplied by some time-dependent coefficients. To introduce the idea, we recall from earlier chapters that a

---

**Op. cit.*, p. 58.

three-dimensional vector $\mathbf{r}$ can be expressed in terms of the cartesian components x, y, z as follows:

$$\mathbf{r}=x\mathbf{i}+y\mathbf{j}+z\mathbf{k} \tag{8.51}$$

where $\mathbf{i}$, $\mathbf{j}$, $\mathbf{k}$ are unit vectors with directions coinciding with the directions of the x, y, z axes, respectively. The components x, y, and z of the vector $\mathbf{r}$ can be obtained by means of the dot products

$$x=\mathbf{i}\cdot\mathbf{r}, \qquad y=\mathbf{j}\cdot\mathbf{r}, \qquad z=\mathbf{k}\cdot\mathbf{r} \tag{8.52}$$

The above resolution of $\mathbf{r}$ in terms of Cartesian components is based on the fact that axes x, y, and z are orthogonal. In fact, the set of unit vectors $\mathbf{i}$, $\mathbf{j}$, and $\mathbf{k}$ is orthonormal, as

$$\begin{aligned} &\mathbf{i}\cdot\mathbf{i}=1, \qquad \mathbf{j}\cdot\mathbf{j}=1, \qquad \mathbf{k}\cdot\mathbf{k}=1 \\ &\mathbf{i}\cdot\mathbf{j}=\mathbf{j}\cdot\mathbf{i}=0, \qquad \mathbf{i}\cdot\mathbf{k}=\mathbf{k}\cdot\mathbf{i}=0, \qquad \mathbf{j}\cdot\mathbf{k}=\mathbf{k}\cdot\mathbf{j}=0 \end{aligned} \tag{8.53}$$

The same operation can be expressed in a notation more consistent with our objective. To this end, let us introduce the vectors

$$\mathbf{e}_1=\begin{bmatrix}1\\0\\0\end{bmatrix}, \qquad \mathbf{e}_2=\begin{bmatrix}0\\1\\0\end{bmatrix}, \qquad \mathbf{e}_3=\begin{bmatrix}0\\0\\1\end{bmatrix} \tag{8.54}$$

Clearly, $\mathbf{e}_1$, $\mathbf{e}_2$, and $\mathbf{e}_3$ are the equivalent of the unit vectors $\mathbf{i}$, $\mathbf{j}$, and $\mathbf{k}$. Indeed, they are orthonormal, as they satisfy

$$\mathbf{e}_s^{\mathrm{T}}\mathbf{e}_r=\delta_{rs}, \qquad r, s=1, 2, 3 \tag{8.55}$$

and we observe that Eqs. (8.55) are the counterpart of Eqs. (8.53). Any three-dimensional vector $\mathbf{r}$ with components x, y, and z can be expressed in terms of the unit vectors $\mathbf{e}_1$, $\mathbf{e}_2$, and $\mathbf{e}_3$ in the form

$$\mathbf{r}=\begin{bmatrix}x\\y\\z\end{bmatrix}=x\mathbf{e}_1+y\mathbf{e}_2+z\mathbf{e}_3 \tag{8.56}$$

where x, y, z are called the coordinates of $\mathbf{r}$ with respect to the set $\mathbf{e}_1$, $\mathbf{e}_2$, $\mathbf{e}_3$. It is easy to see that Eq. (8.56) is entirely equivalent to Eq. (8.51). The coordinates x, y, and z can be obtained by recalling Eqs. (8.55) and writing

$$x=\mathbf{e}_1^{\mathrm{T}}\mathbf{r}, \qquad y=\mathbf{e}_2^{\mathrm{T}}\mathbf{r}, \qquad z=\mathbf{e}_3^{\mathrm{T}}\mathbf{r} \tag{8.57}$$

which are the counterpart of Eqs. (8.52).

The above decomposition of a vector into three components can be interpreted geometrically as expressing a vector in terms of projections along three orthogonal axes. The same idea can be extended to a space of higher dimension than 3, although in the process some of the physical content of the idea must be abandoned in favor of more abstract thinking. Hence, following the same line of thought as above, we can conceive of a set of n orthonormal vectors of the form

$$\mathbf{e}_1 = \begin{bmatrix} 1 \\ 0 \\ \vdots \\ 0 \end{bmatrix}, \quad \mathbf{e}_2 = \begin{bmatrix} 0 \\ 1 \\ \vdots \\ 0 \end{bmatrix}, \quad \ldots, \quad \mathbf{e}_n = \begin{bmatrix} 0 \\ 0 \\ \vdots \\ 1 \end{bmatrix} \tag{8.58}$$

Then, an n-dimensional vector $\mathbf{u}$ can be expressed in terms of components $u_1, u_2, \ldots, u_n$ along these unit vectors as follows:

$$\mathbf{u} = \begin{bmatrix} u_1 \\ u_2 \\ \vdots \\ u_n \end{bmatrix} = u_1 \mathbf{e}_1 + u_2 \mathbf{e}_1 + \cdots + u_n \mathbf{e}_n = \sum_{i=1}^{n} u_i \mathbf{e}_i \tag{8.59}$$

In linear algebra, the vectors $\mathbf{e}_1, \mathbf{e}_2, \ldots, \mathbf{e}_n$ are known as *standard unit vectors.*

The question remains as to how the above developments relate to vibrations. In this regard, we note that the n-dimensional vector $\mathbf{u}$ can be regarded as representing the displacement vector of an n-degree-of-freedom system. But, the decomposition of the displacement vector $\mathbf{u}$ into components along the unit vectors $\mathbf{e}_1, \mathbf{e}_1, \ldots, \mathbf{e}_n$ does not have much practical value in vibrations. However, the decomposition of the displacement vector $\mathbf{u}$ into components along another set of orthogonal vectors has considerable value. This latter set is simply the set of modal vectors $\mathbf{u}_1, \mathbf{u}_1, \ldots, \mathbf{u}_n$. Indeed, letting $\alpha_1, \alpha_2, \ldots, \alpha_n$ be the components of $\mathbf{u}$ along the modal vectors, we can express $\mathbf{u}$ in the form of the linear combination

$$\mathbf{u} = \alpha_1 \mathbf{u}_1 + \alpha_2 \mathbf{u}_2 + \cdots + \alpha_n \mathbf{u}_n = \sum_{r=1}^{n} \alpha_r \mathbf{u}_r \tag{8.60}$$

The modal vectors do not possess the ordinary orthogonality property but orthogonality with respect to the mass matrix. Moreover, we assume that the modal vectors have been normalized so as to satisfy Eqs. (8.49). Then, premultiplying Eq. (8.60) by $\mathbf{u}_s^T M$ and considering Eqs. (8.49), we obtain the components of $\mathbf{u}$ in the form

$$\alpha_r = \mathbf{u}_r^T M \mathbf{u}, \qquad r = 1, 2, \ldots, n \tag{8.61}$$

Equations (8.60) and (8.61) are referred to as the *expansion theorem*, and they form the basis for modal analysis, whereby the response of an n-degree-of-freedom system is expressed in terms of the modal vectors multiplied by time-dependent coefficients known as modal coordinates.

Example 8.3

Verify that the two modal vectors $\mathbf{u}_1$ and $\mathbf{u}_2$ computed in Example 8.2 are orthogonal with respect to the mass and stiffness matrices.

The modal vectors of Example 8.2 are

$$\mathbf{u}_1 = m^{-1/2} \begin{bmatrix} 0.3691 \\ 0.6572 \end{bmatrix}, \qquad \mathbf{u}_2 = m^{-1/2} \begin{bmatrix} -0.9294 \\ 0.2610 \end{bmatrix} \tag{a}$$

and the mass and stiffness matrices are

$$M = m\begin{bmatrix}1 & 0\\ 0 & 2\end{bmatrix}, \qquad K = k\begin{bmatrix}2 & -1\\ -1 & 1\end{bmatrix} \tag{b}$$

Hence, to verify the orthogonality with respect to M, we write

$$\begin{aligned}\mathbf{u}_1^{\mathrm{T}} M \mathbf{u}_2 &= m^{-1/2}\begin{bmatrix}0.3691\\ 0.6572\end{bmatrix}^{\mathrm{T}} m\begin{bmatrix}1 & 0\\ 0 & 2\end{bmatrix} m^{-1/2}\begin{bmatrix}-0.9294\\ 0.2610\end{bmatrix}\\ &= -0.3691\times 0.9294 + 2\times 0.6572\times 0.2610\\ &= -0.3430 + 0.3430 = 0\end{aligned} \tag{c}$$

so that the orthogonality of $\mathbf{u}_1$ and $\mathbf{u}_2$ with respect to M is verified. Similarly, to verify the orthogonality of $\mathbf{u}_1$ and $\mathbf{u}_2$ with respect to K, we form

$$\begin{aligned}\mathbf{u}_1^{\mathrm{T}} K \mathbf{u}_2 &= m^{-1/2}\begin{bmatrix}0.3691\\ 0.6572\end{bmatrix}^{\mathrm{T}} k\begin{bmatrix}2 & -1\\ -1 & 1\end{bmatrix} m^{-1/2}\begin{bmatrix}-0.9294\\ 0.2610\end{bmatrix}\\ &= \frac{k}{m}\begin{bmatrix}0.3691\\ 0.6572\end{bmatrix}^{\mathrm{T}}\begin{bmatrix}-2\times 0.9294 - 0.2610\\ 0.9294 + 0.2610\end{bmatrix} = \frac{k}{m}\begin{bmatrix}0.3691\\ 0.6572\end{bmatrix}^{\mathrm{T}}\begin{bmatrix}-2.1198\\ 1.1904\end{bmatrix}\\ &= \frac{k}{m}(-0.3691\times 2.1198 + 0.6572\times 1.1904)\\ &= \frac{k}{m}(-0.7824 + 0.7824) = 0\end{aligned} \tag{d}$$

so that the orthogonality of $\mathbf{u}_1$ and $\mathbf{u}_2$ with respect to K is also verified.

8.5 RESPONSE OF MULTI-DEGREE-OF-FREEDOM UNDAMPED SYSTEMS BY MODAL ANALYSIS

The response of a multi-degree-of-freedom undamped system is governed by the differential equation (8.24), which represents a set of n simultaneous nonhomogeneous second-order ordinary differential equations. An attempt to solve the equations in the form (8.24) is likely to encounter enormous difficulties, so that the idea of transforming Eq. (8.24) into one that lends itself to an easier solution arises naturally. To this end, the Laplace transformation used so successfully for single-degree-of-freedom systems comes to mind immediately. It turns out that such an approach may be feasible for very low-order systems but becomes prohibitive as n increases. Hence, a different kind of transformation is advisable. Such a transformation does indeed exist, and it represents a coordinate transformation reducing the set of simultaneous equations (8.24) into a set of independent second-order equations, where each of the independent equations resembles that of an undamped single-degree-of-freedom system.

Let us consider the n-dimensional displacement vector $\mathbf{q}(t) = [q_1(t)\ q_2(t) \ldots q_n(t)]^{\mathrm{T}}$. If the time t is frozen for a moment, then the vector $\mathbf{q}$ can be regarded as a constant vector in an n-dimensional vector space. But, in Section 8.4 we have shown that an n-dimensional vector space can be defined in terms of n orthonormal vectors in the same way that a three-dimensional Cartesian space can be defined in terms of the unit vectors **i**, **j**, **k**. For reasons that will become evident later, we choose to define the n-dimensional space in terms of the n orthonormal modal vectors $\mathbf{u}_1$, $\mathbf{u}_2, \ldots, \mathbf{u}_n$, where the vectors satisfy the orthonormality relations (8.49) and (8.50). Denoting the coordinates of $\mathbf{q}$ with respect to the set $\mathbf{u}_1, \mathbf{u}_2, \ldots, \mathbf{u}_n$, by $\eta_1, \eta_2, \ldots, \eta_n$, respectively, we can use the expansion theorem (8.60) and write

$$\mathbf{q}=\eta_1\mathbf{u}_1+\eta_2\mathbf{u}_2+\cdots+\eta_n\mathbf{u}_n=\sum_{r=1}^{n}\eta_r\mathbf{u}_r \tag{8.62}$$

and we note that $\eta_1, \eta_2, \ldots, \eta_n$ correspond to the frozen time t. Next, let us unfreeze the time t and consider a later instant $t+\Delta t$, where Δt is a small time increment. A coordinate transformation similar to Eq. (8.62) can be written between $q_1(t+\Delta t), q_2(t+\Delta t), \ldots, q_n(t+\Delta t)$ and $\eta_1(t+\Delta t), \eta_2(t+\Delta t), \ldots, \eta_n(t+\Delta t)$. By continuity, small changes in $q_1, q_2, \ldots, q_n$ resulting from the time difference Δt lead to small changes in $\eta_1, \eta_2, \ldots, \eta_n$, so that transformation (8.62) is valid whether $\mathbf{q}$ is a constant vector or a time-dependent vector. In view of this, we can rewrite Eq. (8.62) in the form

$$\mathbf{q}(t)=\eta_1(t)\mathbf{u}_1+\eta_2(t)\mathbf{u}_2+\cdots+\eta_n(t)\mathbf{u}_n=\sum_{r=1}^{n}\eta_r(t)\mathbf{u}_r \tag{8.63}$$

Equation (8.63) represents a linear transformation between two sets of generalized coordinates, $q_1(t), q_2(t), \ldots, q_n(t)$ and $\eta_1(t), \eta_2(t), \ldots, \eta_n(t)$. It is the coordinate transformation capable of reducing a set of simultaneous differential equations of motion to a set of independent equations.

Introducing Eq. (8.63) into Eq. (8.24), we obtain

$$M\sum_{r=1}^{n}\ddot{\eta}_r(t)\mathbf{u}_r+K\sum_{r=1}^{n}\eta_r(t)\mathbf{u}_r=\sum_{r=1}^{n}[M\mathbf{u}_r\ddot{\eta}_r(t)+K\mathbf{u}_r\eta_r(t)]=\mathbf{Q}(t) \tag{8.64}$$

Then, premultiplying Eq. (8.64) by $\mathbf{u}_s^{\mathrm{T}}$ and recalling the orthonormality relations (8.49) and (8.50), we can write

$$\sum_{r=1}^{n}[\mathbf{u}_s^{\mathrm{T}}M\mathbf{u}_r\ddot{\eta}_r(t)+\mathbf{u}_s^{\mathrm{T}}K\mathbf{u}_r\eta_r(t)]=\sum_{r=1}^{n}[\delta_{rs}\,\ddot{\eta}_r(t)+\omega_r^2\,\delta_{rs}\,\eta_r(t)]=\mathbf{u}_s^{\mathrm{T}}\mathbf{Q}(t) \tag{8.65}$$

However, by the nature of the Kronecker delta, only one term survives in the series in Eq. (8.65), namely, the term corresponding to $r=s$. Hence, letting s take all the integer values from 1 to n, we see that Eq. (8.65) yields

$$\ddot{\eta}_s(t)+\omega_s^2\eta_s(t)=N_s(t), \qquad s=1, 2, \ldots, n \tag{8.66}$$

where

$$N_s(t)=\mathbf{u}_s^{\mathrm{T}}\mathbf{Q}(t), \qquad s=1, 2, \ldots, n \tag{8.67}$$

are generalized forces corresponding to the coordinates $\eta_s(t)\,(s=1, 2, \ldots, n)$.

Equations (8.66) represent a set of independent equations of motion, so that transformation (8.63) is indeed capable of reducing a set of simultaneous equations of motion to a set of independent equations of motion. In contrast with Eqs. (8.24), which in general use physical coordinates, the coordinates appearing in Eqs. (8.66) tend to be abstract, in the sense that they may not have physical meaning. The coordinates $\eta_1(t), \eta_2(t), \ldots, \eta_n(t)$, corresponding to the set of independent equations, are called *principal coordinates*, or *natural coordinates*. In view of the fact that the modal vectors $\mathbf{u}_r$ $(r=1, 2, \ldots, n)$ have been normalized, $\eta_1(t), \eta_2(t), \ldots, \eta_n(t)$ can also be called *normal coordinates*. Moreover, because $\eta_1(t), \eta_2(t), \ldots, \eta_n(t)$ are the coordinates of $\mathbf{q}(t)$ with respect to the modal vectors, they are often referred to as *modal coordinates*. Correspondingly, $N_1(t), N_2(t), \ldots, N_n(t)$ are called *modal forces*. Unlike the forces $Q_1(t), Q_2(t), \ldots, Q_n(t)$, which tend to be actual forces and moments, $N_1(t), N_2(t), \ldots, N_n(t)$ tend to be abstract forces.

Equations (8.66) have the appearance of the equations of motion of a set of n independent single-degree-of-freedom undamped systems of the type discussed in Chapter 4. The homogeneous solution of such equations has the form of Eq. (4.57) and the particular solution can be written in the form of the convolution integral given by Eq. (4.170). Denoting the initial modal displacements and velocities by $\eta_s(0)$ and $\dot{\eta}_s(0)$, respectively, recognizing that, by letting $m=1$, $\zeta=0$, $\omega_n=\omega_d=\omega_s$, the impulse response for a typical equation in the independent set can be obtained from Eq. (4.162) in the form

$$g_s(t)=\frac{1}{\omega_s} \sin \omega_s t \, \mathscr{u}(t), \qquad s=1, 2, \ldots, n \tag{8.68}$$

where $\mathscr{u}(t)$ is the unit step function, and using Eqs. (4.57) and (4.170), we can write the complete solution of Eqs. (8.66) as follows:

$$\begin{aligned}\eta_s(t)&=\eta_s(0) \cos \omega_s t+\frac{1}{\omega_s} \dot{\eta}_s(0) \sin \omega_s t+\int_0^t N_s(\tau) g_s(t-\tau)\, d\tau \\ &=\eta_s(0) \cos \omega_s t+\frac{1}{\omega_s} \dot{\eta}_s(0) \sin \omega_s t \\ &\quad+\frac{1}{\omega_s} \int_0^t N_s(\tau) \sin \omega_s(t-\tau)\, d\tau, \qquad s=1, 2, \ldots, n\end{aligned} \tag{8.69}$$

The initial modal displacements and velocities can be obtained from Eq. (8.63). Indeed, they are related to the actual initial displacement and velocity vectors $\mathbf{q}(0)$ and $\dot{\mathbf{q}}(0)$ by

$$\mathbf{q}(0)=\sum_{r=1}^{n} \eta_r(0) \mathbf{u}_r, \qquad \dot{\mathbf{q}}(0)=\sum_{r=1}^{n} \dot{\eta}_r(0) \mathbf{u}_r \tag{8.70}$$

Premultiplying Eqs. (8.70) by $\mathbf{u}_s^{\mathrm{T}} M$ and recalling the orthonormality relations (8.49), we obtain

$$\eta_s(0)=\mathbf{u}_s^{\mathrm{T}} M \mathbf{q}(0), \qquad \dot{\eta}_s(0)=\mathbf{u}_s^{\mathrm{T}} M \dot{\mathbf{q}}(0) \tag{8.71}$$

Inserting Eqs. (8.69) into Eq. (8.63) and recalling Eqs. (8.67) and (8.71), we can write the complete solution

$$\begin{aligned}\mathbf{q}(t)&=\sum_{r=1}^{n}\left[\eta_r(0)\cos\omega_r t+\frac{1}{\omega_r}\dot{\eta}_r(0)\sin\omega_r t\right.\\&\quad\left.+\frac{1}{\omega_r}\int_0^t N_r(\tau)\sin\omega_r(t-\tau)\,d\tau\right]\mathbf{u}_r\\&=\sum_{r=1}^{n}\left\{\mathbf{u}_r^{\mathrm{T}}M\left[\mathbf{q}(0)\cos\omega_r t+\frac{1}{\omega_r}\dot{\mathbf{q}}(0)\sin\omega_r t\right]\right.\\&\quad\left.+\frac{1}{\omega_r}\mathbf{u}_r^{\mathrm{T}}\int_0^t \mathbf{Q}(\tau)\sin\omega_r(t-\tau)\,d\tau\right\}\mathbf{u}_r\end{aligned}\tag{8.72}$$

Equation (8.72) gives the system response in terms of the natural frequencies ω_r and normal modes $\mathbf{u}_r$ $(r=1, 2, \ldots, n)$ for any initial conditions $\mathbf{q}(0)$ and $\dot{\mathbf{q}}(0)$ and external excitation $\mathbf{Q}(t)$. The procedure for deriving the system response in terms of the normal modes is known as *modal analysis.*

Although not stated so explicitly, Eq. (8.72) gives the system response for the case of transient excitation. Indeed, by its very nature, the use of the convolution integral is suitable for transient forces, but not for steady-state harmonic forces. Yet, modal analysis is not restricted to any class of excitations and can be used for harmonic excitations as well. Hence, using the approach of Chapters 3 and 4 and letting the external force vector have the form

$$\mathbf{Q}(t)=\mathbf{Q}_0 e^{i\omega t}, \qquad i=\sqrt{-1}\tag{8.73}$$

where $\mathbf{Q}_0$ is a constant vector and ω is the excitation frequency, we can obtain the modal forces from Eqs. (8.67) by writing

$$N_s=\mathbf{u}_s^{\mathrm{T}}\mathbf{Q}(t)=\mathbf{u}_s^{\mathrm{T}}\mathbf{Q}_0 e^{i\omega t}=N_{s0}e^{i\omega t}, \qquad s=1, 2, \ldots, n\tag{8.74}$$

where

$$N_{s0}=\mathbf{u}_s^{\mathrm{T}}\mathbf{Q}_0, \qquad s=1, 2, \ldots, n\tag{8.75}$$

can be identified as amplitudes of the modal forces. Inserting Eqs. (8.74) into Eqs. (8.66), we obtain the modal equations

$$\ddot{\eta}_s(t)+\omega_s^2\eta_s(t)=N_{s0}e^{i\omega t}, \qquad s=1, 2, \ldots, n\tag{8.76}$$

Equations of the type (8.76) were discussed in Section 4.9. Indeed, using results obtained in Section 4.9, we can write the modal response to steady-state harmonic excitation in the form

$$\eta_s(t)=\frac{N_{s0}}{\omega_s^2-\omega^2}e^{i\omega t}=\frac{N_{s0}}{\omega_s^2}\frac{1}{1-(\omega/\omega_s)^2}e^{i\omega t}, \qquad s=1, 2, \ldots, n\tag{8.77}$$

Inserting Eqs. (8.77) into Eq. (8.63) and considering Eqs. (8.75), we obtain the actual response

$$\mathbf{q}(t)=\sum_{s=1}^{n}\eta_s(t)\mathbf{u}_s=\left[\sum_{s=1}^{n}\frac{\mathbf{u}_s^{\mathrm{T}}\mathbf{Q}_0}{\omega_s^2}\frac{1}{1-(\omega/\omega_s)^2}\mathbf{u}_s\right]e^{i\omega t}, \qquad s=1, 2, \ldots, n\tag{8.78}$$

From Eq. (8.78), we conclude that the response increases without bounds every time the driving frequency ω is in the neighborhood of one of the natural frequencies ω_s $(s=1, 2, \ldots, n)$. Hence, *for an n-degree-of-freedom system, there are n resonant frequencies.*

Example 8.4

Consider the two-degree-of-freedom undamped system of Example 8.2, and derive the response to the initial conditions

$$\mathbf{q}(0)=\begin{bmatrix}0\\0\end{bmatrix}, \qquad \dot{\mathbf{q}}(0)=\begin{bmatrix}0\\1\end{bmatrix} \tag{a}$$

by modal analysis.

From Eq. (8.72), we note that to obtain the system response by modal analysis we need the mass matrix M, the natural frequencies ω_r and the normal modes $\mathbf{u}_r$ $(r=1, 2)$. The mass matrix is

$$M=m\begin{bmatrix}1 & 0\\0 & 2\end{bmatrix} \tag{b}$$

Moreover, from Example 8.2, we have

$$\begin{aligned}\omega_1&=0.4682\sqrt{\frac{k}{m}}, \qquad \omega_2=1.5102\sqrt{\frac{k}{m}}\\ \mathbf{u}_1&=m^{-1/2}\begin{bmatrix}0.3691\\0.6572\end{bmatrix}, \qquad \mathbf{u}_2=m^{-1/2}\begin{bmatrix}-0.9294\\0.2610\end{bmatrix}\end{aligned} \tag{c}$$

Using Eqs. (a)–(c), we can form

$$\begin{aligned}&\mathbf{u}_1^{\mathrm{T}}M\mathbf{q}(0)=0\\ &\frac{1}{\omega_1}\mathbf{u}_1^{\mathrm{T}}M\dot{\mathbf{q}}(0)=\frac{m}{0.4682\sqrt{k}}\begin{bmatrix}0.3691\\0.6572\end{bmatrix}^{\mathrm{T}}\begin{bmatrix}1 & 0\\0 & 2\end{bmatrix}\begin{bmatrix}0\\1\end{bmatrix}=2.8073\frac{m}{\sqrt{k}}\\ &\mathbf{u}_2^{\mathrm{T}}M\mathbf{q}(0)=0\\ &\frac{1}{\omega_2}\mathbf{u}_2^{\mathrm{T}}M\dot{\mathbf{q}}(0)=\frac{m}{1.5102\sqrt{k}}\begin{bmatrix}-0.9294\\0.2610\end{bmatrix}^{\mathrm{T}}\begin{bmatrix}1 & 0\\0 & 2\end{bmatrix}\begin{bmatrix}0\\1\end{bmatrix}=0.3457\frac{m}{\sqrt{k}}\end{aligned} \tag{d}$$

so that inserting Eqs. (c) and (d) into Eq. (8.72), we obtain the response

$$\begin{aligned}\mathbf{q}(t)&=\sum_{r=1}^{2}\mathbf{u}_r^{\mathrm{T}}M\left[\mathbf{q}(0)\cos\omega_r t+\frac{1}{\omega_r}\dot{\mathbf{q}}(0)\sin\omega_r t\right]\mathbf{u}_r\\ &=2.8073\sqrt{\frac{m}{k}}\sin 0.4682\sqrt{\frac{k}{m}}t\begin{bmatrix}0.3691\\0.6572\end{bmatrix}\\ &\quad+0.3457\sqrt{\frac{m}{k}}\sin 1.5102\sqrt{\frac{k}{m}}t\begin{bmatrix}-0.9294\\0.2610\end{bmatrix}\end{aligned} \tag{e}$$

which is identical to the response obtained in Example 8.2.

Example 8.5

Derive the response of the two-degree-of-freedom undamped system of Example 8.4 to the external excitation

$$\mathbf{Q}(t)=\begin{bmatrix} Q_0 \mathscr{u}(t) \\ 0 \end{bmatrix} \tag{a}$$

where $\mathscr{u}(t)$ is the unit step function applied to $t=0$. Let the initial conditions be zero.

Introducing Eq. (a) into Eq. (8.72) and letting $\mathbf{q}(0)=\dot{\mathbf{q}}(0)=\mathbf{0}$, we obtain the response in the form of the convolution integral

$$\begin{aligned}\mathbf{q}(t)&=\sum_{r=1}^{2}\left[\frac{1}{\omega_r}\mathbf{u}_r^T\int_0^t \mathbf{Q}(\tau)\sin\omega_r(t-\tau)\,d\tau\right]\mathbf{u}_r\\ &=Q_0\sum_{r=1}^{2}\left\{\frac{1}{\omega_r}\mathbf{u}_r^T\int_0^t\begin{bmatrix}\mathscr{u}(\tau)\\0\end{bmatrix}\sin\omega_r(t-\tau)\,d\tau\right\}\mathbf{u}_r\\ &=Q_0\sum_{r=1}^{2}\frac{u_{r1}}{\omega_r}\int_0^t[\mathscr{u}(\tau)\sin\omega_r(t-\tau)\,d\tau]\,\mathbf{u}_r\end{aligned} \tag{b}$$

where u_{r1} is the first component of the vector $\mathbf{u}_r$. But,

$$\begin{aligned}\int_0^t \mathscr{u}(\tau)\sin\omega_r(t-\tau)\,d\tau&=-\int_0^t\sin\omega_r(\tau-t)\,d\tau\\ &=\frac{\cos\omega_r(\tau-t)}{\omega_r}\bigg|_0^t=\frac{1}{\omega_r}(1-\cos\omega_r t)\end{aligned} \tag{c}$$

so that Eq. (b) becomes

$$\mathbf{q}(t)=Q_0\left[\frac{u_{11}}{\omega_1^2}(1-\cos\omega_1 t)\mathbf{u}_1+\frac{u_{21}}{\omega_2^2}(1-\cos\omega_2 t)\mathbf{u}_2\right]\mathscr{u}(t) \tag{d}$$

where the right side of Eq. (d) has been multiplied by $\mathscr{u}(t)$ in recognition of the fact that the response is zero for $t<0$. Inserting the modal data from Example 8.4 into Eq. (d), we obtain the response

$$\begin{aligned}\mathbf{q}(t)&=\frac{Q_0}{k}\left\{\frac{0.3691}{0.4682^2}\left(1-\cos 0.4682\sqrt{\frac{k}{m}}\,t\right)\begin{bmatrix}0.3691\\0.6572\end{bmatrix}\right.\\ &\qquad\left.-\frac{0.9294}{1.5102^2}\left(1-\cos 1.5102\sqrt{\frac{k}{m}}\,t\right)\begin{bmatrix}-0.9294\\0.2610\end{bmatrix}\right\}\mathscr{u}(t)\\ &=\frac{Q_0}{k}\left\{1.6839\left(1-\cos 0.4682\sqrt{\frac{k}{m}}\,t\right)\begin{bmatrix}0.3691\\0.6572\end{bmatrix}\right.\\ &\qquad\left.-0.4075\left(1-\cos 1.5102\sqrt{\frac{k}{m}}\,t\right)\begin{bmatrix}-0.9294\\0.2610\end{bmatrix}\right\}\mathscr{u}(t)\end{aligned} \tag{e}$$

8.6 DAMPED SYSTEMS

The modal analysis for deriving the response of undamped multi-degree-of-freedom vibrating systems described in Section 8.5 is based on the fact that the modal vectors are orthogonal both with respect to the mass matrix M and the stiffness matrix K. In this section, we wish to explore the possibility of using the same approach in the case of damped systems as well. Hence, inserting Eq. (8.63) into Eq. (8.21) and premultiplying the result by $\mathbf{u}_s^{\mathrm{T}}$, we obtain

$$\sum_{r=1}^{n} [\mathbf{u}_s^{\mathrm{T}} M \mathbf{u}_r \ddot{\eta}_r(t) + \mathbf{u}_s^{\mathrm{T}} C \mathbf{u}_r \dot{\eta}_r(t) + \mathbf{u}_s^{\mathrm{T}} K \mathbf{u}_r \eta_r(t)] = \mathbf{u}_s^{\mathrm{T}} \mathbf{Q}(t), \qquad s = 1, 2, \ldots, n \tag{8.79}$$

Introducing the notation

$$c_{sr} = \mathbf{u}_s^{\mathrm{T}} C \mathbf{u}_r, \qquad r, s = 1, 2, \ldots, n \tag{8.80}$$

considering the orthogonality of the modal vectors with respect to M and K, Eqs. (8.49) and (8.50), and recalling Eq. (8.67), we can reduce Eqs. (8.79) to

$$\ddot{\eta}_s(t) + \sum_{r=1}^{n} c_{sr} \dot{\eta}_r(t) + \omega_s^2 \eta_s(t) = N_s(t), \qquad s = 1, 2, \ldots, n \tag{8.81}$$

In the general case,

$$c_{sr} \neq 0 \qquad \text{for} \quad s \neq r \tag{8.82}$$

so that Eqs. (8.81) represent a set of simultaneous equations of motion. Hence, in the general case, the modal analysis presented for undamped systems in Section 8.5 cannot be used to reduce a set of simultaneous equations describing the motion of a damped system to a set of independent equations.

There are some special cases, however, when the modal analysis described in Section 8.5 can be used to decouple damped systems also. One such case is known as *proportional damping*. In this case, the damping matrix is a linear combination of the mass matrix M and the stiffness matrix K, or

$$C = \alpha M + \beta K \tag{8.83}$$

where α and β are two constant scalars. Introducing Eq. (8.83) into Eq. (8.80) and recalling the orthogonality of the modal vectors with respect to M and K, Eqs. (8.49) and (8.50), we obtain

$$\begin{aligned} c_{sr} &= \mathbf{u}_s^{\mathrm{T}}(\alpha M + \beta K)\mathbf{u}_r = \alpha \mathbf{u}_s^{\mathrm{T}} M \mathbf{u}_r + \beta \mathbf{u}_s^{\mathrm{T}} K \mathbf{u}_r \\ &= (\alpha + \beta\omega_s^2)\, \delta_{sr}, \qquad r, s = 1, 2, \ldots, n \end{aligned} \tag{8.84}$$

where δ_{sr} is the Kronecker delta. It will prove convenient to introduce the notation

$$\alpha + \beta\omega_s^2 = 2\zeta_s \omega_s, \qquad s = 1, 2, \ldots, n \tag{8.85}$$

where ζ_s $(s = 1, 2, \ldots, n)$ are *modal damping factors*. Then, inserting Eqs. (8.84) and (8.85) into Eqs. (8.81), we obtain the set of n independent equations

$$\ddot{\eta}_s(t) + 2\zeta_s \omega_s \dot{\eta}_s(t) + \omega_s^2 \eta_s(t) = N_s(t), \qquad s = 1, 2, \ldots, n \tag{8.86}$$

and we observe that every one of Eqs. (8.86) resembles the equation of motion of a viscously damped single-degree-of-freedom system.

Another case of interest is that of small damping. When the elements of the damping matrix C are small, all the coefficients c_{rs} in Eq. (8.84) are small. Then, it is assumed that the error committed by simply ignoring the coefficients c_{rs} corresponding to $r \neq s$ is not significant. Hence, inserting

$$c_{sr}=2\zeta_s\omega_s\,\delta_{sr}, \qquad r, s=1, 2, \dots, n \tag{8.87}$$

into Eqs. (8.81), we obtain once again the set of independent equations (8.86).

The homogeneous solution of an equation resembling Eqs. (8.86) was discussed in Section 4.6. Moreover, the impulse response of a single-degree-of-freedom damped system was derived in Section 4.13. Using these results, we can write the complete solution of Eqs. (8.86) in the form

$$\begin{aligned}\eta_s(t)=&\frac{\eta_s(0)}{(1-\zeta_s^2)^{1/2}}\,e^{-\zeta_s\omega_s t}\cos(\omega_{ds}t-\psi_s)\\&+\frac{\dot{\eta}_s(0)}{\omega_{ds}}\,e^{-\zeta_s\omega_s t}\sin\omega_{ds}t+\int_0^t N_s(\tau)g_s(t-\tau)\,d\tau, \qquad s=1, 2, \dots, n\end{aligned} \tag{8.88}$$

where

$$\omega_{ds}=(1-\zeta_s^2)^{1/2}\omega_s, \qquad s=1, 2, \dots, n \tag{8.89}$$

is the frequency of damped oscillation in the sth mode and

$$\psi_s=\tan^{-1}\frac{\zeta_s}{(1-\zeta_s^2)^{1/2}}, \qquad s=1, 2, \dots, n \tag{8.90}$$

is the corresponding phase angle. Moreover,

$$g_s(t)=\frac{1}{\omega_s}\,e^{-\zeta_s\omega_s t}\sin\omega_{ds}t\,\mathscr{u}(t), \qquad s=1, 2, \dots, n \tag{8.91}$$

is the impulse response associated with Eqs. (8.86). Note that the above solution assumes small damping. To complete the response of an n-degree-of-freedom system subject to proportional damping, or merely small damping, we insert Eqs. (8.88) into

$$\mathbf{q}(t)=\sum_{s=1}^{n}\eta_s(t)\mathbf{u}_s \tag{8.92}$$

In the general case of damping, no transformation of the type (8.92) is capable of producing the decoupling necessary for the ready derivation of the system response. Indeed, we shall show in Chapter 10 that to obtain the response it is necessary to work with the state vector, a $2n$-dimensional vector defined by the displacements $q_i(t)$ and velocities $\dot{q}_i(t)\,(i=1, 2, \dots, n)$. In fact, a modal analysis exists also for this general case, but the details are different.

As in the undamped case, the modal response to steady-state harmonic excitation is not covered by Eq. (8.88) and must be obtained by a different approach, such as that described in Section 4.9. Hence, considering an excitation in the form (8.73), we obtain the modal equations

$$\ddot{\eta}_s + 2\zeta_s\omega_s\dot{\eta}_s(t) + \omega_s^2\eta_s(t) = N_{s0}e^{i\omega t}, \qquad s = 1, 2, \ldots, n \tag{8.93}$$

where N_{s0} are modal force amplitudes, as given by Eqs. (8.75). Using results obtained in Section 4.9, the particular solution of Eqs. (8.93) can be shown to have the form

$$\eta_s(t) = \frac{N_{s0}}{\omega_s^2 - \omega^2 + i2\zeta_s\omega\omega_s}e^{i\omega t} = \frac{N_{s0}}{\omega_s^2}|G_s(i\omega)|e^{i(\omega t + \phi_s)}, \qquad s = 1, 2, \ldots, n \tag{8.94}$$

where

$$|G_s(i\omega)| = \frac{1}{\{[1-(\omega/\omega_s)^2]^2 + (2\zeta_s\omega/\omega_s)^2\}^{1/2}}, \qquad s = 1, 2, \ldots, n \tag{8.95}$$

$$\phi_s = \tan^{-1}\left[-\frac{2\zeta_s\omega/\omega_s}{1-(\omega/\omega_s)^2}\right], \qquad s = 1, 2, \ldots, n \tag{8.96}$$

are modal magnification factors and modal phase angles, respectively. Inserting Eqs. (8.94) into Eq. (8.92) and considering Eqs. (8.75), we conclude that the response of a multi-degree-of-freedom system to steady-state harmonic excitation has the expression

$$\mathbf{q}(t) = \sum_{s=1}^{n} \eta_s(t)\mathbf{u}_s = \sum_{s=1}^{n} \frac{\mathbf{u}_s^T\mathbf{Q}_0}{\omega_s^2}|G_s(i\omega)|e^{i(\omega t + \phi_s)}\mathbf{u}_s \tag{8.97}$$

Example 8.6

The two-degree-of-freedom system of Fig. 8.3 is the same as that of Fig. 8.2, except that it is damped. Derive the response of the system to the initial conditions

$$\mathbf{q}(0) = \begin{bmatrix} 1 \\ 2 \end{bmatrix}, \qquad \dot{\mathbf{q}}(0) = \begin{bmatrix} 0 \\ 0 \end{bmatrix} \tag{a}$$

for the case in which $c = 0.8\sqrt{km}$.

Rayleigh's dissipation function (Section 8.2) has the form

$$\mathscr{F} = \tfrac{1}{2}[c\dot{q}_1^2 + c(\dot{q}_2 - \dot{q}_1)^2] = \tfrac{1}{2}(2c\dot{q}_1^2 + c\dot{q}_2^2 - 2c\dot{q}_1\dot{q}_2)$$

$$= \frac{1}{2}\begin{bmatrix} \dot{q}_1 \\ \dot{q}_2 \end{bmatrix}^T \begin{bmatrix} 2c & -c \\ -c & c \end{bmatrix} \begin{bmatrix} \dot{q}_1 \\ \dot{q}_2 \end{bmatrix} \tag{b}$$

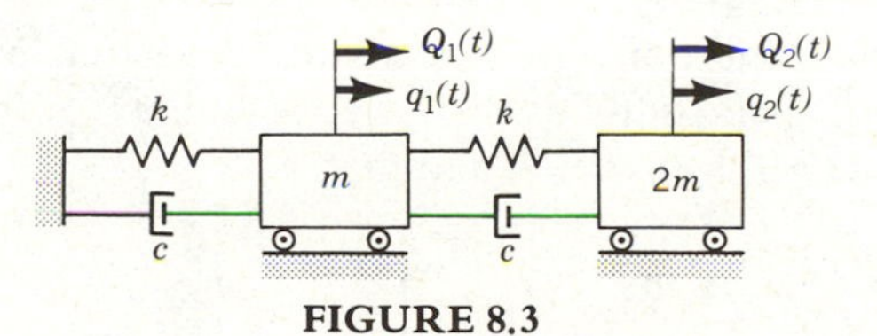

FIGURE 8.3

so that the damping matrix is

$$C = c\begin{bmatrix} 2 & -1 \\ -1 & 1 \end{bmatrix} \tag{c}$$

which has the same form as the stiffness matrix. In fact, we can write

$$C = \frac{c}{k}K \tag{d}$$

so that damping is of the proportional type, with $\alpha = 0$ and $\beta = c/k$. Using Eq. (8.84), we have

$$c_{sr} = (\alpha + \beta\omega_s^2)\,\delta_{sr} = \left(\frac{c}{k}\,\omega_s^2\right)\delta_{sr} = 2\zeta_s\omega_s\,\delta_{sr}, \qquad r, s = 1, 2 \tag{e}$$

so that, recalling that $c = 0.8\sqrt{km}$, we can express the modal damping factors in the form

$$\zeta_s = \frac{c\omega_s}{2k} = 0.4\sqrt{\frac{m}{k}}\,\omega_s, \qquad s = 1, 2 \tag{f}$$

The modal data for the system are given in Example 8.4. Inserting the natural frequencies from Eqs. (c) of Example 8.4 into Eqs. (f), we obtain

$$\begin{aligned} \zeta_1 &= 0.4\sqrt{\frac{m}{k}}\,\omega_1 = 0.4 \times 0.4682 = 0.1873 \\ \zeta_2 &= 0.4\sqrt{\frac{m}{k}}\,\omega_2 = 0.4 \times 1.5102 = 0.6401 \end{aligned} \tag{g}$$

so that from Eqs. (8.89) the frequencies of the damped oscillations are

$$\begin{aligned} \omega_{d1} &= (1 - \zeta_1^2)^{1/2}\omega_1 = (1 - 0.1873^2)^{1/2} \times 0.4682\sqrt{\frac{k}{m}} = 0.4599\sqrt{\frac{k}{m}} \\ \omega_{d2} &= (1 - \zeta_2^2)^{1/2}\omega_2 = (1 - 0.6401^2)^{1/2} \times 1.5102\sqrt{\frac{k}{m}} = 1.2035\sqrt{\frac{k}{m}} \end{aligned} \tag{h}$$

Equations (8.90) imply that the phase angles are related to the damping factors by

$$\psi_s = \sin^{-1}\zeta_s, \qquad s = 1, 2, \ldots, n \tag{i}$$

so that the phase angles are

$$\begin{aligned} \psi_1 &= \sin^{-1}\zeta_1 = \sin^{-1} 0.1873 = 10.795^\circ \\ \psi_2 &= \sin^{-1}\zeta_2 = \sin^{-1} 0.6401 = 39.799^\circ \end{aligned} \tag{j}$$

The modal initial conditions are related to the actual initial conditions by Eqs. (8.71). Inserting Eqs. (a) into Eqs. (8.71) and recalling Eqs. (b) and (c) of Example 8.4, we conclude that $\dot{\eta}_1(0) = \dot{\eta}_2(0) = 0$ and that

$$\eta_1(0) = \mathbf{u}_1^T M\mathbf{q}(0) = m^{1/2}\begin{bmatrix} 0.3691 \\ 0.6572 \end{bmatrix}^T \begin{bmatrix} 1 & 0 \\ 0 & 2 \end{bmatrix}\begin{bmatrix} 1 \\ 2 \end{bmatrix} = 2.9979m^{1/2} \tag{k}$$

$$\eta_2(0)=\mathbf{u}_2^{\mathrm{T}}M\mathbf{q}(0)=m^{1/2}\begin{bmatrix}-0.9294\\0.2610\end{bmatrix}\begin{bmatrix}1&0\\0&2\end{bmatrix}\begin{bmatrix}1\\2\end{bmatrix}=0.1146m^{1/2}$$

Finally, introducing Eqs. (k) into Eqs. (8.88) and using Eq. (8.92), we can write the desired response as follows:

$$\begin{aligned}\mathbf{q}(t)&=\sum_{s=1}^{2}\eta_s(t)\mathbf{u}_s\\&=\frac{2.9979}{0.9823}e^{-0.1873\times0.4682\sqrt{k/m}t}\cos\left(0.4599\sqrt{\frac{k}{m}}t-10.795^\circ\right)\begin{bmatrix}0.3691\\0.6572\end{bmatrix}\\&\quad+\frac{0.1146}{0.7969}e^{-0.6041\times1.5102\sqrt{k/m}t}\cos\left(1.2035\sqrt{\frac{k}{m}}t-39.799^\circ\right)\begin{bmatrix}-0.9294\\0.2610\end{bmatrix}\\&=3.0519\,e^{-0.0877\sqrt{k/m}t}\cos\left(0.4599\sqrt{\frac{k}{m}}t-10.795^\circ\right)\begin{bmatrix}0.3691\\0.6572\end{bmatrix}\\&\quad+0.1438e^{-0.9123\sqrt{k/m}t}\cos\left(1.2035\sqrt{\frac{k}{m}}t-39.799^\circ\right)\begin{bmatrix}-0.9294\\0.2610\end{bmatrix}\end{aligned}\tag{l}$$

We observe from Eq. (l) that the initial participation of the second mode in the response is less than 5% of that of the first mode. This can be explained by the fact that the initial displacement resembles the first mode closely. Moreover, the second mode decays considerably faster than the first mode, as can be concluded from the exponential terms in Eq. (l). This behavior is consistent with the proportional damping assumption. Indeed from Eqs. (e), we conclude that the exponents are proportional to ω_s^2, so that higher modes tend to die out faster than lower modes, the relative rate of decay depending on how large the higher frequencies are compared to the lower ones.

PROBLEMS

8.1 Derive the linearized equations of motion for the system of Fig. 6.27 by assuming small angles θ_1 and θ_2.

8.2 The system depicted in Fig. 8.4 serves as a mathematical model for an automobile, where the rigid beam represents the chassis and body and the spring–damper assemblies represent the suspension and tires. Use the vertical

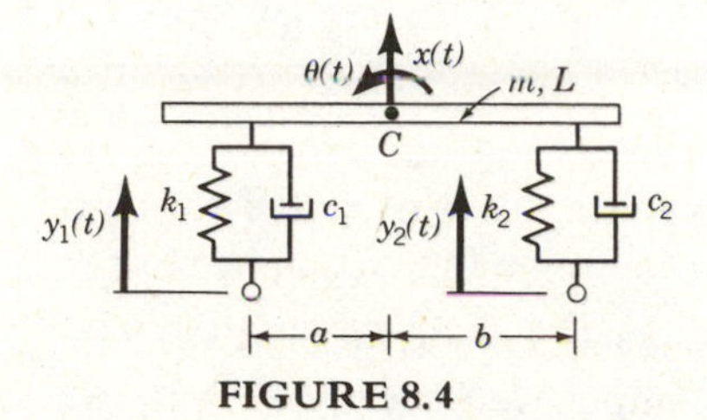

FIGURE 8.4

displacement $x(t)$ of the mass center C and the angular displacement $\theta(t)$ of the beam as generalized coordinates and derive the equations of motion for the system under the assumption that $\theta(t)$ is small. Note that $y_1(t)$ and $y_2(t)$ represent displacements of the support.

8.3 Determine the equilibrium position $x = x_E$, $\theta = \theta_E$ of the system of Fig. 6.28 for the parameters $k_1 = k_2 = k = mg/L$. Note that θ can be arbitrarily large. Then, let $x(t) = x_E + x_1(t)$ and $\theta(t) = \theta_E + \theta_1(t)$, where $x_1(t)$ and $\theta_1(t)$ are small perturbations from equilibrium, and derive the linearized equations of motion about the equilibrium.

8.4 Derive and solve the eigenvalue problem for the system of Problem 8.1 for the case $m_1 = m_2 = m$, $L_1 = L_2 = L$.

8.5 Derive and solve the eigenvalue problem for the system of Problem 8.2 for the case $a = 0.3L$, $b = 0.4L$, $k_1 = k_2 = k$, $c_1 = c_2 = 0$. Assume that the beam is uniform.

8.6 Derive and solve the eigenvalue problem for the airfoil section of Fig. 6.26 for the case $I_E = 0.1mc^2$, $s = 0.2c$, $k_\theta = 0.04c^2 k_x$, where c is the length of the chord. Assume that $\theta = 0$ in the equilibrium position.

8.7 Derive and solve the eigenvalue problem for the system of Fig. 8.1 for the parameters $m_1 = m_2 = m$, $m_3 = 2m$, $c_1 = c_2 = c_3 = c_4 = 0$, $k_1 = k_2 = k$, $k_3 = k_4 = 2k$. Plot the modes.

8.8 Verify the orthogonality with respect to the mass and stiffness matrices of the modes computed in Problem 8.7 and normalize them according to Eq. (8.49).

8.9 Derive and solve the eigenvalue problem for the system of Problem 8.3. Let $M = 2m$.

8.10 The system of Problem 8.4 is subjected to the initial conditions $\theta_1(0) = -\theta_2(0) = \theta_0$, $\dot{\theta}_1(0) = \dot{\theta}_2(0) = 0$. Obtain the system response.

8.11 The system of Problem 8.9 is subject to the initial conditions $x_1(0) = \theta_1(0) = \dot{\theta}_1(0) = 0$, $\dot{x}_1(0) = v_0$. Obtain the system response.

8.12 The system of Problem 8.7 is subject to the initial conditions $q_1(0) = q_3(0) = 0.8$, $q_2(0) = 1$, $\dot{q}_1(0) = \dot{q}_2(0) = \dot{q}_3(0) = 0$. Determine the system response and explain the results (in terms of the participation of the individual modes in the overall response).

8.13 The system of Problem 8.5 undergoes vibration testing during which the supports are imparted the motions $y_1(t) = 2 \sin \omega t$, $y_2(t) = \sin \omega t$. Obtain the system response.

8.14 The system of Problem 8.7 is subjected to the forces $Q_1(t) = Q_0 \cos(\omega t + \alpha)$, $Q_2(t) = Q_3(t) = 0$. Determine the system response.

8.15 The system of Problem 8.4 is struck horizontally at the bottom end of the lower link by the impulsive force $F(t) = \hat{F}\,\delta(t)$, where $\delta(t)$ is the unit impulse

(Section 1.8). Use the virtual work expression to determine the generalized forces and obtain the system response.

8.16 The system of Problem 8.5 is subjected to the displacement excitation $y_1(t) = y_0 u(t)$, $y_2(t) = 0$, where $u(t)$ is the unit step function (Section 1.8). Determine the system response.

8.17 The system of Problem 8.3 is subjected to a horizontal force acting on mass M, where the force has the form of the rectangular pulse shown in Fig. 8.5. Determine the system response.

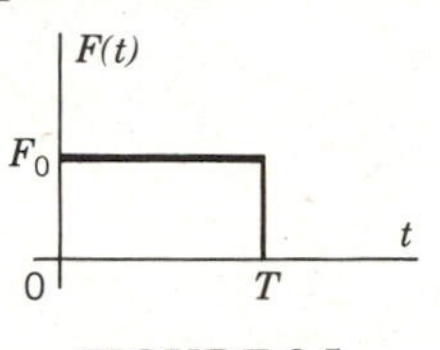

FIGURE 8.5

8.18 The systems of Example 8.2 is subjected to a force $Q_2(t)$ in the form of the triangular pulse shown in Fig. 8.6, while $Q_1(t) = 0$. Determine the system response.

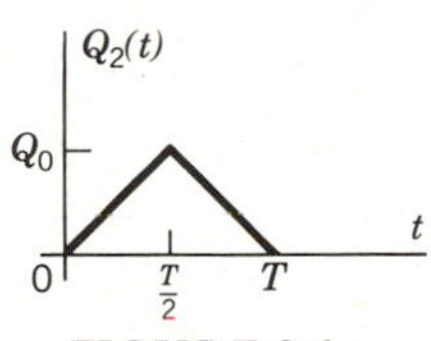

FIGURE 8.6

8.19 The system of Problem 8.6 is subjected to a vertical force applied at C, where the force has the form of the sinusoidal pulse shown in Fig. 8.7. Calculate the system response.

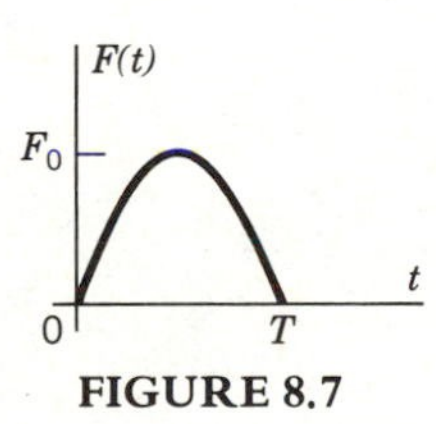

FIGURE 8.7

8.20 The system of Example 8.6 is subjected to the force $Q_2(t) = Q_0 u(t)$, where $u(t)$ is the unit step function (Section 1.8), while $Q_1(t) = 0$. Determine the system response, and ascertain the steady-state value of the displacements (obtained by letting $t \to \infty$). Draw conclusions.

8.21 Consider the system of Problem 8.7, but now assume that the damping coefficients are $c_1 = c_2 = c$, $c_2 = c_4 = 2c$. Obtain the response of the system to the excitation $Q_1(t) = Q_2(t) = 0$, $Q_3(t) = Q_0 r(t)$, where $r(t)$ is the unit ramp function (Section 1.8). Then determine the response for very large time and draw conclusions.

CHAPTER 9
Introduction to System Stability

9.1 INTRODUCTION

The differential equations governing the motion of dynamical systems are quite often very complicated and do not lend themselves to closed-form solution, particularly if the equations are nonlinear. One can always attempt numerical integration of the equations. However, in many cases one is content with only a qualitative description of the motion. In this regard, in this chapter we discuss the description of the motion in the phase space and introduce various concepts of system stability.

9.2 MOTION IN THE PHASE PLANE

Let us consider a particle of mass m in one-dimensional motion under the force $f(q, \dot{q})$, where $q(t)$ denotes the displacement of m, as shown in Fig. 9.1. We assume that the force $f(q, \dot{q})$ does not depend on the time t explicitly. Using Newton's second law, we write the equation of motion

$$m\ddot{q} = f(q, \dot{q}) \tag{9.1}$$

The motion of the particle is known as soon as the differential equation (9.1) is solved, which requires two conditions on q or on $\dot{q}$, or one on q and the second on $\dot{q}$. Ordinarily, these are the initial displacement $q(0) = q_0$ and initial velocity $\dot{q}(0) = \dot{q}_0$.

Quite often Eq. (9.1) does not admit a closed-form solution, which is particularly true when $f(q, \dot{q})$ is a nonlinear function of q and $\dot{q}$, so that our interest is in extracting information concerning the system behavior by other means. To this end, we wish to consider a geometric description of the motion, and in particular one capable of describing the motion for all possible initial conditions. Plots $q(t)$ versus t are not suitable, because curves corresponding to different initial conditions can

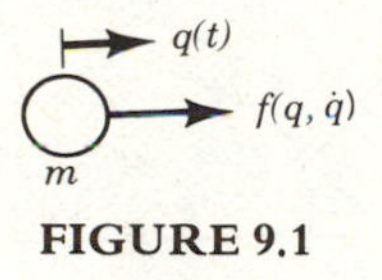

FIGURE 9.1

intersect, so that the description is not unique. Indeed, there can be more than one velocity corresponding to the same displacement. A much more satisfactory picture is obtained by introducing the momentum $p(t)=m\dot{q}(t)$ as an auxiliary variable and describing the motion in the plane p versus q. Such a plane is known as the *phase plane*.

A solution in the phase plane is known as a *trajectory*. To obtain the trajectory equation, we replace the second-order differential equation (9.1) by the two first-order differential equations

$$\dot{q}=p/m \tag{9.2a}$$

$$\dot{p}=f(q, p) \tag{9.2b}$$

Because $f(q, p)$ does not depend on the time t explicitly, the time can be eliminated from Eqs. (9.2). Indeed, dividing Eq. (9.2b) by Eq. (9.2a), we obtain

$$\frac{dp}{dq}=\frac{f(q, p)m}{p} \tag{9.3}$$

which represents the differential equation of the trajectories in the phase plane. For a set of initial conditions $q(0)=q_0$, $p(0)=p_0$, Eq. (9.3) defines a unique trajectory, in the sense that no other trajectory will ever intersect it. Hence, the trajectories resemble the streamlines in fluid flow, as shown in Fig. 9.2. Generally, Eq. (9.3) does not admit a closed-form solution, but trajectories can be obtained by various graphical methods, such as the *method of isoclines*, or better yet, by means of a numerical integration procedure carried out on a computer. The graphical solution is based on the fact that Eq. (9.3) defines the tangent to a given trajectory uniquely, except at critical points, at which the ratio on the right side of Eq. (9.3) is undefined.

A point $q=\alpha_1$, $p=\alpha_2$, where α_1 and α_2 are constants satisfying the equations

$$p=0, \qquad f(\alpha_1, \alpha_2)=0 \tag{9.4}$$

represents a critical point, at which the tangent to a given trajectory is undefined. A critical point is known as an *equilibrium point*, or a *singular point*. Because $p=\alpha_2=0$, *all the equilibrium points lie on the q-axis.* Observing from Eqs. (9.2) that $p=0$, $f=0$ implies that the velocity $\dot{q}$ and acceleration $\ddot{q}$ are zero, we can see that the term equilibrium point becomes obvious. Linear systems have one equilibrium point, and nonlinear systems have generally more than one equilibrium point.

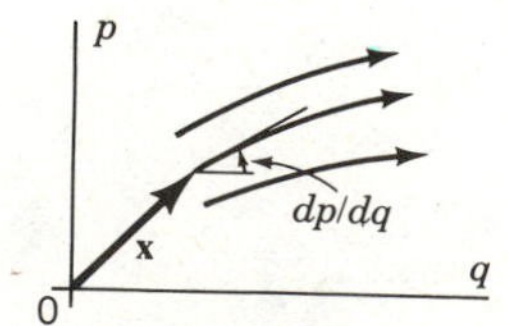

FIGURE 9.2

Next, let us introduce the two-dimensional vector $\mathbf{x}$ with the components q and p, $\mathbf{x}=[x_1\ x_2]^T=[q\ p]^T$, where $\mathbf{x}$ is known as the *phase vector*. Similarly, we can introduce the associated vector $\mathbf{X}=[X_1\ X_2]^T=[p/m\ f]^T=\mathbf{X}(\mathbf{x})$, so that Eqs. (9.2) can be written in the vector form

$$\dot{\mathbf{x}}=\mathbf{X}(\mathbf{x}) \tag{9.5}$$

The magnitude of the vector $\mathbf{x}$ can be defined as the *Euclidean length*

$$\|\mathbf{x}\|=(x_1^2+x_2^2)^{1/2}=(q^2+p^2)^{1/2} \tag{9.6}$$

and it gives the distance of a point in the phase plane from the origin (Fig. 9.2). A circle of radius r and with the center at origin 0 is given by

$$\|\mathbf{x}\|=r \tag{9.7}$$

If $\mathbf{x}(t)=[q(t)\ p(t)]^T$ represents a solution of Eqs. (9.2) such that

$$\|\mathbf{x}(t)\|<r \tag{9.8}$$

for some given radius r, then the solution is said to be *bounded*.

In terms of the phase vector formulation, an equilibrium point is defined as a constant vector $\boldsymbol{\alpha}=[\alpha_1\ \alpha_2]^T$ that renders the right side of Eq. (9.5) equal to zero. Hence, an equilibrium point is a solution of the algebraic equation

$$\mathbf{X}(\boldsymbol{\alpha})=\mathbf{0} \tag{9.9}$$

where we recall that, because $\alpha_2=0$, all the equilibrium points lie on the q-axis. Hence, to determine the vector $\boldsymbol{\alpha}$, it is only necessary to determine the component α_1.

Our interest lies in motions in the neighborhood of equilibrium points. But, because all the equilibrium points lie on the q-axis, the simple coordinate transformation $q=\alpha_1+q_1$ can translate the origin of the phase plane so as to cause it to coincide with an equilibrium point (Fig. 9.3). Hence, we can assume without loss of generality that $\alpha_1=0$, so that the equilibrium point under consideration is defined by $\boldsymbol{\alpha}=\mathbf{0}$. In view of this, Eq. (9.9) reduces to

$$\mathbf{X}(\mathbf{0})=\mathbf{0} \tag{9.10}$$

which implies that Eq. (9.5) admits the *trivial solution*, or the *null solution*, as a special solution.

Quite frequently, one does not use the phase vector to represent the motion, but a different vector, namely, one obtained by replacing the momentum $p=m\dot{q}$ by

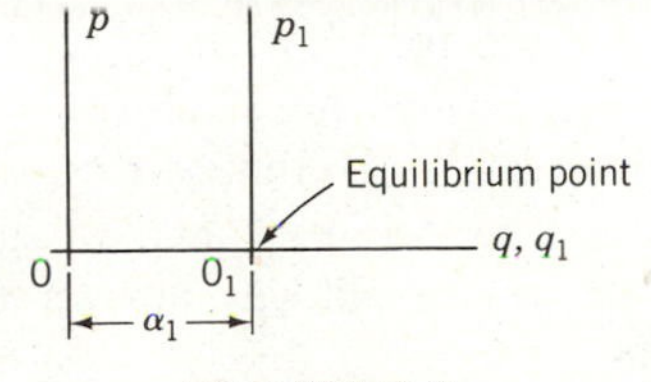

FIGURE 9.3

the velocity $\dot{q}$ as the auxiliary variable. The vector $\mathbf{x}=[q\ \dot{q}]^T$ is known as the *state vector*, and the corresponding plane is called the *state plane*. There is no essential difference between the geometric representations in terms of the state vector and that in terms of the phase vector. In system analysis, the state-vector representation is used almost exclusively.

9.3 STABILITY OF EQUILIBRIUM POINTS

A very important question in dynamics is the *stability of motion* in the neighborhood of equilibrium points. We consider an equilibrium point coinciding with the origin of the phase plane, so that our interest is in the stability of the trivial solution. There are various definitions of stability. One definition requires that the solution $\mathbf{x}(t)$ be bounded for the motion in the neighborhood of the origin to be stable. This definition is too vague, however, and can exclude motions known to be stable. The most satisfying definitions of stability are due to Liapunov.

Let us consider two circles $||\mathbf{x}||=\delta$ and $||\mathbf{x}||=\varepsilon$ in the phase plane, where δ and ε are two positive constants such that $\delta<\varepsilon$. Then:

1. The solution $\mathbf{x}(t)$ is said to be *stable* if for

$$||\mathbf{x}(0)||=||\mathbf{x}_0||<\delta \tag{9.11}$$

the inequality

$$||\mathbf{x}(t)||<\varepsilon, \qquad t>0 \tag{9.12}$$

is satisfied.

2. The solution $\mathbf{x}(t)$ is said to be *asymptotically stable* if it is stable and in addition

$$\lim_{t\to\infty} ||\mathbf{x}(t)||=0 \tag{9.13}$$

3. The solution $\mathbf{x}(t)$ is *unstable* if for

$$||\mathbf{x}(0)||=||\mathbf{x}_0||=\delta \tag{9.14}$$

we have

$$||\mathbf{x}(t_1)||=\varepsilon \tag{9.15}$$

for some finite time t_1.

The geometric implication of the above definitions can be found in Fig. 9.4. If a trajectory initiated inside the circle $||\mathbf{x}||=\delta$ remains inside the circle $||\mathbf{x}||=\varepsilon$ at all times, as in the case of trajectory I, then the motion is stable. If a trajectory initiated inside $||\mathbf{x}||=\delta$ does not touch the circle $||\mathbf{x}||=\varepsilon$ and it eventually reaches the origin, as in the case of trajectory II, then the motion is asymptotically stable. If on the other hand, a trajectory initiated inside $||\mathbf{x}||=\delta$ reaches $||\mathbf{x}||=\varepsilon$, and it actually crosses the circle $||\mathbf{x}||=\varepsilon$ on its way out, as in the case of trajectory III, then the motion is unstable.

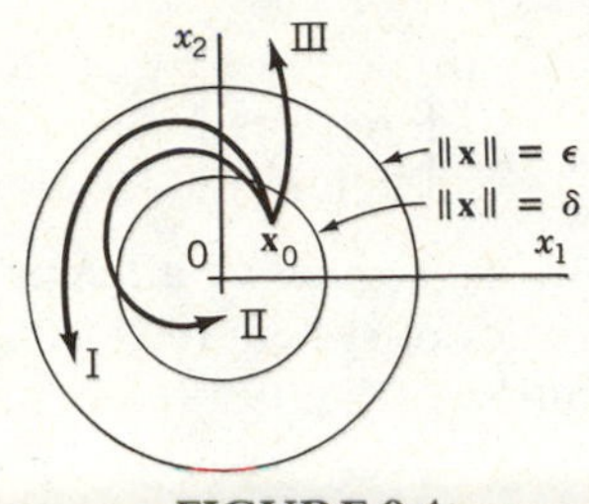

FIGURE 9.4

Physically, motion initiated close to an equilibrium point remains in the neighborhood of the point for mere stability, it tends to the equilibrium point for asymptotic stability, and it moves away from it for instability.

The above definitions are qualitative in nature. In the next section we examine more quantitative ways of assessing stability.

9.4 MOTION IN THE NEIGHBORHOOD OF EQUILIBRIUM POINTS

Let us consider Eqs. (9.2), or Eq. (9.5), and examine the motion in the neighborhood of an equilibrium point coinciding with the origin. We make the so-called small-motions assumption, which implies that the motion is confined to a small neighborhood of the origin.

Expanding Eq. (9.2b) in a Taylor series about the origin, we can write

$$\dot{p} = \left.\frac{\partial f}{\partial q}\right|_{q,p=0} q + \left.\frac{\partial f}{\partial p}\right|_{q,p=0} p + O^2(q, p) \tag{9.16}$$

where $O^2(q, p)$ denotes an expression containing all the nonlinear terms in q and p, i.e., terms such as q^2, qp, p^2, q^3, ... Then, introducing the coefficients

$$\begin{aligned} a_{11} &= 0, \qquad a_{12} = 1/m \\ a_{21} &= \left.\frac{\partial f}{\partial q}\right|_{q,p=0}, \qquad a_{22} = \left.\frac{\partial f}{\partial p}\right|_{q,p=0} \end{aligned} \tag{9.17}$$

recalling the notation used in Eq. (9.5), and ignoring the nonlinear terms, we can rewrite Eqs. (9.2) in the form of the *linearized equation*

$$\begin{aligned} \dot{x}_1 &= a_{11}x_1 + a_{12}x_2 \\ \dot{x}_2 &= a_{21}x_1 + a_{22}x_2 \end{aligned} \tag{9.18}$$

Equations (9.18) are known as *perturbation equations*. They can be written in the compact form

$$\dot{\mathbf{x}} = A\mathbf{x} \tag{9.19}$$

where

$$A=\begin{bmatrix} a_{11} & a_{12} \\ a_{21} & a_{22} \end{bmatrix} \tag{9.20}$$

is the matrix of coefficients. In the sequel we attempt to draw conclusions concerning the motion characteristics of system (9.5) from the linearized equations.

Let us assume that the solution of Eq. (9.19) has the exponential form

$$\mathbf{x}(t)=\mathbf{x}e^{\lambda t} \tag{9.21}$$

where $\mathbf{x}$ is a constant vector and λ is a constant exponent. Introducing Eq. (9.21) into Eq. (9.19) and dividing through by $e^{\lambda t}$, we obtain

$$A\mathbf{x}=\lambda\mathbf{x} \tag{9.22}$$

Equation (9.22) is known as the *algebraic characteristic-value problem*, or *eigenvalue problem.* The problem can be stated as follows: Find the values of the parameter λ such that Eq. (9.22) admits nontrivial solutions. Equation (9.22) represents a set of homogeneous algebraic equations having the explicit form

$$\begin{aligned} (a_{11}-\lambda)x_1+a_{12}x_2&=0 \\ a_{21}x_1+(a_{22}-\lambda)x_2&=0 \end{aligned} \tag{9.23}$$

The set of homogeneous algebraic equations admits a nontrivial solution provided the determinant of the coefficients is equal to zero, or

$$\begin{aligned} \det[A-\lambda I]&=\det\begin{bmatrix} a_{11}-\lambda & a_{12} \\ a_{21} & a_{22}-\lambda \end{bmatrix} \\ &=\lambda^2-(a_{11}+a_{22})\lambda+a_{11}a_{22}-a_{12}a_{21}=0 \end{aligned} \tag{9.24}$$

where I is the 2×2 identity matrix. Equation (9.24) is known as the *characteristic equation.* Its solutions λ_1 and λ_2 are known as *characteristic values*, or *eigenvalues.* The values of the vector $\mathbf{x}=[x_1\ x_2]^T$ corresponding to λ_1 on the one hand and to λ_2 on the other hand form the so-called *characteristic vectors*, or *eigenvectors*, $\mathbf{x}_1=[x_{11}\ x_{21}]^T$ and $\mathbf{x}_2=[x_{12}\ x_{22}]^T$, so that the solution can be written in the form

$$\mathbf{x}(t)=\mathbf{x}_1e^{\lambda_1 t}+\mathbf{x}_2e^{\lambda_2 t} \tag{9.25}$$

The motion characteristics depend on the nature of the eigenvalues λ_1 and λ_2, as discussed in the sequel.

There are basically three cases:

i. The roots of λ_1 and λ_2 are *pure imaginary complex conjugates.* In this case, the solution represents pure harmonic oscillation about the equilibrium point, so that the motion of the linearized system is *stable.*

ii. The roots λ_1 and λ_2 are either *real and negative* or they are *complex conjugates with negative real part.* In this case the solution approaches the equilibrium point as time increases, so that the motion is *asymptotically stable.*

iii. The roots λ_1 and λ_2 are such that at least one is *real and positive* or they are *complex conjugates with positive real part*. In this case the solution increases exponentially with time, so that the motion is *unstable*.

In Case i, the linearized system is said to possess *critical behavior*, and in Cases ii and iii the linearized system is said to possess *significant behavior*. If the system possesses significant behavior, then the linearized system has the same motion characteristics as the original nonlinear system. In the case of critical behavior, the linearized system does not necessarily have the same motion characteristics as the original nonlinear system (although it can have), and one must examine the effect of the nonlinear terms. Of course, for a linear system no such distinction is necessary.

The stability criteria based on the nature of the eigenvalues can be best visualized by plotting the eigenvalues in a complex plane referred to as the λ-plane. Then, the left half-plane represents the region of asymptotic stability, the imaginary axis the region of mere stability, and the right half-plane the region of instability (Fig. 9.5). The λ-plane is used widely in automatic controls, in which the control forces are chosen so as to drive the eigenvalues into the left half-plane from the imaginary axis or from the right half-plane, thus rendering the system asymptotically stable.

Example 9.1

Let us consider the mass–damper–spring system of Section 4.2. In the absence of external forces, the differential equation of motion is

$$m\ddot{q}(t)+c\dot{q}(t)+kq(t)=0 \tag{a}$$

where $q(t)$ is the displacement of mass m. Equation (a) can be replaced by

$$\dot{q}=p/m, \qquad \dot{p}=f(q,p)=-kq-(c/m)p \tag{b}$$

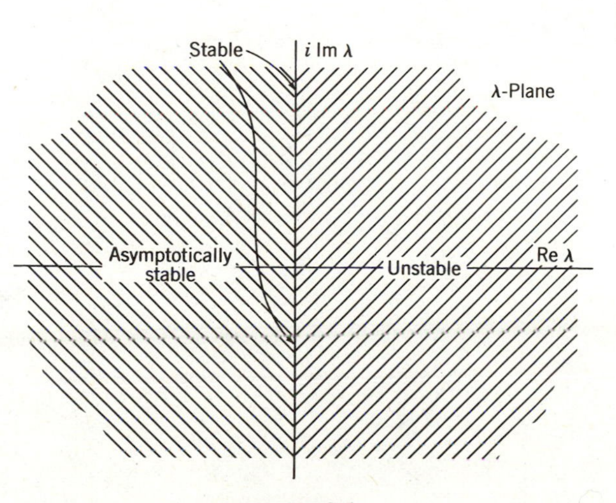

FIGURE 9.5

Clearly, the origin of the phase plane is the only equilibrium point. From Eqs. (b) and (9.17), we obtain the coefficients

$$a_{11}=0, \qquad a_{12}=1/m$$
$$a_{21}=\left.\frac{\partial f}{\partial q}\right|_{q,p=0}=-k, \qquad a_{22}=\left.\frac{\partial f}{\partial p}\right|_{q,p=0}=-c/m \tag{c}$$

so that using Eq. (9.24), the characteristic equation becomes

$$\lambda^2+(c/m)\lambda+k/m=0 \tag{d}$$

which can be rewritten in the more familiar form

$$\lambda^2+2\zeta\omega_n\lambda+\omega_n^2=0 \tag{e}$$

where ζ is the damping factor and ω_n is the frequency of undamped oscillation (Section 4.6). The solution of the characteristic equation yields the eigenvalues

$$\left.\begin{matrix}\lambda_1\\ \lambda_2\end{matrix}\right\}=-\zeta\omega_n\pm\sqrt{\zeta^2-1}\,\omega_n \tag{f}$$

If $\zeta>1$, the eigenvalues λ_1 and λ_2 are real and negative. If $\zeta<1$, the eigenvalues are complex conjugates with negative real part. Hence, the equilibrium point $q=p=0$ is *asymptotically stable.*

Example 9.2

Consider a nonlinear system consisting of a mass m and a softening spring, where the force exerted by the spring on the mass is given by

$$f(q)=-kq\left[1-\left(\frac{q}{2a}\right)^2\right] \tag{a}$$

The plot $f(q)$ versus q is shown in Fig. 9.6. Identify the equilibrium points and test their stability.

The equations of motion can be written in the form

$$\dot{q}=p/m$$
$$\dot{p}=-kq\left[1-\left(\frac{q}{2a}\right)^2\right] \tag{b}$$

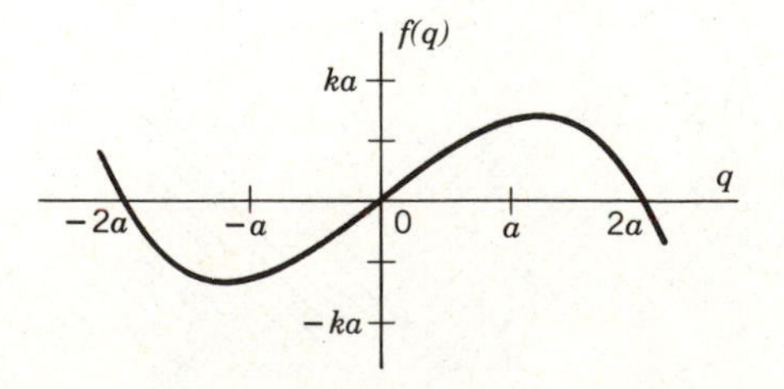

FIGURE 9.6

All the equilibrium points are on the q-axis, i.e., $p=0$. Their location is obtained by solving

$$-kq\left[1-\left(\frac{q}{2a}\right)^2\right]=0 \tag{c}$$

which yields

$$q=0, \qquad q=2a, \qquad q=-2a \tag{d}$$

so that there are three equilibrium points.

To test the stability of the trivial equilibrium point $q=p=0$, we first calculate the coefficients a_{ij} $(i, j=1, 2)$

$$\begin{aligned} &a_{11}=0, \qquad a_{12}=1/m \\ &a_{21}=\left.\frac{\partial f}{\partial q}\right|_{q,p=0}=-k, \qquad a_{22}=\left.\frac{\partial f}{\partial p}\right|_{q,p=0}=0 \end{aligned} \tag{e}$$

so that the characteristic equation is

$$\lambda^2+k/m=0 \tag{f}$$

yielding the eigenvalues

$$\left.\begin{matrix}\lambda_1\\ \lambda_2\end{matrix}\right\}=\pm i\omega_{\text{n}} \tag{g}$$

where $\omega_{\text{n}}=\sqrt{k/m}$. Hence, in the neighborhood of the origin, the linearized system acts like a harmonic oscillator. The motion is merely *stable*. Because this implies critical behavior, we must examine the effect of the nonlinear terms to determine the exact nature of the equilibrium point. This is done in the next section.

Next, let us examine the system behavior in the neighborhood of the equilibrium point $q=2a$, $p=0$. It is not necessary to check the equilibrium point $q=-2a$, $p=0$, as the behavior is the same as for $q=2a$, $p=0$. The coefficients a_{11}, a_{12}, and a_{22} are the same as for $q=p=0$. On the other hand,

$$a_{21}=\left.\frac{\partial f}{\partial q}\right|_{q=2a,\,p=0}=2k \tag{h}$$

so that in this case the characteristic equation is

$$\lambda^2-2k/m=0 \tag{i}$$

yielding the eigenvalues

$$\left.\begin{matrix}\lambda_1\\ \lambda_2\end{matrix}\right\}=\pm\sqrt{2k/m} \tag{j}$$

Because one of the roots is real and positive, the equilibrium point $q=2a$, $p=0$ is *unstable*. This implies significant behavior, so that even when the nonlinear terms are considered the same conclusion is reached.

9.5 CONSERVATIVE SYSTEMS

In the case of conservative systems, the force does not depend on the velocity and the equation of motion admits an integral in the form of the total energy E.

Let us assume that the system differential equation of motion has the form

$$m\ddot{q} = f(q) \tag{9.26}$$

From differential calculus, however, we can write

$$\ddot{q} = \frac{d\dot{q}}{dt} = \frac{d\dot{q}}{dq}\frac{dq}{dt} = \dot{q}\frac{d\dot{q}}{dq} \tag{9.27}$$

so that Eq. (9.26) can be rewritten in the form

$$m\dot{q}\,d\dot{q} - f(q)\,dq = 0 \tag{9.28}$$

Integrating Eq. (9.28), we obtain

$$T + V = E = \text{const} \tag{9.29}$$

where

$$T = \frac{1}{2}m\dot{q}^2 = \frac{1}{2m}p^2 \tag{9.30}$$

is the kinetic energy and

$$V = -\int_0^q f(s)\,ds = \int_q^0 f(s)\,ds = V(q) \tag{9.31}$$

is the potential energy. Equation (9.29) expresses the conservation of energy principle. Mathematically, it represents an integral of the motion.

Because the energy is conserved, we conclude that it is impossible for a trajectory initiated at an arbitrary point in the phase plane for which the total energy is different from zero to reach the origin of the phase plane, where the total energy is zero. Hence, *for conservative systems asymptotic stability is not possible*, so that the equilibrium points can be either merely stable or unstable.

Equations (9.29) and (9.30) yield

$$p = \pm\sqrt{2m[E - V(q)]} \tag{9.32}$$

so that, for a given total energy E and potential energy function $V(q)$ Eq. (9.32) yields the momentum p. Hence, Eq. (9.32) can be used to plot the trajectories in the phase plane.

As an illustration, let us consider the system of Example 9.2, which is recognized as being conservative. The potential energy is

$$V(q) = -\int_q^0 ks\left[1 - \left(\frac{s}{2a}\right)^2\right]ds = \frac{k}{2}q^2\left(1 - \frac{q^2}{8a^2}\right) \tag{9.33}$$

The plot $V(q)$ versus q is shown in Fig. 9.7a and trajectories in the phase plane are shown in Fig. 9.7b for three different values of the total energy. Points for which

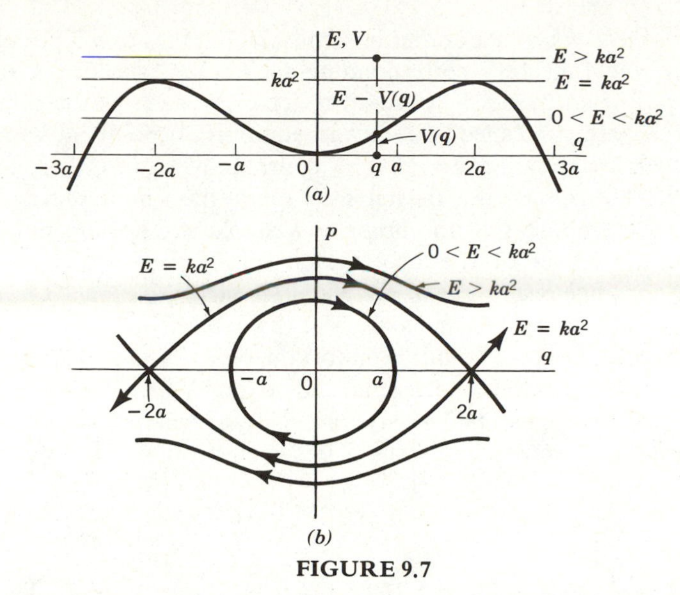

FIGURE 9.7

$f = -dV/dq = 0$ are equilibrium points. At the equilibrium point $q = 0$ the potential energy is zero, and at the equilibrium point $q = 2a$ the potential energy has the value $V(2a) = ka^2$. The type of motion depends on the total energy of the system. If $0 < E < ka^2$, then the trajectories are closed, as shown in Figs. 9.7*b* and 9.8. From Fig. 9.8, we conclude that the motion never reaches the circle $||\mathbf{x}|| = 2a$, so that the origin is a *stable* equilibrium point. Note that this conclusion is the same as that reached in Example 9.2 for the linearized system. On the other hand, for $E > ka^2$ we conclude from Figs. 9.7*b* and 9.9 that there is no circle $||\mathbf{x}|| = \varepsilon$ that the trajectories will not cross in their way out, so that the equilibrium point $q = 2a$, $p = 0$ is

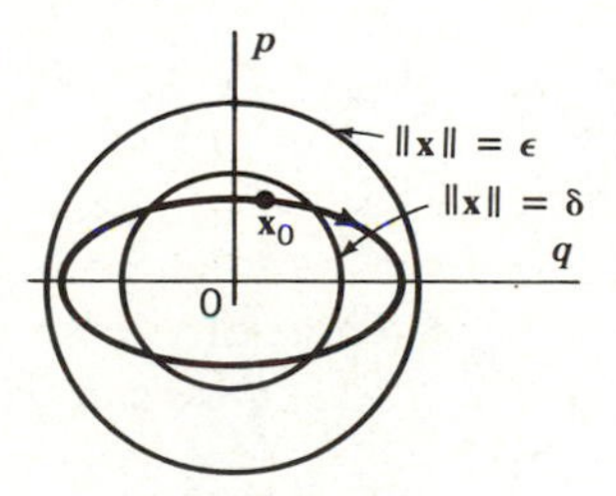

FIGURE 9.8

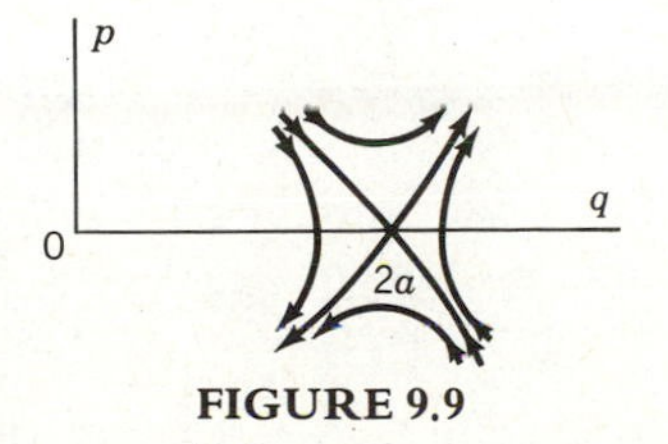

FIGURE 9.9

unstable. The case $E=ka^2$ is a borderline case dividing the two types of motion. Appropriately, the trajectory corresponding to $E=ka^2$ is called a *separatrix*. Equilibrium points surrounded by closed trajectories, such as the origin in the case considered here, are called *centers*, and equilibrium points at which trajectories meet, such as the points $q=\pm 2a$, $p=0$, are known as *saddle points*.

From Fig. 9.7*a*, we observe that the potential energy has a minimum at the origin, and we established above that the origin is a stable equilibrium point. This is essentially the idea expressed by a theorem due to Lagrange: *An equilibrium point corresponding to a minimum value of the potential energy is stable*. On the other hand, the potential energy has a maximum at the equilibrium point $q=2a$, $p=0$, and we established above that this equilibrium point is unstable. This case is covered by a theorem due to Liapunov that can be stated as follows: *If the potential energy has no minimum at an equilibrium point, then the equilibrium point is unstable.* Both theorems can be proved rigorously, but such proofs lie beyond the scope of this text.

9.6 MOTION IN THE STATE SPACE

In Sections 9.2–9.5, we discussed the geometric interpretation of the motion of a single-degree-of-freedom in the phase plane, or alternatively the motion in the state plane. In this section, we wish to extend some of the concepts introduced in Sections 9.2–9.5 to multi-degree-of-freedom systems.

In Section 9.2, we showed how a second-order differential equation describing the motion of a single-degree-of-freedom system, Eq. (9.1), can be replaced by two first-order differential equations, Eqs. (9.2). This was accomplished by introducing the momentum as an auxiliary variable, so that the problem formulation was in terms of two dependent variables, the coordinate $q(t)$ and the momentum $p(t)$, where q and p define the phase vector $\mathbf{x}=[q \;\; p]^T$. Alternatively, the velocity $\dot{q}(t)$ can be taken as the auxiliary variable, instead of the momentum $p(t)$, giving rise to a formulation in terms of the state vector $\mathbf{x}=[q \;\; \dot{q}]^T$, replacing the phase vector.

The approach described above can be used also for multi-degree-of-freedom systems. Indeed, considering an n-degree-of-freedom system described by the generalized coordinates $q_i(t)$ $(i=1, 2, \ldots, n)$, we can use as auxiliary variables the generalized momenta defined by

$$p_i=\frac{\partial L}{\partial \dot{q}_i}, \qquad i=1, 2, \ldots, n \tag{9.34}$$

where L is the Lagrangian, first introduced in Section 7.5. Then, using a well established theory of analytical dynamics, we can derive the so-called Hamilton equations, a set of $2n$ first-order differential equations in terms of the generalized coordinates $q_i(t)$ and generalized momenta $p_i(t)$ $(i=1, 2, \ldots, n)$. The generalized coordinates define an n-dimensional vector

$$\mathbf{q}(t)=[q_1(t) \quad q_2(t) \quad \ldots \quad q_n(t)]^T \tag{9.35}$$

known as the *configuration vector*. Moreover, introducing the n-dimensional vector of generalized momenta

$$\mathbf{p}(t) = [p_1(t) \quad p_2(t) \quad \dots \quad p_n(t)]^T \tag{9.36}$$

we can define the $2n$-dimensional *phase vector*

$$\mathbf{x}(t) = \begin{bmatrix} \mathbf{q} \\ \mathbf{p} \end{bmatrix} \tag{9.37}$$

The $2n$-dimensional space defined by the phase vector $\mathbf{x}(t)$ is called the *phase space*. The motion characteristics in the phase space are the same as in the phase plane. However, whereas one can envision trajectories in the phase space, it is not feasible to plot them. We will not pursue this subject here, but turn to an analogous description of the motion, one based on the state vector.

Lagrange's equations for an n-degree-of-freedom can be written in the form

$$\ddot{q}_i(t) = f_i[q_1(t), q_2(t), \dots, q_n(t), \dot{q}_1(t), \dot{q}_2(t), \dots, \dot{q}_n(t)], \qquad i = 1, 2, \dots, n \tag{9.38}$$

where f_i are in general nonlinear functions of the generalized coordinates and generalized velocities. To derive the state equations of motion, we supplement Eqs. (9.37) by the identities

$$\dot{q}_i(t) = \dot{q}_i(t), \qquad i = 1, 2, \dots, n \tag{9.39}$$

Next, we consider the substitutions

$$\begin{aligned} q_i &= x_i, \qquad \dot{q}_i = x_{i+n}, \\ x_{i+n} &= X_i, \qquad f_i = X_{i+n}, \end{aligned} \qquad i = 1, 2, \dots, n \tag{9.40}$$

and introduce the $2n$-dimensional state vector

$$\mathbf{x}(t) = \begin{bmatrix} \mathbf{q}(t) \\ \dot{\mathbf{q}}(t) \end{bmatrix} = [x_1(t) \quad x_2(t) \quad \dots \quad x_{2n}(t)]^T \tag{9.41}$$

as well as the associated $2n$-dimensional vector

$$\mathbf{X}(t) = \begin{bmatrix} \dot{\mathbf{q}}(t) \\ \mathbf{f}(t) \end{bmatrix} = [X_1(t) \quad X_2(t) \quad \dots \quad X_{2n}(t)]^T \tag{9.42}$$

where $\mathbf{X}(t)$ depends explicitly on $\mathbf{x}(t)$ and only implicitly on time, $\mathbf{X} = \mathbf{X}(\mathbf{x})$. Then, using Eqs. (9.40), we can write Eqs. (9.38) and (9.39) in the compact vector form

$$\dot{\mathbf{x}} = \mathbf{X}(\mathbf{x}) \tag{9.43}$$

Equation (9.43) represents the *state equations of motion* for the system. The $2n$-dimensional space defined by the state vector is called the *state space*. The state space is topologically equivalent to the phase space. It follows by implication that all the concepts introduced in Sections 9.2–9.5 in connection with the phase plane are equally valid for the state space. In fact, we observe that the $2n$ state equations of motion, Eq. (9.43), have the same appearance as the two phase equations of motion, Eq. (9.5). Hence, Eq. (9.43) can be treated in a manner similar to that of the second-order system given by Eq. (9.5).

The equilibrium positions are defined by constant state vectors $\boldsymbol{\alpha}$ satisfying the equation

$$\mathbf{X}(\boldsymbol{\alpha})=\mathbf{0} \tag{9.44}$$

Considering Eqs. (9.40), we conclude that the vectors $\boldsymbol{\alpha}$ must be such that

$$\dot{q}_i=\alpha_{i+n}=0, \qquad i=1, 2, \ldots, n \tag{9.45a}$$

$$f_i=f_i(\alpha_1, \alpha_2, \ldots, \alpha_n)=0, \qquad i=1, 2, \ldots, n \tag{9.45b}$$

Equations (9.45a) imply that the equilibrium points lie in the configuration space, namely, the $\mathbf{q}$-space. On the other hand, Eqs. (9.45b) represent algebraic equations that can be solved for the constants α_i ($i=1, 2, \ldots, n$). If f_i are linear functions, then there is only one equilibrium point, and if f_i are nonlinear functions, then there are as many equilibrium points as there are distinct solutions of $f_i=0$ ($i=1, 2, \ldots, n$).

To derive the perturbation equations, we expand the functions f_i in Taylor series about the equilibrium positions. Assuming, as in the two-dimensional case, that the origin of the state space has been translated so as to coincide with an equilibrium point, $\boldsymbol{\alpha}=\mathbf{0}$, we can write

$$f_i(x_1, x_2, \ldots, x_{2n})=f_i(0, 0, \ldots, 0)+\sum_{j=1}^{2n} \left.\frac{\partial f_i}{\partial x_j}\right|_{\mathbf{x}=0} x_i+O(\mathbf{x}^2), \qquad i=1, 2, \ldots, n \tag{9.46}$$

where $O(\mathbf{x}^2)$ denotes nonlinear terms in $\mathbf{x}$. But the first term on the right side of Eqs. (9.46) is zero by virtue of Eqs. (9.45b). Hence, the linearized perturbation equations are

$$\dot{\mathbf{x}}=A\mathbf{x} \tag{9.47}$$

where the elements of the coefficient matrix A are

$$\begin{aligned} &a_{ij}=0, \qquad i, j=1, 2, \ldots, n \\ &a_{ij}=\delta_{n+i,j}, \qquad i=1, 2, \ldots, n, \qquad j=n+1, n+2, \ldots, 2n \\ &a_{ij}=\left.\frac{\partial f_{i-n}}{\partial x_j}\right|_{\mathbf{x}=0}, \qquad i=n+1, n+2, \ldots, 2n, \qquad j=1, 2, \ldots, 2n \end{aligned} \tag{9.48}$$

The solution of Eq. (9.47) has the form

$$\mathbf{x}(t)=e^{\lambda t}\mathbf{x} \tag{9.49}$$

where λ is a constant scalar and $\mathbf{x}$ is an m-dimensional constant vector, $m=2n$. Introducing Eq. (9.49) into Eq. (9.47), we obtain the *eigenvalue problem*

$$[A-\lambda I]\mathbf{x}=\mathbf{0} \tag{9.50}$$

in which I is the $m \times m$ identity matrix. Similarly, the *characteristic equation* can be written in the form

$$\det[A-\lambda I]=0 \tag{9.51}$$

The determinant in Eq. (9.51) represents a polynomial of degree m in λ, known as the *characteristic polynomial*. The characteristic polynomial has m roots, where the roots are the system *characteristic values*, or *eigenvalues*, $\lambda_1, \lambda_2, \ldots, \lambda_m$.

The motion characteristics of the linearized system depend on the eigenvalues. We distinguish the following three cases:

i. If all the eigenvalues $\lambda_1, \lambda_2, \ldots, \lambda_m$ are pure imaginary, then the system is *stable*.

ii. If the eigenvalues $\lambda_1, \lambda_2, \ldots, \lambda_m$ are either real and negative or they are complex with negative real parts, or any combination thereof, the system is *asymptotically stable*.

iii. If at least one of the roots $\lambda_1, \lambda_2, \ldots, \lambda_m$ is real and positive or complex with positive real part, then the system is *unstable*.

The validity of the stability conclusions reached on the basis of the linearized system remains as described at the end of Section 9.4.

In deriving the state equations by means of the procedure suggested by Eqs. (9.40), there is the implication that the number of state equations is always even. This is indeed the case when one begins with a simultaneous set of second-order differential equations. There are cases, however, when the order of the original system can be either even or odd. In this regard, we recall that in Section 1.12 we transformed a single equation of order n to n first-order state equations, where n was an arbitrary integer, not necessarily even. The set of n first-order equations derived in Section 1.12 was expressed in a matrix form entirely analogous to that of Eq. (9.47). In discussing the nature of the solution following the derivation of Eq. (9.47), the order of the system played no particular role, so that the discussion is as valid for odd values as for even values of m.

Example 9.3

Derive the state equations of motion for the two-degree-of-freedom system of Example 6.5. Then, let $F=0$, identify the equilibrium positions, linearize the equations of motion about the equilibrium positions, and check the stability of these positions by solving the eigenvalue problem and checking the eigenvalues.

The system of Example 6.5 is shown in Fig. 6.8*a*. It consists of a point mass constrained to move horizontally and a thin rigid bar with one end hinged to the point mass. The motion of the system is described by the translation $x=x(t)$ of the point mass and the rotation $\theta=\theta(t)$ of the rigid bar about the hinge. The equations of motion, derived in Example 6.5, are

$$\begin{gathered}(M+m)\ddot{x}+ma(\ddot{\theta}\cos\theta-\dot{\theta}^2\sin\theta)+kx=F\\ ma\ddot{x}\cos\theta+\tfrac{4}{3}ma^2\ddot{\theta}+mga\sin\theta=2Fa\cos\theta\end{gathered}\tag{a}$$

To derive the state equations, we introduce the notation

$$x=x_1,\qquad \theta=x_2,\qquad \dot{x}=x_3,\qquad \dot{\theta}=x_4\tag{b}$$

Then, Eqs. (a) can be written in the first-order form

$$\begin{aligned} (M+m)\dot{x}_3 + ma\cos x_2 \dot{x}_4 &= F + max_4^2 \sin x_2 - kx_1 \\ ma\cos x_2 \dot{x}_3 + \tfrac{4}{3}ma^2 \dot{x}_4 &= 2Fa\cos x_2 - mga\sin x_2 \end{aligned} \tag{c}$$

Equations (c) are supplemented by the identities

$$\dot{x}_1 = x_3, \qquad \dot{x}_2 = x_4 \tag{d}$$

Equations (c) and (d) represent the four state equations of motion. Solving Eqs. (c) for $\dot{x}_2$ and $\dot{x}_4$ and using Eqs. (d), the state equations can be written in the form (9.43), in which $\mathbf{x}$ is a four-dimensional state vector. Moreover, the vector $\mathbf{X}$ on the right side of Eq. (9.43) can be shown to have the components

$$\begin{aligned} X_1 &= x_3 \\ X_2 &= x_4 \\ X_3 &= \frac{1}{4(M+m)ma^2 - 3(ma\cos x_2)^2} \\ &\quad \times [4ma^2(F + max_4^2 \sin x_2 - kx_1) - 3ma\cos x_2(2Fa\cos x_2 - mga\sin x_2)] \\ X_4 &= \frac{3}{4(M+m)ma^2 - 3(ma\cos x_2)^2} \\ &\quad \times [-ma\cos x_2(F + max_4^2 \sin x_2 - kx_1) + (M+m)(2Fa\cos x_2 - mga\sin x_2)] \end{aligned} \tag{e}$$

Letting $F=0$, we conclude that the components X_1 and X_2 remain the same and that the components X_3 and X_4 reduce to

$$\begin{aligned} X_3 &= \frac{1}{4(M+m)ma^2 - 3(ma\cos x_2)^2} \\ &\quad \times [4ma^2(max_4^2 \sin x_2 - kx_1) + 3(ma)^2 g \sin x_2 \cos x_2] \\ X_4 &= \frac{3}{4(M+m)ma^2 - 3(ma\cos x_2)^2} \\ &\quad \times [-ma\cos x_2(max_4^2 \sin x_2 - kx_1) - (M+m)mga\sin x_2] \end{aligned} \tag{f}$$

The equilibrium positions are constant solutions satisfying $\mathbf{X}=\mathbf{0}$. From the first two of Eqs. (e) and Eqs. (f), we conclude that the equilibrium positions must satisfy

$$x_1 = x_3 = x_4 = 0, \qquad \sin x_2 = 0 \tag{g}$$

Whereas mathematically $\sin x_2 = 0$ has the infinity of solutions $x_2 = r\pi$, $r=0, \pm 1, \pm 2, \ldots$, physically only two of these solutions represent distinct positions, namely, $x_2=0$ and $x_2=\pi$. The first represents the case in which the bar is at rest pointing downward and the second the case in which the bar is at rest in the upright position. We denote them by

$$\begin{aligned} &E_1: x_1 = x_2 = x_3 = x_4 = 0 \\ &E_2: x_1 = x_3 = x_4 = 0, \qquad x_2 = \pi \end{aligned} \tag{h}$$

Of course, E_1 represents the trivial solution.

To linearize about E_1, we write simply $\sin x_2 \cong x_2$, $\cos x_2 \cong 1$, neglect nonlinear terms and reduce X_3 and X_4 to

$$\begin{aligned} X_3 &= \frac{1}{4M+m}(-4kx_1 + 3mgx_2) \\ X_4 &= \frac{3}{(4M+m)a}[kx_1 - (M+m)gx_2] \end{aligned} \tag{i}$$

Hence, the state equations can be written in the linearized form (9.47) in which the matrix of coefficients has the form

$$A = \begin{bmatrix} 0 & 0 & 1 & 0 \\ 0 & 0 & 0 & 1 \\ -\dfrac{4k}{4M+m} & \dfrac{3mg}{4M+m} & 0 & 0 \\ \dfrac{3k}{(4M+m)a} & -\dfrac{3(M+m)g}{(4M+m)a} & 0 & 0 \end{bmatrix} \tag{j}$$

In the case of the equilibrium point E_2, we replace x_2 by $\pi + x_2$ in Eqs. (f) where x_2 is small. Then, making the approximation $\sin(\pi + x_2) \cong -x_2$, $\cos(\pi + x_2) \cong -1$, we can reduce X_3 and X_4 to

$$\begin{aligned} X_3 &= \frac{1}{4M+m}(-4kx_1 + 3mgx_2) \\ X_4 &= \frac{3}{(4M+m)a}[-kx_1 + (M+m)gx_2] \end{aligned} \tag{k}$$

so that the matrix of coefficients for the equilibrium point E_2 is

$$A = \begin{bmatrix} 0 & 0 & 1 & 0 \\ 0 & 0 & 0 & 1 \\ -\dfrac{4k}{4M+m} & \dfrac{3mg}{4M+m} & 0 & 0 \\ -\dfrac{3k}{(4M+m)a} & \dfrac{3(M+m)g}{(4M+m)a} & 0 & 0 \end{bmatrix} \tag{l}$$

The characteristic equation associated with the motion in the neighborhood of the equilibrium point E_1 is given by

$$\det[A-\lambda I]=\begin{bmatrix} -\lambda & 0 & 1 & 0 \\ 0 & -\lambda & 0 & 1 \\ -\dfrac{4k}{4M+m} & \dfrac{3mg}{4M+m} & -\lambda & 0 \\ \dfrac{3k}{(4M+m)a} & -\dfrac{3(M+m)g}{(4M+m)a} & 0 & -\lambda \end{bmatrix}$$

$$=\lambda^4+\frac{3(M+m)g+4ka}{(4M+m)a}\lambda^2+\frac{3kg}{(4M+m)a}=0 \qquad \text{(m)}$$

Introducing the notation

$$\frac{3(M+m)g+4ka}{(4M+m)a}=2b, \qquad \frac{3kg}{(4M+m)a}=c \qquad \text{(n)}$$

we can write the roots of the characteristic equation in the form

$$\lambda^2=-b\pm\sqrt{b^2-c} \qquad \text{(o)}$$

But,

$$b^2-c=\frac{1}{2(4M+m)a}\{[3(M+m)g+4ka]^2-12kg(4M+m)a\}$$

$$=\frac{1}{2(4M+m)a}[(3Mg-4ka)^2+9(2M+m)mg^2+12mgka]>0 \qquad \text{(p)}$$

It follows that both roots in Eq. (o) are real and negative, so that all four eigenvalues are pure imaginary. Hence, the equilibrium point E_1 is *stable* (in the linear sense).

The characteristic equation associated with E_2 is

$$\det[A-\lambda I]=\begin{bmatrix} -\lambda & 0 & 1 & 0 \\ 0 & -\lambda & 0 & 1 \\ -\dfrac{4k}{4M+m} & \dfrac{3mg}{4M+m} & -\lambda & 0 \\ -\dfrac{3k}{(4M+m)a} & \dfrac{3(M+m)g}{(4M+m)a} & 0 & -\lambda \end{bmatrix}$$

$$=\lambda^4+\frac{-3(M+m)g+4ka}{(4M+m)a}\lambda^2-\frac{3kg}{(4M+m)a}=0 \qquad \text{(q)}$$

Now, letting

$$\frac{-3(M+m)g+4ka}{(4M+m)a}=2b, \qquad -\frac{3kg}{(4M+m)a}=c \qquad \text{(r)}$$

we conclude that the roots retain the form given by Eq. (o). In this case, however, because c is negative

$$b^2 - c > b^2 \tag{s}$$

so that one of the roots in Eq. (o) is positive and the other is negative, independently of the sign of b. It follows that two of the eigenvalues are pure imaginary, the third one is real and negative, and the fourth one is real and positive. Hence, the equilibrium point E_2 is *unstable.*

The above conclusions conform to expectations. Indeed, it is not difficult to envision that the equilibrium position in which the bar hangs straight down is stable and that in which the bar is upright is unstable.

9.7 ROUTH-HURWITZ CRITERION

As pointed out in Section 9.6, the stability of a linear system is determined by the eigenvalues of the matrix A of the coefficients. From Fig. 9.5, it is clear that it is the real part of the eigenvalues that controls the stability of the solution. For a system of order m, the eigenvalues can be obtained by solving the characteristic equation

$$a_0\lambda^m + a_1\lambda^{m-1} + a_2\lambda^{m-2} + \cdots + a_{m-1}\lambda + a_m = 0 \tag{9.52}$$

For relatively high-order systems, solving the characteristic equations for all the m eigenvalues is likely to be a difficult task computationally. Hence, it appears desirable to be able to make a statement concerning the system stability without actually solving the characteristic equation. Because the imaginary parts of the eigenvalues do not affect the system stability, only the information concerning the real parts is necessary, and in particular the sign of the real parts.

There are two conditions necessary for none of the roots $\lambda_1, \lambda_2, \ldots, \lambda_m$ of Eq. (9.52) to have positive real part. The conditions are:

1. All the coefficients $a_0, a_1, \ldots, a_m$ of the characteristic polynomial must have the same sign.
2. None of the coefficients vanishes.

Assuming that $a_0 > 0$, the conditions imply that all the coefficients must be positive.

The above conditions are only necessary, and their satisfaction does not guarantee stability. The conditions can be used, however, to identify unstable systems by inspection. Necessary and sufficient conditions for asymptotic stability are provided by the *Routh–Hurwitz criterion.*

The coefficients a_i $(i = 0, 1, 2, \ldots, m)$ of the characteristic polynomial on the left side of Eq. (9.52) can be used to construct the determinants

$$\Delta_1 = a_1$$

$$\Delta_2 = \begin{vmatrix} a_1 & a_0 \\ a_3 & a_2 \end{vmatrix}$$

$$\Delta_3 = \begin{vmatrix} a_1 & a_0 & 0 \\ a_3 & a_2 & a_1 \\ a_5 & a_4 & a_3 \end{vmatrix} \tag{9.53}$$

$$\vdots$$

$$\Delta_m = \begin{vmatrix} a_1 & a_0 & 0 & \cdots & 0 \\ a_3 & a_2 & a_1 & \cdots & 0 \\ a_5 & a_4 & a_3 & \cdots & 0 \\ \vdots & \vdots & \vdots & & \vdots \\ a_{2m-1} & a_{2m-2} & a_{2m-3} & \cdots & a_m \end{vmatrix}$$

All the entries in the determinants $\Delta_1, \Delta_2, \ldots, \Delta_m$ corresponding to subscripts r such that $r > m$ or $r < 0$ are to be replaced by zero. Then, assuming that $a_0 > 0$, the Routh–Hurwitz criterion can be stated as follows: *The necessary and sufficient conditions for all the solutions λ_j $(j = 1, 2, \ldots, m)$ of the characteristic equation* (9.52) *to have negative real parts is that all the determinants $\Delta_1, \Delta_2, \ldots, \Delta_m$ be positive.* We observe that the last two determinants are related by

$$\Delta_m = a_m \Delta_{m-1} \tag{9.54}$$

so that it is only necessary to check the sign of the first $m-1$ determinants and of the coefficient a_m.

For relatively large-order systems, application of the Routh–Hurwitz criterion tends to be very laborious, as the computation of the large-order determinants in Eq. (9.53) involves a large number of multiplications. Fortunately, the computation of large-order determinants can be circumvented by introducing the *Routh array*

$$\begin{array}{l|lllll} \lambda^m & a_0 & a_2 & a_4 & a_6 & \cdots \\ \lambda^{m-1} & a_1 & a_3 & a_5 & a_7 & \cdots \\ \lambda^{m-2} & c_1 & c_2 & c_3 & c_4 & \cdots \\ \lambda^{m-3} & d_1 & d_2 & d_3 & d_4 & \cdots \\ \vdots & \vdots & \vdots & \vdots & \vdots & \\ \lambda^1 & m_1 & 0 & 0 & 0 & \\ \lambda^0 & n_1 & 0 & 0 & 0 & \end{array}$$

where $a_0, a_1, \ldots, a_m$ are the coefficients of the characteristic polynomial and

$$c_1 = -\frac{1}{a_1}\begin{vmatrix} a_0 & a_2 \\ a_1 & a_3 \end{vmatrix}$$

$$c_2 = -\frac{1}{a_1}\begin{vmatrix} a_0 & a_4 \\ a_1 & a_5 \end{vmatrix} \tag{9.55a}$$

$$c_3 = -\frac{1}{a_1}\begin{vmatrix} a_0 & a_6 \\ a_1 & a_7 \end{vmatrix}$$
$$\vdots$$

are the entries in the row corresponding to λ^{m-2},

$$d_1 = -\frac{1}{c_1}\begin{vmatrix} a_1 & a_3 \\ c_1 & c_2 \end{vmatrix}$$
$$d_2 = -\frac{1}{c_1}\begin{vmatrix} a_1 & a_5 \\ c_1 & c_3 \end{vmatrix} \tag{9.55b}$$
$$d_3 = -\frac{1}{c_1}\begin{vmatrix} a_1 & a_7 \\ c_1 & c_4 \end{vmatrix}$$
$$\vdots$$

are the entries in the row corresponding to λ^{m-3}, etc. Then, the Routh–Hurwitz criterion can be stated in terms of the Routh array as follows: *All the solutions λ_j ($j = 1, 2, \ldots, m$) of the characteristic equation (9.52) have negative real parts if all the entries in the first column of the Routh array have the same sign.*

The Routh–Hurwitz criterion requires the coefficients $a_0, a_1, \ldots, a_m$, and hence the derivation of the characteristic polynomial. Whereas this may be an easy task for low-order systems, for relatively high-order systems the task may not be so easy. In such cases, it may be more expeditious to check the system stability by computing the eigenvalues of the coefficient matrix A as outlined in Section 9.6.

Example 9.4

The system shown in Fig. 9.10 is the same as that of Example 9.3, except that there is a viscous damper between the supporting wall and the mass M. Use the Routh–Hurwitz criterion to show that the equilibrium point E_1 is asymptotically stable.

Adding the damping term $c\dot{x}$ to the equation for the coordinate x in Eqs. (a) of Example 9.3, we can write the equations of motion

$$\begin{aligned} (M+m)\ddot{x} + c\dot{x} + ma(\ddot{\theta}\cos\theta - \dot{\theta}^2\sin\theta) + kx &= F \\ ma\ddot{x}\cos\theta + \tfrac{4}{3}ma^2\ddot{\theta} + mga\sin\theta &= 2Fa\cos\theta \end{aligned} \tag{a}$$

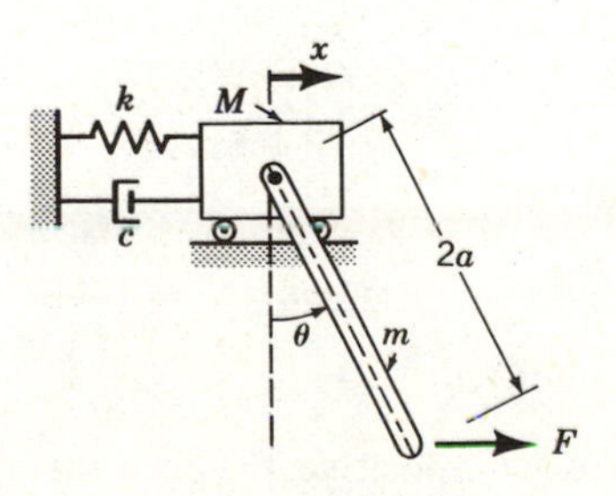

FIGURE 9.10

Letting $F=0$ and linearizing about $x=0$, $\theta=0$, we obtain

$$\begin{aligned}(M+m)\ddot{x}+c\dot{x}+kx+ma\ddot{\theta}&=0\\ ma\ddot{x}+\tfrac{4}{3}ma^2\ddot{\theta}+mga\theta&=0\end{aligned} \tag{b}$$

Assuming a solution of Eqs. (b) in the form

$$x(t)=Xe^{\lambda t}, \qquad \theta(t)=\Theta e^{\lambda t} \tag{c}$$

and dividing through by $e^{\lambda t}$, we have

$$\begin{aligned}[(M+m)\lambda^2+c\lambda+k]X+ma\lambda^2\Theta&=0\\ ma\lambda^2X+(\tfrac{4}{3}ma^2\lambda^2+mga)\Theta&=0\end{aligned} \tag{d}$$

Setting the determinant of the coefficients equal to zero, we obtain the characteristic equation

$$\begin{aligned}&\begin{vmatrix}(M+m)\lambda^2+c\lambda+k & ma\lambda^2\\ ma\lambda^2 & \tfrac{4}{3}ma^2\lambda^2+mga\end{vmatrix}\\ &=[(M+m)\lambda^2+c\lambda+k](\tfrac{4}{3}ma^2\lambda^2+mga)-(ma\lambda^2)^2\\ &=\tfrac{1}{3}(4M+m)a^2\lambda^4+\tfrac{4}{3}ma^2c\lambda^3+[(M+m)mga+\tfrac{4}{3}ma^2k]\lambda^2+mgac\lambda+mgak=0\end{aligned} \tag{e}$$

so that

$$\begin{aligned}&a_0=\tfrac{1}{3}(4M+m)ma^2, \qquad a_1=\tfrac{4}{3}ma^2c\\ &a_2=(M+m)mga+\tfrac{4}{3}ma^2k, \qquad a_3=mgac, \qquad a_4=mgak\end{aligned} \tag{f}$$

Clearly, the coefficients a_i ($i=0, 1, \ldots, 4$) are all positive, so that the necessary conditions for asymptotic stability are satisfied. To check the satisfaction of the necessary and sufficient conditions, we turn to the Routh–Hurwitz criterion. Inserting Eqs. (f) into Eqs. (9.53), we obtain the determinants

$$\begin{aligned}\Delta_1&=a_1=\tfrac{4}{3}ma^2c>0\\ \Delta_2&=\begin{vmatrix}a_1 & a_0\\ a_3 & a_2\end{vmatrix}=\begin{vmatrix}\tfrac{4}{3}ma^2c & \tfrac{1}{3}(4M+m)ma^2\\ mgac & (M+m)mga+\tfrac{4}{3}ma^2k\end{vmatrix}\\ &=(\tfrac{4}{3}ma^2)^2ck+(ma)^3cg>0\\ \Delta_3&=\begin{vmatrix}a_1 & a_0 & 0\\ a_3 & a_2 & a_1\\ 0 & a_4 & a_3\end{vmatrix}=-a_1\begin{vmatrix}a_1 & a_0\\ 0 & a_4\end{vmatrix}+a_3\,\Delta_2\\ &=-(\tfrac{4}{3}ma^2c)^2mgak+mgac[(\tfrac{4}{3}ma^2)^2ck+(ma)^3cg]=(ma)^4(cg)^2>0\\ \Delta_4&=a_4\,\Delta_3=mgak(ma)^4(cg)^2>0\end{aligned} \tag{g}$$

Because all four determinants are positive, the equilibrium position E_1 is asymptotically stable.

Before abandoning this example, it will prove of interest to check the stability of E_1 by the procedure based on the Routh array. In our case, the Routh array is

$$\begin{array}{c|ccc} \lambda^4 & a_0 & a_2 & a_4 \\ \lambda^3 & a_1 & a_3 & 0 \\ \lambda^2 & c_1 & c_2 & 0 \\ \lambda^1 & d_1 & 0 & 0 \\ \lambda^0 & e_1 & 0 & 0 \end{array}$$

where, from the first two of Eqs. (9.55a),

$$\begin{aligned} c_1 &= -\frac{1}{a_1}\begin{vmatrix} a_0 & a_2 \\ a_1 & a_3 \end{vmatrix} = \frac{\Delta_2}{a_1} = \frac{(\frac{4}{3}ma^2)^2ck + (ma)^3cg}{\frac{4}{3}ma^2c} > 0 \\ c_2 &= -\frac{1}{a_1}\begin{vmatrix} a_0 & a_4 \\ a_1 & 0 \end{vmatrix} = a_4 = mgak \end{aligned} \tag{h}$$

and from the first of Eqs. (9.55b),

$$\begin{aligned} d_1 &= -\frac{1}{c_1}\begin{vmatrix} a_1 & a_3 \\ c_1 & c_2 \end{vmatrix} = \frac{c_1a_3 - c_2a_1}{c_1} \\ &= \frac{\{[(\frac{4}{3}ma^2)^2ck + (ma)^3cg]mgac/\frac{4}{3}ma^2c\} - mgak\frac{4}{3}ma^2c}{c_1} \\ &= \frac{(ma)^4(cg)^2}{\frac{4}{3}ma^2cc_1} > 0 \end{aligned} \tag{i}$$

Moreover, by induction,

$$e_1 = -\frac{1}{d_1}\begin{vmatrix} c_1 & c_2 \\ d_1 & 0 \end{vmatrix} = c_2 = mgak > 0 \tag{j}$$

where the value of c_2 was obtained from (h). It is easy to see that all the entries in the first column of the Routh array are positive, so that we conclude once again that the equilibrium position E_1 is asymptotically stable.

PROBLEMS

9.1 The differential equation describing the longitudinal motion of an aircraft has the form

$$m\dot{v} = T - D$$

where m is the mass of the aircraft, v the speed, T the engine thrust, and D the drag. The thrust is related to the power P by $Tv = P$ and the drag is given by $D = Av^2 + B/v^2$, where A and B are constant coefficients. Give the equations for the cruising speed of the aircraft for the two cases: (1) constant power and (2) constant thrust.

9.2 Consider a mass–damper–spring system of the type examined in Example 9.1. By contrast, however, the spring considered here is nonlinear, exhibiting the restoring force $f(q) = -k[q - (\pi/2)\sin q]$. Derive the state equations of motion and determine the equilibrium positions.

9.3 A bead of mass m is free to slide along a circular hoop of radius R. The hoop rotates about a vertical diametrical axis with the constant angular velocity Ω (Fig. 9.11). Derive the state equations of motion and determine the equilibrium positions.

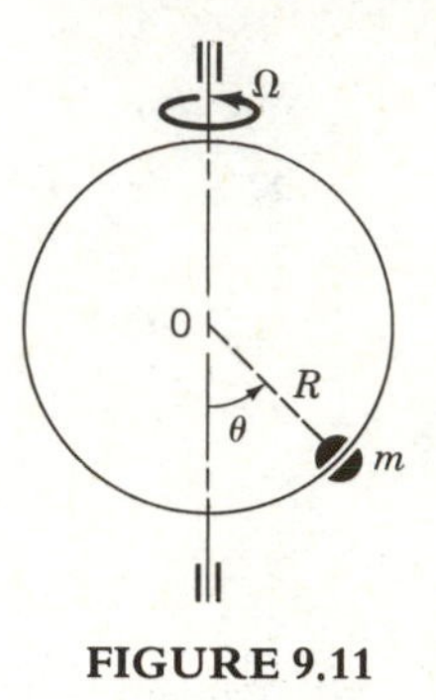

FIGURE 9.11

9.4 Determine the cruising speed for the aircraft of Problem 9.1 for the two cases: (1) $P = 4\times10^6$ W = const and (2) $T = 4\times10^4$ N = const. The drag coefficients are $A = 1.3693\ \mathrm{N\cdot s^2/m^2}$, $B = 2.0835\times10^8\ \mathrm{N\cdot m^2/s^2}$. Then, derive the linearized perturbation equation for the two cases, and check the stability of each cruising speed by examining the eigenvalue.

9.5 The equations of motion of a particle subjected to a force according to Newton's inverse square law are given by Eqs. (3.70) and (3.71). Use Eq. (3.71) to eliminate $\dot{\theta}$ from Eq. (3.70), thus obtaining an equation for the radial distance r alone. Then, write the state equations of motion, determine the equilibrium position, derive the linearized perturbation equations, and check the stability of the equilibrium position by examining the system eigenvalues.

9.6 Consider the system of Problem 9.2, assume that the damping is zero, and use the approach described in Section 9.5 to plot trajectories in the neighborhood of all the equilibrium positions and to check the existence of a separatrix.

9.7 Repeat Problem 9.6 for the system of Problem 9.3. Consider the cases: (1) $\Omega = 0$ and (2) $\Omega = 2\sqrt{g/R}$.

9.8 Consider the system of Example 9.3, except that here the spring is nonlinear, exhibiting the restoring force $f(x) = -kx[1-(x/2b)^2]$. Derive the state equations and identify the equilibrium positions.

9.9 Derive the state equations for the gyroscopic system described by Eqs. (6.96) and identify the equilibrium positions.

9.10 Determine the stability of the system of Problem 1.13 by solving the eigenvalue problem and examining the eigenvalues.

9.11 Consider the system of Problem 9.2, derive the linearized perturbation equations corresponding to each of the equilibrium positions, and determine the stability of the equilibrium positions by solving the associated eigenvalue problems.

9.12 Repeat Problem 9.11 for the system of Problem 9.3.

9.13 Repeat Problem 9.11 for the system of Problem 9.8.

9.14 Determine the stability of the equilibrium positions for the system of Problem 1.13 by means of the two approaches discussed in Section 9.7.

9.15 Repeat Problem 9.14 for the system of Problem 9.11.

CHAPTER 10
Computational Techniques for the Response

10.1 INTRODUCTION

In earlier chapters, such as Chapters 3, 4, and 8, we discussed ways of deriving the response of single-degree-of-freedom and multi-degree-of-freedom systems. The emphasis in previous discussions has been on general techniques for obtaining closed-form solutions. In this chapter, the emphasis is placed on numerical evaluation of the response, and in particular on evaluation on a digital computer.

For the most part, the response of a multi-degree-of-freedom vibrating system can be derived by formulating the problem in terms of a set of second-order differential equations and using an approach called modal analysis. This approach works for undamped systems and for some special cases of damping, such as proportional damping, but it fails in the more general case of nonproportional damping. Of course, the approach is not applicable to systems described by differential equations whose structure differs from that of multi-degree-of-freedom vibrating systems. Indeed, to obtain the response of general dynamical systems, it is necessary to formulate the problem in terms of a set of first-order state equations. Then, for linear systems, the response can be obtained by a procedure based on the so-called transition matrix, and we note in passing that systems with general nonproportional damping can be regarded as special cases of the general class of linear dynamical systems amenable to treatment by the transition matrix. The solution based on the transition matrix represents a closed-form solution involving a convolution integral, which implies a linear system with constant coefficients. However, even though a solution in terms of a convolution integral represents a closed-form solution, in the case of complicated excitation the integrals still require numerical evaluation. No closed-form solution is possible in the case of nonlinear systems. In fact, for nonlinear systems the only solutions possible involve direct numerical integration of the system differential equations.

In attempting to produce a numerical solution on a digital computer, one difficulty becomes apparent immediately. In all the problems encountered to this point, the excitation and response were functions of time, and the time was regarded as continuous. Such systems are known as continuous-time systems. Digital computers, however, are incapable of handling continuous-time systems. Hence, if a numerical solution on a digital computer is the object, it becomes necessary to treat the time as a discrete variable, which gives rise to discrete-time systems. In reducing a continuous-time system to a discrete-time system, differential equations become difference equations. Moreover, convolution integrals become convolution sums. Nevertheless, most of the concepts developed for continuous-time systems have counterparts in discrete-time systems. Indeed, the approach based on the transition matrix can still be used for discrete-time systems, albeit in a somewhat different form. As with the response, the various concepts of stability introduced in Chapter 9 in conjunction with continuous-time systems also apply to discrete-time systems, although the definitions assume a somewhat different form. Numerical procedures for the response of nonlinear systems tend to be incremental procedures, which implies discretization in time by definition.

This chapter begins with a discussion of the solution of the state equations for continuous-time systems by the approach based on the transition matrix, with emphasis on efficient computational algorithms. The approach is specialized subsequently to the problem of general damped systems. Then, the subject of discrete-time systems is presented in detail. In particular, procedures are developed for evaluating the response on a digital computer and for testing the stability of linear discrete-time systems. Finally, the Euler method and the Runge–Kutta methods for the response of nonlinear systems are presented.

10.2 SOLUTION OF THE STATE EQUATIONS FOR LINEAR SYSTEMS BY THE TRANSITION MATRIX

In Chapter 8, we discussed ways of deriving the response of linear multi-degree-of-freedom vibrating systems. In particular, we presented a very efficient procedure for deriving the response of undamped systems, known as modal analysis, whereby a set of n simultaneous second-order differential equations is transformed into a set of n independent second-order differential equations. The approach is made possible by the structure of the differential equations, i.e., a structure characterized by symmetric mass matrix M and stiffness matrix K. Indeed, in this case the orthogonality of the modal vectors with respect to the matrices M and K permits decoupling of the differential equations of motion. The same procedure can also be used in the case of proportional damping, or in the case of small damping, but it fails in the general case of damping, as in the latter case the modal vectors are not orthogonal with respect to M and K. Moreover, in the case of controlled systems, unless a certain method for the design of control forces is used, the structure of the differential equations governing the behavior of the closed-loop system tends to be different from that of the open-loop system (see Chapter 11), so that the decoupling approach presented in Chapter 8 is not applicable. Hence, a more general approach to the solution of the system differential equations is highly desirable.

In this section, we develop an approach capable of producing the response of general linear systems with constant coefficients. The approach is formulated on the basis of state equations, which are of first order, and can be used for any type of dynamical systems and for any number of equations, even or odd. In fact, it can treat general damped systems as a special case. The basic ideas involved in the approach can be introduced conveniently by means of a first-order system.

The differential equations describing the behavior of first-order mechanical and electrical systems were derived in Sections 4.2 and 4.3. As pointed out in Section 4.3, the two equations possess similar structure, and the only difference lies in the notation for the system parameters, the response, and the excitation. To unify the notation, we write the differential equation of a typical first-order system in the form

$$\dot{x}(t) = ax(t) + bu(t) \tag{10.1}$$

where $x(t)$ and $u(t)$ are scalar functions of time representing the response and excitation, respectively, and a and b are constant scalars. The solution of Eq. (10.1) can be obtained by the Laplace transformation method (see the Appendix). Transforming both sides of Eq. (10.1), we obtain

$$sX(s) - x(0) = aX(s) + bU(s) \tag{10.2}$$

where $X(s)$ and $U(s)$ are the transforms of $x(t)$ and $u(t)$, respectively, and $x(0)$ is the initial condition. A solution of Eq. (10.2) yields the transformed response

$$X(s) = \frac{x(0)}{s-a} + \frac{b}{s-a} U(s) \tag{10.3}$$

But, from the table of Laplace transforms (Section A.7), we can write the inverse transformation

$$\mathscr{L}^{-1} \frac{1}{s-a} = e^{at} \tag{10.4}$$

Moreover, using the convolution theorem (Section A.6), we obtain the response

$$x(t) = e^{at} x(0) + \int_0^t e^{a(t-\tau)} bu(\tau)\, d\tau \tag{10.5}$$

which contains the response to the initial condition $x(0)$ and the excitation $u(t)$. The first can be identified as the homogeneous solution and the second as the particular solution.

Next, let us consider a system of order m and write the differential equations describing the behavior of the system in the matrix form

$$\dot{\mathbf{x}}(t) = A\mathbf{x}(t) + B\mathbf{u}(t) \tag{10.6}$$

where $\mathbf{x}(t)$ is the m-dimensional state vector denoting the response, $\mathbf{u}(t)$ is an l-dimensional vector representing the excitation, and A and B are $m \times m$ and $m \times l$ matrices of coefficients, respectively. Equation (10.6) resembles Eq. (10.1) entirely, except that Eq. (10.6) is a matrix equation and Eq. (10.1) is a scalar equation. Because of the similar structure, one can expect the solution of Eq. (10.6) to resemble the solution of Eq. (10.1), as given by Eq. (10.5).

We derive the solution of Eq. (10.6) in two stages, first the homogeneous solution and then the particular solution, where the second is based on the first. Letting $\mathbf{u}(t)=\mathbf{0}$ in Eq. (10.6), we obtain the homogeneous equation

$$\dot{\mathbf{x}}(t)=A\mathbf{x}(t) \tag{10.7}$$

Using the analogy with the homogeneous solution of the first-order system, as given by the first term on the right side of Eq. (10.5), we can write the solution of Eq. (10.7) in the form

$$\mathbf{x}(t)=e^{At}\mathbf{x}(0) \tag{10.8}$$

where $\mathbf{x}(0)$ is the initial state vector and e^{At} is an $m\times m$ matrix, which can be obtained from the coefficient matrix A by means of the series

$$e^{At}=I+tA+\frac{t^2}{2!}A^2+\frac{t^3}{3!}A^3+\cdots \tag{10.9}$$

Indeed, from Eq. (10.9), the time derivative of e^{At} is

$$\begin{aligned}\frac{d}{dt}e^{At}&=A+tA^2+\frac{t^2}{2!}A^3+\frac{t^3}{3!}A^4+\cdots\\&=A\left(I+tA+\frac{t^2}{2!}A^2+\frac{t^3}{3!}A^3+\cdots\right)=Ae^{At}\end{aligned} \tag{10.10}$$

so that Eq. (10.8) is verified to represent the solution of Eq. (10.7).

The task of deriving the particular solution of Eq. (10.6) is more involved. As a first step, we premultiply Eq. (10.6) by the time-dependent $m\times m$ matrix $K(t)$, so that

$$K(t)\dot{\mathbf{x}}(t)=K(t)A\mathbf{x}(t)+K(t)B\mathbf{u}(t) \tag{10.11}$$

Then, we consider

$$\frac{d}{dt}\{K(t)\mathbf{x}(t)\}=\dot{K}(t)\mathbf{x}(t)+K(t)\dot{\mathbf{x}}(t) \tag{10.12}$$

Inserting Eq. (10.6) into Eq. (10.12), we obtain

$$\frac{d}{dt}\{K(t)\mathbf{x}(t)\}=\dot{K}(t)\mathbf{x}(t)+K(t)A\mathbf{x}(t)+K(t)B\mathbf{u}(t) \tag{10.13}$$

Next, let us assume that the matrix $K(t)$ satisfies the equation

$$\dot{K}(t)=-AK(t) \tag{10.14}$$

so that, by analogy with Eqs. (10.7) and (10.8), we conclude that a matrix $K(t)$ satisfying Eq. (10.14) has the form

$$K(t)=e^{-At}K(0) \tag{10.15}$$

where

$$e^{-At}=I-tA+\frac{t^2}{2!}A^2-\frac{t^3}{3!}A^3+\cdots \tag{10.16}$$

For convenience, we choose

$$K(0)=I \tag{10.17}$$

so that Eq. (10.15) reduces to

$$K(t)=e^{-At} \tag{10.18}$$

Hence, from Eq. (10.16), we conclude that the matrix $K(t)$ commutes with the matrix A, or

$$AK(t)=K(t)A \tag{10.19}$$

from which it follows that $K(t)$ also satisfies

$$\dot{K}(t)=-K(t)A \tag{10.20}$$

In view of Eq. (10.20), Eq. (10.13) reduces to

$$\frac{d}{dt}\{K(t)\mathbf{x}(t)\}=K(t)B\mathbf{u}(t) \tag{10.21}$$

Equation (10.21) can be integrated readily, with the result

$$\begin{aligned} K(t)\mathbf{x}(t) &= K(0)\mathbf{x}(0)+\int_0^t K(\tau)B\mathbf{u}(\tau)\,d\tau \\ &= \mathbf{x}(0)+\int_0^t K(\tau)B\mathbf{u}(\tau)\,d\tau \end{aligned} \tag{10.22}$$

Premultiplying Eq. (10.22) through by $K^{-1}(t)$ and recalling Eq. (10.18), we obtain

$$\begin{aligned} \mathbf{x}(t) &= K^{-1}(t)\mathbf{x}(0)+K^{-1}(t)\int_0^t K(\tau)B\mathbf{u}(\tau)\,d\tau \\ &= e^{At}\mathbf{x}(0)+\int_0^t e^{A(t-\tau)}B\mathbf{u}(\tau)\,d\tau \end{aligned} \tag{10.23}$$

and we observe that Eq. (10.23) contains both the homogeneous solution and the particular solution. Of course, the particular solution has the same form as that given by Eq. (10.8). Comparing Eqs. (10.5) and (10.23), we conclude that the response of the system of order m has the same structure as the response of a first-order system, except that in the first case the excitation and response are represented by vectors and in the second case by scalars. Moreover, the matrices A and B replace the scalars a and b. This analogy is entirely consistent with the differential equations (10.1) and (10.6), in which the same analogy exists. Note that both solutions, Eqs. (10.5) and (10.23), involve the evaluation of convolution integrals.

The matrix

$$\Phi(t,\tau)=e^{A(t-\tau)} \tag{10.24}$$

is commonly known as the *transition matrix*. It is obtained from Eq. (10.9) by replacing t by $t-\tau$.

The transition matrix can be evaluated, perhaps more directly, by means of the

Laplace transformation method. Indeed, taking the Laplace transform of Eq. (10.7), we can write

$$sX(s) - x(0) = AX(s) \tag{10.25}$$

where $\mathbf{X}(s)$ is the Laplace transform of $\mathbf{x}(t)$. Equation (10.25) can be solved for $\mathbf{X}(s)$ as follows:

$$\mathbf{X}(s) = [sI - A]^{-1}\mathbf{x}(0) \tag{10.26}$$

where $[sI - A]^{-1}$ is the inverse of the matrix $[sI - A]$, in which I is the identity matrix of order m. Taking the inverse Laplace transformation of Eq. (10.26), we obtain

$$\mathbf{x}(t) = \mathscr{L}^{-1}\{[sI - A]^{-1}\}\mathbf{x}(0) \tag{10.27}$$

so that, comparing Eq. (10.27) to Eq. (10.8) and recalling Eq. (10.24), we conclude that

$$\Phi(t, 0) = \mathscr{L}^{-1}\{[sI - A]^{-1}\} \tag{10.28}$$

Moreover, $\Phi(t, \tau)$ can be obtained from Eq. (10.28) by replacing t by $t - \tau$. The approach described above involves the inverse of the matrix $[sI - A]$, which can be a drawback. Indeed, the approach is quite effective for relatively low-order systems, but it becomes impractical as the order of the system increases.

Example 10.1

Derive the transition matrix for an undamped single-degree-of-freedom system. Then, use the approach based on the transition matrix to derive the response to the impulsive force

$$f(t) = \hat{f}\,\delta(t) \tag{a}$$

where $\hat{f}$ is the magnitude of the impulse.

Introducing Eq. (a) into Eq. (4.8) and letting $c = 0$, we obtain the equation of motion

$$m\ddot{x}(t) + kx(t) = \hat{f}\,\delta(t) \tag{b}$$

Before proceeding with the derivation of the transition matrix, we must calculate the matrix A of the coefficients. To this end, let us introduce the notation

$$x(t) = x_1(t), \qquad \dot{x}(t) = x_2(t) \tag{c}$$

so that, inserting Eqs. (c) into Eq. (b) and dividing through by m, we can write the state equations in the form

$$\dot{x} = x_2, \qquad \dot{x}_2 = -\omega_n^2 x_1 + \frac{\hat{f}}{m}\,\delta(t) \tag{d}$$

where $\omega_n = \sqrt{k/m}$ is the natural frequency of the system. The state equations can be written in the matrix form (10.6), where the state and excitation vectors are

$$\mathbf{x}(t) = [x_1(t) \quad x_2(t)]^T, \qquad \mathbf{u}(t) = \hat{f}\,\delta(t) \tag{e}$$

in which $\mathbf{u}(t)$ is not really a vector but a scalar, and the coefficient matrices are

$$A=\begin{bmatrix} 0 & 1 \\ -\omega_n^2 & 0 \end{bmatrix}, \qquad B=\begin{bmatrix} 0 \\ 1/m \end{bmatrix} \tag{f}$$

Next, let us expand the series

$$\begin{aligned} e^{At} &= I+tA+\frac{t^2}{2!}A^2+\frac{t^3}{3!}A^3+\cdots \\ &= \begin{bmatrix} 1 & 0 \\ 0 & 1 \end{bmatrix}+t\begin{bmatrix} 0 & 1 \\ -\omega_n^2 & 0 \end{bmatrix}-\frac{(\omega_n t)^2}{2!}\begin{bmatrix} 1 & 0 \\ 0 & 1 \end{bmatrix}-\frac{\omega_n^2 t^3}{3!}\begin{bmatrix} 0 & 1 \\ -\omega_n^2 & 0 \end{bmatrix} \\ &\quad +\frac{(\omega_n t)^4}{4!}\begin{bmatrix} 1 & 0 \\ 0 & 1 \end{bmatrix}+\frac{\omega_n^4 t^5}{5!}\begin{bmatrix} 0 & 1 \\ -\omega_n^2 & 0 \end{bmatrix}-\cdots \\ &= \begin{bmatrix} 1-\dfrac{(\omega_n t)^2}{2!}+\dfrac{(\omega_n t)^4}{4!}-\cdots & \dfrac{1}{\omega_n}\left[\omega_n t-\dfrac{(\omega_n t)^3}{3!}+\dfrac{(\omega_n t)^5}{5!}-\cdots\right] \\ -\omega_n\left[\omega_n t-\dfrac{(\omega_n t)^2}{3!}+\dfrac{(\omega_n t)^5}{5!}-\cdots\right] & 1-\dfrac{(\omega_n t)^2}{2!}+\dfrac{(\omega_n t)^4}{4!}-\cdots \end{bmatrix} \end{aligned} \tag{g}$$

But

$$\begin{aligned} 1-\frac{(\omega_n t)^2}{2!}+\frac{(\omega_n t)^4}{4!}-\cdots &= \cos\omega_n t \\ \omega_n t-\frac{(\omega_n t)^3}{3!}+\frac{(\omega_n t)^5}{5!}-\cdots &= \sin\omega_n t \end{aligned} \tag{h}$$

so that Eq. (g) can be rewritten as

$$e^{At}=\begin{bmatrix} \cos\omega_n t & \omega_n^{-1}\sin\omega_n t \\ -\omega_n\sin\omega_n t & \cos\omega_n t \end{bmatrix} \tag{i}$$

which represents the transition matrix for $\tau=0$.

Letting the initial state $\mathbf{x}(0)$ be equal to zero and considering Eqs. (e), (f), and (i), we can obtain the response of the system from Eq. (10.23) as follows:

$$\begin{aligned} \begin{bmatrix} x_1(t) \\ x_2(t) \end{bmatrix} &= \int_0^t \begin{bmatrix} \cos\omega_n(t-\tau) & \omega_n^{-1}\sin\omega_n(t-\tau) \\ -\omega_n\sin\omega_n(t-\tau) & \cos\omega_n(t-\tau) \end{bmatrix}\begin{bmatrix} 0 \\ 1/m \end{bmatrix}\hat{f}\,\delta(\tau)\,d\tau \\ &= \begin{bmatrix} \hat{f}(m\omega_n)^{-1}\sin\omega_n t \\ \hat{f}m^{-1}\cos\omega_n t \end{bmatrix} \end{aligned} \tag{j}$$

and we observe that the response given by Eq. (j) contains the displacement $x_1(t)=x(t)$ as well as the velocity $x_2(t)=\dot{x}(t)$.

Example 10.2

Calculate the transition matrix for the system of Example 10.1 by means of Eq. (10.28).

The coefficient matrix A is given by the first of Eqs. (f) in Example 10.1. Hence, we can write

$$sI - A = \begin{bmatrix} s & -1 \\ \omega_n^2 & s \end{bmatrix} \tag{a}$$

The inverse of $sI - A$ is simply

$$[sI - A]^{-1} = \frac{1}{s^2 + \omega_n^2} \begin{bmatrix} s & 1 \\ -\omega_n^2 & s \end{bmatrix} = \begin{bmatrix} \dfrac{s}{s^2+\omega_n^2} & \dfrac{1}{s^2+\omega_n^2} \\ -\dfrac{\omega_n^2}{s^2+\omega_n^2} & \dfrac{s}{s^2+\omega_n^2} \end{bmatrix} \tag{b}$$

Taking the inverse Laplace transformation of Eq. (b) and making use of the table of Laplace transform pairs in Section A.7, we obtain

$$\Phi(t, 0) = \mathscr{L}^{-1}\{[sI - A]^{-1}\} = \mathscr{L}^{-1} \begin{bmatrix} \dfrac{s}{s^2+\omega_n} & \dfrac{1}{s^2+\omega_n^2} \\ -\dfrac{\omega_n^2}{s^2+\omega_n^2} & \dfrac{s}{s^2+\omega_n^2} \end{bmatrix}$$

$$= \begin{bmatrix} \cos \omega_n t & \omega_n^{-1} \sin \omega_n t \\ -\omega_n \sin \omega_n t & \cos \omega_n t \end{bmatrix} \tag{c}$$

which is identical to the matrix given by Eq. (i) of Example 10.1.

10.3 COMPUTATIONAL ASPECTS OF THE RESPONSE BY THE TRANSITION MATRIX

Closed-form solutions for the response by means of the transition matrix can be obtained only in very simple cases, such as the single-degree-of-freedom system of Example 10.1. For more involved cases, it is necessary to evaluate the transition matrix numerically. To this end, we recall that the transition matrix can be expressed as a series of the type (10.9), which involves raising the matrix A to various powers. Series (10.9) is infinite, however, so that for practical reasons it must be truncated. If the series is truncated so as to include terms through the nth power only, or

$$\Phi(t, 0) \cong \Phi_n(t) = I + tA + \frac{t^2}{2!} A^2 + \cdots + \frac{t^n}{n!} A^n \tag{10.29}$$

then the approximation Φ_n requires a total of $n(n-1)/2$ matrix multiplications. It turns out that the computation of Φ_n can be carried out in a more economical form. Indeed, series (10.29) can be rewritten as

$$\Phi_n = I + tA\left(I + \frac{tA}{2}\left(I + \frac{tA}{3}\left(I + \cdots + \frac{tA}{n-1}\left(I + \frac{tA}{n}\right)\right)\cdots\right)\right) \tag{10.30}$$

and can be computed recursively as follows:

$$\begin{aligned}
\psi_1 &= I + \frac{tA}{n} \\
\psi_2 &= I + \frac{tA}{n-1}\psi_1 \\
\psi_3 &= I + \frac{tA}{n-2}\psi_2 \\
&\vdots \\
\Phi_n &= I + tA\psi_{n-1}
\end{aligned} \tag{10.31}$$

Computation of Φ_n by means of Eqs. (10.31) requires only $n-1$ matrix multiplications.

The question remains as to how many terms must be included in the series for the computation of the transition matrix to achieve convergence for a given level of accuracy. Clearly, the answer to this question depends on the magnitude of the elements of the matrix A. It also depends on the interval $(t, \tau) = t - \tau$. But, it is not difficult to verify that the transition matrix satisfies

$$\Phi(t, \tau) = \Phi(t, t_1)\Phi(t_1, t_2) \cdots \Phi(t_{k-1}, t_k)\Phi(t_k, \tau) \tag{10.32}$$

Hence, by breaking the interval (t, τ) into the smaller intervals (t, t_1), $(t_1, t_2), \ldots,$ (t_{k-1}, t_k), (t_k, τ), we can accelerate convergence of the series and actually achieve convergence with fewer terms. We shall consider this approach later in this chapter.

Calculating the response by means of the transition matrix involving the matrix A may still not be the most efficient one computationally, so that we wish to explore an alternative approach. The eigenvalue problem associated with the matrix A has the form

$$A\mathbf{x}_i = \lambda_i \mathbf{x}_i, \qquad i = 1, 2, \ldots, m \tag{10.33}$$

where λ_i and $\mathbf{x}_i$ $(i = 1, 2, \ldots, m)$ are the eigenvalues and eigenvectors, respectively, both complex in general. Unlike the case in which A is symmetric, in the case in which A is not symmetric the eigenvectors are not mutually orthogonal. Nevertheless, the eigenvectors $\mathbf{x}_i$ $(i = 1, 2, \ldots, m)$ do possess some kind of orthogonality property, as shown in the following.

Consider the eigenvalue problem associated with A^T

$$A^T \mathbf{y}_j = \lambda_j \mathbf{y}_j, \qquad j = 1, 2, \ldots, m \tag{10.34}$$

which is known as the *adjoint* eigenvalue problem. Because

$$\det A^T = \det A \tag{10.35}$$

the eigenvalues of A^T are the same as the eigenvalues of A. On the other hand, the eigenvectors of A^T are different from the eigenvectors A. The set of eigenvectors $\mathbf{y}_j$ $(j = 1, 2, \ldots, m)$ is known as the *adjoint* of the set of eigenvectors $\mathbf{x}_i$ $(i = 1, 2, \ldots, m)$.

The eigenvalue problem (10.34) can be written in the transposed form

$$\mathbf{y}_j^T A = \lambda_j \mathbf{y}_j^T, \qquad j=1, 2, \ldots, m \tag{10.36}$$

Because of their position relative to the matrix A in eigenvalue problems (10.33) and (10.36), the eigenvectors $\mathbf{x}_i$ $(i=1, 2, \ldots, m)$ are called *right eigenvectors* of A and the eigenvectors $\mathbf{y}_j$ $(j=1, 2, \ldots, m)$ are referred to as *left eigenvectors* of A.

Next, premultiply Eq. (10.33) by $\mathbf{y}_j^T$ and postmultiply Eq. (10.36) by $\mathbf{x}_i$, so that

$$\mathbf{y}_j^T A \mathbf{x}_i = \lambda_i \mathbf{y}_j^T \mathbf{x}_i \tag{10.37a}$$

$$\mathbf{y}_j^T A \mathbf{x}_i = \lambda_j \mathbf{y}_j^T \mathbf{x}_i \tag{10.37b}$$

Subtracting Eq. (10.37b) from Eq. (10.37a), we obtain

$$0 = (\lambda_i - \lambda_j) \mathbf{y}_j^T \mathbf{x}_i \tag{10.38}$$

from which we conclude that, if $\lambda_i \neq \lambda_j$, then

$$\mathbf{y}_j^T \mathbf{x}_i = 0, \qquad i \neq j, \qquad i, j = 1, 2, \ldots, m \tag{10.39}$$

or the left eigenvectors of A are orthogonal to the right eigenvectors of A. This type of orthogonality is known as *biorthogonality*. Inserting Eq. (10.39) into Eqs. (10.37) we conclude that the biorthogonality is also with respect to A, or

$$\mathbf{y}_j^T A \mathbf{x}_i = 0, \qquad i \neq j, \qquad i, j = 1, 2, \ldots, m \tag{10.40}$$

When $j=i$, the products $\mathbf{y}_i^T \mathbf{x}_i$ and $\mathbf{y}_i^T A \mathbf{x}_i$ are not zero. It is convenient to normalize the right and left eigenvectors of A so as to satisfy

$$\mathbf{y}_i^T \mathbf{x}_i = 1, \qquad i = 1, 2, \ldots, m \tag{10.41}$$

Letting $j=i$ in Eqs. (10.37), we conclude that

$$\mathbf{y}_i^T A \mathbf{x}_i = \lambda_i, \qquad i = 1, 2, \ldots, m \tag{10.42}$$

In this case the left eigenvectors of A are said to be *biorthonormal* to the right eigenvectors of A. Introducing the $m \times m$ matrices of right and left eigenvectors of A

$$X = [\mathbf{x}_1 \quad \mathbf{x}_2 \quad \ldots \quad \mathbf{x}_m], \qquad Y = [\mathbf{y}_1 \quad \mathbf{y}_2 \quad \ldots \quad \mathbf{y}_m] \tag{10.43a, b}$$

as well as the $m \times m$ matrix of eigenvalues

$$\Lambda = \text{diag}[\lambda_1 \quad \lambda_2 \quad \ldots \quad \lambda_m] \tag{10.44}$$

we can rewrite Eqs. (10.41) and (10.42) in the compact form

$$Y^T X = I, \qquad Y^T A X = \Lambda \tag{10.45a, b}$$

Equations (10.45) can be used to express the transition matrix in a very useful form. From Eq. (10.45a), we can write

$$Y^{\mathrm{T}}=X^{-1}, \qquad X=(Y^{\mathrm{T}})^{-1} \tag{10.46}$$

so that premultiplying Eq. (10.45a) by X and postmultiplying it by Y^{T}, we obtain

$$XY^{\mathrm{T}}=I \tag{10.47}$$

Moreover, premultiplying Eq. (10.45b) by X and postmultiplying it by Y^{T}, we have

$$A=X\Lambda Y^{\mathrm{T}} \tag{10.48}$$

Equations (10.47) and (10.48) can be used to express the transition matrix in the desired form. Indeed, introducing Eqs. (10.47) and (10.48) into Eq. (10.9), we obtain

$$\begin{aligned}
e^{At}&=XY^{\mathrm{T}}+tX\Lambda Y^{\mathrm{T}}+\frac{t^2}{2!}X\Lambda Y^{\mathrm{T}}X\Lambda Y+\frac{t^3}{3!}X\Lambda Y^{\mathrm{T}}X\Lambda Y^{\mathrm{T}}X\Lambda Y^{\mathrm{T}}+\cdots\\
&=XY^{\mathrm{T}}+tX\Lambda Y^{\mathrm{T}}+\frac{t^2}{2!}X\Lambda^2Y^{\mathrm{T}}+\frac{t^3}{3!}X\Lambda^3Y^{\mathrm{T}}+\cdots\\
&=X\left(I+t\Lambda+\frac{t^2}{2!}\Lambda^2+\frac{t^3}{3!}\Lambda^3+\cdots\right)Y^{\mathrm{T}}=Xe^{\Lambda t}Y^{\mathrm{T}}
\end{aligned} \tag{10.49}$$

where use has been made of Eq. (10.45a). Moreover, the series in parentheses on the right side of Eq. (10.49) has been recognized as $e^{\Lambda t}$. Hence it follows from Eq. (10.24) that the transition matrix can be expressed in the form

$$\Phi(t,\tau)=Xe^{\Lambda(t-\tau)}Y^{\mathrm{T}} \tag{10.50}$$

The form (10.50) has important implications in the computation of the transition matrix when the order of the system is high. When the order of the system is low, the form (10.50) has no particular computational advantage. In fact, when the order of the system is low, it may be more advantageous to evaluate the transition matrix by means of Eq. (10.28).

The form (10.50) replaces the coefficient matrix A in the transition matrix by a triple matrix product involving the eigensolution of A. The object is to permit a more efficient computation of the series for $e^{A(t-\tau)}$. The form (10.50) provides an explanation for the statement made earlier in this section to the effect that the number of terms required for the convergence of the computation of the transition matrix depends on the matrix A. Indeed, from Eqs. (10.49) and (10.50), we conclude that the number of terms depends on the product of the eigenvalue of A of largest modulus and the time interval $(t,\tau)=t-\tau$. Of course, the time interval can be divided into smaller subintervals, as pointed out earlier.

The response can be expressed in terms of the transition matrix in the form (10.50). Indeed, inserting Eq. (10.49) into Eq. (10.23), we obtain

$$\mathbf{x}(t)=Xe^{\Lambda t}Y^{\mathrm{T}}\mathbf{x}(0)+\int_0^t Xe^{\Lambda(t-\tau)}Y^{\mathrm{T}}B\mathbf{u}(\tau)\,d\tau \tag{10.51}$$

The procedure consisting of solving the eigenvalue problem for A and A^{T} and using Eq. (10.51) to complete the system response is the equivalent of a modal analysis for general dynamic linear systems. Clearly, this approach requires expressing the equations of motion in state form.

10.4 RESPONSE OF GENERAL DAMPED SYSTEMS BY THE TRANSITION MATRIX

The approach based on the transition matrix is valid for any linear system reducible to the form (10.6). As such, it can be used to derive the response in the general case of damping, i.e., in the case in which the modal vectors associated with the undamped system cannot be used to decouple the system equations of motion. This necessitates the transformation of the equations to state form.

The equations of motion of an n-degree-of-freedom damped linear system was shown in Chapter 8 to have the form [see Eq. (8.21)]

$$M\ddot{\mathbf{q}}(t)+C\dot{\mathbf{q}}(t)+K\mathbf{q}(t)=\mathbf{Q}(t) \tag{10.52}$$

To derive the state equations, we introduce the notation

$$q_i(t)=x_i(t), \qquad \dot{q}_i(t)=x_{i+n}(t), \qquad i=1, 2, \ldots, n \tag{10.53}$$

and supplement the equation of motion, Eq. (10.52), by the auxiliary equations

$$\dot{x}_i(t)=x_{i+n}(t), \qquad i=1, 2, \ldots, n \tag{10.54}$$

Multiplying Eq. (10.52) through by M^{-1}, rearranging, and considering Eqs. (10.54), we can write the state equations in the form (10.6), in which

$$\mathbf{x}(t)=\left[\begin{array}{c}\mathbf{q}(t)\\ \hline \dot{\mathbf{q}}(t)\end{array}\right], \qquad \mathbf{u}(t)=\left[\begin{array}{c}\mathbf{0}\\ \hline \mathbf{Q}(t)\end{array}\right] \tag{10.55}$$

are the $2n$-dimensional state vector and associated excitation vector, respectively, and

$$A=\left[\begin{array}{c|c}0 & I\\ \hline -M^{-1}K & -M^{-1}C\end{array}\right], \qquad B=\left[\begin{array}{c|c}0 & 0\\ \hline 0 & M^{-1}\end{array}\right] \tag{10.56}$$

are $2n\times 2n$ coefficient matrices.

The response of the damped system for any initial excitation $\mathbf{x}(0)$ and/or external excitation $\mathbf{u}(t)$ can be obtained directly by using Eq. (10.23). It can also be obtained by solving first the eigenvalue problems associated with A and A^{T} for the matrix Λ of eigenvalues and the matrices X and Y of right and left eigenvectors, respectively. Then, the response can be calculated by means of Eq. (10.51).

Example 10.3

Derive the state equations for the two-degree-of-freedom damped system of Example 9.4 (linearized about the equilibrium point E_1) and obtain the response to the excitation

$$F(t)=\hat{F}\,\delta(t) \tag{a}$$

To avoid confusion with the matrix M, denote the masses by $M=m_1$, $m=m_2$. The system parameters are

$$\begin{aligned} &m_1=5 \quad \text{kg}, \qquad m_2=15 \quad \text{kg}, \qquad c=5 \quad \text{N}\cdot\text{s/m}, \\ &k=40 \quad \text{N/m}, \qquad a=2 \quad \text{m}, \qquad \hat{F}=0.2 \quad \text{N}\cdot\text{s} \end{aligned} \tag{b}$$

From Example 9.4, the linearized equations of motion are

$$\begin{aligned} (m_1+m_2)\ddot{x}+c\dot{x}+kx+m_2a\ddot{\theta}&=F \\ m_2a\ddot{x}+\tfrac{4}{3}m_2a^2\ddot{\theta}+m_2ga\theta&=2Fa \end{aligned} \tag{c}$$

which can be written in the matrix from (10.52), in which the configuration and force vectors are

$$\mathbf{q}=\begin{bmatrix} x \\ \theta \end{bmatrix}, \qquad \mathbf{Q}=\begin{bmatrix} F \\ 2Fa \end{bmatrix} \tag{d}$$

and the coefficient matrices are

$$M=\begin{bmatrix} m_1+m_2 & m_2a \\ m_2a & \tfrac{4}{3}m_2a^2 \end{bmatrix}, \qquad C=\begin{bmatrix} c & 0 \\ 0 & 0 \end{bmatrix}, \qquad K=\begin{bmatrix} k & 0 \\ 0 & m_2ga \end{bmatrix} \tag{e}$$

Hence, the state equations have the form (10.6), in which the state and excitation vectors have the form

$$\mathbf{x}=\begin{bmatrix} x \\ \theta \\ \dot{x} \\ \dot{\theta} \end{bmatrix}, \qquad \mathbf{u}=\begin{bmatrix} 0 \\ 0 \\ F \\ 2Fa \end{bmatrix} \tag{f}$$

Moreover, inserting Eqs. (b) and (e) into Eqs. (10.56), we obtain the coefficient matrices

$$\begin{aligned} A&=\begin{bmatrix} 0 & 0 & 1 & 0 \\ 0 & 0 & 0 & 1 \\ -\dfrac{4k}{4m_1+m_2} & \dfrac{3m_2g}{4m_1+m_2} & -\dfrac{4c}{4m_1+m_2} & 0 \\ \dfrac{3k}{(4m_1+m_2)a} & -\dfrac{3(m_1+m_2)g}{(4m_1+m_2)a} & \dfrac{3c}{(4m_1+m_2)a} & 0 \end{bmatrix} \\ &=\begin{bmatrix} 0 & 0 & 1 & 0 \\ 0 & 0 & 0 & 1 \\ -4.5714 & 12.6128 & -0.5714 & 0 \\ 1.7143 & -8.4085 & 0.2143 & 0 \end{bmatrix} \end{aligned} \tag{g}$$

and

$$B=\begin{bmatrix} 0 & 0 & 0 & 0 \\ 0 & 0 & 0 & 0 \\ 0 & 0 & \dfrac{4}{4m_1+m_2} & -\dfrac{3}{(4m_1+m_2)a} \\ 0 & 0 & -\dfrac{3}{(4m_1+m_2)a} & \dfrac{3(m_1+m_2)}{(4m_1+m_2)m_2a^2} \end{bmatrix}$$

$$=\begin{bmatrix} 0 & 0 & 0 & 0 \\ 0 & 0 & 0 & 0 \\ 0 & 0 & 0.1143 & -0.0429 \\ 0 & 0 & -0.0429 & 0.0286 \end{bmatrix} \tag{h}$$

The response was obtained by means of Eq. (10.51). To this end, the eigenvalue problems for A and A^{T} were solved. The eigenvalues are

$$\left.\begin{matrix}\lambda_1\\ \lambda_2\end{matrix}\right\}=-0.0632\pm i1.2100, \qquad \left.\begin{matrix}\lambda_3\\ \lambda_4\end{matrix}\right\}=-0.2225\pm i3.3773 \tag{i}$$

The matrices X and Y of right and left eigenvectors, respectively, were also computed, but are not listed here for brevity. The truncated series for the transition matrix, Eq. (10.50), was computed by using a recursive algorithm for $\exp \Lambda t$ similar to that described by Eqs. (10.31). Plots of $x(t)$ versus t and $\theta(t)$ versus t are shown in Fig. 10.1.

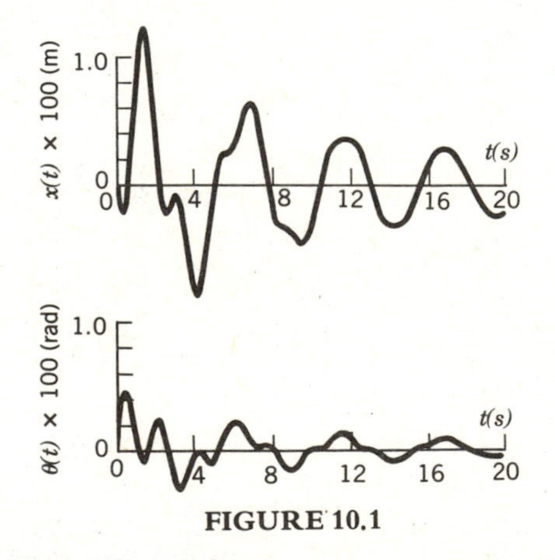

FIGURE 10.1

10.5 DISCRETE-TIME SYSTEMS. THE CONVOLUTION SUM

In Chapter 8, we discussed methods for deriving the response of vibrating linear systems. In the case of external excitation, the task reduces essentially to the evaluation of convolution integrals. Then, in Section 10.2 we discussed a procedure for deriving the response of general dynamic systems by means of the transition matrix. This latter procedure also required the evaluation of a convolution integral. Except for relatively simple excitation functions, the evaluation of convolution integrals cannot be carried out in closed form and must be carried out numerically. This implies evaluation on a computer, and in particular on a digital computer.

In carrying out numerical solutions on a digital computer, it is necessary to reduce the information to a form acceptable to the computer. To discuss this process, it is desirable to introduce a number of concepts. Functions of time such as the excitation and response, also known as input and output, respectively, are referred to in system theory as *signals*. In the signals discussed so far the time t was a continuous variable, and a digital computer cannot handle such signals. Indeed, a digital computer can work only with signals defined for discrete values of time. To distinguish between the two types of signals, the first are referred to as *continuous-time signals* and the second as *discrete-time signals*. Accordingly, a system involving continuous-time signals is called a *continuous-time system*, and one involving discrete-time signals is referred to as a *discrete-time system*.

Although discrete-time systems can arise naturally, our interest in discrete-time systems results from our desire to process information on a digital computer. Because the input and output are continuous-time signals, it is necessary to convert continuous-time signals into discrete-time signals and vice versa. The process consists of three cascaded operations, as shown in Fig. 10.2. The conversion from continuous time to discrete time is carried out by means of a *sampler*, which changes a continuous-time signal into a sequence of numbers corresponding to the values of the signal at the times $t_n = nT$ $(n = 1, 2, \ldots)$, where T is the *sampling period*. The continuous-time to discrete-time conversion is denoted symbolically by C/D. The sampler can be represented by a switch, as shown in Fig. 10.3, where the switch is open for all times except at the sampling instances t_n, when it closes instantaneously to let the signal pass through. The conversion from discrete time to continuous time is carried out by a *data hold circuit*. The simplest and the most frequently used hold is the *zero-order hold*, defined mathematically by

$$x(t) = x(t_n) = x(n), \qquad nT \leqslant t \leqslant nT + T \tag{10.57}$$

The zero-order hold generates a *staircase* signal of the type shown in Fig. 10.4.

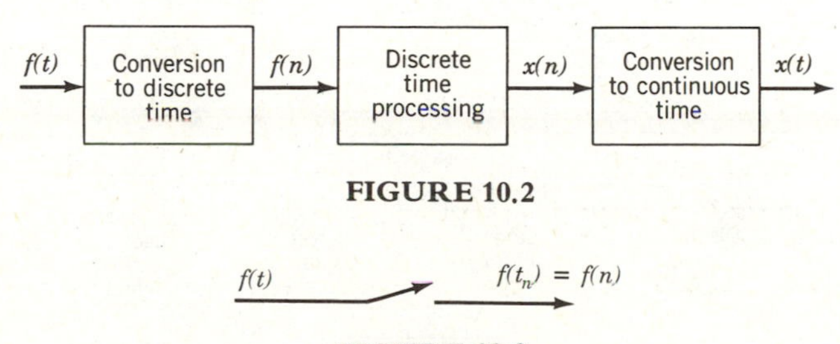

FIGURE 10.2

FIGURE 10.3

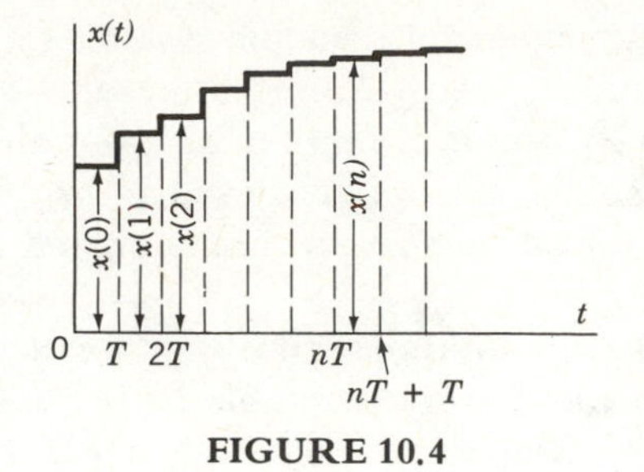

FIGURE 10.4

The conversion from discrete time to continuous time is denoted symbolically by D/C.

Strictly speaking, if the information processing is to be carried out by a digital computer, then some intermediate steps are necessary. Indeed, both the continuous-time and discrete-time signals are *analog signals*, i.e., signals whose amplitudes are not restricted to any particular values. By contrast, a *digital signal* is a signal whose amplitude is restricted to a given set of values. A digital computer can accept only digital signals, ordinarily encoded in a binary code. Hence, to use a digital computer, we must change the format of a signal from discrete analog to discrete digital. This task is carried out by means of an *analog-to-digital converter*, denoted symbolically by A/D. Of course, after the computations on a digital computer have been performed, the digital signal can be converted back to an analog signal by means of a *digital-to-analog converter*, denoted symbolically by D/A. The complete process is shown schematically in Fig. 10.5.

The conversion from discrete analog signals to discrete digital signals involves certain *quantization*, which implies that the analog signal is rounded so as to coincide with the closest value from the restricted set. Hence, this quantization introduces some error, where the error depends on the number of quantization levels. Indeed, the error decreases as the number of quantization levels increases, and vice versa. The number of quantization levels depends on the number of bits in the binary word used by the digital computer, and it increases with the binary word length. For computers using binary words containing 12 bits or more, the quantization error tends to be mathematically insignificant. In such cases, the A/D and D/A operations can be omitted from Fig. 10.5. Assuming that the A/D and D/A conversions are mathematically insignificant, we shall make no particular distinction between discrete analog and discrete digital signals and refer to them simply as discrete-time signals.

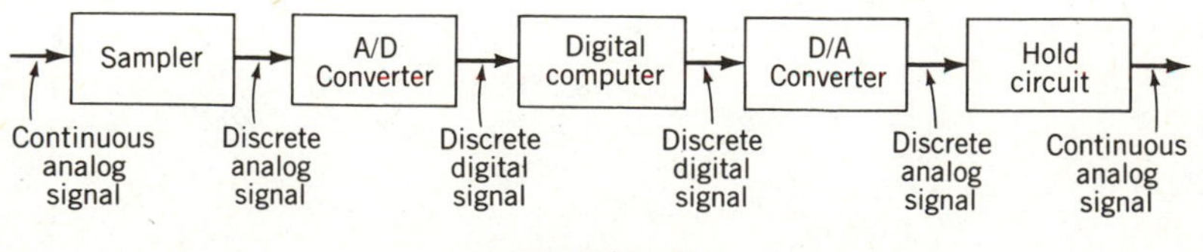

FIGURE 10.5

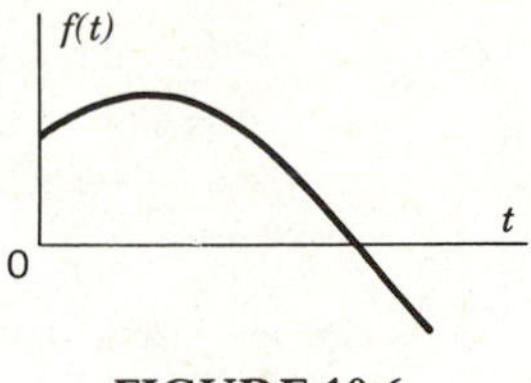

FIGURE 10.6

Now, let us turn our attention to the mathematical description of discrete-time systems. All signals involved in discrete-time systems are regarded as sequences of sample values resulting from sampling continuous-time signals. Hence, assuming that a continuous-time signal $f(t)$ (Fig. 10.6) is sampled every T seconds beginning at $t=0$, the discrete-time signal $f(nT)=f(n)$ consists of the sequence

$$\{f(n)\}=f(0),\ f(1),\ f(2),\ldots \tag{10.58}$$

To describe the sequence mathematically, it is convenient to introduce the *discrete-time unit impulse*, or *unit sample*, as the Kronecker delta

$$\delta(n-k)=\begin{cases}1, & n=k\\ 0, & n\neq k\end{cases} \tag{10.59}$$

The unit impulse is shown in Fig. 10.7. Then, the discrete-time signal $f(n)$ can be represented mathematically in the form

$$f(n)=\sum_{k=0}^{\infty} f(k)\,\delta(n-k) \tag{10.60}$$

The discrete-time signal $f(n)$ is displayed in Fig. 10.8.

Next, we propose to derive the response of discrete-time systems to an excitation $f(n)$. By analogy with continuous-time systems, we can define the *discrete-time impulse response $g(n)$ as the response of a linear discrete-time system to a discrete-time impulse $\delta(n)$ applied at $k=0$, with all initial conditions equal to zero.* The excita-

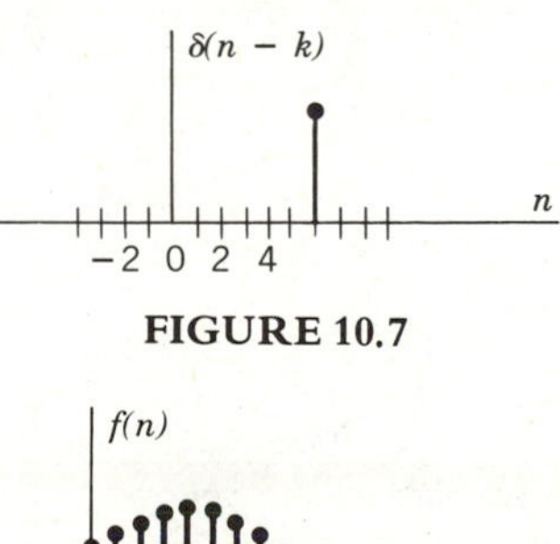

FIGURE 10.7

FIGURE 10.8

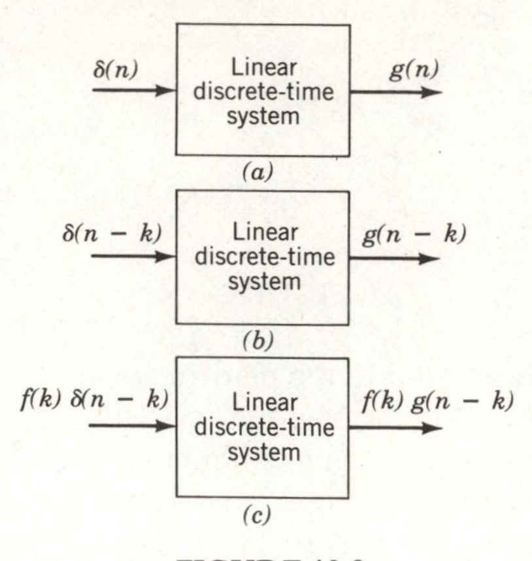

FIGURE 10.9

tion–response relation is shown schematically in the block diagram of Fig. 10.9*a*. We observe in passing that $g(n)=0$ for $n<0$, because there cannot be any response before the system has been excited. If the unit impulse is delayed by k sampling periods, so that the excitation is now $\delta(n-k)$, the response must also be delayed by k periods. Hence, the response to $\delta(n-k)$ is $g(n-k)$, as shown in Fig. 10.9*b*, and we note that $g(n-k)=0$ for $n<k$. Moreover, because the system is linear, if the excitation has the form of an impulse of magnitude $f(k)$ at $n=k$, then the response is simply $f(k)g(n-k)$, as shown in Fig. 10.9*c*. But, the excitation $f(n)$ can be regarded as an infinite sum of impulses $f(k)\,\delta(n-k)$, as indicated by Eq. (10.60). Hence, denoting the response to $f(n)$ by $x(n)$ and recalling that $g(n-k)=0$ for $k>n$, we can write simply

$$x(n)=\sum_{k=0}^{\infty} f(k)g(n-k)=\sum_{k=0}^{n} f(k)g(n-k) \tag{10.61}$$

Equation (10.61) expresses the response of a linear discrete-time system in the form of a *convolution sum*, and it represents the discrete-time counterpart of the convolution integral, Eq. (4.170).

Example 10.4

Obtain the response of a mass–spring system to the excitation

$$f(t)=f_0 \mathscr{r}(t) \tag{a}$$

where $\mathscr{r}(t)$ is the unit ramp function. Note that f_0 has units N/s. Treat the system as a discrete-time system, and obtain the response by means of the convolution sum, Eq. (10.61).

Letting $t=nT$, we can write the excitation in the form of Eq. (10.60), in which

$$f(k)=f_0 kT, \qquad k=0, 1, 2, \ldots \tag{b}$$

Moreover, the impulse response of the equivalent discrete-time system is (see

Example 10.5)

$$g(k) = g(kT) = \frac{T}{m\omega_n} \sin \omega_n kT, \qquad k = 0, 1, 2, \ldots \tag{c}$$

where m is the mass and ω_n the natural frequency. Inserting Eqs. (b) and (c) into Eq. (10.61), we obtain the response sequence

$$\begin{aligned}
x(1) &= f(0)g(1) + f(1)g(0) = 0 \\
x(2) &= f(0)g(2) + f(1)g(1) + f(2)g(0) = \frac{f_0 T^2}{m\omega_n} \sin \omega_n T \\
x(3) &= f(0)g(3) + f(1)g(2) + f(2)g(1) + f(3)g(0) \\
&= \frac{f_0 T^2}{m\omega_n} (\sin 2\omega_n T + 2 \sin \omega_n T)
\end{aligned} \tag{d}$$

10.6 DISCRETE-TIME SOLUTION OF THE STATE EQUATIONS

In Section 10.2, we derived the solution of the state equations by means of the transition matrix, as indicated by Eq. (10.23). The solution is in continuous time, and, for evaluation on a digital computer, it is desirable to derive the discrete-time counterpart. This can be done through a discretization in time of the continuous-time solution.

Let us assume that we are interested in the solution only at the sampling times $t_k = kT$ $(k = 1, 2, \ldots)$. Letting $t = kT$ in Eq. (10.23), we obtain

$$\mathbf{x}(kT) = e^{kTA}\mathbf{x}(0) + \int_0^{kT} e^{(kT-\tau)A} B\mathbf{u}(\tau)\, d\tau \tag{10.62}$$

at the next sampling time, the solution is

$$\mathbf{x}(kT+T) = e^{(kT+T)A}\mathbf{x}(0) + \int_0^{kT+T} e^{(kT+T-\tau)A} B\mathbf{u}(\tau)\, d\tau \tag{10.63}$$

which can be rewritten in the form

$$\begin{aligned}
\mathbf{x}(kT+T) = e^{TA}&\left[e^{kTA}\mathbf{x}(0) + \int_0^{kT} e^{(kT-\tau)A} B\mathbf{u}(\tau)\, d\tau \right] \\
&+ \int_{kT}^{kT+T} e^{(kT+T-\tau)A} B\mathbf{u}(\tau)\, d\tau
\end{aligned} \tag{10.64}$$

If the sampling period T is small, it is reasonable to assume that $\mathbf{u}(\tau)$ does not change very much during the interval $kT \leqslant \tau \leqslant kT+T$, so that we can make the approximation

$$\mathbf{u}(\tau) \cong \mathbf{u}(kT), \qquad kT \leqslant \tau \leqslant kT+T \tag{10.65}$$

Hence, we can write

$$\int_{kT}^{kT+T} e^{(kT+T-\tau)A} B\mathbf{u}(\tau)\,d\tau \cong \left[\int_{kT}^{kT+T} e^{(kT+T-\tau)A}\,d\tau\right] B\mathbf{u}(kT)$$
$$= \left[\int_{T}^{0} e^{tA}(-dt)\right] B\mathbf{u}(kT)$$
$$= \left[\int_{0}^{T} e^{tA}\,dt\right] B\mathbf{u}(kT) \tag{10.66}$$

Then, recognizing that the expression in brackets in Eq. (10.64) is $\mathbf{x}(kT)$ and omitting T from the argument, we can rewrite Eq. (10.64) as follows:

$$\mathbf{x}(k+1) = \Phi\mathbf{x}(k) + \Gamma\mathbf{u}(k), \qquad k = 0, 1, 2, \ldots \tag{10.67}$$

where

$$\Phi = e^{TA}, \qquad \Gamma = \left(\int_0^T e^{tA}\,dt\right) B \tag{10.68}$$

The first of Eqs. (10.68) represents the transition matrix for the discrete-time model. It can be computed by the algorithm given by Eqs. (10.31) by simply replacing t by the sampling period T. Equations (10.67) are a set of *difference equations*, and they represent the counterpart of the differential equation (10.6) for continuous-time systems.

Equations (10.67) can be solved recursively. Indeed, if the initial vector $\mathbf{x}(0)$ and the excitation vector $\mathbf{u}(k)$ at the sampling times $t = kT$ ($k = 0, 1, 2, \ldots$) are known, then the response can be obtained in sequence by writing Eqs. (10.67) in the form

$$\begin{aligned}
\mathbf{x}(1) &= \Phi\mathbf{x}(0) + \Gamma\mathbf{u}(0) \\
\mathbf{x}(2) &= \Phi\mathbf{x}(1) + \Gamma\mathbf{u}(1) = \Phi^2\mathbf{x}(0) + \Phi\Gamma\mathbf{u}(0) + \Gamma\mathbf{u}(1) \\
&\vdots \\
\mathbf{x}(n+1) &= \Phi\mathbf{x}(n) + \Gamma\mathbf{u}(n) = \Phi^{n+1}\mathbf{x}(0) + \sum_{k=0}^{n} \Phi^{n-k}\Gamma\mathbf{u}(k)
\end{aligned} \tag{10.69}$$

The last of Eqs. (10.69) represents the general solution $\mathbf{x}(n+1)$ for the discrete-time model, and we observe that the sum in this last equation is a convolution sum. In computing the response on a digital computer, however, it is more convenient to carry out the computations recursively by updating old values of the response vector as soon as new values have been computed, which amounts to using the first expressions on the right side of Eqs. (10.69) rather than the second ones.

In the case of high-order systems, the computation of the transition matrix by means of the first of Eqs. (10.68) can cause difficulties. In this case, it may prove advantageous to solve first the eigenvalue problems associated with A and A^{T} for the matrix Λ of the eigenvalues and the matrices X and Y of right and left eigenvectors, respectively. Then, the matrices Φ and Γ given by Eqs. (10.68) can be expressed in the form

$$\Phi = e^{TA} = X\Theta Y^{\mathrm{T}}, \qquad \Gamma = \left(\int_0^T e^{tA}\, dt \right) B = X\,\Delta Y^{\mathrm{T}} B \tag{10.70}$$

where

$$\Theta = e^{T\Lambda}, \qquad \Delta = \int_0^T e^{t\Lambda}\, dt \tag{10.71}$$

Note that, if the eigenvalues are distinct, then Θ and Δ are diagonal matrices. In view of Eqs. (10.70), Eqs. (10.67) can be replaced by

$$\mathbf{y}(k+1) = \Theta \mathbf{y}(k) + \Delta \mathbf{z}(k), \qquad k = 0, 1, 2, \ldots \tag{10.72}$$

where

$$\mathbf{y}(k) = Y^{\mathrm{T}} \mathbf{x}(k), \qquad \mathbf{z}(k) = Y^{\mathrm{T}} B \mathbf{u}(k), \qquad k = 0, 1, 2, \ldots \tag{10.73}$$

Of course, to recover the actual response, we must write

$$\mathbf{x}(k) = X \mathbf{y}(k), \qquad k = 0, 1, 2, \ldots \tag{10.74}$$

Note that to compute $\Theta = e^{T\Lambda}$ one can use the algorithm described by Eqs. (10.31), which simplifies considerably for diagonal matrices.

Example 10.5

Derive the discrete-time impulse response for the mass–spring system of Example 10.4. Compare the result with the impulse response used in Example 10.4 and provide an explanation for the apparent discrepancy.

The response is obtained by means of Eqs. (10.67), rewritten in the form

$$\mathbf{x}_{k+1} = \Phi \mathbf{x}_k + \Gamma \mathbf{u}_k, \qquad k = 0, 1, 2, \ldots \tag{a}$$

In the case at hand,

$$\begin{aligned} &\mathbf{x}_0 = \mathbf{0} \\ &\mathbf{u}_0 = \delta(n) = 1, \qquad \mathbf{u}_1 = \mathbf{u}_2 = \cdots = 0 \end{aligned} \tag{b}$$

and from Example 10.1 and the first of Eqs. (10.68),

$$B = \begin{bmatrix} 0 \\ 1/m \end{bmatrix}, \quad \Phi = e^{TA} = \begin{bmatrix} \cos \omega_{\mathrm{n}} T & \omega_{\mathrm{n}}^{-1} \sin \omega_{\mathrm{n}} T \\ -\omega_{\mathrm{n}} \sin \omega_{\mathrm{n}} T & \cos \omega_{\mathrm{n}} T \end{bmatrix} \tag{c}$$

Moreover, from the second of Eqs. (10.68),

$$\begin{aligned} \Gamma &= \left(\int_0^T e^{tA}\, dt \right) B = \int_0^T \begin{bmatrix} \cos \omega_{\mathrm{n}} t & \omega_{\mathrm{n}}^{-1} \sin \omega_{\mathrm{n}} t \\ -\omega_{\mathrm{n}} \sin \omega_{\mathrm{n}} t & \cos \omega_{\mathrm{n}} t \end{bmatrix} dt \begin{bmatrix} 0 \\ 1/m \end{bmatrix} \\ &= \frac{1}{m\omega_{\mathrm{n}}} \begin{bmatrix} -\omega_{\mathrm{n}}^{-1} \cos \omega_{\mathrm{n}} t \\ \sin \omega_{\mathrm{n}} t \end{bmatrix} \Bigg|_0^T \\ &= \frac{1}{m\omega_{\mathrm{n}}^2} \begin{bmatrix} 1 - \cos \omega_{\mathrm{n}} T \\ \omega_{\mathrm{n}} \sin \omega_{\mathrm{n}} T \end{bmatrix} \end{aligned} \tag{d}$$

Letting $k=0$ in Eqs. (a) and using Eqs. (b) and (d), we obtain the sequence

$$\mathbf{x}_1=\frac{1}{m\omega_n^2}\begin{bmatrix}1-\cos\omega_n T\\ \omega_n\sin\omega_n T\end{bmatrix}$$

$$\mathbf{x}_2=\Phi\mathbf{x}_1=\frac{1}{m\omega_n^2}\begin{bmatrix}\cos\omega_n T & \omega_n^{-1}\sin\omega_n T\\ -\omega_n\sin\omega_n T & \cos\omega_n T\end{bmatrix}\begin{bmatrix}1-\cos\omega_n T\\ \omega_n\sin\omega_n T\end{bmatrix}$$

$$=\frac{1}{m\omega_n^2}\begin{bmatrix}\cos\omega_n T-\cos 2\omega_n T\\ -\omega_n(\sin\omega_n T-\sin 2\omega_n T)\end{bmatrix} \tag{e}$$

$$\mathbf{x}_3=\Phi\mathbf{x}_2=\frac{1}{m\omega_n^2}\begin{bmatrix}\cos\omega_n T & \omega_n^{-1}\sin\omega_n T\\ -\omega_n\sin\omega_n T & \cos\omega_n T\end{bmatrix}\begin{bmatrix}\cos\omega_n T-\cos 2\omega_n T\\ -\omega_n(\sin\omega_n T-\sin 2\omega_n T)\end{bmatrix}$$

$$=\frac{1}{m\omega_n^2}\begin{bmatrix}\cos 2\omega_n T-\cos 3\omega_n T\\ -\omega_n(\sin 2\omega_n T-\sin 3\omega_n T)\end{bmatrix}$$

$$\vdots$$

which can be expressed in the compact form

$$\mathbf{x}_k=\frac{1}{m\omega_n^2}\begin{bmatrix}\cos(k-1)\omega_n T-\cos k\omega_n T\\ -\omega_n[\sin(k-1)\omega_n T-\sin k\omega_n T]\end{bmatrix},\qquad k=1,2,\ldots \tag{f}$$

The discrete-time impulse response is the upper component of the vector $\mathbf{x}_k$, or

$$g(k)=\frac{1}{m\omega_n^2}[\cos(k-1)\omega_n T-\cos k\omega_n T],\qquad k=1,2,\ldots \tag{g}$$

Comparing Eqs. (g) with Eqs. (c) of Example 10.4, we are tempted to conclude that the two sequences are different, and if the sampling period T is relatively large, then they are indeed. However, for the discrete-time model to yield reasonably accurate results, the sampling period must be small, i.e., T must be such that $\omega_n T$ is a small number. In particular, $\omega_n T$ must be sufficiently small that

$$\begin{aligned}\cos(k-1)\omega_n T&=\cos k\omega_n T\cos\omega_n T+\sin k\omega_n T\sin\omega_n T\\ &\cong\cos k\omega_n T+\omega_n T\sin k\omega_n T\end{aligned} \tag{h}$$

Introducing Eq. (h) into Eqs. (g), we obtain the discrete-time impulse response

$$g(k)=\frac{T}{m\omega_n}\sin k\omega_n T,\qquad k=0,1,2,\ldots \tag{i}$$

where, from the first of Eqs. (b), we included $g(0)=0$ in the sequence. Now, if we compare Eqs. (i) with Eqs. (c) of Example 10.4, we conclude that the two sequences are identical.

This simple example permits us to reach an important conclusion concerning the accuracy of discrete-time models. Indeed, if the system exhibits periodic behavior, then the sampling period T must be a small fraction of the period of the response for the discrete-time model to simulate the continuous-time system with acceptable accuracy.

10.7 STABILITY OF DISCRETE-TIME SYSTEMS

In the absence of external excitation, the equations governing the behavior of a discrete-time system are

$$\mathbf{x}(k+1)=\Phi\mathbf{x}(k), \qquad k=0, 1, 2, \ldots \tag{10.75}$$

where Φ is the transition matrix, as given by the first of Eqs. (10.68). To examine the stability of the system, we use Eqs. (10.72) instead of Eqs. (10.75), or

$$\mathbf{y}(k+1)=\Theta\mathbf{y}(k), \qquad k=0, 1, 2, \ldots \tag{10.76}$$

in which Θ and $\mathbf{y}(k)$ are given by the first of Eqs. (10.71) and (10.73). For distinct eigenvalues of A, the matrix Θ is diagonal,* so that Eqs. (10.76) can be written in the scalar form

$$y_j(k+1)=\Theta_j y_j(k), \qquad j=1, 2, \ldots, m, \qquad k=0, 1, 2, \ldots \tag{10.77}$$

where we note from the first of Eqs. (10.70) that Θ_j are the eigenvalues of the transition matrix Φ.

In general the scalars Θ_j $(j=1, 2, \ldots, m)$ are complex quantities. They can be expressed in the form

$$\Theta_j=\gamma_j+i\delta_j=|\Theta_j|e^{i\psi_j}, \qquad j=1, 2, \ldots, m \tag{10.78}$$

where $|\Theta_j|$ is the magnitude of Θ_j and ψ_j is a phase angle given by

$$\psi_j=\tan^{-1} \delta_j/\gamma_j, \qquad j=1, 2, \ldots, m \tag{10.79}$$

Hence, Eqs. (10.77) can be rewritten as follows:

$$y_j(k+1)=|\Theta_j|e^{i\psi_j}y_j(k), \qquad j=1, 2, \ldots, m, \qquad k=0, 1, 2, \ldots \tag{10.80}$$

But, $e^{i\psi_j}$ is a complex vector of unit magnitude, so that it has no effect on the magnitude of $y_j(k+1)$. On the other hand, $|\Theta_j|$ is a real scalar whose magnitude affects directly the magnitude of $y_j(k+1)$. In particular, we distinguish the cases: (i) if $|\Theta_j|<1$, then $y_j(k+1)$ is smaller in magnitude relative to $y_j(k)$, (ii) if $|\Theta_j|=1$, then $y_j(k+1)$ has the same magnitude as $y_j(k)$, and (iii) if $|\Theta_j|>1$, then $y_j(k+1)$ is larger in magnitude relative to $y_j(k)$. These observations permit us to make the following stability statements:

i. The system is *asymptotically stable if all the eigenvalues of* Φ *are such that* $|\Theta_j|<1$ $(j=1, 2, \ldots, m)$.

ii. The system is merely *stable if some eigenvalues of* Φ *are such that* $|\Theta_j|=1$ *and the remaining ones are such that* $|\Theta_l|<1$, $l\neq j$.

iii. The system is *unstable if at least one eigenvalue of* Φ *is such that* $|\Theta_j|>1$.

The stability characteristics can be best visualized by considering the Θ-plane, as shown in Fig. 10.10. For asymptotic stability, all the eigenvalues must lie inside

---

*See L. Meirovitch, *Computational Methods in Structural Dynamics*, Sijthoff-Noordhoff International Publishers. The Netherlands, 1980, p. 71.

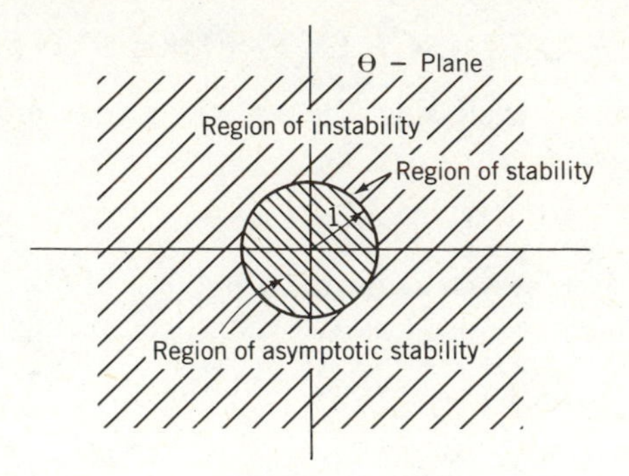

FIGURE 10.10

the unit circle with center at the origin. For mere stability, some of the eigenvalues must lie on the circle, and the balance of the eigenvalues must lie inside the circle. For instability, at least one eigenvalue must lie outside the unit circle.

Returning to the first of Eqs. (10.71), we observe that for distinct eigenvalues $\lambda_j (j=1, 2, \ldots, m)$ of A, we can write

$$\Theta_j = e_j^{T\lambda_j}, \qquad j=1, 2, \ldots, m \tag{10.81}$$

Letting

$$\lambda_j = \alpha_j + i\beta_j, \qquad j=1, 2, \ldots, m \tag{10.82}$$

we obtain

$$\Theta_j = e^{T\lambda_j} = e^{T\alpha_j} e^{iT\beta_j}, \qquad j=1, 2, \ldots, m \tag{10.83}$$

But, $e^{iT\beta_j}$ is a complex vector of unit magnitude, so that

$$|\Theta_j| = e^{T\alpha_j}, \qquad j=1, 2, \ldots, m \tag{10.84}$$

For stability of the continuous-time system, all α_j must be nonpositive,

$$\alpha_j \leqslant 0, \qquad j=1, 2, \ldots, m \tag{10.85}$$

which implies that all λ_j must lie in the left half of the complex λ-plane (see Fig. 9.5). From Eqs. (10.84), we conclude that if all α_j are nonpositive, then all $|\Theta_j|$ are smaller than or equal to unity,

$$|\Theta_j| \leqslant 1, \qquad j=1, 2, \ldots, m \tag{10.86}$$

so that the left half of the λ-plane, including the imaginary axis, maps into the circular region $|\Theta| \leqslant 1$ in the Θ-plane. Hence, the equivalent discrete-time system is stable regardless of the value of T.

The fact that the sampling period T does not change the stability characteristics of the system does not imply that the response of the discrete-time system converges to the response of the continuous-time system for all T. Indeed, both responses can be stable, but if T is too large, then the simulated response using discrete-time equations can drift away from the response of the actual continuous-time system.

10.8 EULER'S METHOD

In Section 10.2 we have shown how the response can be obtained by means of a procedure based on the transition matrix. Then, in Section 10.6 we have shown how the procedure can be carried out in discrete time. Unfortunately, the approach of Sections 10.2 and 10.6 is restricted to linear systems, so that an approach applicable to nonlinear systems appears desirable. Because the solution of nonlinear equations implies numerical integration, we consider a discrete-time solution directly. Numerical integration is carried out most conveniently in terms of first-order equations, which implies a state form. To introduce the ideas, we first consider a first-order system and then extend the results to a system of order m.

Let us consider a nonlinear first-order system described by the differential equation

$$\dot{x}(t) = f[x(t), t] \tag{10.87}$$

where f is a nonlinear function of $x(t)$ and t. Expanding the solution of Eq. (10.87) in a Taylor series in the neighborhood of t, we can write

$$x(t+T) = x(t) + T\dot{x}(t) + \frac{T^2}{2!}\ddot{x}(t) + \cdots \tag{10.88}$$

where T is a small time increment. But $\dot{x} = f$ according to Eq. (10.87). Moreover,

$$\ddot{x} = \frac{\partial f}{\partial x}\dot{x} + \frac{\partial f}{\partial t} = f\frac{\partial f}{\partial x} + \frac{\partial f}{\partial t} \tag{10.89}$$

so that Eq. (10.88) becomes

$$x(t+T) = x(t) + Tf + \frac{T^2}{2!}\left(f\frac{\partial f}{\partial x} + \frac{\partial f}{\partial t}\right) + \cdots \tag{10.90}$$

Equation (10.90) forms the basis for our numerical algorithm.

To derive the solution of Eq. (10.87) in discrete time, we introduce the times $t = t_k$, $t + T_k = t_{k+1}$, where T_k is a small time increment $(k = 0, 1, 2, \ldots)$. Hence, letting

$$x(t) = x(t_k) = x(k), \qquad x(t+T) = x(t_{k+1}) = x(k+1) \tag{10.91a}$$

$$f[x(t), t] = f[x(t_k), t_k] = f(k) \tag{10.91b}$$

we can rewrite Eq. (10.90) in the form

$$x(k+1) = x(k) + T_k f(k) + \frac{T_k^2}{2!}\left[f(k)\frac{\partial f(k)}{\partial x} + \frac{\partial f(k)}{\partial t}\right] + \cdots, \qquad k = 0, 1, 2, \ldots \tag{10.92}$$

Equations (10.92) can be used to derive solutions to any desired order.

The simplest approximation to Eq. (10.87) is obtained by retaining the first-order term in the Taylor series (10.90). The equivalent approximation for discrete-time systems is obtained from Eqs. (10.92) in the form

$$x(k+1) = x(k) + T_k f(k), \qquad k = 0, 1, 2, \ldots \tag{10.93}$$

The method of computing the first-order approximation by means of Eqs. (10.93) is known as *Euler's method.*

Euler's method can be extended to higher-order systems. Indeed, for a system of order m, the state equations can be written in the vector form

$$\dot{\mathbf{x}}(t)=\mathbf{f}[\mathbf{x}(t), t] \tag{10.94}$$

where $\mathbf{x}(t)$ and $\mathbf{f}[\mathbf{x}(t), t]$ are m-dimensional vectors. Then, the first-order approximation can be written for discrete-time systems as follows:

$$\mathbf{x}(k+1)=\mathbf{x}(k)+T_k\mathbf{f}(k), \qquad k=0, 1, 2, \ldots \tag{10.95}$$

where

$$\mathbf{f}(k)=\mathbf{f}[\mathbf{x}(t_k), t_k], \qquad k=0, 1, 2, \ldots \tag{10.96}$$

In most cases, the time increments are taken equal to one another, so that $T_k=T$ and $t_k=kT$ $(k=0, 1, 2, \ldots)$, where T can be identified as the sampling period.

The algorithm based on Euler's method works for both linear and nonlinear systems, and it is easy to program on a digital computer. Because it is only a first-order approximation, the question arises as to the error produced by the discretization process. The error induced by the discretization process in integrating a differential equation across one time increment T is referred to as *local truncation error*. On the other hand, the difference between the computed value $y(k)$ and the true value $y(t_k)$ is called *total truncation error* by some and *global truncation error* by others. In the case of Euler's method the local truncation error is of order T^2 and the total truncation error is of order T. This is relatively low accuracy, for which reason Euler's method is not used widely.

Example 10.6

The oscillation of a simple pendulum is governed by the nonlinear differential equation

$$\ddot{\theta}(t)+\omega_{\mathrm{n}}^2 \sin \theta(t)=0, \qquad \omega_{\mathrm{n}}=\sqrt{g/L} \tag{a}$$

where $\theta(t)$ is the angular displacement, ω_{n} is a constant having units of frequency, g is the gravitational constant, and L is the length of the pendulum. Use Euler's method to obtain the response to the initial conditions $\theta(0)=\pi/4$, $\dot{\theta}(0)=0$. Let $\omega_{\mathrm{n}}=2$ rad/s.

Introducing the notation

$$\theta(t)=x_1(t), \qquad \dot{\theta}(t)=x_2(t) \tag{b}$$

Eq. (a) can be replaced by the state equations

$$\dot{x}_1=x_2, \qquad \dot{x}_2=-\omega_{\mathrm{n}}^2 \sin x_1 \tag{c}$$

so that the components of the vector $\mathbf{f}$ in Eq. (10.96) are

$$f_1=x_2, \qquad f_2=-\omega_{\mathrm{n}}^2 \sin x_1 \tag{d}$$

Inserting Eqs. (d) into Eqs. (10.95), we obtain the equations defining the Euler method in the form

$$\begin{aligned} x_1(k+1) &= x_1(k) + Tf_1(k) = x_1(k) + Tx_2(k) \\ x_2(k+1) &= x_2(k) + Tf_2(k) = x_2(k) - T\omega_n^2 \sin x_1(k) \end{aligned} \qquad k = 0, 1, 2, \ldots \tag{e}$$

where T is the increment of time used in the approximation, assumed to be the same for every step. Equations (e) are subject to

$$x_1(0) = \pi/4, \qquad x_2(0) = 0 \tag{f}$$

The response corresponding to $\omega_n = 2$ rad/s was obtained using the sampling period $T = 0.01$ s. The angular displacement $\theta(k) = x_1(k)$ is displayed in Fig. 10.11. For comparison, the true solution is plotted in dashed line. It is clear that the solution obtained by Euler's method deviates from the true solution, and to reduce the error it is necessary to reduce the sampling period T. We shall not pursue this subject, as the use of Euler's method is not really advocated.

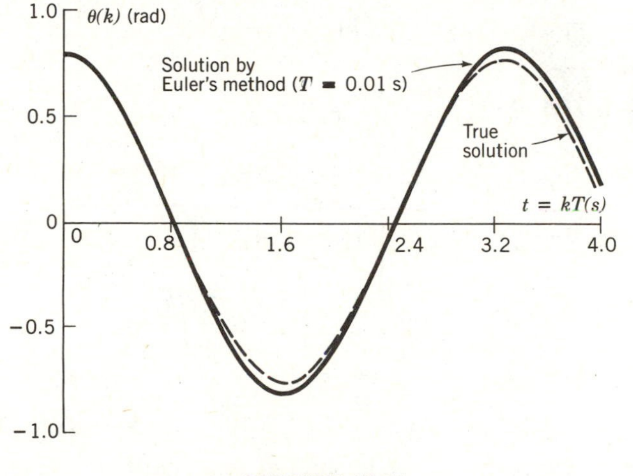

FIGURE 10.11

10.9 THE RUNGE-KUTTA METHODS

The Euler method was based on the first two terms on the right side of the Taylor series (10.88), so that the Euler method represents a first-order approximation. This implies that the series was *truncated* after the first derivative term. In fact, the poor results produced by Euler's method can be attributed to a large extent to the truncation error. To reduce this error, it is necessary to retain more terms in series (10.88). This is basically the idea behind the *Runge–Kutta methods.* As the name indicates, there is more than one method, the difference between them lying in the order of approximation. The most popular of the Runge–Kutta methods is the fourth-order method. To introduce the ideas, we propose to derive the second-order method and only extend the results to the fourth-order method.

As with Euler's method, we begin by considering the case of a single first-order differential equation, namely, Eq. (10.87). The discrete-time counterpart of Eq. (10.87) is Eq. (10.92). To develop the second-order Runge–Kutta method, we assume an approximation of the form

$$x(k+1)=x(k)+c_1 g_1+c_2 g_2, \qquad k=0, 1, 2, \ldots \tag{10.97}$$

where c_1 and c_2 are constants and

$$g_1=T_k f[x(t_k), t_k]=T_k f(k) \tag{10.98a}$$

$$g_2=T_k f[x(t_k)+\alpha_2 g_1, t_k+\beta_2 T_k] \tag{10.98b}$$

in which α_2 and β_2 are constants. The constants c_1, c_2, α_2, and β_2 are determined so that Eqs. (10.92) and (10.97) agree through terms of second order in T_k. From Eqs. (10.98), we can write the expansion

$$\begin{aligned} g_2/T_k &= f[x(t_k)+\alpha_2 g_1, t_k+\beta_2 T_k]=f[x(t_k)+\alpha_2 T_k f(k), t_k+\beta_2 T_k] \\ &= f[x(t_k), t_k]+\alpha_2 T_k f(k)\frac{\partial f[x(t_k), t_k]}{\partial x}+\beta_2 T_k \frac{\partial f[x(t_k), t_k]}{\partial t}+\cdots \\ &= f(k)+\alpha_2 T_k f(k)\frac{\partial f(k)}{\partial x}+\beta_2 T_k\frac{\partial f(k)}{\partial t}+\cdots \end{aligned} \tag{10.99}$$

so that, inserting Eqs. (10.98) and (10.99) into Eq. (10.97), we obtain

$$\begin{aligned} x(k+1) &= x(k)+c_1 T_k f(k)+c_2 T_k\left[f(k)+\alpha_2 T f(k)\frac{\partial f(k)}{\partial x}+\beta_2 T\frac{\partial f(k)}{\partial t}+\cdots\right] \\ &= x(k)+(c_1+c_2)T_k f(k)+c_2 T_k^2\left[\alpha_2 f(k)\frac{\partial f(k)}{\partial x}+\beta_2\frac{\partial f(k)}{\partial t}\right]+\cdots \\ &\qquad k=0, 1, 2, \ldots \end{aligned} \tag{10.100}$$

Comparing terms through second order in T_k in Eqs. (10.92) and (10.100), we conclude that the constants, c_1, c_2, α_2, and β_2 must satisfy the equations

$$c_1+c_2=1, \qquad c_2\alpha_2=\tfrac{1}{2}, \qquad c_2\beta_2=\tfrac{1}{2} \tag{10.101}$$

Equations (10.101) constitute three equations in four unknowns, so that there is no unique solution. Indeed, the equations can be satisfied in an infinite number of ways by taking

$$c_2=1-c_1, \qquad \alpha_2=\beta_2=\frac{1}{2(1-c_1)} \tag{10.102}$$

where $c_1 \neq 1$, but otherwise can be chosen arbitrarily. One satisfactory choice is $c_1=1/2$, which yields

$$c_1=c_2=\tfrac{1}{2}, \qquad \alpha_2=\beta_2=1 \tag{10.103}$$

Inserting Eqs. (10.103) into Eqs. (10.97) and (10.98), we obtain the algorithm defining the *second-order Runge–Kutta method* in the form

$$x(k+1)=x(k)+\tfrac{1}{2}(g_1+g_2), \qquad k=0,1,2,\ldots \tag{10.104}$$

where

$$g_1=T_k f(k) \tag{10.105a}$$

$$g_2=T_k f[x(k)+g_1,\ t_k+T_k] \tag{10.105b}$$

and we note that the choice given by Eqs. (10.103) yields a symmetric form for the algorithm.

The same procedure can be used to derive the higher-order Runge–Kutta methods. In general, the method of order p requires terms through order p in T_k in Eq. (10.92) and p functions, $g_1, g_2, \ldots, g_p$, in Eq. (10.97). In the process, the derivation becomes progressively tedious, so that presentation of the details serves no useful purpose. Perhaps the most widely used of the Runge–Kutta methods is the *fourth-order method* defined by the algorithm

$$x(k+1)=x(k)+\tfrac{1}{6}(g_1+2g_2+2g_3+g_4), \qquad k=0,1,2,\ldots \tag{10.106}$$

where

$$\begin{aligned} g_1&=T_k f(k)\\ g_2&=T_k f[x(k)+g_1/2,\ t_k+T_k/2]\\ g_3&=T_k f[x(k)+g_2/2,\ t_k+T_k/2]\\ g_4&=T_k f[x(k)+g_3,\ t_k+T_k] \end{aligned} \tag{10.107}$$

As with Euler's method, the Runge–Kutta methods can be extended to systems of higher order. For a system of order m, described by the vector state equation (10.94), the fourth-order Runge–Kutta method is defined by the algorithm

$$\mathbf{x}(k+1)=\mathbf{x}(k)+\tfrac{1}{6}(\mathbf{g}_1+2\mathbf{g}_2+2\mathbf{g}_3+\mathbf{g}_4) \tag{10.108}$$

where

$$\begin{aligned} \mathbf{g}_1&=T_k\mathbf{f}(k)\\ \mathbf{g}_2&=T_k\mathbf{f}[\mathbf{x}(k)+\mathbf{g}_1/2,\ t_k+T_k/2]\\ \mathbf{g}_2&=T_k\mathbf{f}[\mathbf{x}(k)+\mathbf{g}_2/2,\ t_k+T_k/2]\\ \mathbf{g}_4&=T_k\mathbf{f}[\mathbf{x}(k)+\mathbf{g}_3,\ t_k+T_k] \end{aligned} \tag{10.109}$$

are m-dimensional vectors. For the most part, the time increments are taken equal to one another, so that $T_k=T$ and $t_k=kT$ $(k=0,1,2,\ldots)$, where T is the sampling period.

The fourth-order Runge–Kutta method requires a large amount of computer time, as for each integration step it requires four evaluations of the vector $\mathbf{f}[x(t_k), t_k]$. However, the method is extremely accurate. Indeed, the local truncation error (see Section 10.8) is of order T^5. The total truncation error (see Section 10.8) is difficult to assess, but it can be assumed conservatively to be of order T^4.

Example 10.7

Solve the problem of Example 10.6 by the fourth-order Runge–Kutta method, compare results, and draw conclusions.

The state equations are given by Eqs. (c) of Example 10.6 and the components of the vector **f** by Eqs. (d) of the same example. To set up the computational algorithm for the fourth-order Runge–Kutta method, we use these equations in conjunction with Eqs. (10.108) and (10.109). The algorithm is defined by the equations

$$\begin{aligned} x_1(k+1) &= x_1(k) + \tfrac{1}{6}[g_{11}(k) + 2g_{21}(k) + 2g_{31}(k) + g_{41}(k)] \\ x_2(k+1) &= x_2(k) + \tfrac{1}{6}[g_{12}(k) + 2g_{22}(k) + 2g_{32}(k) + g_{42}(k)] \\ k &= 0, 1, 2, \ldots \end{aligned} \tag{a}$$

where g_{1i}, g_{2i}, g_{3i}, and g_{4i} $(i=1, 2)$ are the components of the vectors $\mathbf{g}_1$, $\mathbf{g}_2$, $\mathbf{g}_3$, and $\mathbf{g}_4$, respectively. They are given by the formulas

$$\begin{aligned} g_{11}(k) &= Tf_1[x_1(k), x_2(k)] = Tx_2(k) \\ g_{12}(k) &= Tf_2[x_1(k), x_2(k)] = -T\omega_n^2 \sin x_1(k) \\ g_{21}(k) &= Tf_1[x_1(k) + 0.5g_{11}(k), x_2(k) + 0.5g_{12}(k)] = T[x_2(k) + 0.5g_{12}(k)] \\ g_{22}(k) &= Tf_2[x_1(k) + 0.5g_{11}(k), x_2(k) + 0.5g_{12}(k)] \\ &= -T\omega_n^2 \sin[x_1(k) + 0.5g_{11}(k)] \\ g_{31}(k) &= Tf_1[x_1(k) + 0.5g_{21}(k), x_2(k) + 0.5g_{22}(k)] = T[x_2(k) + 0.5g_{22}(k)] \\ g_{32}(k) &= Tf_2[x_1(k) + 0.5g_{21}(k), x_2(k) + 0.5g_{22}(k)] \\ &= -T\omega_n^2 \sin[x_1(k) + 0.5g_{21}(k)] \\ g_{41}(k) &= Tf_1[x_1(k) + g_{31}(k), x_2(k) + g_{32}(k)] = T[x_2(k) + g_{32}(k)] \\ g_{42}(k) &= Tf_2[x_1(k) + g_{31}(k), x_2(k) + g_{32}(k)] = -T\omega_n^2 \sin[x_1(k) + g_{31}(k)] \\ k &= 0, 1, 2, \ldots \end{aligned} \tag{b}$$

where it was assumed that the time increment is the same for every step, $T_k = T$ $(k=0, 1, 2, \ldots)$. Equations (a) and (b) are subject to

$$x_1(0) = \pi/4, \qquad x_2(0) = 0 \tag{c}$$

The response corresponding to $\omega_n = 2$ rad/s was obtained using the sampling period $T = 0.1$ s. The angular displacement $\theta(k) = x_1(k)$ is shown in Fig. 10.12. For comparison, the true solution and the solution obtained in Example 10.6 by Euler's method are also plotted. The Runge–Kutta and the true solutions virtually coincide. On the other hand, the solution obtained by Euler's method, shown here in dashed line, deviates from the true solution. Hence, the Runge–Kutta method yielded far superior results to those obtained by the Euler method. This in spite of the fact that the sampling time used in the Runge–Kutta method was ten times as large as that used in the Euler method. Consistent with this, the computer execution

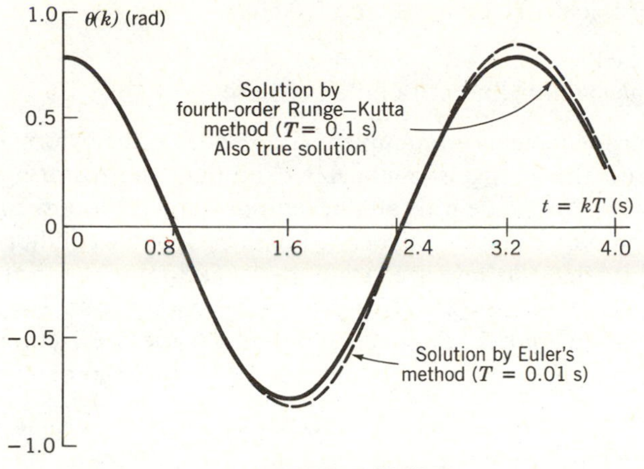

FIGURE 10.12

time in the Runge–Kutta method was one-fourth the execution time in the Euler method.

PROBLEMS

10.1 Calculate the transition matrix for a damped single-degree-of-freedom system by means of Eq. (10.24).

10.2 Calculate the transition matrix for a damped single-degree-of-freedom system by means of Eq. (10.28).

10.3 Calculate the transition matrix for the system of Problem 1.13 by means of Eq. (10.24).

10.4 Calculate the transition matrix for the system of Problem 1.13 by means of Eq. (10.28).

10.5 Obtain the step response of a damped single-degree-of-freedom system by treating it as a discrete-time system and using the convolution sum, Eq. (10.61). Consider the case in which $\omega_n = 1$ rad/s and $\zeta = 0.5$. Plot the response.

10.6 Obtain the ramp response for the system of Problem 10.5. Plot the response.

10.7 Obtain the step response of the system of Problem 1.13 by treating it as a discrete-time system and using Eq. (10.61). Plot the response.

10.8 Obtain the ramp response for the system of Problem 10.7. Plot the response.

10.9 Solve Problem 10.5 by using Eqs. (10.69).

10.10 Solve Problem 10.6 by using Eqs. (10.69).

10.11 Solve Problem 10.7 by using Eqs. (10.69).

10.12 Solve Problem 10.8 by using Eqs. (10.69).

10.13 Discretize the linearized equation for the system of Problem 9.4 in time and check the stability for each of the two cases.

10.14 Consider the system of Example 9.4, except that the external force is zero, $F=0$, and the spring is nonlinear, exhibiting the restoring force $f(x)=-kx[1-(x/2b)^2]$. Identify all the equilibrium positions, and derive the linearized state equations associated with each of the equilibrium positions. Then, discretize the linearized state equations in time, let the system parameters have the values $M=5$ kg, $m=15$ kg, $c=5$ N·s/m, $k=40$ N·s/m, $a=2$ m, $b=0.01$ m, plot the eigenvalues in the Θ-plane (see Section 10.7) for each equilibrium position, and determine the stability of the equilibrium positions.

10.15 Solve the differential equation

$$\ddot{x}+x[1+(x/2)^2]=0, \qquad x(0)=0, \qquad \dot{x}(0)=1$$

by Euler's method using a sampling period $T=0.1$ s. Plot the response.

10.16 Solve Problem 10.15 by the fourth-order Runge–Kutta method. Compare the response with that obtained in Problem 10.15 and draw conclusions.

10.17 The bob of the simple pendulum of Example 10.6 is subjected to a horizontal force as shown in Fig. 10.13. Derive and plot the response by Euler's method using a sampling period $T=0.1$ s. The pendulum is initially at rest.

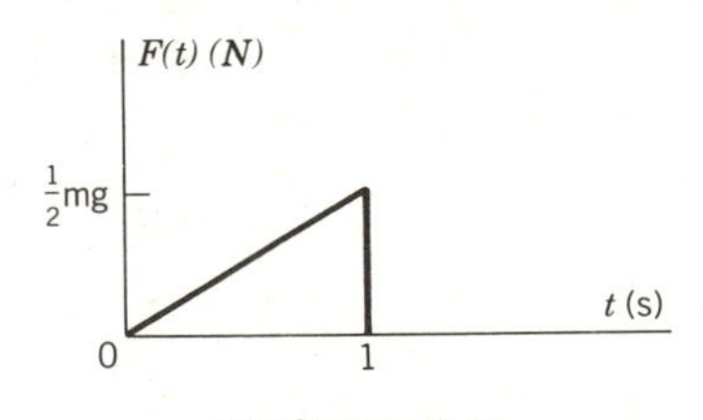

FIGURE 10.13

10.18 Solve Problem 10.17 by the fourth-order Runge–Kutta method. Compare the response with that obtained in Problem 10.17 and draw conclusions.

10.19 Derive the response of the system of Problem 10.14 to the initial excitation $x(0)=\dot{x}(0)=\dot{\theta}(0)=0$, $\theta(0)=\pi/4$ by the fourth-order Runge–Kutta method using a sampling period $T=0.1$ s. Note that the problem is concerned with the nonlinear system and not the linearized one. Plot the response.

10.20 The system of Problem 10.19 is subjected to a force $F=F(t)$ as shown in Fig. 10.13. Let the initial conditions be zero, and derive the response by the fourth-order Runge–Kutta method.

CHAPTER 11

Feedback Control Systems

11.1 INTRODUCTION

The general idea of system control was introduced in a general way in Chapter 1 without going into details. In the same chapter, various concepts and techniques commonly used in control theory, such as frequency response, transfer function, impulse response, and the convolution integral, were also presented. Other concepts and techniques pertinent to control theory, such as system stability and the transition matrix, were discussed in later chapters. This chapter is the final link in our study of dynamical systems, and it provides an introduction to system control.

The field of control can be divided broadly into *classical control* and *modern control*. Classical control is concerned primarily with low-order systems, particularly with systems characterized by a single input and a single output. The basic tools of analysis are the transfer function and the frequency response. On the other hand, modern control is concerned with higher-order systems, described by multiple inputs and multiple outputs. The analysis is carried out mostly in the time domain, as the use of transfer functions becomes impractical. One of the basic tools of analysis in modern control is linear algebra, and in particular matrix theory. Classical control tends to be more physical in nature, and modern control tends to be more mathematical. Classical control and modern control complement each other and can be used together to design control systems.

As pointed out earlier, various techniques of linear system theory were already introduced. In this chapter, we concentrate on techniques of special interest in feedback control, such as the root-locus method, the method of Nyquist, and Bode diagrams. These techniques are basic to classical control. The chapter concludes with an elementary introduction to modern control.

11.2 FEEDBACK CONTROL SYSTEMS

The simplest type of control is *open-loop control*. The term open-loop derives from the comparison with another type of control known as *closed-loop control*. As stated in Chapter 1, in an open-loop control system the input is not influenced by the output.

A typical open-loop control system can be described by the block diagram shown

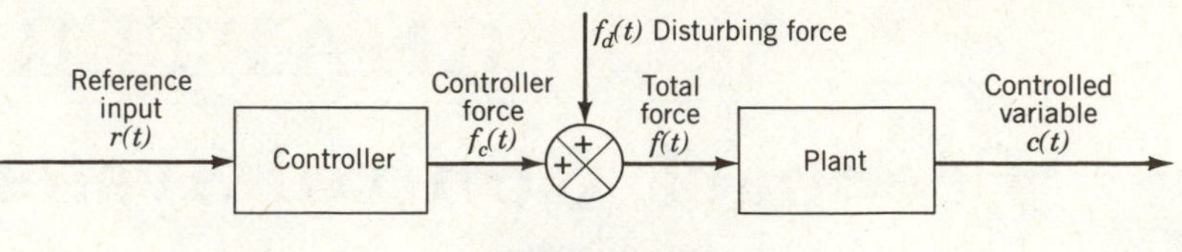

FIGURE 11.1

in Fig. 11.1. The *reference input*, denoted $r(t)$, represents the *desired output*. The *controller* represents the brain of the control systems in the sense that it converts the reference input into a controlling force $f_c(t)$, which combines with the disturbing force $f_d(t)$ to give the total force $f(t)$ acting on the *plant*. The effect of the force $f(t)$ acting on the plant is the *actual output* $c(t)$, which plays the role of the *controlled variable*.

In the case of a closed-loop system, there are two additional elements, namely, a *sensor* and an *error detector*. A typical block diagram of a closed-loop control system is shown in Fig. 11.2. Note that, in addition to the variables defined for the open-loop system, we have the *feedback signal* $b(t)$ and the *actuating signal* $e(t) = r(t) - b(t)$, where $e(t)$ is a measure of the error. Such a control system is known as a *feedback control system*.

As demonstrated in Chapter 1, the analysis of control systems is carried out conveniently in terms of Laplace transformed variables. Moreover, the controller and the plant are quite often regarded as a single entity, so that a typical block diagram of a feedback control system is as shown in Fig. 11.3, where $R(s) = \mathscr{L}r(t)$,

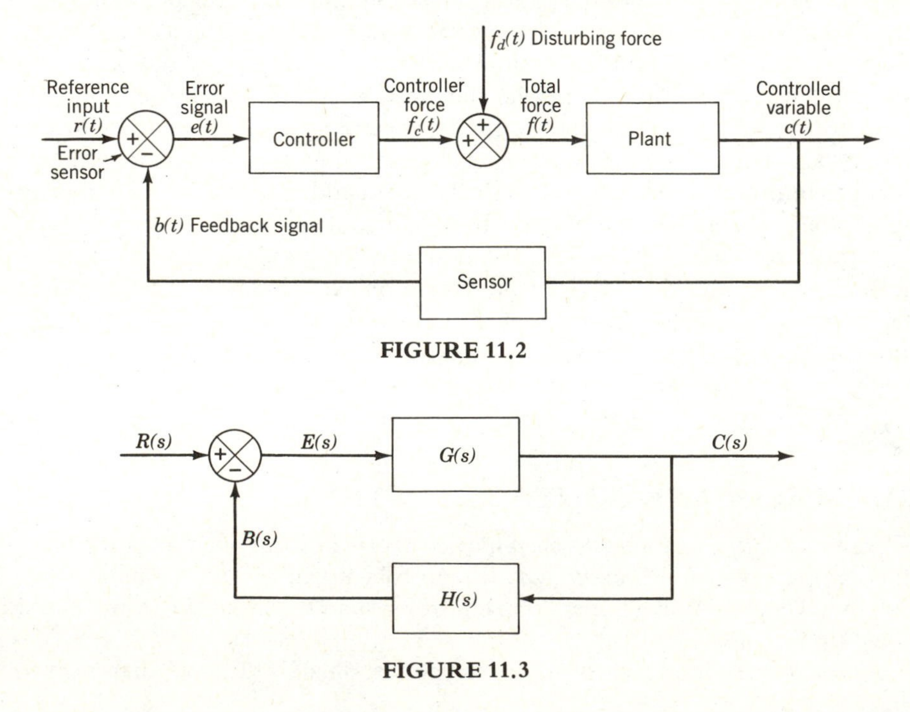

FIGURE 11.2

FIGURE 11.3

$C(s) = \mathscr{L}c(t)$, $B(s) = \mathscr{L}b(t)$, and $E(s) = \mathscr{L}e(t)$. In addition, $G(s)$ is known as the *forward-path transfer function* and $H(s)$ as the *feedback-path transfer function.*

From Fig. 11.3, we obtain the following relations

$$C(s) = G(s)E(s) \tag{11.1}$$

and

$$B(s) = H(s)C(s) \tag{11.2}$$

The *open-loop transfer function* is defined as the ratio of the transformed feedback signal $B(s)$ to the transformed actuating signal $E(s)$. Hence, from Eqs. (11.1) and (11.2), we can write

$$\frac{B(s)}{E(s)} = G(s)H(s) \tag{11.3}$$

The term open-loop transfer function derives from the fact that $G(s)H(s)$ would be the system transfer function if the transformed signal $B(s)$ were not fed back, so that there would be no closing of the loop. When the feedback-path transfer function is equal to unity, $H(s) = 1$, the open-loop and the forward-path transfer functions coincide.

Considering Fig. 11.3 once again, we observe that the transform of the actuating signal is

$$E(s) = R(s) - B(s) \tag{11.4}$$

Introducing Eqs. (11.2) and (11.4) into Eq. (11.1), we obtain

$$C(s) = G(s)R(s) - G(s)H(s)C(s) \tag{11.5}$$

from which we can write

$$\frac{C(s)}{R(s)} = M(s) = \frac{G(s)}{1 + G(s)H(s)} \tag{11.6}$$

where $M(s)$ can be identified as the *closed-loop transfer function*, also known as the *control ratio*, or the *overall transfer function.* Because Eq. (11.6) is equivalent to

$$C(s) = \frac{G(s)}{1 + G(s)H(s)} R(s) \tag{11.7}$$

we conclude that the output depends on the forward-path transfer function, feedback-path transfer function, and the reference input. From Eq. (11.1), we can write the equation for the transformed actuating signal

$$E(s) = \frac{C(s)}{G(s)} = \frac{1}{1 + G(s)H(s)} R(s) \tag{11.8}$$

The above derivations are for a typical single input, single output system. In carrying out the inverse transformation $c(t) = \mathscr{L}^{-1}C(s)$, the details are likely to differ from system to system. Quite often the open-loop transfer function involves the product of several transfer functions, so that the actual calculations can become quite involved.

To evaluate the control system performance, it is customary to use given reference inputs, such as the unit step function or the unit ramp function. Then, the system performance is judged by the manner in which the output follows the reference input. In the next section, we illustrate the idea by means of a simple example.

11.3 PERFORMANCE OF CONTROL SYSTEMS

The system of Fig. 11.4*a* represents a position control system. The plant consists of a wheel of polar mass moment of inertia J mounted on a shaft and supported at both ends by means of ball-bearing sleeves. The sleeves are assumed to provide viscous damping, with a viscous damping coefficient c. The object is to position the wheel by means of a torque applied by a motor. The torque is proportional to the error $e(t)$, with the constant of proportionality of the motor being equal to K_m. The equation of motion of the wheel is

$$J\ddot{\theta}_o(t) + c\dot{\theta}_o(t) = T(t) \tag{11.9}$$

where $\theta_o(t)$ is the wheel output angle, and the relation between the error $e(t)$ and the torque $T(t)$ is

$$T(t) = K_m e(t) \tag{11.10}$$

Moreover, the error is

$$e(t) = \theta_i(t) - \theta_o(t) \tag{11.11}$$

where $\theta_i(t)$ is the reference input angle.

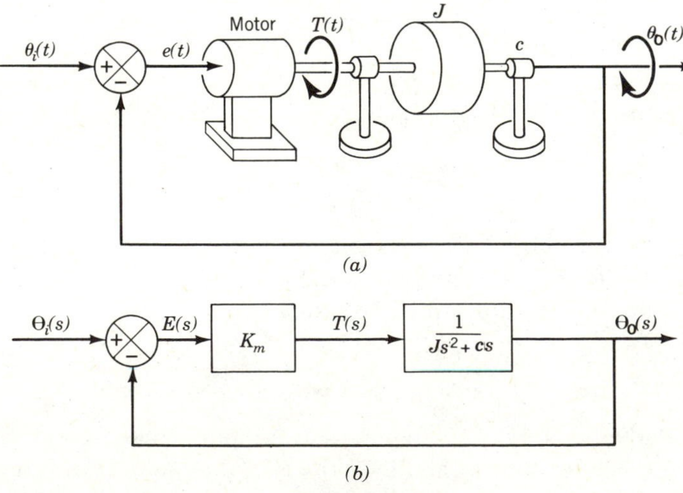

FIGURE 11.4

Laplace transforming Eq. (11.9), we obtain

$$(Js^2+cs)\Theta_o(s)=T(s) \tag{11.12}$$

where $\Theta_o(s)=\mathscr{L}\theta_o(t)$ and $T(s)=\mathscr{L}T(t)$, so that

$$\frac{\Theta_o(s)}{T(s)}=\frac{1}{Js^2+cs} \tag{11.13}$$

Moreover, transforming Eqs. (11.10) and (11.11), we can write

$$T(s)=K_mE(s) \tag{11.14}$$

and

$$E(s)=\Theta_i(s)-\Theta_o(s) \tag{11.15}$$

Equations (11.13)–(11.15) are shown in the block diagram of Fig. 11.4*b*. The forward-path transfer function is given by

$$G(s)=\frac{\Theta_o(s)}{E(s)} \tag{11.16}$$

so that, considering Eqs. (11.13) and (11.14), we obtain

$$G(s)=\frac{K_m}{Js^2+cs} \tag{11.17}$$

Moreover, recognizing that $H(s)=1$, the open-loop transfer function is equal to the forward-path transfer function, so that the relation between the transformed output and transformed reference input can be obtained from Eq. (11.7) in the form

$$\Theta_o(s)=\frac{G(s)}{1+G(s)}\,\Theta_i(s)=\frac{K_m}{Js^2+cs+K_m}\,\Theta_i(s) \tag{11.18}$$

Equation (11.18) has an interesting physical interpretation. Indeed, the closed-loop transfer function is

$$M(s)=\frac{\Theta_o(s)}{\Theta_i(s)}=\frac{K_m}{Js^2+cs+K_m} \tag{11.19}$$

which, except for the factor K_m in the numerator, is identical to the transfer function of a single-degree-of-freedom damped system. Hence, the effect of closing the loop is to add the term K_m in the denominator, thus providing an effective spring of stiffness K_m.

Equation (11.18) can be used to check the control system performance. To this end, it is necessary to select a reference input $\theta_i(t)$, insert its Laplace transform $\Theta_i(s)$ into Eq. (11.18), and then derive the system output $\theta_o(t)$ by inverting Eq. (11.18). One check of particular interest is how well the system can effect an abrupt change in position. This check calls for a reference input in the form of the unit step function (see Section 1.8)

$$\theta_i(t)=\mathscr{u}(t) \tag{11.20}$$

which has the Laplace transform

$$\Theta_i(s) = \frac{1}{s} \tag{11.21}$$

Hence, inserting Eq. (11.21) into Eq. (11.18), we can write the output Laplace transform in the form

$$\Theta_o(s) = \frac{K_m}{J} \frac{1}{s(s^2 + 2\zeta\omega_n s + \omega_n^2)} \tag{11.22}$$

where

$$2\zeta\omega_n = c/J, \qquad \omega_n^2 = K_m/J \tag{11.23}$$

The response of a system to a unit step function is known as the step response (Section 1.10). To determine the step response, we must obtain the inverse transform of Eq. (11.22). Because the function $(s + 2\zeta\omega)/(s^2 + 2\zeta\omega s + \omega^2)$ is a Laplace transform that can be found in the table of Laplace transform pairs in Section A.7, let us consider the partial fractions expansion

$$\begin{aligned} \frac{1}{s(s^2 + 2\zeta\omega_n s + \omega_n^2)} &= \frac{c_1}{s} + \frac{c_2 s + c_3}{s^2 + 2\zeta\omega_n s + \omega_n^2} \\ &= \frac{c_1(s^2 + 2\zeta\omega_n s + \omega_n^2) + (c_2 s + c_3)s}{s(s^2 + 2\zeta\omega_n s + \omega_n^2)} \\ &= \frac{(c_1 + c_2)s^2 + (2\zeta\omega_n c_1 + c_3)s + c_1\omega_n^2}{s(s^2 + 2\zeta\omega_n s + \omega_n^2)} \end{aligned} \tag{11.24}$$

where c_1, c_2, and c_3 are constants. The satisfaction of Eq. (11.24) requires that the factors multiplying s^2 and s be equal to zero and the remaining term be equal to unity. Hence, we must have

$$c_1 + c_2 = 0, \qquad 2\zeta\omega_n c_1 + c_3 = 0, \qquad c_1\omega_n^2 = 1 \tag{11.25}$$

Solving Eqs. (11.25), we obtain the constants

$$c_1 = \frac{1}{\omega_n^2}, \qquad c_2 = -\frac{1}{\omega_n^2}, \qquad c_3 = -\frac{2\zeta}{\omega_n} \tag{11.26}$$

so that the partial fractions expansion takes the form

$$\frac{1}{s(s^2 + 2\zeta\omega_n s + \omega_n^2)} = \frac{1}{\omega_n^2}\left(\frac{1}{s} - \frac{s + 2\zeta\omega_n}{s^2 + 2\zeta\omega_n s + \omega_n^2}\right) \tag{11.27}$$

Inserting Eq. (11.27) into Eq. (11.22) and using the table of Laplace transform pairs in Section A.7, we obtain

$$\begin{aligned} \theta_o(t) = \mathscr{L}^{-1}\Theta_o(s) &= \frac{K_m}{J\omega_n^2} \mathscr{L}^{-1}\left(\frac{1}{s} - \frac{s + 2\zeta\omega_n}{s^2 + 2\zeta\omega_n s + \omega_n^2}\right) \\ &= \left[1 - e^{-\zeta\omega_n t}\left(\cos \zeta\omega_d t + \frac{\zeta\omega_n}{\omega_d} \sin \omega_d t\right)\right] u(t) \end{aligned} \tag{11.28}$$

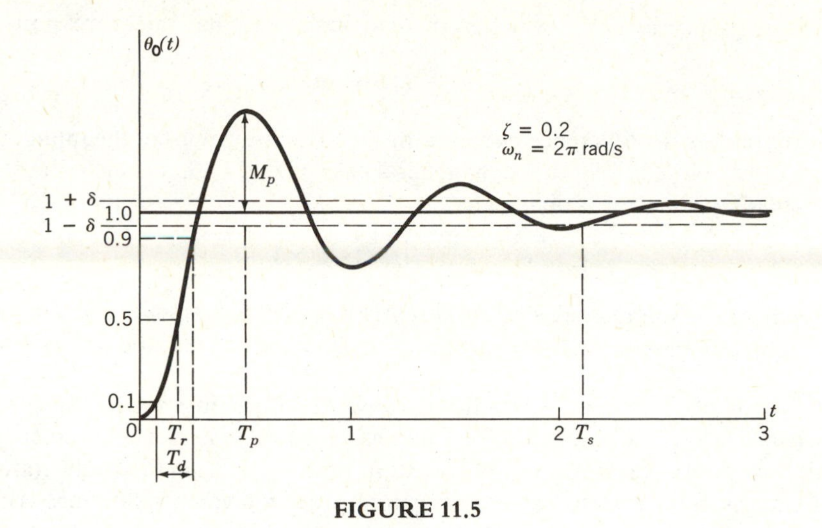

FIGURE 11.5

where $\omega_d = (1-\zeta^2)^{1/2}\omega_n$. This implies that $\zeta < 1$, which is ordinarily the case. From Eq. (11.28), we conclude that the output consists of a unit step function plus some exponentially decaying oscillation, so that the output does not reproduce the reference input exactly. The performance of the control system depends on various design parameters, such as the damping factor ζ and the natural frequency ω_n of the undamped oscillation.

To judge the control system performance, let us plot the step response $\theta_o(t)$ versus t, with ζ and ω_n as parameters. A typical plot is shown in Fig. 11.5. The characteristics of the step response of the control system is defined in terms of a number of criteria. In particular, we identify the following:

1. *Peak time* T_p, defined as the time necessary for the step response to reach its peak value.
2. *Maximum overshoot* M_p, defined as the value of the peak step response minus the final value of the response. An alternative definition of the peak response is in terms of the *percent overshoot* P.O., given by

$$\text{P.O.} = \frac{\theta_o(T_p) - \theta_o(\infty)}{\theta_o(\infty)} \times 100\%$$

3. *Delay time* T_d, defined as the time needed for the step response to reach 50% of its final value.
4. *Rise time* T_r, which is the time required for the step reponse to rise from 10% to 90%, or from 5% to 95%, or from 0% to 100% of its final value.
5. *Settling time* T_s, defined as the time required for the step response to decrease to and stay within a certain percentage δ of the final value. A commonly used figure is 5%.

6. *Steady-state error* e_{ss}, which in this particular case has the expression

$$e_{ss} = \theta_i(\infty) - \theta_o(\infty)$$

represents the difference between the steady-state values of the input and the output. Note that in the more general case of Fig. 11.3, in which $H(s) \neq 1$, the steady-state error is defined as

$$e_{ss} = r(\infty) - b(\infty)$$

where r is the reference input and b is the feedback signal. The steady-state error can be determined by means of the final-value theorem (see Section A.5).

Some of the performance criteria listed above are more important than others. In particular, good performance of a feedback control system is characterized by small maximum overshoot M_p, low settling time T_s, and small steady-state error e_{ss}. Changes in the system parameters can improve a given performance criterion. Care must be exercised, however, as the same changes can degrade another criterion. The object of control system design is to select the parameters so that the important performance criteria are met.

11.4 THE ROOT-LOCUS METHOD

In Chapter 9, we studied methods for determining the stability of systems. In particular, we showed that the stability characteristics depend on the system eigenvalues. Hence, to determine the stability characteristics, it is necessary to solve the eigenvalue problem. But, the eigenvalue problem is basically a numerical problem, so that we must assign numerical values to the system parameters before attempting a solution. Quite often, a mere statement as to whether the system is stable or not suffices. To obtain such a statement, we can use the Routh–Hurwitz criterion, which requires the characteristic equation. To use the Routh–Hurwitz criterion, once again we must assign numerical values to the system parameters, except in some simple cases.

Quite frequently, the designer is interested not merely in a stability statement but also in the response characteristics, and in particular how the response changes as the system parameters change. To answer this question, several graphical methods have been developed. One of these methods is known as the *root-locus method*. The method consists of plotting the roots of the characteristic equation in the s-plane as a certain parameter of the system changes. It turns out that the roots lie on smooth curves, known as *loci*, and the plots themselves are known as *root-locus plots*. It should be pointed out that we already encountered a root-locus plot in Section 4.6, in which the locus of the roots of the characteristic equation for a damped second-order system was plotted using the damping factor as a parameter (see Fig. 4.12).

In Section 11.2, we derived the closed-loop transfer function in the form

$$M(s)=\frac{C(s)}{R(s)}=\frac{G(s)}{1+G(s)H(s)} \tag{11.29}$$

where $G(s)$ is the forward-path transfer function and $G(s)H(s)$ is the open-loop transfer function. The characteristic equation is simply

$$1+G(s)H(s)=0 \tag{11.30}$$

so that the root loci are plots of the roots of Eq. (11.30) as functions of a given parameter. To this end, it is convenient to write the open-loop transfer function in the form

$$G(s)H(s)=KF(s) \tag{11.31}$$

where K is a parameter. Inserting Eq. (11.31) into Eq. (11.30), we can rewrite the characteristic equation as follows:

$$1+KF(s)=0 \tag{11.32}$$

Then, *the root loci are defined as the loci in the s-plane of the roots of Eq. (11.32) as K varies from 0 to* ∞. The loci obtained by letting K vary from 0 to $-\infty$ are called *complementary root loci.* We shall be concerned only with positive values of K.

To establish the conditions under which Eq. (11.32) is satisfied, let us write

$$F(s)=-\frac{1}{K}=|F(s)|e^{i\psi_F} \tag{11.33}$$

where $|F(s)|$ is the magnitude and ψ_F is the phase angle of $F(s)$. Hence, the characteristic equation is satisfied if both equations

$$|F(s)|=\frac{1}{K} \tag{11.34}$$

and

$$\psi_F=\underline{/F(s)}=(2k+1)\pi, \qquad k=0, \pm 1, \pm 2, \ldots \tag{11.35}$$

are satisfied simultaneously, where $\underline{/F(s)}$ is the customary notation for the phase angle of the function $F(s)$. To construct the root loci, we must find all the points in the s-plane that satisfy Eq. (11.35). Then, the value of K for any s on the loci can be determined by means of Eq. (11.34).

It should be pointed out that the roots of the characteristic equation, Eq. (11.32), are not the only poles of the closed-loop transfer function $M(s)$. Indeed, from Eq. (11.29), we observe that $M(s)$ has a pole also where $G(s)$ has a pole and $G(s)H(s)$ is finite, or zero, or has a pole of lower order. Generally, these latter poles can be found by inspection of $G(s)$ and $H(s)$. The root-locus method permits the location of the poles of $M(s)$ that cannot be found by inspection. Note that the poles of $M(s)$ are the *closed-loop poles* of the system.

The use of Eqs. (11.34) and (11.35) to find the roots of the characteristic equation analytically is generally not more advantageous than the use of Eq. (11.32). The construction of the root loci is basically a graphical problem, however, and this graphical construction is made easier by the use of Eqs. (11.34) and (11.35). This is due to the special nature of the function $F(s)$ as a rational function of s. Hence, let us write

$$F(s)=\frac{N(s)}{D(s)}=\frac{(s-z_1)(s-z_2)\cdots(s-z_m)}{(s-p_1)(s-p_2)\cdots(s-p_n)}=\frac{\prod_{i=1}^m (s-z_i)}{\prod_{j=1}^n (s-p_j)}=-\frac{1}{K} \tag{11.36}$$

where $z_i\,(i=1,2,\ldots,m)$ are the *zeros* of $F(s)$ and $p_j\,(j=1,2,\ldots,n)$ are the *poles* of $F(s)$. Inserting Eq. (11.36) into Eq. (11.34), we have simply

$$|F(s)|=\frac{\prod_{i=1}^m |s-z_i|}{\prod_{j=1}^n |s-p_j|}=\frac{1}{K} \tag{11.37}$$

Moreover, denoting the phase angles of $s-z_i$ by θ_{zi} and the phase angles of $s-p_j$ by θ_{pj} and considering Eq. (11.36), we obtain from Eq. (11.35)

$$\psi_F=\underline{/F(s)}=\sum_{i=1}^{m}\theta_{zi}-\sum_{j=1}^{n}\theta_{pj}=(2k+1)\pi, \qquad k=0, \pm 1, \pm 2, \ldots \tag{11.38}$$

As a simple illustration, let us consider the function

$$F(s)=\frac{s-z_1}{(s-p_1)(s-p_2)} \tag{11.39}$$

The vectors $s-z_1$, $s-p_1$, and $s-p_2$ and the corresponding angles θ_{z1}, θ_{p1}, and θ_{p2} are shown in the s-plane depicted in Fig. 11.6. The point s belongs to a root locus if the angles satisfy

$$\underline{/F(s)}=\theta_{z1}-\theta_{p1}-\theta_{p2}=(2k+1)\pi, \qquad k=0, \pm 1, \pm 2, \ldots \tag{11.40}$$

This can be established by trial and error. Then, if a point s has been verified as belonging to a root locus, the magnitude of the parameter K can be determined by writing

$$K=\frac{|s-p_1||s-p_2|}{|s-z_1|} \tag{11.41}$$

The preceding discussion concerning the graphical construction of root loci is predicated upon prior knowledge of the zeros and poles of $F(s)$, which are the

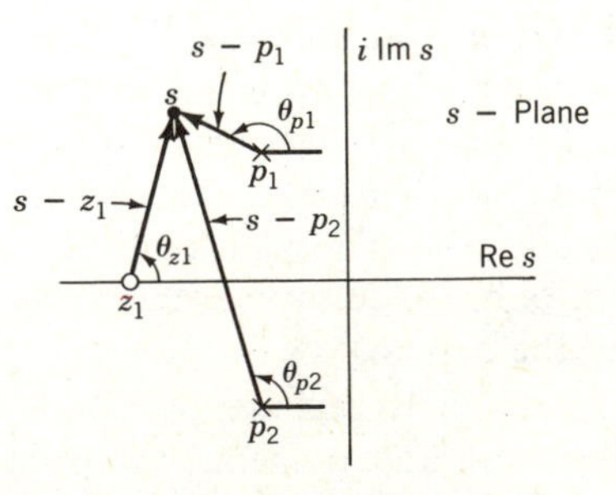

FIGURE 11.6

same as the zeros and poles of the open-loop transfer function $G(s)H(s)$. If the zeros and poles of the function $F(s)$ are not known in advance, then the root loci can be plotted by means of a digital computer program written expressly for this purpose. In fact, the same program can be used even when the zeros and poles are known.

Quite often, particularly in preliminary design, accurate root-locus plots are not really necessary, and rough sketches are sufficient. To help with these sketches, it is possible to formulate a number of rules of construction. These rules not only can help with the construction of root loci but they can also enhance the understanding of the technique. The rules are based on the relation between the zeros and poles of the open-loop transfer function and the roots of the characteristic equation. They are as follows:

Rule 1. *The root loci start ($K=0$) at the poles of the open-loop transfer function, and there are as many loci as there are poles of the open-loop transfer function,* provided the number of poles is greater than the number of zeros.

From Eq. (11.36), it is easy to see that as K approaches zero, $F(s)$ approaches infinity, so that s approaches the poles p_j $(j=1, 2, \ldots, n)$ of $F(s)$, or the poles of $G(s)H(s)$. Because the open-loop transfer function has n poles, there must be n root loci.

Rule 2. *The root loci terminate ($K=\infty$) either at the zeros of the open-loop transfer function or at $s=\infty$*, where the number of root loci terminating at $s=\infty$ is equal to the excess number of poles over zeros.

From Eq. (11.36), we conclude that as K approaches infinity $F(s)$ approaches zero. This can happen in two ways. In the first place, $F(s)$ approaches zero as s approaches the zeros of $F(s)$, or the zeros of $G(s)H(s)$. Second, $F(s)$ approaches zero, and hence $G(s)H(s)$ approaches zero as s approaches infinity.

Rule 3. *The number of branches of the root loci is equal to either the number of poles or the number of zeros of the open-loop transfer function, whichever is larger.*

Using Eq. (11.36), the characteristic equation can be written in the form

$$D(s)+KN(s)=\prod_{j=1}^{n}(s-p_j)+K\prod_{i=1}^{m}(s-z_i)=0 \tag{11.42}$$

The number of roots of the characteristic equation is equal to the degree of the characteristic polynomial, which is equal to the larger of n and m. In turn, the number of roots of the characteristic equation determines the number of branches of the root loci, which verifies the rule. In engineering systems, the number n of poles is ordinarily greater than the number m of zeros, so that there are n branches.

Rule 4. *The root loci are symmetric with respect to the real axis.*

Because the coefficients of the characteristic polynomial are real, all complex roots appear in conjugate pairs. Hence, it is only necessary to plot root loci in the upper half of the s-plane.

Rule 5. *A root locus always exists on a given section of the real axis when the total number of real poles and real zeros of the open-loop transfer function to the right of the section is odd.*

This rule is best demonstrated by a sketch. Figure 11.7 shows the s-plane containing three poles and two zeros of the open-loop transfer function $G(s)H(s)$. As customary, poles are denoted by crosses and zeros by circles. In Fig. 11.7*a*, the test point s lies between the pole p_2 and the zero z_1. The angles to the vectors from p_1 and p_2 to s are $\theta_{p1}=\pi$ and $\theta_{p2}=\pi$, respectively. On the other hand, the angles to the vectors from z_1, p_3, and z_2 to s are all zero. Hence, the phase angle of $F(s)$ corresponding to the chosen test point is

$$\underline{/F(s)}=\sum_{i=1}^{2}\theta_{zi}-\sum_{j=1}^{3}\theta_{pj}=0+0-\pi-\pi-0=-2\pi \tag{11.43}$$

Because $\underline{/F(s)}$ is not an odd multiple of π, the point s in Fig. 11.7*a* does not lie on a root locus. In Fig. 11.7*b* the test point s lies between the zero z_1 and the pole p_3. Following the same pattern, we can write for the phase angle of $F(s)$ corresponding to the latter test point

$$\underline{/F(s)}=\sum_{i=1}^{2}\theta_{zi}-\sum_{j=1}^{3}\theta_{pj}=\pi+0-\pi-\pi-0=-\pi \tag{11.44}$$

so that this test point does lie on a root locus. In the case of Fig. 11.7*a* there are two poles to the right of the section $z_1<s<p_2$, and in the case of Fig. 11.7*b* there are two poles and one zero to the right of the section $p_3<s<z_1$, which verifies the rule. Note that complex poles and zeros contribute multiples of 2π to the phase angle of $F(s)$, because complex poles and zeros occur in pairs of complex conjugates. Hence, the presence of complex poles and zeros does not affect the above conclusions.

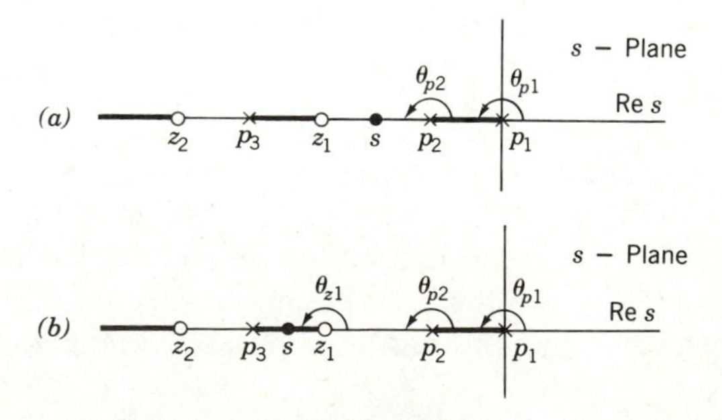

FIGURE 11.7

Rule 6. *For large values of s, the root loci approach straight lines asymptotically, where the angles of the asymptotes are given by*

$$\theta_k = \frac{(2k+1)\pi}{n-m}, \qquad k=0, 1, \ldots, |n-m|-1 \tag{11.45}$$

Using Eq. (11.36), we can write

$$\lim_{s\to\infty} F(s) = \lim_{s\to\infty} \frac{\prod_{i=1}^{m}(s-z_i)}{\prod_{j=1}^{n}(s-p_j)} = \frac{1}{s^{n-m}} \tag{11.46}$$

Moreover, letting the angle of s be $\psi_s = \theta$, we conclude that the angle of $F(s)$ as $s \to \infty$ is

$$\lim_{s\to\infty} \psi_F = -(n-m)\psi_s = -(n-m)\theta \tag{11.47}$$

But, if s is on a root locus, the angle ψ_F must satisfy Eq. (11.38). Hence, combining Eqs. (11.38) and (11.47), we obtain Eq. (11.45). Note that according to Rules 2 and 3 there can be only $n-m$ asymptotes, which explains the upper limit on the integer k in Eq. (11.45). If k exceeds the upper limit, the asymptotes simply repeat themselves.

Rule 7. *The $|n-m|$ asymptotes intersect on the real axis of the s-plane at a distance*

$$c = \frac{\sum_{j=1}^{n} p_j - \sum_{i=1}^{m} z_i}{n-m} \tag{11.48}$$

from the origin of the s-plane, where the intersection of the asymptotes is sometimes called the *centroid*.

Because the root loci are symmetric with respect to the real axis, the asymptotes must intersect on the real axis. To verify Eq. (11.48), let us consider Eq. (11.36) and write it in the form

$$\begin{aligned} F(s) &= \frac{(s-z_1)(s-z_2)\cdots(s-z_m)}{(s-p_1)(s-p_2)\cdots(s-p_n)} \\ &= \frac{s^m + (-\sum_{i=1}^{m} z_i)s^{m-1} + \cdots + (-1)^m \prod_{i=1}^{m} z_i}{s^n + (-\sum_{j=1}^{n} p_j)s^{n-1} + \cdots + (-1)^n \prod_{j=1}^{n} p_j} \\ &= \left[s^{n-m} + \left(\sum_{i=1}^{m} z_i - \sum_{j=1}^{n} p_j \right) s^{n-m-1} + \cdots \right]^{-1} \end{aligned} \tag{11.49}$$

Regarding the asymptotes as genuine root loci, we conclude that corresponding root locus plots appear as in Fig. 11.8 and can be described mathematically by

$$F_a(s) = \frac{1}{(s-c)^{n-m}} = [s^{n-m} - c(n-m)s^{n-m-1} + \cdots]^{-1} \tag{11.50}$$

But, for very large s, Eqs. (11.49) and (11.50) must be identical, so that the coefficients of s^{n-m-1} must be the same in both equations. Equating the coefficients of s^{n-m-1}

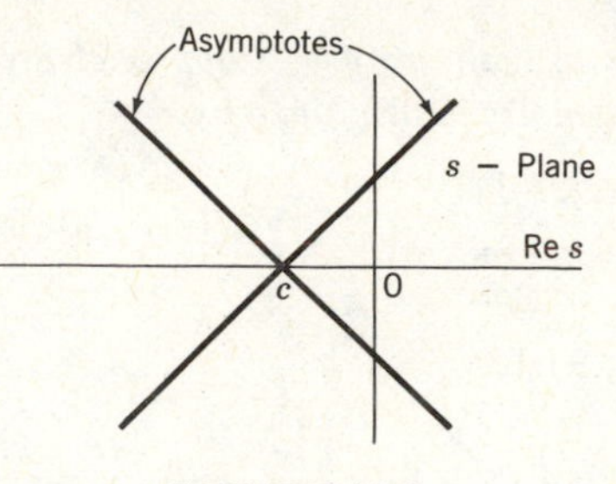

FIGURE 11.8

in Eqs. (11.49) and (11.50), we obtain Eq. (11.48). Note that only finite poles and zeros are included in the sums in Eq. (11.48).

Rule 8. *The breakaway points $s=s_b$ are points on the root loci at which multiple roots occur, and they must satisfy the equation*

$$\left.\frac{dF(s)}{ds}\right|_{s=s_b}=0 \tag{11.51}$$

In the first place, breakaway points or saddle points are defined as points on the root loci at which two (or more) branches meet. Note that these saddle points have an entirely different physical meaning from the saddle points encountered in Section 9.5 in connection with system stability. The difference lies in the fact that here the curves represent root loci in the s-plane and in Section 9.5 the curves represented trajectories in the phase plane. Assuming that at a breakaway point $s=s_b$ there are k roots, we conclude that the characteristic equation must have the form

$$f(s, k)=1+KF(s)=(s-s_b)^k A(s)=0 \tag{11.52}$$

where $A(s_b)\neq 0$. Taking the derivative of Eq. (11.52) with respect to s, we obtain

$$\frac{df}{ds}=\frac{\partial f}{\partial s}+\frac{\partial f}{\partial K}\frac{dK}{ds}=\frac{\partial f}{\partial s}+F\frac{dK}{ds}=0 \tag{11.53}$$

But, from the right side of Eq. (11.52), we can write

$$\left.\frac{\partial f}{\partial s}\right|_{s=s_b}=\left.\left[k(s-s_b)^{k-1}A(s)+(s-s_b)^k\frac{dA(s)}{ds}\right]\right|_{s=s_b}=0 \tag{11.54}$$

Hence, because $F(s_b)\neq 0$, Eq. (11.53) yields

$$\left.\frac{dK}{ds}\right|_{s=s_b}=0 \tag{11.55}$$

so that the parameter K has a stationary value at a breakaway point. From Eq. (11.52), we can also write

$$\frac{dK}{ds}F+K\frac{dF}{ds}=0 \tag{11.56}$$

from which we conclude that the satisfaction of Eq. (11.55) implies the satisfaction of Eq. (11.51).

If there are two branches meeting at a breakaway point, then they make a 180° angle with respect to each other. Moreover, if the breakaway point is on the real axis, then because of the symmetry of the root loci, the tangent to the root loci must make 90° and 270° angles with respect to the real axis. Note that Eq. (11.51), or Eq. (11.55), represents only a necessary condition. In addition, a breakaway point must satisfy the characteristic equation, Eq. (11.32), which determines the value of K corresponding to the point $s=s_b$.

It should be pointed out that the root-locus techniques described above can be used to study the behavior of the roots of polynomials in general and are not confined to problems of feedback control. To this end, we must rewrite the polynomial in the form given by the left side of Eq. (11.30). This form depends on the particular coefficient of the polynomial chosen to play the role of the parameter K.

Example 11.1

Consider the position control system of Fig. 11.4*a*, but with the torque being proportional to the error and the integral of the error, or

$$T(t)=K_1 e(t)+K_2 \int_0^t e(\tau)\, d\tau \tag{a}$$

instead of being merely proportional to the error, as in Eq. (11.10). Note that such control is called proportional and integral control. Draw a block diagram to replace the block diagram of Fig. 11.4*b*, and derive the open-loop transfer function for the system. Then, sketch the root loci using $K=K_1/J$ as a parameter and for the values

$$K_2/K_1=0.3, \qquad c/J=0.5 \tag{b}$$

In the first place, we wish to determine the transfer function between the error and the torque. To this end, we must evaluate the Laplace transform of the integral in Eq. (a). Using integration by parts, we can write

$$\begin{aligned}\mathscr{L}\left[\int_0^t e(\tau)\, d\tau\right] &= \int_0^\infty \left[\int_0^t e(\tau)\, d\tau\right] e^{-st}\, dt \\ &= \left[\int_0^t e(\tau)\, d\tau\right] \frac{e^{-st}}{-s}\Bigg|_0^\infty + \frac{1}{s}\int_0^\infty e^{-st} e(t)\, dt\end{aligned} \tag{c}$$

But, the first term on the right side of Eq. (c) vanishes because at the upper limit

$$\lim_{t\to\infty} e^{-st} \int_0^t e(\tau)\, d\tau = 0 \tag{d}$$

and at the lower limit $\int_0^0 e(\tau)\, d\tau=0$. Hence,

$$\mathscr{L}\left[\int_0^t e(\tau)\, d\tau\right] = \frac{E(s)}{s} \tag{e}$$

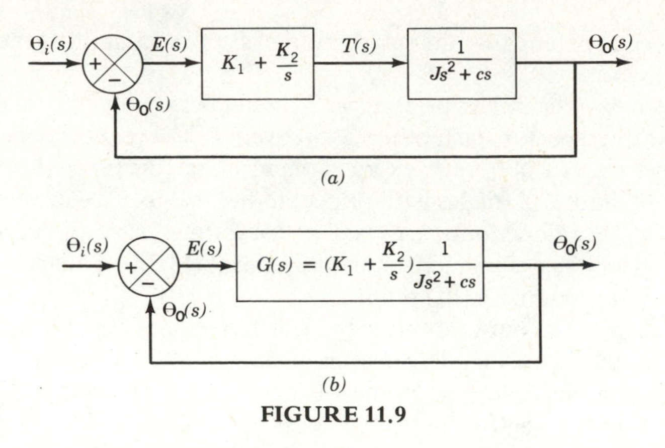

FIGURE 11.9

where $E(s)$ is the Laplace transform of $e(t)$. Transforming both sides of Eq. (a), we obtain

$$T(s)=\left(K_1+\frac{K_2}{s}\right)E(s) \tag{f}$$

where $T(s)$ is the Laplace transform of $T(t)$. The block diagram for the system is shown in Fig. 11.9*a*. The blocks for the torquer and the wheel can be combined as shown in Fig. 11.9*b*, where $G(s)$ is the forward-path transfer function. Inserting Eq. (f) into Eq. (11.13), we can write

$$\Theta_o(s)=\frac{T(s)}{Js^2+cs}=\left(K_1+\frac{K_2}{s}\right)\frac{1}{Js^2+cs}E(s) \tag{g}$$

so that the forward-path transfer function is

$$G(s)=\frac{\Theta_o(s)}{E(s)}=\left(K_1+\frac{K_2}{s}\right)\frac{1}{Js^2+cs} \tag{h}$$

Because $H(s)=1$, $G(s)$ is also the open-loop transfer function. The open-loop transfer function can be given the more convenient form

$$G(s)=\frac{K_1}{J}\frac{s+K_2/K_1}{s^2(s+c/J)}=K\frac{s-z_1}{s^2(s-p_3)} \tag{i}$$

where

$$z_1=-K_2/K_1, \qquad p_1=p_2=0, \qquad p_3=-c/J \tag{j}$$

are the zero and the poles of the open-loop transfer function. Hence, there is one zero, $m=1$, and three poles, $n=3$. Using Eqs. (b), we obtain the numerical values

$$z_1=-0.3, \qquad p_1=p_2=0, \qquad p_3=-0.5 \tag{k}$$

To sketch the root loci, we seek guidance from the various rules discussed earlier, as well as from Eq. (11.38). They permit us to reach the following conclusions:

i. By Rule 1, there are three root loci, one root locus starting from each pole.

ii. Because there is only one zero, by Rule 2, one root locus ends at the zero and the other two end at infinity.

iii. The number of zeros is $m=1$ and the number of poles is $n=3$. Hence, by Rule 3, the number of branches is the larger of the two numbers, i.e., there are three branches.

iv. By Rule 5, there is one root locus on the real axis in the section between p_3 and z_1. This section is a root locus beginning at p_3 and ending at z_1.

v. By Rule 6, there are two asymptotes with the angles $\theta_1=\pi/2$ and $\theta_2=3\pi/2$.

vi. By Rule 7, the intersection of the asymptotes with the real axis is

$$c=\frac{p_1+p_2+p_3-z_1}{2}=\frac{-0.5+0.3}{2}=-0.1 \tag{l}$$

The above conclusions are sufficient to sketch the root loci, except in the vicinity of the poles $p_1=p_2=0$. To complete the sketch, we turn to Eq. (11.38) to determine the angles of departure of the root loci starting at $p_1=p_2$. Equation (11.38) has the specific form

$$\underline{/F(s)}=\theta_{z1}-\theta_{p1}-\theta_{p2}-\theta_{p3}=(2k+1)\pi, \qquad k=0, \pm 1, \pm 2, \ldots \tag{m}$$

Choosing a point s very close to the origin, as shown in Fig. 11.10, the phase angles of the vectors from the zero and poles to s are

$$\theta_{z1}\cong 0, \qquad \theta_{p1}=\theta_{p2}=\text{unknown}, \qquad \theta_{p3}\cong 0 \tag{n}$$

so that

$$\underline{/F(s)}=-2\theta_{p1}=-2\theta_{p2}=(2k+1)\pi \tag{o}$$

There are only two distinct angles, or

$$\begin{aligned}\theta_{p1}=\theta_{p2}&=\pi/2 \qquad \text{for} \quad k=-1\\ \theta_{p1}=\theta_{p2}&=3\pi/2 \qquad \text{for} \quad k=-2\end{aligned} \tag{p}$$

so that the two root loci starting at $p_1=p_2=0$ are tangent to the imaginary axis, one to the positive and the second to the negative imaginary axis. The root loci are sketched in Fig. 11.11. It is clear that the root loci are symmetric with respect to the real axis, so that Rule 4 is verified. The magnitude of the parameter K at

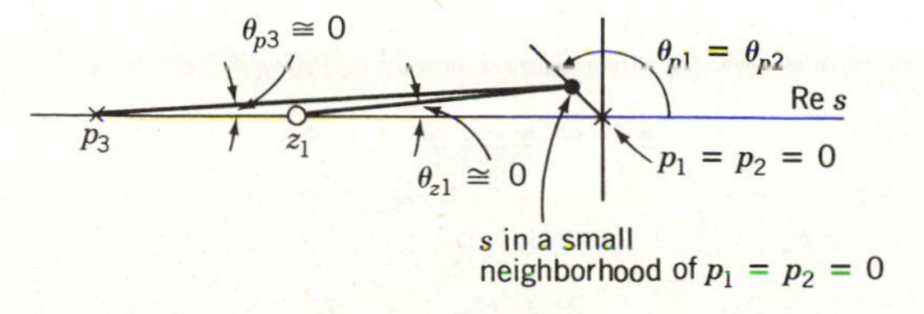

FIGURE 11.10

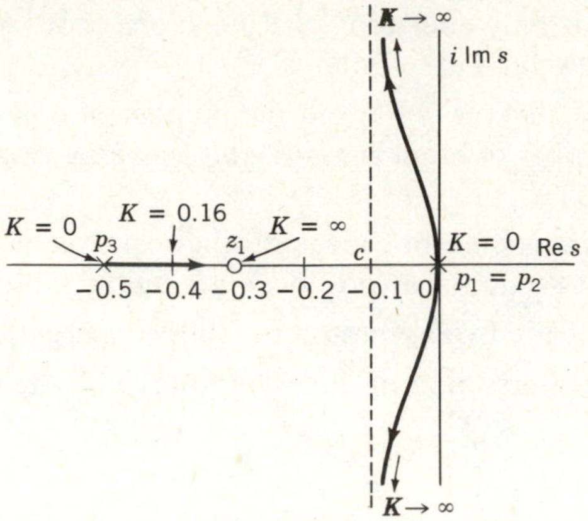

FIGURE 11.11

any point s on a root locus can be obtained by using Eq. (11.37) and writing

$$K = \frac{|s-p_1|\,|s-p_2|\,|s-p_3|}{|s-z_1|} \tag{q}$$

For example, for the point $s=-0.4$, we obtain

$$K = \frac{|s|\,|s|\,|s+0.5|}{|s+0.3|} = \frac{|-0.4|\,|-0.4|\,|-0.4+0.5|}{|-0.4+0.3|} = 0.16 \tag{r}$$

Example 11.2

Plot the root loci for a system with the open-loop transfer function

$$G(s)H(s) = \frac{K}{s(s^2+0.8s+1)} \tag{a}$$

Before proceeding with the sketching of the root loci, we observe that the open-loop transfer function has no zeros and three poles. The poles are

$$p_1 = 0, \qquad p_2 = -0.4000 + i0.9165, \qquad p_3 = -0.4000 - i0.9165 \tag{b}$$

Application of the rules is as follows:

i. By Rule 1, there are three root loci, one root locus starting from each of the poles indicated by Eqs. (b).

ii. Because the open-loop transfer function has no zeros, all three root loci terminate at infinity.

iii. By Rule 3, there are three branches.

iv. There is only one pole on the real axis, namely, at the origin. Because there are no zeros, by Rule 5 the entire negative real axis is a root locus, so that the root locus begins at the origin and ends at $s=-\infty$.

v. By Rule 6, there are three asymptotes with the angles $\theta_1 = \pi/3$, $\theta_2 = \pi$, and $\theta_3 = -\pi/3$.

vi. By Rule 7, the intersection of the asymptotes with the real axis is

$$c = \frac{p_1 + p_2 + p_3}{3} = \frac{-0.4000 + i0.9165 - 0.4000 - i0.9165}{3}$$
$$= -0.2667 \qquad \text{(c)}$$

The above information is not sufficient to sketch the root loci. Additional information can be obtained by considering the angle of departure from p_2 and the intersection with the imaginary axis. To determine the angle of departure from p_2, we denote the angles of the vectors from p_1 and p_2 to an arbitrary point s in the neighborhood of p_2 by θ_{p1} and by θ_{p2}, respectively. Moreover, the angle from p_3 to s is $\theta_{p3} = \pi/2$. Hence, Eq. (11.38) yields

$$\underline{/F(s)} = -\theta_{p1} - \theta_{p2} - \theta_{p3} = -\theta_{p1} - \theta_{p2} - \pi/2 = (2k+1)\pi, \quad k = 0, \pm 1, \pm 2, \ldots \qquad \text{(d)}$$

But, the angle θ_{p1} is known, so that the only value of θ_{p2} that fits the situation is obtained for $k = -1$, or

$$\theta_{p2} = (\pi/2) - \theta_{p1} \qquad \text{(e)}$$

For the intersection with the imaginary axis, we substitute $s = i\omega$ in the characteristic equation, with the result

$$1 + G(i\omega)H(i\omega) = 1 + \frac{K}{i\omega[(i\omega)^2 + 0.8i\omega + 1]} = 0 \qquad \text{(f)}$$

which can be rewritten as follows:

$$(i\omega)^3 + 0.8(i\omega)^2 + i\omega + K = 0 \qquad \text{(g)}$$

Equating the imaginary part of Eq. (g) to zero, we obtain

$$(i\omega)^3 + i\omega = 0 \rightarrow \omega = \pm 1 \qquad \text{(h)}$$

so that the root loci intersect the imaginary axis at $s = \pm i$. To determine the value of the gain constant at the intersections, we equate the real part of Eq. (g) to zero and let $\omega = \pm 1$. This yields $K = 0.8$. The root loci are sketched in Fig. 11.12, and we note that the second asymptote is not a genuine asymptote but a root locus.

Example 11.3

Plot the root loci of the equation

$$s^2 + 2\zeta\omega_n s + \omega_n^2 = 0 \qquad \text{(a)}$$

as ζ varies from 0 to ∞.

In the first place, we must recast Eq. (a) in the form of Eq. (11.32). It is not difficult to verify that Eq. (a) can be reduced to Eq. (11.32) by letting

$$K = 2\zeta\omega_n, \qquad F(s) = \frac{s}{s^2 + \omega_n^2} \qquad \text{(b)}$$

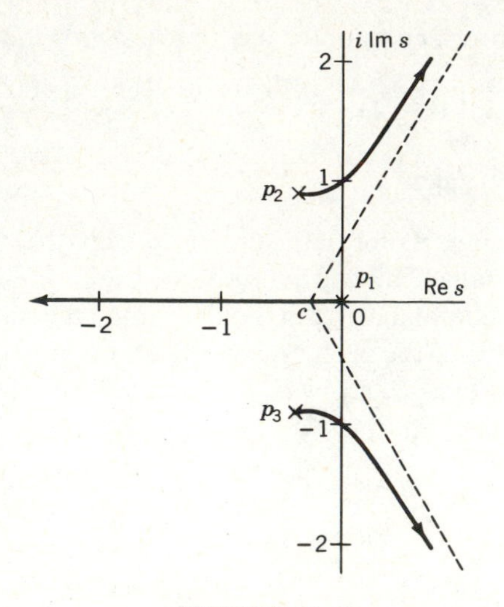

FIGURE 11.12

Hence, $F(s)$ has one zero, $m=1$, and two poles, $n=2$, or

$$z_1=0, \qquad p_1=i\omega_n, \qquad p_2=-i\omega_n \tag{c}$$

Application of the rules is carried out as follows:

i. By Rule 1, there are two root loci, one root locus starting from each pole.

ii. Because there is only one zero, by Rule 2, one root locus ends at the zero and the other one ends at infinity.

iii. The number of zeros is $m=1$ and the number of poles is $n=2$. Hence, by Rule 3, there are two branches.

iv. There are no poles and only one zero on the real axis, namely, at the origin. Hence, by Rule 5, the entire real axis to the left of the origin is a root locus. The nature of the root locus on the real axis is examined by means of a different rule.

v. Application of Rule 6 yields one asymptote coinciding with the negative real axis. But the negative real axis is a root locus, so that this is not a genuine asymptote.

vi. To check for the applicability of Rule 8, we write

$$\frac{dF(s)}{ds}=\frac{s^2+\omega_n^2-2s^2}{(s^2+\omega_n^2)^2}=\frac{\omega_n^2-s^2}{(s^2+\omega_n^2)^2}=0 \tag{d}$$

which yields

$$s_b=\pm\omega_n \tag{e}$$

Inserting $s_b=\omega_n$ in Eq. (11.32) and considering the second of Eqs. (b), we obtain

$$1+KF(\omega_n)=1+\frac{K}{2\omega_n}=0 \tag{f}$$

which requires that

$$K=-2\omega_n \to \zeta=-1 \tag{g}$$

Because ζ is a positive quality, the possibility $s_b=\omega_n$ must be discarded. On the other hand, $s_b=-\omega_n$ yields

$$K=2\omega_n \to \zeta=1 \tag{h}$$

so that there is a breakaway point and a double root at $s=-\omega_n$ corresponding to $\zeta=1$. Because there are two branches, the angle between the two branches is 180°. Hence, because of the symmetry of the root loci, one branch makes a 90° angle with respect to the real axis and the other branch makes a 270° angle. Note, in passing, that the breakaway point corresponds to critical damping (Section 4.6).

It remains to establish the angles of the root loci at the poles. To this end, we consider a trial point s in the neighborhood of p_1, as shown in Fig. 11.13, and use Eq. (11.38) to write the formula for the phase angle of $F(s)$

$$\underline{/F(s)}=\theta_{z1}-\theta_{p1}-\theta_{p2}=(2k+1)\pi, \qquad k=0, \pm 1, \pm 2, \ldots \tag{i}$$

From Fig. 11.13, we can write

$$\theta_{z1}\cong\frac{\pi}{2}, \qquad \theta_{p1}=\text{unknown}, \qquad \theta_{p2}\cong\frac{\pi}{2} \tag{j}$$

There is only one angle that fits the situation, or

$$\theta_{p1}=\pi \qquad \text{for} \quad k=-1 \tag{k}$$

so that the root locus starting at $p_1=i\omega_n$ is parallel to the negative real axis. Because of symmetry, the root locus starting at $p_2=-i\omega_n$ is also parallel to the negative real axis.

To complete the root loci between the poles and the breakaway point, one must use trial and error. Noticing that the tangents at the poles are normal to the imaginary axis and the tangents at the breakaway point are normal to the real axis and that the distance from the origin to the poles and to the breakaway point is the same, we have reasons to believe that the root loci between the poles and

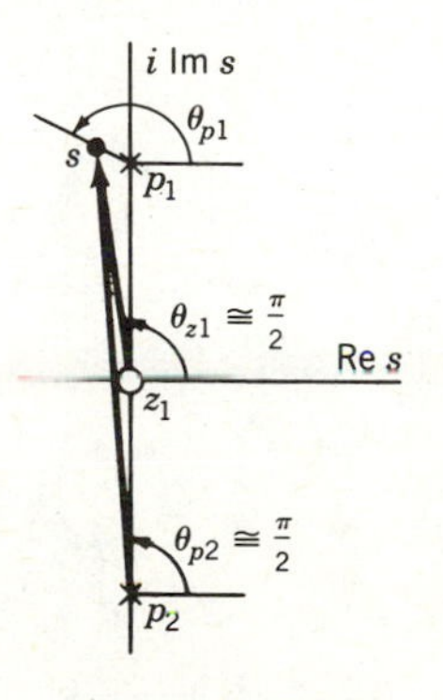

FIGURE 11.13

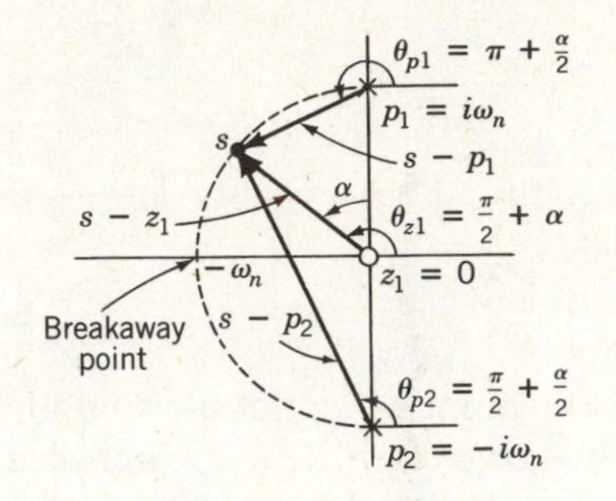

FIGURE 11.14

the breakaway point are quarter-circles. This belief turns out to be correct, as demonstrated by the construction in Fig. 11.14. From Fig. 11.14, the phase angles of $s-z_1$, $s-p_1$, and $s-p_2$ are

$$\theta_{z1}=\frac{\pi}{2}+\alpha, \qquad \theta_{p1}=\pi+\frac{\alpha}{2}, \qquad \theta_{p2}=\frac{\pi}{2}+\frac{\alpha}{2} \tag{l}$$

so that the phase angle of $F(s)$ is

$$\underline{/F(s)}=\theta_{z1}-\theta_{p1}-\theta_{p2}=\frac{\pi}{2}+\alpha-\left(\pi+\frac{\alpha}{2}\right)-\left(\frac{\pi}{2}+\frac{\alpha}{2}\right)=-\pi \tag{m}$$

Hence, Eq. (11.38) is satisfied, and s is indeed a point on the root locus. The root loci are plotted in Fig. 11.15.

It should be pointed out that the root loci of Fig. 11.15 were actually plotted earlier. Indeed, essentially the same plot is shown in Fig. 4.12, but in an entirely different context. In Fig. 4.12, the object was to examine how the roots of the characteristic equation vary as the damping varies. In Fig. 11.15, the object is to demonstrate how root-locus techniques can be used to study the behavior of the roots of a polynomial as one of the coefficients changes.

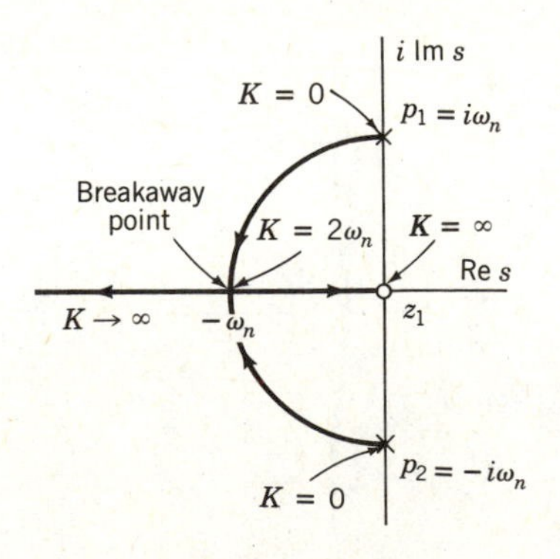

FIGURE 11.15

11.5 THE METHOD OF NYQUIST

The root-locus method is a procedure for the determination of the closed-loop poles based on the knowledge of the poles and zeros of the open-loop transfer function $G(s)H(s)$. At times, the function $G(s)H(s)$ is not available and must be determined experimentally. The Nyquist method is a technique permitting the determination of the degree of stability of a closed-loop system from frequency response data, obtained by subjecting the open-loop system to a sinusoidal force. The method can be explained conveniently by means of concepts from conformal mapping.

Let us consider a single-valued rational function $F(s)$ that is analytic everywhere in a given region in the s-plane, except at a finite number of points. In addition to the s-plane, we consider an $F(s)$-plane, so that to each value s_i in the complex s-plane there is a corresponding value $F(s_i)$ in the complex $F(s)$-plane $(i=1, 2, \ldots)$. If s moves clockwise around a closed contour C_s in the s-plane, then $F(s)$ moves around a closed contour C_F in the $F(s)$-plane in either the clockwise or the counter-clockwise sense, depending on the nature of $F(s)$. Figures 11.16*a* and 11.16*b* show the s-plane and $F(s)$-plane, respectively. In the contour C_F of Fig. 11.16*b*, $F(s)$ moves in the counterclockwise sense, but this is merely for the sake of illustration. Figures 11.16*a* and 11.16*b* represent a *mapping* of a contour in the s-plane to one in the $F(s)$-plane.

Next, consider the rational function

$$F(s)=\frac{(s-z_1)(s-z_2)\cdots(s-z_m)}{(s-p_1)(s-p_2)\cdots(s-p_n)} \tag{11.57}$$

where z_i $(i=1, 2, \ldots, m)$ are the zeros and p_j $(j=1, 2, \ldots, n)$ are the poles of $F(s)$. The object is to establish how the angle of $F(s)$ changes as s moves clockwise around a closed contour C_s in the s-plane. Figure 11.17 shows such a closed contour. The change in the angle of $F(s)$ depends on the location of the zeros and poles relative to the closed contour C_s. To demonstrate this, we consider one zero and one pole inside C_s and one zero and one pole outside C_s, as shown in Fig. 11.17. As s moves once clockwise around C_s, the vectors $s-z_1$ and $s-p_1$ undergo a clockwise angle change of 2π each. At the same time, the vectors $s-z_2$ and $s-p_2$ undergo no net angle change. The question is how to determine the corresponding change in the angle of $F(s)$. To this end, we recall from Section 11.4 that the complex function $F(s)$ can be expressed in the form

$$F(s)=|F(s)|e^{i\psi_F} \tag{11.58}$$

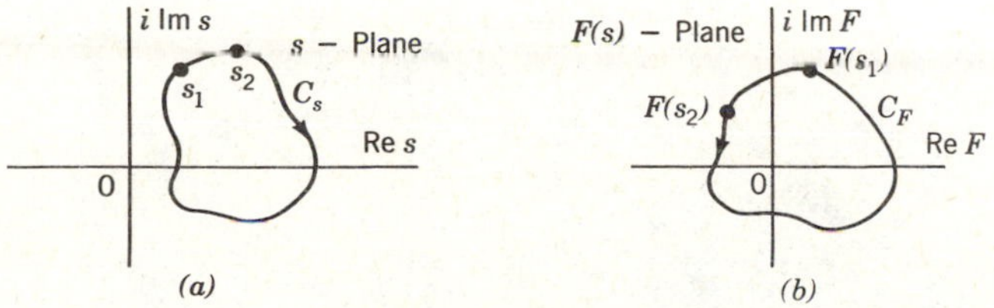

FIGURE 11.16

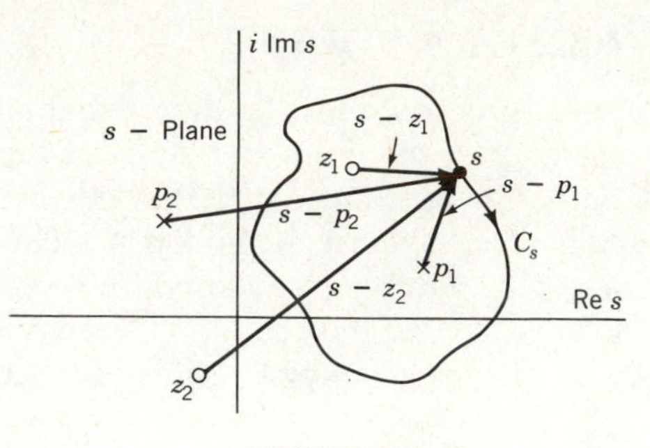

FIGURE 11.17

where $|F(s)|$ is the magnitude and ψ_F the phase angle of $F(s)$. This representation of $F(s)$ is displayed in Fig. 11.18. But, according to Eq. (11.38), the phase angle of $F(s)$ has the expression

$$\psi_F = \sum_{i=1}^{m} \theta_{zi} - \sum_{j=1}^{n} \theta_{pj} \tag{11.59}$$

where θ_{zi} and θ_{pj} are the phase angles of the vectors $s-z_i$ and $s-p_j$, respectively. To an increment Δs along C_s there are corresponding phase angle increments $\Delta\theta_{zi}\,(i=1,2,\ldots,m)$ and $\Delta\theta_{pj}\,(j=1,2,\ldots,n)$, so that the incremental change in the phase angle ψ_F is

$$\Delta\psi_F = \sum_{i=1}^{m} \Delta\theta_{zi} - \sum_{j=1}^{n} \Delta\theta_{pj} \tag{11.60}$$

As s completes one clockwise circuit around C_s, the phase angle change is $\Delta\theta_{zi}=2\pi$ if the zero z_i is inside C_s and $\Delta\theta_{zi}=0$ if z_i is outside C_s. Similarly, $\Delta\theta_{pj}=2\pi$ if the pole p_j is inside C_s and $\Delta\theta_{pj}=0$ if p_j is outside C_s. Hence, *if $F(s)$ has Z zeros and P poles inside a closed contour and s moves once clockwise around the contour, then $F(s)$ experiences a clockwise angle change of $2\pi(Z-P)$.* The preceding statement is known as the *principle of the argument*. Clearly, if $P>Z$, then the angle change experienced by $F(s)$ is in the counterclockwise sense.

Letting s move clockwise around a closed contour in the s-plane, we can determine the difference between the number of zeros and poles of $F(s)$ inside the contour by measuring the angle change in $F(s)$. Because an angle change of 2π

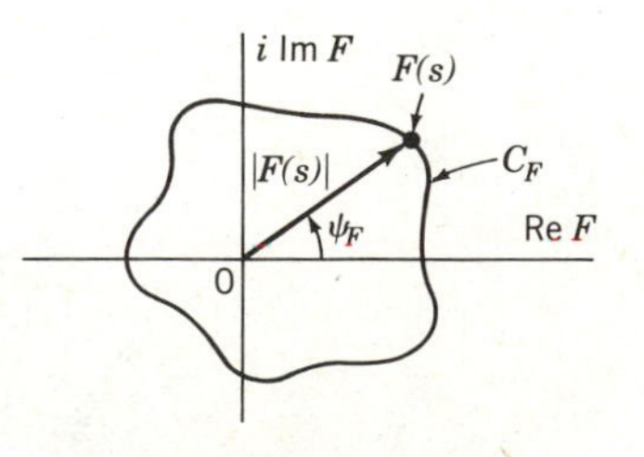

FIGURE 11.18

represents a clockwise encirclement of the origin in the $F(s)$-plane, we can determine the difference between the number of zeros and poles of $F(s)$ inside the closed contour by counting the number of encirclements of the origin in the $F(s)$-plane. This suggests a procedure for checking the stability of a system by using as a mapping the rational function

$$F_1(s) = 1 + KF(s) = 1 + G(s)H(s) \tag{11.61}$$

and by choosing a clockwise contour C_s enclosing the entire right half of the s-plane. Note that according to Eq. (11.6), $F_1(s)$ represents the denominator of the closed-loop transfer function, where $G(s)H(s)$ is the open-loop transfer function. Then, assuming that the number P of poles of $F_1(s)$ in the right half of the s-plane is known, we can obtain the number Z of zeros of $F_1(s)$ in the right half of the s-plane by using the principle of the argument and writing

$$Z = P + N \tag{11.62}$$

where N is the number of clockwise encirclements of the origin of the $F_1(s)$-plane. The $F_1(s)$-plane is shown in Fig. 11.19. But, the zeros of $F_1(s)$ are the poles of the closed-loop transfer function, so that Eq. (11.62) allows us to ascertain whether the system is stable or unstable, as discussed later in this section.

Instead of working with the $F_1(s)$-plane, it is more convenient to work with the $G(s)H(s)$-plane, shown in Fig. 11.20. From Eq. (11.61), we conclude that the $G(s)H(s)$-plane differs from the $F_1(s)$-plane by a translation only. Indeed, the origin of the $F_1(s)$-plane corresponds to the point $-1+i0$ in the $G(s)H(s)$-plane. Hence, stability can be checked by counting the number of encirclements of the point $-1+i0$ in the $G(s)H(s)$-plane instead of the number of encirclements of the origin of the $F_1(s)$-plane.

The method discussed above requires the open-loop transfer function $G(s)H(s)$, so that in this regard it has no advantage over the root-locus method. In fact, it provides less information about the system behavior, as the root-locus method provides not only a stability statement but also the exact location of the poles. The advantage of the approach discussed here is that by choosing a certain contour

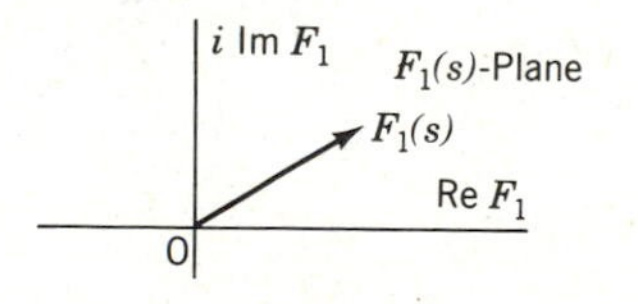

FIGURE 11.19

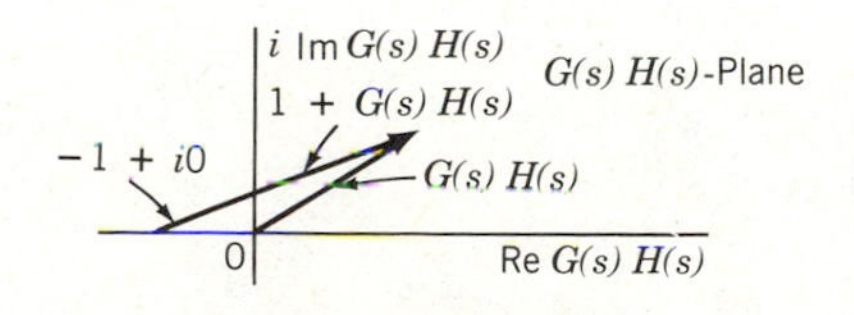

FIGURE 11.20

in the s-plane it is possible to apply the method when only the frequency response of the open-loop system is available. As pointed out earlier, this frequency response can be obtained experimentally by measuring the frequency response of the open-loop system to a sinusoidal excitation. Note that the frequency response represents the same concept as that discussed in Chapter 4 in conjunction with the vibration of a single-degree-of-freedom system.

Let us consider a contour C_s extending along the imaginary axis from the point $-iR$ to iR and then along the semicircle of radius R in the right half of the s-plane, as shown in Fig. 11.21. By letting $R \rightarrow \infty$, the closed contour C_s will enclose the entire right half-plane. The contour C_s is known as a *Nyquist path*. If we substitute the values of s along the contour C_s in $G(s)H(s)$, then the resulting plot in the $G(s)H(s)$-plane is known as a *Nyquist plot*. The Nyquist path of Fig. 11.21 implies that $G(s)H(s)$ has no poles on the imaginary axis. If $G(s)H(s)$ does have poles on the imaginary axis, then the contour C_s must be modified so as to exclude these poles, which can be done by means of small semicircles around the poles. The resulting path is shown in Fig. 11.22*a*.

The Nyquist plot can be expedited by considering certain features of the contour C_s in the s-plane. In the first place, the contour C_s is symmetric with respect to the real axis, so that to any value s in the upper half of the s-plane there corresponds the complex conjugate $\bar{s}$ in the lower half of the s-plane. Because $G(\bar{s})H(\bar{s})$ is the

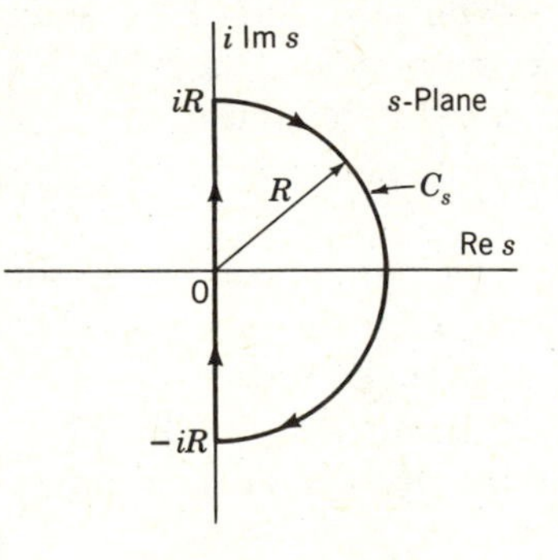

FIGURE 11.21

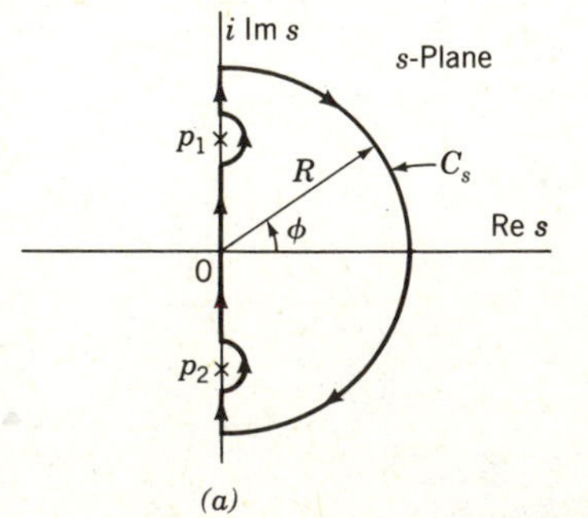

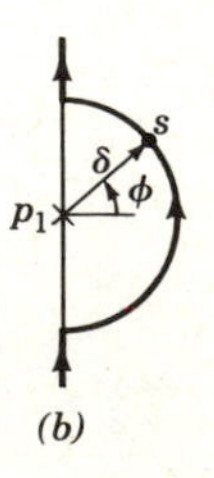

FIGURE 11.22

complex conjugate of $G(s)H(s)$, the Nyquist plot is also symmetric with respect to the real axis.

Next, we consider the behavior of $G(s)H(s)$ as s moves on the large semicircle. Recalling Eq. (11.31) and using Eq. (11.57), we can write

$$G(s)H(s)=KF(s)=K\frac{(s-z_1)(s-z_2)\cdots(s-z_m)}{(s-p_1)(s-p_2)\cdots(s-p_n)}=K\frac{s^m+\cdots}{s^n+\cdots} \tag{11.63}$$

On the large semicircle $s=R\exp i\phi$, where $R=|s|$ is the magnitude and ϕ is the phase angle of s, as shown in Fig. 11.22a. Because R is very large, corresponding to a typical point on the large semicircle, we have

$$G(s)H(s)\cong K\frac{R^m e^{im\phi}}{R^n e^{in\phi}}=KR^{m-n}e^{i(m-n)\phi} \tag{11.64}$$

But, if the number of poles exceeds the number zeros, $n>m$, we can write

$$\lim_{R\to\infty} G(s)H(s)=0 \tag{11.65}$$

Even in the case in which $n=m$, we obtain

$$\lim_{R\to\infty} G(s)H(s)=K=\text{const} \tag{11.66}$$

Hence, as s moves around the large semicircle, the value of $G(s)H(s)$ remains constant in both cases. It follows that the large semicircle maps as a single point in the $G(s)H(s)$-plane, so that *when the number of zeros of $G(s)H(s)$ does not exceed the number of poles, the large semicircle in the s-plane can be ignored.*

Another thing to consider is the behavior of $G(s)H(s)$ as s moves on the small semicircle around a pole. Let us assume that the pole p_1 shown in Fig. 11.22a has multiplicity r. Then, in the neighborhood of $s=p_1$, the open-loop transfer function can be approximated by

$$G(s)H(s)\cong\frac{K}{(s-p_1)^r} \tag{11.67}$$

Figure 11.22b shows the neighborhood of $s=p_1$ in greater detail. Letting δ be the radius of the small semicircle and ϕ the angle to the vector from p_1 to a typical point s on the semicircle, we can write

$$s=p_1+\delta e^{i\phi} \tag{11.68}$$

Hence, for a point s on the small semicircle around p_1, we have

$$G(s)H(s)\simeq\frac{K}{\delta^r}e^{-ir\phi} \tag{11.69}$$

as $\delta\to 0$, the magnitude of $G(s)H(s)$ becomes increasingly large. Moreover, as the angle in the s-plane changes counterclockwise from $\phi=-\pi/2$ to $\phi=\pi/2$, the angle of $G(s)H(s)$ changes clockwise from $r\pi/2$ to $-r\pi/2$.

Finally, over the imaginary axis the value of s is $i\omega$, where ω can take both positive and negative values. Hence, corresponding to the segments on the imaginary axis in the s-plane, we plot simply $G(i\omega)H(i\omega)$, which can be recognized as the frequency response of the open-loop system. As pointed out in the beginning of this section, $G(i\omega)H(i\omega)$ can be obtained experimentally by subjecting the open-loop system to a sinusoidal force. It turns out that the plots of $G(i\omega)H(i\omega)$ are the most significant parts of the Nyquist plot as far as the determination of the encirclements of the point $-1+i0$ is concerned.

Now that we have examined various aspects of the Nyquist plot, let us return to the question of system stability. For the system to be stable, the number of zeros of $1+G(s)H(s)$ in the right half of the s-plane must be zero, in which case Eq. (11.62) reduces to

$$P=-N \tag{11.70}$$

where P is the number of poles of $G(s)H(s)$ in the right half of the s-plane and N is the number of clockwise encirclements of the point $-1+i0$ in the $G(s)H(s)$-plane. But, a clockwise encirclement is the negative of a counterclockwise encirclement. Hence, Eq. (11.70) states that: *A closed-loop system is stable if the Nyquist plot of* $G(s)H(s)$ *encircles the point* $-1+i0$ *in a counterclockwise sense as many times as there are poles of* $G(s)H(s)$ *in the right half of the s-plane*. The preceding statement is known as the *Nyquist criterion*.

Example 11.4

Draw the Nyquist plot of the system of Example 11.1 and apply the Nyquist criterion to determine the stability of the system.

From Example 11.1, the open-loop transfer function has the form

$$G(s)H(s)=K\frac{s+0.3}{s^2(s+0.5)} \tag{a}$$

so that there is a double pole on the imaginary axis. Hence, the C_s contour in the s-plane is as shown in Fig. 11.23. For identification purposes, the segment $-R<\omega<-\delta$ on the imaginary axis is denoted by 1, the semicircle of radius δ and with the center at the origin by 2, the segment $\delta<\omega<R$ on the imaginary axis by 3, and the semicircle of radius R and with the center at the origin by 4. The mappings of these portions of C_s in the $G(s)H(s)$-plane will be denoted by the same numbers.

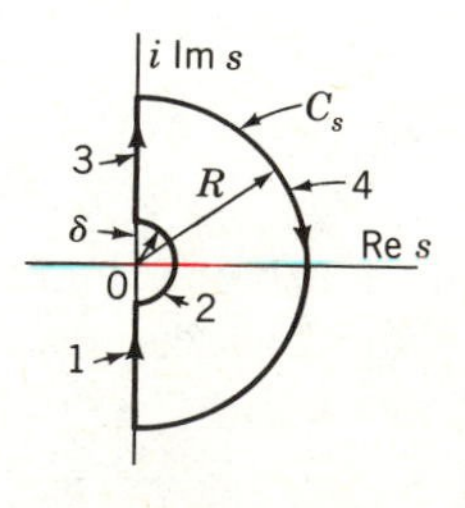

FIGURE 11.23

For any point on the imaginary axis, we replace s by $i\omega$ in Eq. (a) and obtain

$$G(i\omega)H(i\omega) = K\,\frac{i\omega+0.3}{(i\omega)^2(i\omega+0.5)} = -K\,\frac{(i\omega+0.3)(-i\omega+0.5)}{\omega^2(i\omega+0.5)(-i\omega+0.5)}$$

$$= -K\,\frac{\omega^2+0.15+i0.2\omega}{\omega^2(\omega^2+0.25)} \tag{b}$$

Inserting negative values of ω in Eq. (b), we obtain curve 1, and inserting positive values of ω in Eq. (b), we obtain curve 3, as shown in Fig. 11.24. On the small semicircle, we have $s=\delta \exp i\phi$, so that Eq. (a) reduces to

$$G(s)H(s)\bigg|_{\text{on } 2} = \lim_{\delta\to 0} K\,\frac{\delta e^{i\phi}+0.3}{\delta^2 e^{i2\phi}(\delta e^{i\phi}+0.5)} = \lim_{\delta\to 0}\frac{0.6K}{\delta^2}\,e^{-i2\phi} = \infty e^{-i2\phi} \tag{c}$$

As ϕ varies from $-\pi/2$ to $\pi/2$, a point on the small semicircle maps into a point on a circle of very large radius with the angle varying from π to $-\pi$. This large circle is denoted by 2 in Fig. 11.24. Finally, the semicircle of large radius R in the s-plane maps into a point at the origin in Fig. 11.24, which completes the Nyquist plot. Clearly, the point $-1+i0$ is not encircled by the Nyquist plot, so that the closed-loop system of Example 11.1 is stable. In fact, it is asymptotically stable.

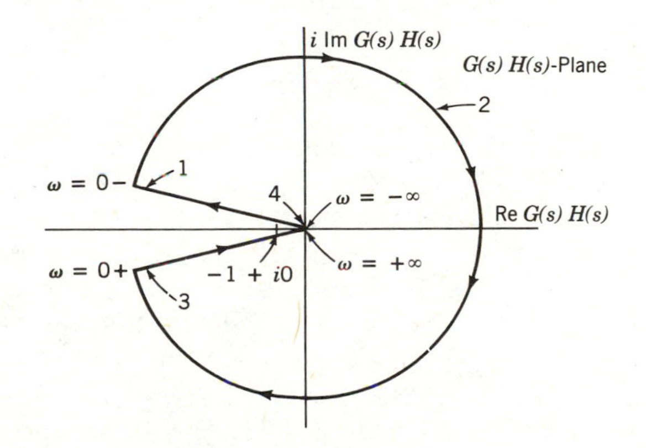

FIGURE 11.24

Example 11.5

Repeat the problem of Example 11.4 but with the zero at $z_1=-0.8$.

The open-loop transfer function in this case has the expression

$$G(s)H(s) = K\,\frac{s+0.8}{s^2(s+0.5)} \tag{a}$$

The contour C_s in the s-plane remains the same as in Fig. 11.23 and the same can be said about much of the Nyquist plot shown in Fig. 11.24, with one significant difference, namely, the curves 1 and 3 corresponding to the respective segments on the imaginary axis in the s-plane. To derive the new curves 1 and 3, we let $s=i\omega$

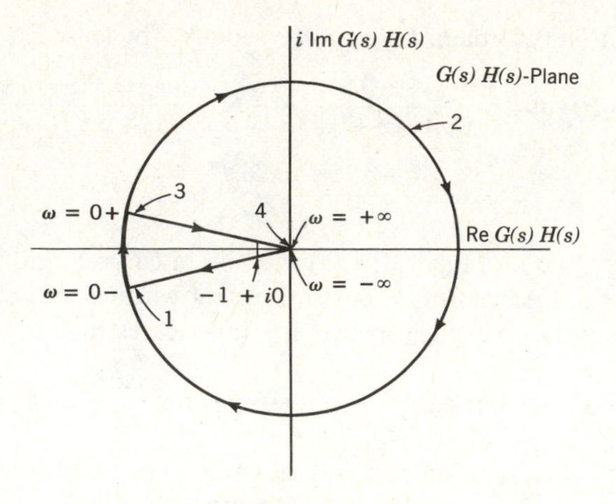

FIGURE 11.25

in Eq. (a) and obtain

$$G(i\omega)H(i\omega)=K\frac{i\omega+0.8}{(i\omega)^2(i\omega+0.5)}=-K\frac{(i\omega+0.8)(-i\omega+0.5)}{\omega^2(i\omega+0.5)(-i\omega+0.5)}$$

$$=-K\frac{\omega^2+0.4-i0.3\omega}{\omega^2(\omega^2+0.25)} \qquad \text{(b)}$$

The new curves 1 and 3 are obtained from Eq. (b) by using negative and positive values of ω, respectively. The new Nyquist plot is shown in Fig. 11.25. This time, however, the plot does encircle the point $-1+i0$, so that we cannot use the Nyquist criterion to ascertain stability. In fact, the system is unstable. To demonstrate this, we turn to Eq. (11.62). From Eq. (a), we conclude that $G(s)H(s)$, and hence $1+G(s)H(s)$, has no poles on the right half of the s-plane, so that $P=0$. Because the Nyquist plot encircles the point $-1+i0$ twice in a clockwise sense, $N=2$ and it follows from Eq. (11.62) that $1+G(s)H(s)$ has two zeros in the right half of the s-plane, so that the system is unstable. Hence, too much integral control can render the system unstable.

It is easy to verify that for stability we must have

$$0\leqslant K_2<0.5K_1 \qquad \text{(c)}$$

where K_1 and K_2 have been defined in Example 11.1. Note that for $K_2=0.5K_1$, which implies a zero with the value $z_1=-0.5$, there is a pole–zero cancellation, and the transfer function reduces to

$$G(s)H(s)=\frac{K}{s^2} \qquad \text{(d)}$$

It follows that for $K_2=0.5K_1$ the system is merely stable, which can be regarded as the borderline case separating asymptotic stability and instability.

11.6 BODE DIAGRAMS

In Chapters 1 and 4, we observed that the steady-state response of systems to sinusoidal excitation is also sinusoidal and has the same frequency as the excitation. Indeed, the differences between the input and output lie only in amplitude and phase angle. We recall from Chapter 4 that it is more convenient to characterize steady-state response not in terms of functions of time but in terms of functions of the excitation frequency. This characterization of the response consists of plots of the magnitude versus frequency and of the phase angle versus frequency. Such plots are referred to as *frequency response*. The frequency response plots used in Chapter 4 were in terms of rectangular coordinates. On the other hand, the Nyquist plots of the open-loop transfer function discussed in Section 11.5 and corresponding to the imaginary axis in the s-plane represent frequency response plots in terms of polar coordinates. Frequency response can also be plotted in terms of logarithmic coordinates. Logarithmic plots of the frequency response are known as *Bode plots*. As in Chapter 4, there are two such plots, namely, magnitude versus frequency and phase angle versus frequency. There are several advantages to logarithmic plots of the frequency response compared to rectangular plots. One advantage is that logarithmic plots permit a much larger range of the excitation frequency. Another advantage is that logarithmic plots are easier to construct. Indeed, ordinarily the frequency response is a rational function with the numerator and denominator consisting of products of various factors. In plotting the logarithm of the frequency response each factor can be plotted separately and then added or subtracted, depending on whether the factor is in the numerator or denominator, respectively. As with the root-locus plots and Nyquist plots, Bode plots also enable us to draw conclusions about the stability of a closed-loop system by working with the open-loop transfer function.

Let us consider the case in which the feedback-path transfer function $H(s)$ is equal to unity, in which case the open-loop transfer function $G(s)H(s)$ reduces to the forward-path transfer function $G(s)$. The frequency response is obtained by replacing s in $G(s)$ by $i\omega$. Because the frequency response is a complex function, it can be written in the form

$$G(i\omega)=|G(i\omega)|e^{i\psi(\omega)} \tag{11.71}$$

where $|G(i\omega)|$ is the magnitude, often referred to as *gain*, and $\psi(\omega)$ is the phase angle. The gain and the phase angle have the expressions

$$|G(i\omega)|=\sqrt{\mathrm{Re}^2[G(i\omega)]+\mathrm{Im}^2[G(i\omega)]}, \qquad \psi(\omega)=\tan^{-1}\frac{\mathrm{Im}\ G(i\omega)}{\mathrm{Re}\ G(i\omega)} \tag{11.72a, b}$$

The natural logarithm of $G(i\omega)$ is

$$\ln G(i\omega)=\ln|G(i\omega)|+i\psi(\omega) \tag{11.73}$$

where $\ln|G(i\omega)|$ is the magnitude. It is customary to work with logarithms to the base 10 instead of natural logarithms, where the first is denoted by log, so that

Eq. (11.73) is replaced by

$$\log G(i\omega) = \log|G(i\omega)| + i0.434\psi(\omega) \tag{11.74}$$

where $\log|G(i\omega)|$ is called the *logarithmic gain*. The factor 0.434 is ordinarily omitted from the complex part of Eq. (11.74). The logarithmic gain is commonly expressed in *decibels*, where the decibel (dB) represents a unit, so that for a given number N

$$N_{\mathrm{dB}} = 20 \log N \quad \text{(dB)} \tag{11.75}$$

Hence, the logarithmic gain in decibels is

$$|G(i\omega)|_{\mathrm{dB}} = 20 \log|G(i\omega)| \quad \text{(dB)} \tag{11.76}$$

There are two units used to express the frequency band from ω_1 to ω_2 or the ratio ω_2/ω_1 of the two frequencies. The frequencies ω_1 and ω_2 are separated by an *octave* if $\omega_2/\omega_1 = 2$. Then, the number of octaves between any two arbitrary frequencies ω_1 and ω_2 is

$$\frac{\log(\omega_2/\omega_1)}{\log 2} = \frac{1}{0.301} \log \frac{\omega_2}{\omega_1} = 3.32 \log \frac{\omega_2}{\omega_1} \quad \text{(octaves)} \tag{11.77}$$

Similarly, the frequencies ω_1 and ω_2 are separated by a *decade* if $\omega_2/\omega_1 = 10$. Then, the number of decades between any two arbitrary frequencies ω_1 and ω_2 is

$$\frac{\log(\omega_2/\omega_1)}{\log 10} = \log \frac{\omega_2}{\omega_1} \quad \text{(decades)} \tag{11.78}$$

The units of octave and decade are commonly used to express the slope of straight lines in Bode plots.

Next, let us consider a frequency response function in the general form

$$G(i\omega) = K \frac{\prod_k (1 + i\omega\tau_k)}{(i\omega)^l \prod_m (1 + i\omega\tau_m) \prod_n [1 - (\omega/\omega_n)^2 + i2\zeta\omega/\omega_n]} \tag{11.79}$$

where K is a positive real gain constant, τ_k and τ_m are real time constants, ζ is a damping factor, and ω_n are natural frequencies. Then, the logarithmic gain in decibels has the simple form

$$\begin{aligned} |G(i\omega)|_{\mathrm{dB}} = 20[\log K &+ \sum_k \log|1 + i\omega\tau_k| - l \log|i\omega| \\ &- \sum_m \log|1 + i\omega\tau_m| - \sum_n \log|1 - (\omega/\omega_n)^2 + i2\zeta\omega/\omega_n|] \quad \text{(dB)} \end{aligned} \tag{11.80}$$

so that the logarithmic gain consists of a summation of the logarithms of the individual factors. Moreover, the phase angle of $G(i\omega)$ has the expression

$$\begin{aligned} \psi(\omega) = \angle G(i\omega) = \angle K &+ \sum_k \angle 1 + \omega\tau_k - l\angle i\omega - \sum_m \angle 1 + i\omega\tau_m \\ &- \sum_n \angle 1 - (\omega/\omega_n)^2 + i2\zeta\omega/\omega_n \end{aligned} \tag{11.81}$$

In plotting the logarithmic gain one can plot the factors separately and then add or subtract them according to their sign in the summation. From Eq. (11.80), we observe that the logarithmic gain has only four types of factors: (i) a constant factor K, (ii) poles at the origin, $i\omega$, (iii) poles or zeros at $\omega \neq 0$, $(1+i\omega\tau)^{\pm 1}$, and (iv) complex poles, $[1-(\omega/\omega_n)^2+i2\zeta\omega/\omega_n]^{-1}$. Hence, a Bode plot can be constructed with relative ease. It is customary to use $\log \omega$ as the abscissa. In the following, we construct Bode plots for the four types of factors.

i. Constant Factor

Because $K = \text{const}$,

$$K_{\text{dB}} = 20 \log K = \text{const} \quad (\text{dB}) \tag{11.82}$$

so that the plot of the logarithmic gain is simply a horizontal straight line. Under the earlier assumption that K is positive, the constant logarithmic gain raises the logarithmic gain of the complete transfer function by that amount. Under the same assumption, if we use Eq. (11.72b), we conclude that the phase angle plot is the horizontal straight line

$$\underline{/K} = \tan^{-1} \frac{0}{K} = \tan^{-1} 0 = 0° \tag{11.83}$$

ii. Poles at the Origin, $i\omega$

The logarithmic gain of $i\omega$ in decibels is

$$|i\omega|_{\text{dB}} = 20 \log|i\omega| = 20 \log \omega \quad (\text{dB}) \tag{11.84}$$

and it represents a straight line passing through point 0 dB at $\omega = 1$. The slope of the line is

$$\frac{d(20 \log \omega)}{d \log \omega} = 20 \quad (\text{dB/decade}) \tag{11.85}$$

From Eqs. (11.77) and (11.78), we conclude that

$$\text{one} \quad \text{decade} = \frac{1}{0.301} \quad \text{octaves} \tag{11.86}$$

so that

$$20 \quad \text{dB/decade} = 20 \times 0.301 = 6.02 \quad \text{dB/octave} \tag{11.87}$$

Using Eq. (11.72b), the phase angle is simply

$$\underline{/i\omega} = \tan^{-1} \frac{\omega}{0} = \tan^{-1} \infty = 90° \tag{11.88}$$

In Eq. (11.78), we considered the factor $(i\omega)^{-l}$. For such a factor, the logarithmic gain is

$$|(i\omega)^{-l}|_{\text{dB}} = -20l \log|i\omega| = -20l \log \omega \quad (\text{dB}) \tag{11.89}$$

which is a straight line through the point 0 dB at $\omega = 1$ with the slope

$$\frac{d(-20l \log \omega)}{d \log \omega} = -20l \quad \text{(dB/decade)} \tag{11.90}$$

The phase angle of $(i\omega)^{-l}$ is

$$\underline{/(i\omega)^{-l}} = -l \underline{/i\omega} = -l \times 90^\circ \tag{11.91}$$

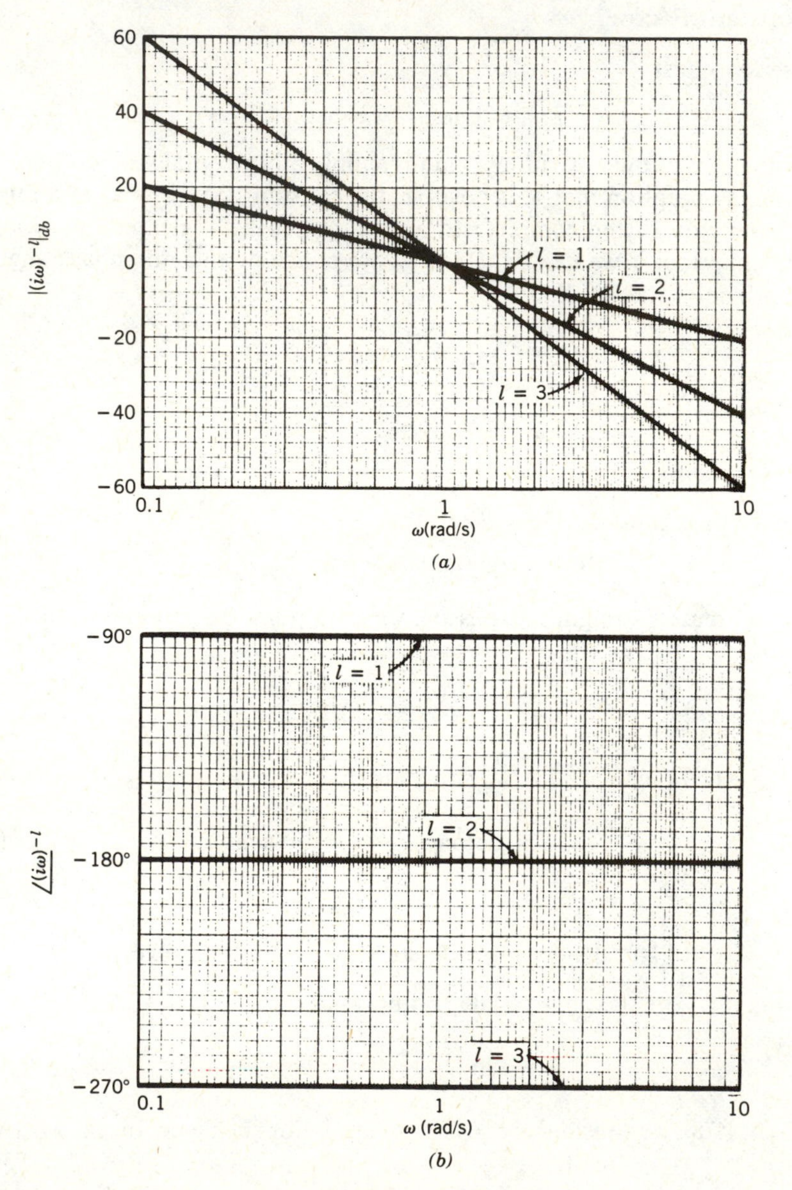

FIGURE 11.26

The logarithmic gain and phase angle of $i\omega$ and $(i\omega)^{-l}$ for several values of l are shown in Figs. 11.26*a* and 11.26*b*. Note that the plots use semilog coordinates.

iii. Factors $(1+i\omega\tau)^{\pm 1}$

The logarithmic gain of the factor $1+i\omega\tau$ is

$$|1+i\omega\tau|_{\text{dB}} = 20\log|1+i\omega\tau| = 20\log\sqrt{1+\omega^2\tau^2} \quad \text{(dB)} \tag{11.92}$$

To construct the Bode plot, it is convenient to consider two limiting cases. In the first case $\omega\tau$ is very small, and in the second case $\omega\tau$ is very large. The curves corresponding to the two cases are straight lines representing asymptotes for the curve. Hence, we have

$$\lim_{\omega\tau\to 0} |1+i\omega\tau|_{\text{dB}} = \lim_{\omega\tau\to 0} 20\log\sqrt{1+\omega^2\tau^2} = 20\log 1 = 0 \quad \text{(dB)} \tag{11.93}$$

and

$$\lim_{\omega\to\infty} |1+i\omega\tau|_{\text{dB}} = \lim_{\omega\tau\to\infty} 20\log\sqrt{1+\omega^2\tau^2} = 20\log\sqrt{\omega^2\tau^2} = 20\log\omega\tau \quad \text{(dB)} \tag{11.94}$$

From Eq. (11.93), the asymptote corresponding to very small $\omega\tau$ is a horizontal straight line passing through the origin. It is convenient to plot the logarithmic gain as a function of $\omega\tau$ instead of ω. In this case, we conclude from Eq. (11.94) that the asymptote corresponding to very large $\omega\tau$ is a straight line with the slope 20 dB/decade. The intersection of the two asymptotes can be obtained by equating Eqs. (11.93) and (11.94). Hence, the intersection is given by

$$\omega_{\text{cf}}\tau = 1 \tag{11.95}$$

where

$$\omega_{\text{cf}} = \frac{1}{\tau} \tag{11.96}$$

is known as the *corner frequency*. A rough approximation of the curve consists of the two asymptotes. Figure 11.27*a* shows both the approximate and the exact Bode plots for $\tau = 1$.

The logarithmic gain of the factor $(1+i\omega\tau)^{-1}$ is simply the negative of that given by Eq. (11.92). As a result, the Bode plot of the factor $(1+i\omega\tau)^{-1}$ is the mirror image of the Bode plot of the factor $1+i\omega\tau$, as shown in Fig. 11.27*a*.

The phase angle of the factor $1+i\omega\tau$ is obtained from Eq. (11.72b) in the form

$$\underline{/1+i\omega\tau} = \tan^{-1}\omega\tau \tag{11.97}$$

The plot of the phase angle can also be approximated by straight lines. Figure 11.27*b* shows an approximated plot consisting of three lines, a horizontal line at $\underline{/1+i\omega\tau} = 0°$ for $\omega\tau < 0.1$, a straight line connecting the point $\underline{/1+i\omega\tau} = 0°$, $\omega\tau = 0.1$ with the point $\underline{/1+i\omega\tau} = 90°$, $\omega\tau = 10$, and a horizontal line at $\underline{/1+i\omega\tau} = 90°$ for $\omega\tau > 10$. The exact plot is also shown in Fig. 11.27*b*. Note that the phase angle has the value of 45° at the corner frequency.

The phase angle of the factor $(1+i\omega\tau)^{-1}$ is the negative of the phase angle of $1+i\omega\tau$. It is plotted in Fig. 11.27*b* as well.

The frequency plots of the factor $(1+i\omega\tau)^{-1}$ of Figs. 11.27*a* and 11.27*b* are not entirely new, although one may not be able to recognize them immediately. Indeed, the factor $(1+i\omega\tau)^{-1}$ represents the frequency response of a first-order

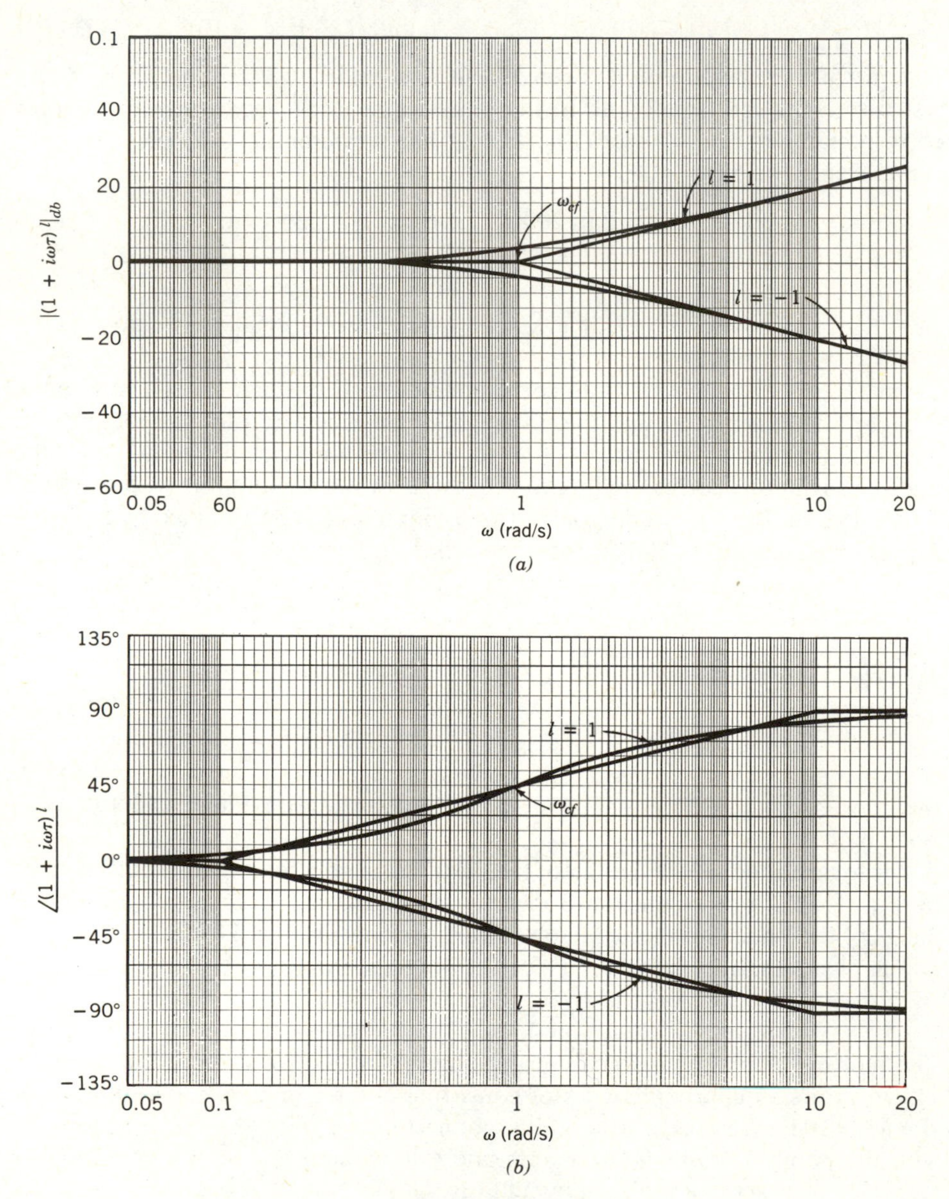

FIGURE 11.27

system and was first discussed in Section 4.8, where we also plotted the magnitude and phase angle of the factor versus $\omega\tau$. In Section 4.8, however, frequency response plots were in terms of ordinary rectangular coordinates, instead of the logarithmic coordinates used here, as can be seen in Figs. 4.16 and 4.17. We also note that in Section 4.8 the phase angle was defined as the negative of the one defined here.

iv. Factor $[1-(\omega/\omega_n)^2+i2\zeta\omega/\omega_n]^{-1}$

We must distinguish between two cases, $\zeta>1$ and $\zeta<1$. In the case $\zeta>1$, we can write

$$\frac{1}{1-(\omega/\omega_n)^2+i2\zeta\omega/\omega_n}=\frac{1}{(\zeta-\sqrt{\zeta^2-1}+i\omega/\omega_n)(\zeta+\sqrt{\zeta^2-1}+i\omega/\omega_n)} \tag{11.98}$$

so that the quadratic factor $[1-(\omega/\omega_n)^2+i2\zeta\omega/\omega_n]^{-1}$ can be expressed as the product of two first-order factors of the type just discussed. When $\zeta<1$, the factor $[1-(\omega/\omega_n)^2+i2\zeta\omega/\omega_n]^{-1}$ can once again be expressed as the product of two first-order factors, but they are more complicated than the type discussed earlier, so that it is more advantageous to consider the original second-order factor without factorization.

The logarithmic gain of $[1-(\omega/\omega_n)^2+i2\zeta\omega/\omega_n]^{-1}$ is

$$\begin{aligned}|[1-(\omega/\omega_n)^2+i2_\zeta\omega/\omega_n]^{-1}|_{\text{dB}}&=-20\log|1-(\omega/\omega_n)^2+i2\zeta\omega/\omega_n|\\&=-20\log\{[1-(\omega/\omega_n)^2]^2+(2\zeta\omega/\omega_n)^2\}^{1/2}\quad(\text{dB})\end{aligned} \tag{11.99}$$

The logarithmic plot has two asymptotes corresponding to small values and large values of ω/ω_n. For small ω/ω_n, we can write

$$\begin{aligned}&\lim_{\omega/\omega_n\to 0}|[1-(\omega/\omega_n)^2+i2\zeta\omega/\omega_n]^{-1/2}|_{\text{dB}}\\&\quad=\lim_{\omega/\omega_n\to 0}[-20\log\{[1-(\omega/\omega_n)^2]^2+(2\zeta\omega/\omega_n)^2\}^{1/2}]\\&\quad=-20\log 1=0\quad(\text{dB})\end{aligned} \tag{11.100}$$

and for large ω/ω_n, we obtain

$$\begin{aligned}&\lim_{\omega/\omega_n\to\infty}|[1-(\omega/\omega_n)^2+i2\zeta\omega/\omega_n]^{-1/2}|_{\text{dB}}\\&\quad=\lim_{\omega/\omega_n\to\infty}[-20\{\log[1-(\omega/\omega_n)^2]^2+(2\zeta\omega/\omega_n)^2\}^{1/2}]\\&\quad=-20\log[(\omega/\omega_n)^4]^{1/2}=-40\log\omega/\omega_n\quad(\text{dB})\end{aligned} \tag{11.101}$$

Hence, the low-frequency asymptote is a horizontal straight line through the origin and the high-frequency asymptote is a straight line with the slope of -40 dB/decade, where the latter is obtained by taking the derivative of Eq. (11.101) with respect to $\log\omega/\omega_n$. The corner frequency lies at the intersection of the two asymptotes. Equating Eqs. (11.100) and (11.101), we obtain the corner frequency

$$\omega_{cf}=\omega_n \tag{11.102}$$

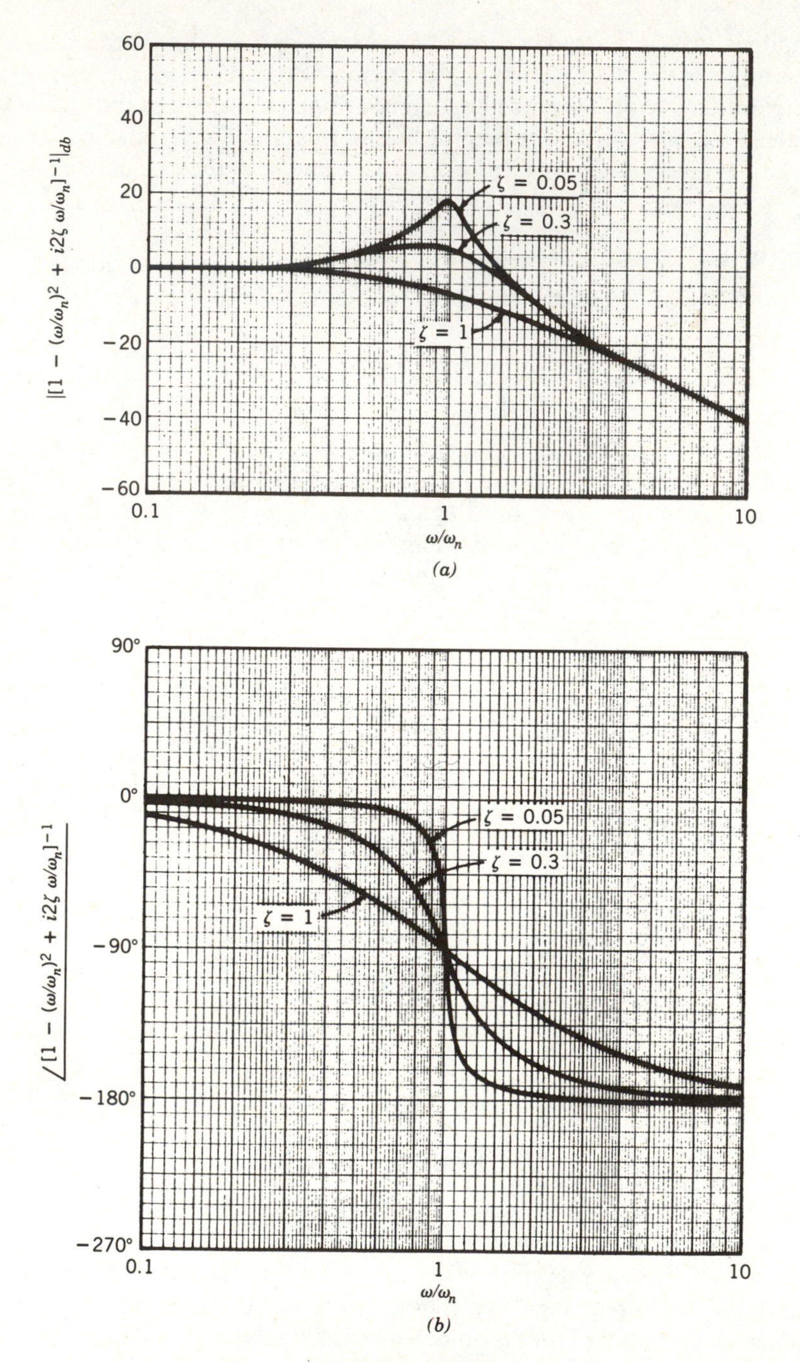
60
40
20
0
−20
−40
−60
$|[1 - (\omega/\omega_n)^2 + i2\zeta\,\omega/\omega_n]^{-1}|_{db}$
$\zeta = 0.05$
$\zeta = 0.3$
$\zeta = 1$
0.1
1
10
ω/ω_n
(a)
90°
0°
−90°
−180°
−270°
$\angle [1 - (\omega/\omega_n)^2 + i2\zeta\,\omega/\omega_n]^{-1}$
$\zeta = 0.05$
$\zeta = 0.3$
$\zeta = 1$
0.1
1
10
ω/ω_n
(b)

FIGURE 11.28

The asymptotes do not play such an important role in the case of the quadratic factor as for first-order factors. Indeed, for small ζ, the Bode plots deviate substantially from the asymptotes in the neighborhood of $\omega=\omega_n$, so that the asymptotes are more an item of academic interest than an aid for the construction of Bode plots. This statement can be verified by examining the Bode plots shown in Fig. 11.28*a* for a variety of values of the parameter ζ.

To obtain the phase angle of the factor $[1-(\omega/\omega_n)^2+i2\zeta\omega/\omega_n]^{-1}$, we write the factor in the form

$$\frac{1}{1-(\omega/\omega_n)^2+i2\zeta\omega/\omega_n}=\frac{1-(\omega/\omega_n)^2-i2\zeta\omega/\omega_n}{\{[1-(\omega/\omega_n)^2]^2+(2\zeta\omega/\omega_n)^2\}^{1/2}} \tag{11.103}$$

so that using Eq. (11.72b) we obtain

$$\begin{aligned}\underline{/[1-(\omega/\omega_n)^2+i2\zeta\omega/\omega_n]^{-1}}&=\tan^{-1}\frac{-2\zeta\omega/\omega_n}{1-(\omega/\omega_n)^2}\\&=-\tan^{-1}\frac{2\zeta\omega/\omega_n}{1-(\omega/\omega_n)^2}\end{aligned} \tag{11.104}$$

The phase angle is plotted in Fig. 11.28*b* for various values of the parameter ζ.

As with the first-order factor discussed above, the frequency response plots of Figs. 11.28*a* and 11.28*b* have been encountered earlier in a somewhat different form. In particular, the frequency response for a second-order system was discussed in Section 4.9, and the magnitude and phase angle were plotted in terms of rectangular coordinates in Figs. 4.18 and 4.19. Once again we observe that the phase angle used in Section 4.9 is the negative of the one defined here.

At this point it is perhaps appropriate to examine the role of the gain constant K in determining the system stability by the various graphical procedures discussed. In the case of root-locus plots, the constant K played the role of a parameter, taking values from 0 to ∞. Hence, the root-locus method presents a broad picture of the system stability. To obtain this picture, one must use trial and error to locate the roots. Of course, the problem is obviated by the use of a digital computer program to obtain the root loci. In the case of Nyquist diagrams, there are cases in which the gain constant K plays no role at all, as can be concluded from Examples 11.4 and 11.5. In Example 11.4, the plot of $G(s)H(s)$ did not intersect the negative real axis, and in Example 11.5 it intersected the axis at $-\infty$. In cases in which the intersection is not far from the point $-1+i0$, a change in K can bring about a change in the stability characteristics, as we shall see in the next section. Bode diagrams are the simplest to draw, but the diagrams are for a specific value of K. This limitation is not so severe as it may seem, because by increasing K one simply shifts the entire plot of the logarithmic gain upward and vice versa. In this regard, the role of K in determining the system stability is quite clear.

Example 11.6

Draw the Bode plots for the system of Example 11.1 corresponding to $K=10$.

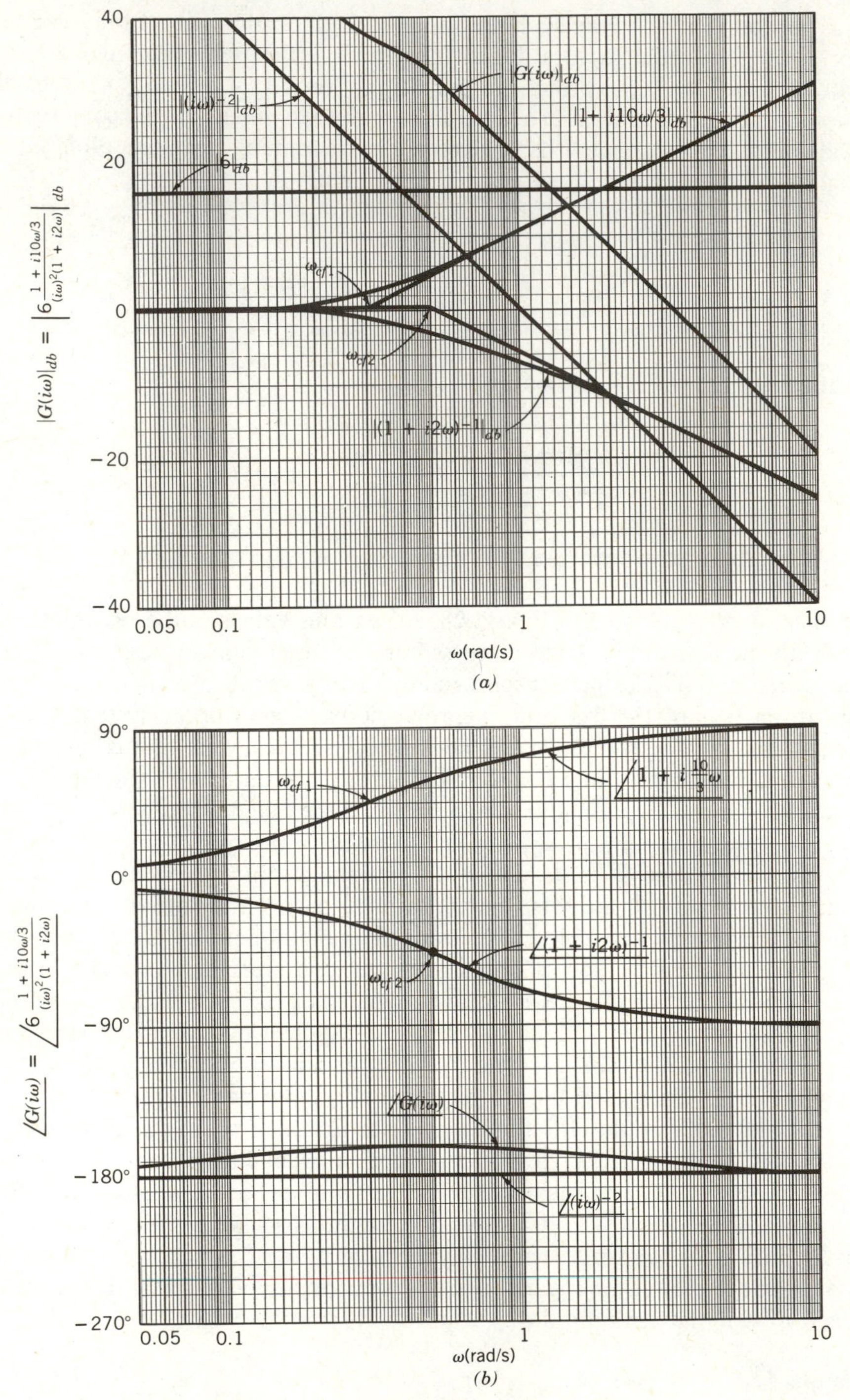
40
20
0
-20
-40
0.05
0.1
1
10
ω(rad/s)
(a)
$|G(i\omega)|_{db}$
$|(i\omega)^{-2}|_{db}$
$|1+i10\omega/3|_{db}$
$|6|_{db}$
ω_{cf1}
ω_{cf2}
$|(1+i2\omega)^{-1}|_{db}$
$|G(i\omega)|_{db} = \left|6\frac{1+i10\omega/3}{(i\omega)^2(1+i2\omega)}\right|_{db}$
90°
0°
-90°
-180°
-270°
0.05
0.1
1
10
ω(rad/s)
(b)
ω_{cf1}
$\angle 1+i\frac{10}{3}\omega$
$\angle(1+i2\omega)^{-1}$
ω_{cf2}
$\angle G(i\omega)$
$\angle(i\omega)^{-2}$
$\angle G(i\omega) = \angle 6\frac{1+i10\omega/3}{(i\omega)^2(1+i2\omega)}$

FIGURE 11.29

From Example 11.1, we obtain the transfer function

$$G(s)=10\frac{s+0.3}{s^2(s+0.5)} \tag{a}$$

Introducing $s=i\omega$ into Eq. (a), we obtain the frequency response

$$G(i\omega)=6\frac{1+(1/0.3)i\omega}{(i\omega)^2[1+(1/0.5)i\omega]}=6\frac{1+i(10/3)\omega}{(i\omega)^2(1+i2\omega)} \tag{b}$$

so that we distinguish four factors: the constant factor 6, the double pole at the origin $(i\omega)^{-2}$, the first-order factor $1+i(10/3)\omega$, and the first-order factor $(1+i2\omega)^{-1}$. We first plot the curves for the four factor separately and then combine them into one diagram.

The magnitude plot of the constant factor 6 is a horizontal line at 20 log 6 dB = 15.5630 dB. According to Eqs. (11.89) and (11.90), the magnitude plot of the factor $(i\omega)^{-2}$ is a straight line passing through 0 dB and $\omega=1$ and with the slope of -40 dB/decade. The magnitude plot of the factor $1+i(10/3)\omega$ has a corner frequency at $\omega_{cf1}=0.3$ and two asymptotes. The first asymptote coincides with the 0 dB axis for $\omega<\omega_{cf1}=0.3$ and the second asymptote is a straight line passing through the point 0 dB, $\omega=0.3$ and with the slope of 20 dB/decade. Finally, the magnitude plot of the factor $(1+i2\omega)^{-1}$ has a corner frequency at $\omega_{cf2}=0.5$ and two asymptotes, one coinciding with the 0-dB axis for $\omega<\omega_{cf2}=0.5$ and the second being a straight line passing through the point 0 dB, $\omega=0.5$ and having the slope of -20 dB/decade. The individual magnitude plots and the combined plot are shown in Fig. 11.29*a*.

The phase angle of the constant factor 6 is a horizontal line at 0°. According to Eq. (11.91), the phase angle of $(i\omega)^{-2}$ is a horizontal line at $-180°$. From Eq. (11.97), the phase angle of the factor $1+i(10/3)\omega$ varies as $\tan^{-1}(10/3)\omega$ between the horizontal lines at 0° and 90° and has the value of 45° at the corner frequency $\omega_{cf1}=0.3$. On the other hand, the phase angle varies as $-\tan^{-1} 0.5\omega$ between the horizontal lines at 0° and $-90°$ and has the value $-45°$ at the corner frequency $\omega_{cf2}=0.5$. The individual phase angle plots and the combined plot are shown in Fig. 11.29*b*.

11.7 POSITION SERVOMECHANISM

In Section 11.3, we investigated a simplified model of a position control system in which the torque applied by the motor was assumed to be proportional to the error signal. The model is shown in Fig. 11.4*a*. In this section, we consider the same position control system, but with one exception. In particular, we replace the simple model of the motor by a more elaborate one. The new model is shown schematically in Fig. 11.30, and it represents a field-controlled dc motor.

According to Kirchhoff's voltage law, the sum of the voltage drops in the elements must be equal to the field voltage (see Section 4.3). Denoting the field voltage by $v_f(t)$ and the field current by $i_f(t)$ and referring to Section 4.3, we can write

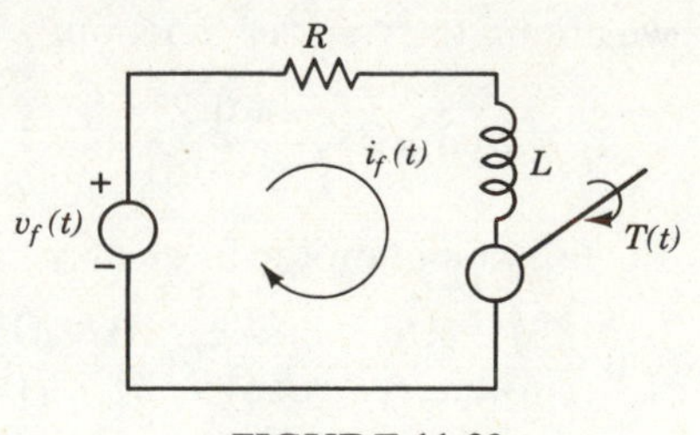

FIGURE 11.30

$$L\frac{di_f(t)}{dt}+Ri_f(t)=v_f(t) \tag{11.105}$$

where L is the inductance and R the resistance. The relation between the error signal $e(t)$ and the field voltage is assumed to have the simple linear form

$$v_f(t)=K_v e(t) \tag{11.106}$$

where K_v is the amplifier constant. Inserting Eq. (11.106) into Eq. (11.105), we obtain the differential equation

$$L\frac{di_f(t)}{dt}+Ri_f(t)=K_v e(t) \tag{11.107}$$

which relates the field current to the error signal.

The torque produced by the motor is proportional to the air-gap flux, which in turn is proportional to the field current. Hence, we can write the relation

$$T(t)=K_i i_f(t) \tag{11.108}$$

where $T(t)$ is the torque and K_i is the torque constant. It remains to provide a relation between the motor torque and the output angle $\theta_o(t)$. This relation is the moment equation for the rotational motion about a fixed axis. Denoting by J the combined mass polar moment of inertia of the wheel and the motor rotor and by c the combined coefficient of viscous damping due to friction in the sleeve and in the motor, the moment equation has the form

$$J\frac{d^2\theta_o(t)}{dt^2}+c\frac{d\theta_o(t)}{dt}=T(t) \tag{11.109}$$

which, except for a redefinition of the parameters J and c, is entirely analogous to Eq. (11.9). Introducing Eq. (11.108) into Eq. (11.109), we obtain

$$J\frac{d^2\theta_o(t)}{dt^2}+c\frac{d\theta_o(t)}{dt}=K_i i_f(t) \tag{11.110}$$

Equations (11.107) and (11.110) relate the output angle $\theta_o(t)$ to the error signal $e(t)$. The equations can be used to derive the forward-path transfer function.

Taking the Laplace transforms of Eqs. (11.107) and (11.110) and ignoring the initial conditions, we can write

$$(Ls+R)I_f(s)=K_v E(s)$$
$$s(Js+c)\Theta_o(s)=K_i I_f(s) \qquad (11.111)$$

where $I_f(s)=\mathscr{L}i_f(t)$, $E(s)=\mathscr{L}e(t)$, and $\Theta_o(s)=\mathscr{L}\theta_o(t)$. Eliminating the transformed current $I_f(s)$ from Eqs. (11.111), we obtain the forward-path transfer function

$$G(s)=\frac{\Theta_o(s)}{E(s)}=\frac{K}{s(s+c/J)(s+R/L)} \qquad (11.112)$$

where

$$K=\frac{K_i K_v}{JL} \qquad (11.113)$$

is an overall gain constant. We observe that the transfer function given by Eq. (11.112) differs from the transfer function for the simplified model developed in Section 11.3 by the multiplying factor $K_v/L(s+R/L)$. Letting the feedback-path transfer function $H(s)$ be equal to unity and using Eq. (11.112), we can write the closed-loop transfer function for the system in the form

$$M(s)=\frac{\Theta_o(s)}{\Theta_i(s)}=\frac{G(s)}{1+G(s)}=\frac{K}{s(s+c/J)(s+R/L)+K} \qquad (11.114)$$

where $\Theta_i(s)=\mathscr{L}\theta_i(t)$.

Although the model of Section 11.3 may not be comparable quantitatively to the one discussed in this section, it may prove of interest to compare the step response for the two cases qualitatively. Recalling Eq. (11.21), we can write the Laplace transform of the step response from Eq. (11.114) as follows:

$$\Theta_o(s)=\frac{K}{s[s(s+c/J)(s+R/L)+K]} \qquad (11.15)$$

which can be rewritten in the factored form

$$\Theta_o(s)=\frac{K}{(s-s_1)(s-s_2)(s-s_3)(s-s_4)} \qquad (11.116)$$

where s_1, s_2, s_3, and s_4 are the roots of the characteristic equation

$$s[s(s+c/J)(s+R(L)+K]=0 \qquad (11.117)$$

Of course, $s_1=0$. To evaluate the inverse of Eq. (11.116), we consider the partial fractions expression

$$\Theta_o(s)=\frac{c_1}{s}+\frac{c_2}{s-s_2}+\frac{c_3}{s-s_3}+\frac{c_4}{s-s_4} \qquad (11.118)$$

where, from Section A.3, the coefficients c_k $(k=1, 2, 3, 4)$ have the values

$$
\begin{aligned}
c_1 &= \lim_{s\to 0} s\Theta_o(s) = -\frac{K}{s_2 s_3 s_4} \\
c_2 &= \lim_{s\to s_2} (s-s_2)\Theta_o(s) = \frac{K}{s_2(s_2-s_3)(s_2-s_4)} \\
c_3 &= \lim_{s\to s_3} (s-s_3)\Theta_o(s) = \frac{K}{s_3(s_3-s_2)(s_3-s_4)} \\
c_4 &= \lim_{s\to s_4} (s-s_4)\Theta_o(s) = \frac{K}{s_4(s_4-s_2)(s_4-s_3)}
\end{aligned}
\tag{11.119}
$$

The step response is the inverse Laplace transform of Eq. (11.118), which from Section A.3 is

$$\theta_o(t) = \mathscr{L}^{-1}\Theta_o(s) = (c_1 + c_2 e^{s_2 t} + c_3 e^{s_3 t} + c_4 e^{s_4 t})\,u(t) \tag{11.120}$$

where $u(t)$ is the unit step function.

As an illustration, let us consider the numerical values

$$K=4, \qquad c/J=1, \qquad R/L=2 \tag{11.121}$$

yielding the characteristic equation

$$s[s(s+1)(s+2)+4]=0 \tag{11.122}$$

which has the roots

$$
\begin{aligned}
&s_1=0, \qquad s_2=-2.7964 \\
&s_3=-0.1018+i1.1917, \qquad s_4=-0.1018-i1.1917
\end{aligned}
\tag{11.123}
$$

Inserting Eq. (11.123) into Eqs. (11.119), we obtain the coefficients

$$
\begin{aligned}
&c_1=1, \qquad c_2=-0.1648 \\
&c_3=-0.4176+i0.2290, \qquad c_4=-0.4176-i0.2290
\end{aligned}
\tag{11.124}
$$

Hence, introducing Eqs. (11.123) and (11.124) into Eq. (11.120), we obtain the step response

$$
\begin{aligned}
\theta_o(t) = [1 &- 0.1648e^{-2.7964t} \\
&- e^{-0.1018t}(0.8352\cos 1.1917t + 0.4580\sin 1.1917t)]\,u(t)
\end{aligned}
\tag{11.125}
$$

The response is plotted in Fig. 11.31. Comparing Fig. 11.31 with Fig. 11.5, we conclude that the step response of the model developed in this section resembles the step response of the model of Section 11.3, at least for the values of parameters given by Eq. (11.121). The explanation lies in the fact that the root $s=s_2$ is real and negative and its magnitude is quite large compared to the real part of s_3 and s_4. As a result, the solution component corresponding to $s=s_2$ decays very fast, so that for the most part the system acts like a second-order system rather than a third-order system.

From Fig. 11.31, we conclude that the performance of the position servo-

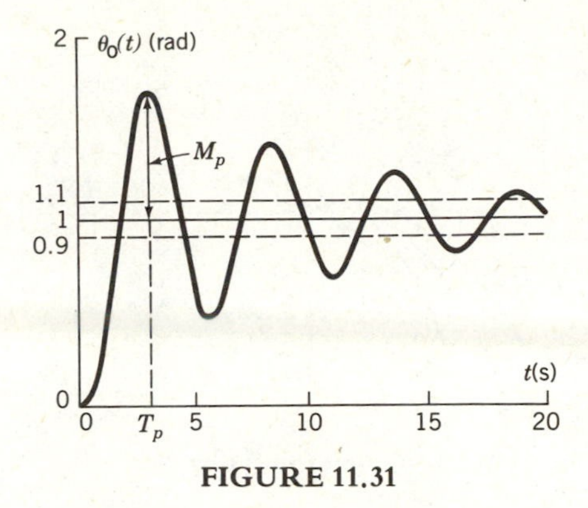

FIGURE 11.31

mechanism is not very good. Indeed, the maximum overshoot is approximately $M_p \cong 0.7$ rad, which is relatively large. This peak response occurs at $T_p \cong 3$ s. Moreover, defining the settling time T_s as the time required for the step response to settle within 10% of the final value, we conclude that $T_s > 20$ s. On the other hand, one good performance criterion is the steady-state error, which is zero. Later in this chapter, we discuss ways of improving the performance.

11.8 RELATIVE STABILITY. GAIN MARGIN AND PHASE MARGIN

In Chapter 9, we studied various concepts of stability and techniques for checking the stability of systems analytically. In this section, we wish to return to the subject of stability and examine it from a different point of view. In particular, the interest lies in additional information concerning the system stability. For example, the Routh–Hurwitz criterion discussed in Section 9.7 can be used only to establish whether all the roots of a characteristic polynomial have negative real parts, i.e., whether the system is stable in an absolute sense. Assuming that the system is stable, a question of interest is how close each root is to the imaginary axis in the s-plane, because this has a direct effect on the settling time. The closeness of a root to the imaginary axis determines the *degree of stability* or the *relative stability* of the root. One way of ascertaining the relative stability is to shift the imaginary axis to the left by a given amount and use the Routh–Hurwitz criterion again, but this is a tedious task. Another way is by means of the use of the Nyquist plot, or alternatively by means of the Bode plots.

The Nyquist criterion defines the stability of a closed-loop system in terms of the number of encirclements of the point $-1+i0$ in the $G(s)H(s)$-plane, where $G(s)H(s)$ is the open-loop transfer function. Actually, in using the Nyquist criterion, it is the open-loop frequency response $G(i\omega)H(i\omega)$ that is of interest, rather than the transfer function $G(s)H(s)$. Hence, the proximity of the $G(i\omega)H(i\omega)$ plot to the point $-1+i0$ determines the degree of stability of the system. In many cases, the degree of stability depends on the gain constant K. To illustrate this, we resort to a specific example.

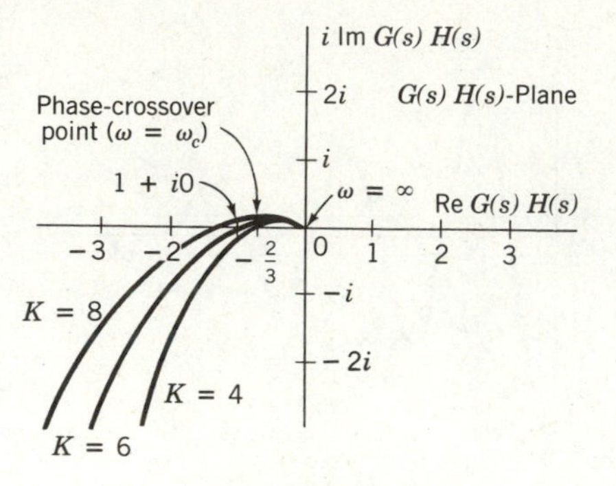

FIGURE 11.32

Let us consider the transfer function of Section 11.7, or

$$G(s)H(s)=\frac{K}{s(s+1)(s+2)}=\frac{K}{s(s^2+3s+2)} \tag{11.126}$$

so that the frequency response is

$$G(i\omega)H(i\omega)=\frac{K}{i\omega[(i\omega)^2+3i\omega+2]}=K\,\frac{-3\omega^2-i\omega(2-\omega^2)}{\omega^2[(2-\omega^2)^2+9\omega^2]} \tag{11.127}$$

The essential portion of the frequency response is plotted in Fig. 11.32 for $K=4$, which is the same value as that used in Section 11.7, as well as for $K=6$ and $K=8$. Clearly, the plot corresponding to $K=4$ does not encircle the point $-1+i0$, so that the system is stable for $K=4$. The point at which the plot intersects the real axis is called the *phase-crossover point*, and it defines a value of ω known as the *phase-crossover frequency* and denoted by ω_c. Then, the distance between the phase-crossover point and the point $-1+i0$ represents a quantitative measure of the relative stability. This quantity is called the *gain margin* and is defined mathematically as

$$\text{gain margin}=20\log\frac{1}{|G(i\omega_c)H(i\omega_c)|}\quad(\text{dB}) \tag{11.128}$$

By increasing the gain constant K, the phase-crossover point approaches the point $-1+i0$. Hence, *the gain margin is the amount in decibels by which the gain can be increased before the closed-loop system reaches instability*. In the case at hand, the phase-crossover frequency is $\omega_c=\sqrt{2}$ rad/s and the gain margin is

$$20\log\frac{1}{|G(i\sqrt{2})H(i\sqrt{2})|}=20\log\frac{1}{2/3}=3.5214\quad\text{dB} \tag{11.129}$$

which implies that the gain can be increased by 3.5214 dB or by 50% before instability is reached. Indeed, the Nyquist plot for $K=6$ passes through the point $-1+i0$. Of course, for $K>6$ the Nyquist plot encircles the point $-1+i0$ twice in the clockwise sense, and there is no pole of $G(s)H(s)$ in the right half of the s-plane.

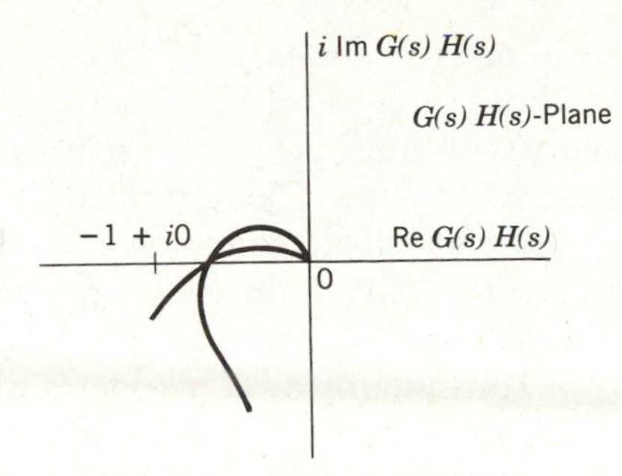

FIGURE 11.33

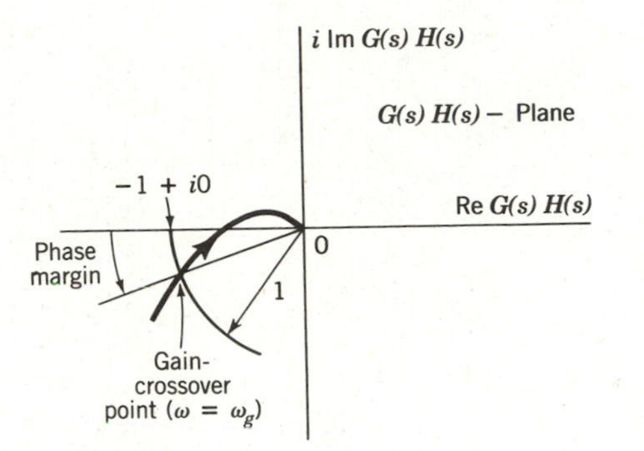

FIGURE 11.34

Hence, according to Eq. (11.62), there must be two roots of $1+G(s)H(s)$ in the right half of the s-plane, so that the system is unstable.

There are cases in which the gain margin defined above is not a good measure of the relative stability of a closed-loop system. For example, there can be two systems with the same gain margin but with different relative stability. Indeed, although the Nyquist plots for two different systems shown in Fig. 11.33 have the same gain margin, one plot comes appreciably closer to the point $-1+i0$ than the other. Hence, a supplementary way of ascertaining relative stability is desirable. To this end, we define the *gain-crossover point* as a point in the Nyquist plot of $G(s)H(s)$ for which the magnitude is unity. The point lies at the intersection of the Nyquist plot and a circle of unit radius, as shown in Fig. 11.34. Then, another quantitative measure of relative stability is the *phase margin* defined as

$$\text{phase margin} = \underline{/G(i\omega_g)H(i\omega_g)} - 180° \tag{11.130}$$

where ω_g is the frequency corresponding to the gain-crossover point and known as the *gain-crossover frequency*. Hence, *the phase margin is the angle in degrees through which the $G(i\omega)H(i\omega)$ plot must be rotated about the origin to bring the gain-crossover point into coincidence with the point $-1+i0$.* The phase margin can be interpreted as the additional phase lag necessary for the system to become unstable. The phase margin is shown in Fig. 11.34.

The gain margin and phase margin are easy to define in terms of Nyquist diagrams. They can also be defined in terms of Bode plots. In fact, because Bode plots are easier to draw than Nyquist plots, the gain margin and phase margin can be evaluated more easily from Bode plots. The critical point for stability is the point $-1+i0$ in the $G(i\omega)H(i\omega)$-plane. This point corresponds to a logarithmic gain of 0 dB and a phase angle of $-180°$ in the Bode diagram. Hence, the gain margin is the portion between the logarithmic gain curve and the 0-dB line corresponding to a phase angle of $-180°$ and the phase margin is the portion between the phase angle curve and the $-180°$ line corresponding to a logarithmic gain of 0 dB. The gain margin and phase margin are shown in Fig. 11.35, which represents

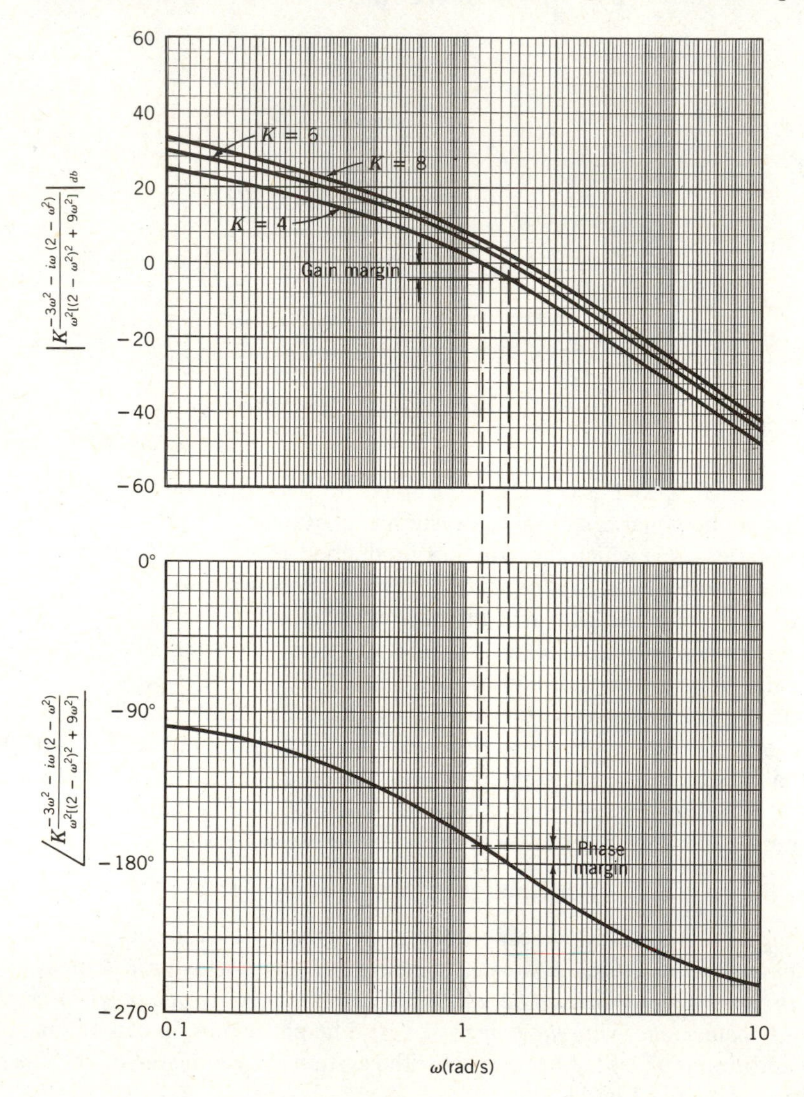

FIGURE 11.35

the Bode diagrams for the frequency response given by Eq. (11.127) and for $K=4$. From Fig. 11.35, we conclude that the gain margin is approximately 3.5 dB and the phase margin is approximately 11.5°. We recall from Section 11.6 that the effect of increasing the gain constant K is to shift the entire logarithmic gain plot upward. From Fig. 11.35, we observe that as K increases the gain margin decreases. At the same time, the gain-crossover point moves to the right, which causes the phase margin to decrease. When K becomes equal to 6, the gain margin reduces to zero and so does the phase margin.

11.9 COMPENSATORS

The performance of a feedback control system was examined in Section 11.3 and again in Section 11.7. In particular, the performance is considered as good if the system exhibits small maximum overshoot M_p, low settling time T_s, and small steady-state error e_{ss}. Of course, it is assumed that the control system is stable. Achieving good performance is the main goal in the design of a control system. If the performance is not entirely satisfactory, then we can consider modifying the system parameters. It turns out that this is not always a viable option, as quite often there are severe limits on the system parameters. In such cases, it is advisable to redesign the control system by inserting in the control loop a device capable of eliminating deficiencies from the system performance. The process of redesigning the control system to bring the system performance to an acceptable level is known as *compensation*, and the device is called a *compensator*. More often than not the compensator is an electrical circuit.

A commonly used compensator is the electrical circuit shown in Fig. 11.36. To derive a relation between the output voltage $v_o(t)$ and the input voltage $v_i(t)$, we consider the *Kirchhoff current law*, which can be stated as follows: *The sum of the currents entering a node must be equal to the sum of the currents leaving the node.* Using the notation of Fig. 11.36, we can write

$$i_1(t)+i_2(t)=i(t) \tag{11.131}$$

where repeated use of Kirchhoff's voltage law yields

$$C\left[\frac{dv_i(t)}{dt}-\frac{dv_o(t)}{dt}\right]=i_1(t) \tag{11.132a}$$

$$\frac{1}{R_1}[v_i(t)-v_o(t)]=i_2(t) \tag{11.132b}$$

$$\frac{1}{R_2}v_o(t)=i(t) \tag{11.132c}$$

Introducing Eqs. (11.132) into Eq. (11.131) and rearranging, we obtain

$$C\frac{dv_i}{dt}+\frac{1}{R_1}v_i=C\frac{dv_o}{dt}+\left(\frac{1}{R_1}+\frac{1}{R_2}\right)v_o=C\frac{dv_o}{dt}+\frac{R_1+R_2}{R_1R_2}v_o \tag{11.133}$$

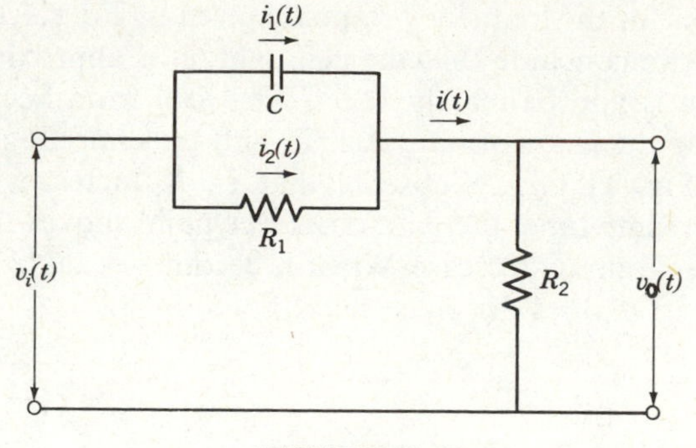

FIGURE 11.36

which represents a relation between the output voltage and the input voltage.

Equation (11.133) can be used to derive the transfer function of the circuit. Taking the Laplace transform of both sides of Eq. (11.133) and ignoring the initial conditions, we have

$$\left(Cs+\frac{1}{R_1}\right)V_i(s)=\left(Cs+\frac{R_1+R_2}{R_1R_2}\right)V_o(s) \tag{11.134}$$

where $V_i(s)=\mathscr{L}v_i(t)$ and $V_o(s)=\mathscr{L}v_o(t)$, so that the transfer function of the compensator is

$$G_c(s)=\frac{V_o(s)}{V_i(s)}=\frac{Cs+1/R_1}{Cs+(R_1+R_2)/R_1R_2} \tag{11.135}$$

Introducing the notation

$$\frac{1}{CR_1}=a, \qquad \frac{R_1+R_2}{CR_1R_2}=b \tag{11.136a, b}$$

the transfer function can be written in the form

$$G_c(s)=\frac{s+a}{s+b} \tag{11.137}$$

The frequency response is obtained by substituting $s=i\omega$ in Eq. (11.137), or

$$G_c(i\omega)=\frac{a}{b}\frac{1+i\omega/a}{1+i\omega/b}=\frac{a}{b}\frac{1+(\omega^2/ab)+i\omega(b-a)/ab}{1+\omega^2/b^2} \tag{11.138}$$

From Eqs. (11.136), we observe that $b>a$. Hence, recalling Eq. (11.72b), we conclude that the phase angle of the compensator is positive. For this reason the compensator shown in Fig. 11.36 is referred to as a *phase-lead compensator*, or simply as a *lead compensator*.

Next, let us use the compensator of Fig. 11.36 to improve the performance of the position servomechanism discussed in Section 11.7. The block diagram for the

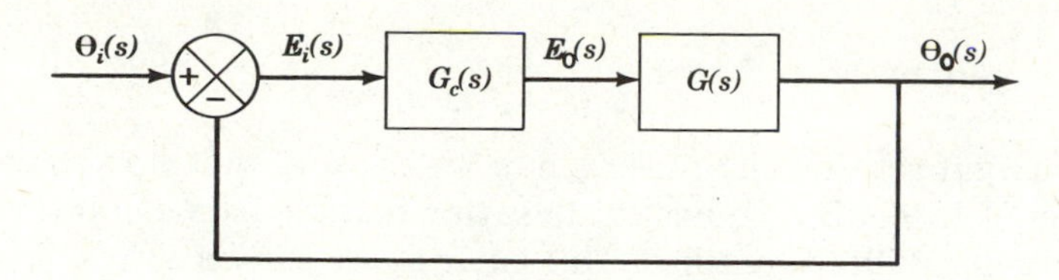

FIGURE 11.37

compensated system is shown in Fig. 11.37. The corresponding open-loop transfer function is

$$\frac{\Theta_o(s)}{E_i(s)} = G_c(s)G(s) \tag{11.139}$$

so that, inserting Eqs. (11.112) and (11.137) into Eq. (11.139), we obtain

$$\begin{aligned} G_c(s)G(s) &= \frac{s+a}{s+a}\,\frac{K}{s(s+c/J)(s+R/L)} \\ &= \frac{aKJL}{bcR}\,\frac{1+s/a}{s(1+s/b)(1+Js/c)(1+Ls/R)} \end{aligned} \tag{11.140}$$

Hence, the effect of using the compensator is to add one zero and one pole to the open-loop transfer function. The problem reduces to the design of the compensator so as to achieve satisfactory performance, which amounts to selecting the parameters a and b.

As an example, let us design a lead compensator so that the compensated system will have a gain margin exceeding 15 dB and a phase margin exceeding 70°. We propose to solve the problem first by means of Nyquist plots. To this end, we construct Nyquist plots for a variety of parameter pairs. One parameter pair satisfying the performance criteria is

$$a = 1/2, \qquad b = 2 \tag{11.141}$$

Indeed, introducing Eqs. (11.121) and (11.141) into Eq. (11.140), we have

$$G_c(s)G(s) = \frac{1}{2}\,\frac{1+2s}{s(1+0.5s)^2(1+s)}$$

The open-loop frequency response is obtained by inserting $s = i\omega$ in Eq. (11.142). The result is

$$G_c(i\omega)G(i\omega) = \frac{1}{2}\,\frac{1+i2\omega}{i\omega(1+i0.5\omega)^2(1+i\omega)} = -\frac{1}{8}\,\frac{9\omega^3 + i(4+11\omega^2-2\omega^4)}{\omega(1+0.25\omega^2)^2(1+\omega^2)} \tag{11.143}$$

The meaningful portion of the Nyquist plot of $G_c(i\omega)G(i\omega)$ is shown in Fig. 11.38. The crossover frequency can be obtained by equating the imaginary part in Eq. (11.143) to zero. This yields the crossover frequency $\omega_c = 2.4171$ rad/s, so that using Eqs. (11.128) and (11.143), the gain margin is

$$\text{gain margin} = 20 \log \frac{1}{|G_c(i2.4171)G(i2.4171)|} = 15.99 \quad \text{dB} \tag{11.144}$$

The gain margin satisfies the performance criterion in that it exceeds 15 dB. A gain margin of 15.99 dB is equivalent to saying that the gain can be increased by 530% before instability is reached. This latter figure can be arrived at by taking the reciprocal of $|G_c(i2.4171)G(i2.4171)|$, subtracting 1, and multiplying by 100. Note that the value of $|G_c(i2.4171)G(i2.4171)|$ can be read directly from Fig. 11.38. We observe that the gain could be increased only by 50% in the uncompensated case. The phase margin can also be read directly from Fig. 11.38. It has the value of 75°, which satisfies the phase margin performance criterion in that it exceeds 70°. Note that a phase margin of 75° is a substantial improvement over the value of 11.5° for the uncompensated system. Figure 11.38 also shows plots for the pair of parameters $a=1/2$ and $b=1$ and for the pair $a=1/2$ and $b=5$. Although the parameters $a=1/2$ and $b=5$ give larger gain and phase margins that $a=1/2$ and $b=2$, they require a resistance ratio R_1/R_2 of 9 compared to 3 for $a=1/2$ and $b=2$. Because the parameters $a=1/2$ and $b=2$ yield gain and phase margins exceeding the performance criteria, we consider this choice as satisfactory. The step response of the compensated system is presented in the next section.

The compensator design can be carried out, perhaps more effectively, by means of Bode plots. The reason for this is that the Bode diagram of the compensator can be added graphically to that of the uncompensated system. Hence, by plotting Bode magnitude and phase angle diagrams corresponding to different compensator parameters, it is possible to check graphically which parameters yield satisfactory results. From Eq. (11.138), the logarithmic gain and phase angle of the lead compensator are

$$|G_c(i\omega)|_{\text{dB}} = 20 \log|G_c(i\omega)| = 20 \log \frac{a}{b}\sqrt{\frac{1+\omega^2/a^2}{1+\omega^2/b^2}} \tag{11.145}$$

and

$$\psi(\omega) = \tan^{-1}\frac{\text{Im } G_c(i\omega)}{\text{Re } G_c(i\omega)} = \tan^{-1}\frac{\omega(b-a)/ab}{1+\omega^2/ab} \tag{11.146}$$

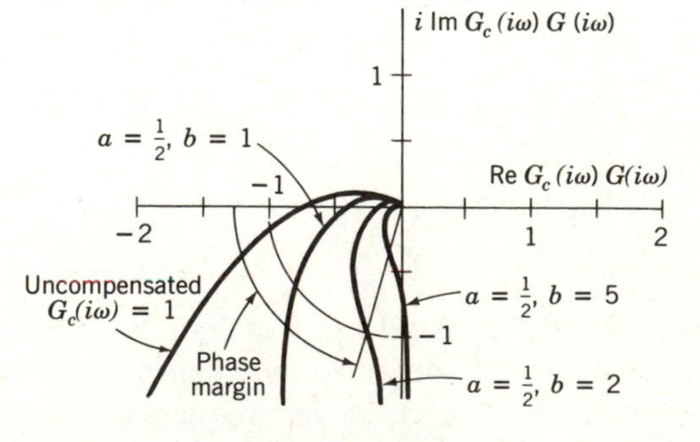

FIGURE 11.38

They are plotted in Fig. 11.39 as functions of ω/a for various values of a/b, where a/b is known as the *lead ratio*. From Fig. 11.39, we conclude that the addition of a lead compensator to a system lowers the logarithmic gain curve in the region of low frequencies and raises the phase angle curve in the low- to intermediate-frequency region. The amount of low-frequency attenuation and phase lead produced by the compensator depends on the lead ratio. The maximum phase lead occurs at the frequency $\omega_m = \sqrt{ab}$ and it has the value

$$\psi_{max} = 90 - 2\tan^{-1}\sqrt{a/b} \quad \text{deg} \tag{11.147}$$

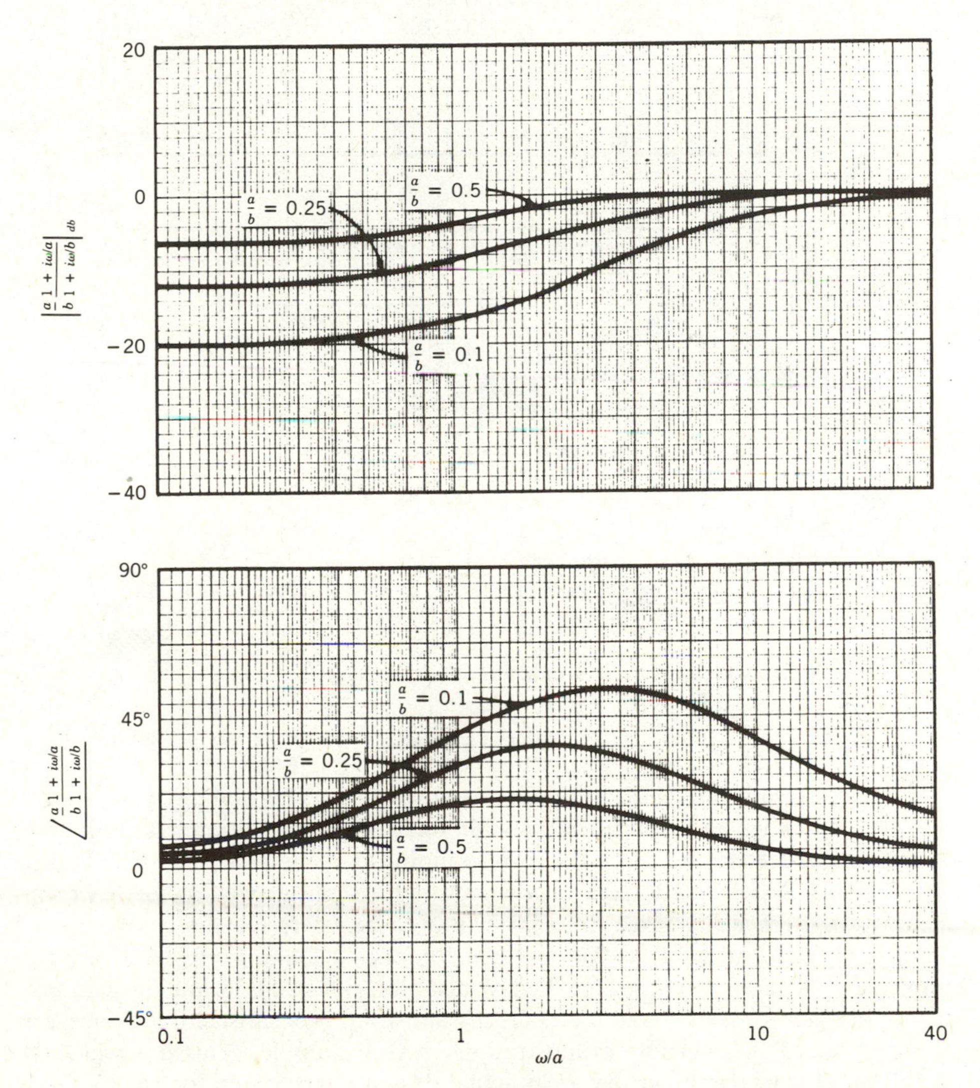

FIGURE 11.39

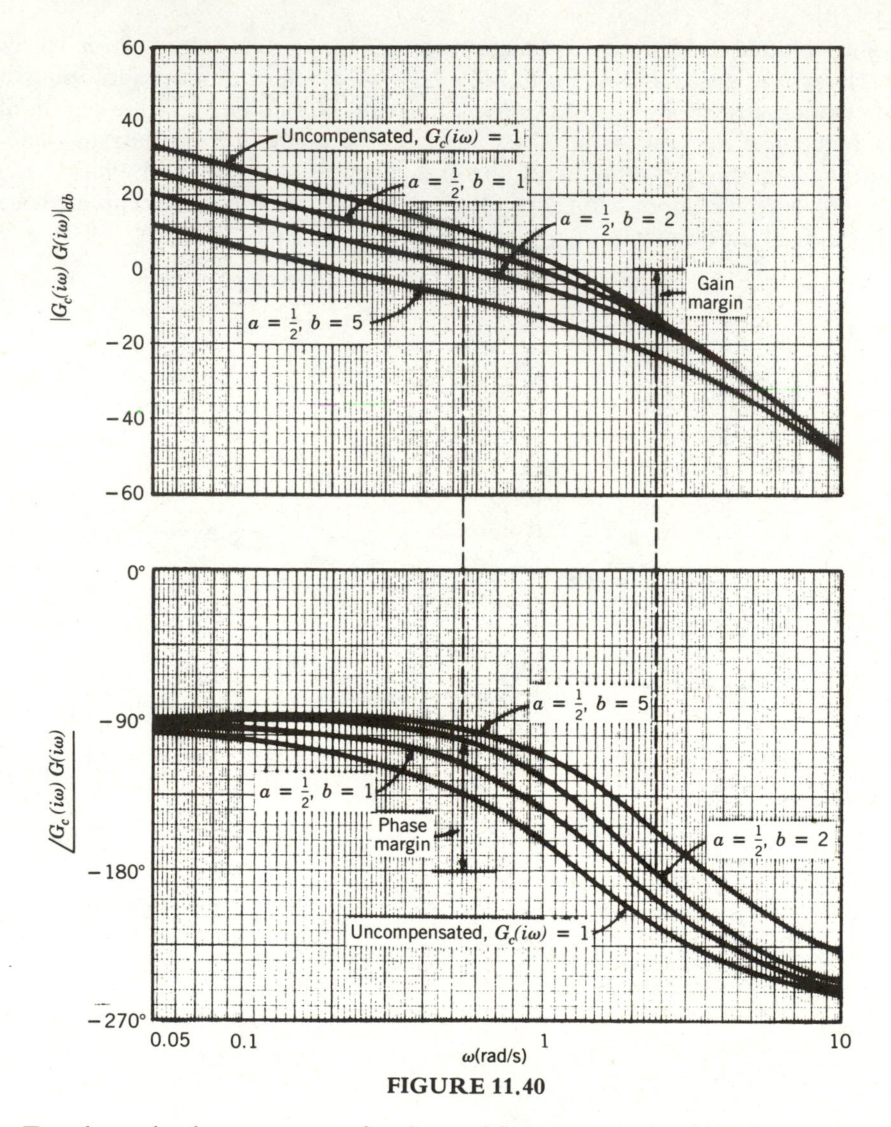

FIGURE 11.40

To select a lead compensator for the position servomechanism discussed earlier in this section, we add compensators with various system parameters to the uncompensated system and check the corresponding gain and phase margins. Figure 11.40 shows Bode plots for the uncompensated system, as well as for the compensated system using three different sets of parameters for the compensator, namely, $a=1/2$ and $b=1$, $a=1/2$ and $b=2$, and $a=1/2$ and $b=5$. These are the same pairs of parameters used for the Nyquist plots of Fig. 11.38, and the gain and phase margins for $a=1/2$ and $b=2$, already shown to be a satisfactory choice, are pointed out for the purpose of illustration. In this example we used a relatively small number of parameter pairs; in actual design a larger number are likely to be investigated.

There are other types of compensating networks. One network worthy of mention is shown in Fig. 11.41. It can be shown that the transfer function for the compensator of Fig. 11.41 is

$$G_c(s) = \frac{a}{b}\frac{s+b}{s+a} \tag{11.148}$$

where a and b are as defined in Eqs. (11.136). This compensator is known as a *lag compensator*. Another compensator is the *lead-lag compensator*, shown in Fig. 11.42. The transfer function of the lead-lag compensator can be shown to be

$$G_c(s) = \frac{(s+a_1)(s+b_2)}{(s+a_2)(s+b_1)} \tag{11.149}$$

where $a_1 = 1/R_1C_1$, $b_2 = 1/R_2C_2$, $b_1 + a_2 = a_1 + b_2 + 1/R_2C_1$, $a_2b_1 = a_1b_2$, and we note that $b_1 > a_1$, $b_2 > a_2$.

In the compensated system shown in Fig. 11.37 the compensator is inserted in the forward path, thus adding zeros and poles to $G(s)$. This process is known as *series compensation*. If the compensator is inserted in the feedback path, then zeros and poles are added to $H(s)$. Such compensation is called *parallel compensation*.

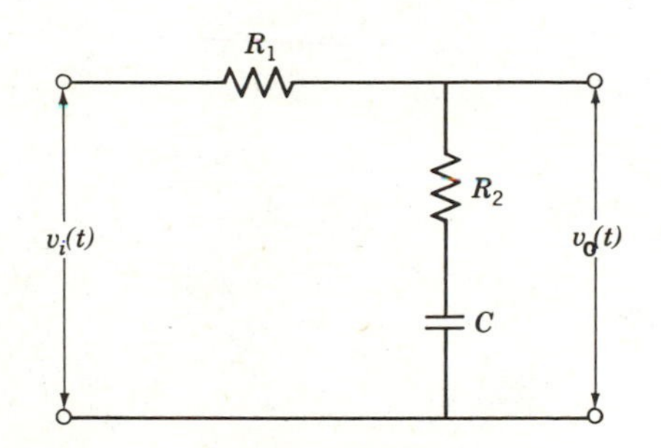

FIGURE 11.41

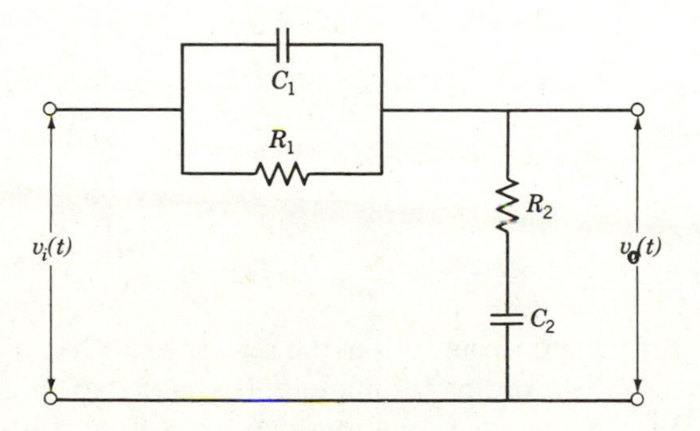

FIGURE 11.42

11.10 DERIVATION OF THE SYSTEM RESPONSE FROM THE TRANSFER FUNCTION

The gain margin and phase margin provide information concerning the system stability. To evaluate the performance of a control system design, criteria such as maximum overshoot or settling time are highly desirable. This information, however, can be obtained only from the time response and not from the frequency response. In many cases, the control design takes place in the s-domain, as shown in Section 11.9, so that the interest lies in deriving the time response of a system given the transfer function. Of course, the time response can be obtained by resolving the transformed response into partial fractions and inverting a sum of simple functions. As the order of the system increases, however, this procedure becomes progressively more tedious, so that a more systematic approach is desirable.

Let us consider a feedback control system and denote the open-loop transfer function by $G(s)$, which implies that $H(s)=1$. If there is a compensator in the loop, then its effect can be included in $G(s)$. We assume that the transfer function has the general form

$$G(s)=K\frac{\prod_{j=1}^{m}(s-z_j)}{\prod_{i=1}^{n}(s-p_i)} \tag{11.150}$$

where $z_j\,(j=1,2,\ldots,m)$ are the zeros and $p_i\,(i=1,2,\ldots,n)$ are the poles of the open-loop transfer function. The closed-loop transfer function is

$$M(s)=\frac{C(s)}{R(s)}=\frac{G(s)}{1+G(s)}=\frac{K\prod_{j=1}^{m}(s-z_j)}{\prod_{i=1}^{n}(s-p_i)+K\prod_{j=1}^{m}(s-z_j)} \tag{11.151}$$

in which $C(s)$ and $R(s)$ are the transformed output and input, respectively. Introducing the notation

$$\begin{aligned}\prod_{i=1}^{n}(s-p_i)+K\prod_{j=1}^{m}(s-z_j)&=s^n+a_1s^{n-1}+\cdots+a_n\\ K\prod_{j=1}^{m}(s-z_j)&=d_0s^m+d_1s^{m-1}+\cdots+d_m\end{aligned} \tag{11.152}$$

and using Eq. (11.151), we can write

$$(s^n+a_1s^{n-1}+\cdots+a_n)C(s)=(d_0s^m+d_1s^{m-1}+\cdots+d_m)R(s) \tag{11.153}$$

Equation (11.153) can be regarded as the Laplace transformation of the differential equation

$$\frac{d^nc(t)}{dt^n}+a_1\frac{d^{n-1}c(t)}{dt^{n-1}}+\cdots+a_nc(t)=d_0\frac{d^mr(t)}{dt^m}+d_1\frac{d^{m-1}r(t)}{dt^{m-1}}+\cdots+d_mr(t) \tag{11.154}$$

where $c(t)$ is the time response and $r(t)$ is the reference input.

Equation (11.154) is an ordinary differential equation of order n. Its solution can be obtained by replacing the equation by a set of first-order equations. Ordinarily, this is a simple procedure, similar to that for the derivation of the state equations. But, because the transfer function has m zeros, the right side of Eq.

(11.154) contains derivatives of the reference input through order m, which complicates matters. The problem created by the presence of the derivatives of the reference input can be obviated by redefining the state variables. The formulation reduces to the relatively simple one for the case in which the transfer function has no zeros (see Section 1.12).

Let us consider the transformation

$$
\begin{aligned}
c(t) &= x_1(t) \\
\frac{dc(t)}{dt} &= \frac{dx_1(t)}{dt} = x_2(t) \\
\frac{d^2c(t)}{dt^2} &= \frac{dx_2(t)}{dt} = x_3(t) \\
&\vdots \\
\frac{d^{n-m-1}c(t)}{dt^{n-m-1}} &= \frac{dx_{n-m-1}(t)}{dt} = x_{n-m}(t) \\
\frac{d^{n-m}c(t)}{dt^{n-m}} &= \frac{dx_{n-m}(t)}{dt} = x_{n-m+1}(t) + b_{n-m}r(t) \\
\frac{d^{n-m+1}c(t)}{dt^{n-m+1}} &= \frac{dx_{n-m+1}(t)}{dt} + b_{n-m}\frac{dr(t)}{dt} = x_{n-m+2}(t) + b_{n-m+1}r(t) + b_{n-m}\frac{dr(t)}{dt} \\
&\vdots \\
\frac{d^{n-1}c(t)}{dt^{n-1}} &= \frac{dx_{n-1}(t)}{dt} + b_{n-2}\frac{dr(t)}{dt} + \cdots + b_{n-m+1}\frac{d^{m-2}r(t)}{dt^{m-2}} + b_{n-m}\frac{d^{m-1}r(t)}{dt^{m-1}} \\
&= x_n(t) + b_{n-1}r(t) + b_{n-2}\frac{dr(t)}{dt} + \cdots + b_{n-m+1}\frac{d^{m-2}r(t)}{dt^{m-2}} + b_{n-m}\frac{d^{m-1}r(t)}{dt^{m-1}} \\
\frac{d^nc(t)}{dt^n} &= \frac{dx_n(t)}{dt} + b_{n-1}\frac{dr(t)}{dt} + b_{n-2}\frac{d^2r(t)}{dt^2} + \cdots + b_{n-m+1}\frac{d^{m-1}r(t)}{dt^{m-1}} + b_{n-m}\frac{d^mr(t)}{dt^m}
\end{aligned}
\tag{11.155}
$$

where $b_{n-m}, b_{n-m+1}, \ldots, b_{n-2}, b_{n-1}$ are constant coefficients. They can be determined by writing Eq. (11.154) in the form

$$
\frac{d^nc(t)}{dt^n} = -a_1\frac{d^{n-1}c(t)}{dt^{n-1}} - \cdots - a_nc(t) + d_0\frac{d^mr(t)}{dt^m} + d_1\frac{d^{m-1}r(t)}{dt^{m-1}} + \cdots + d_mr(t) \tag{11.156}
$$

Considering Eqs. (11.155), we can rewrite Eq. (11.156) as follows:

$$
\begin{aligned}
&\frac{dx_n(t)}{dt} + b_{n-1}\frac{dr(t)}{dt} + b_{n-2}\frac{d^2r(t)}{dt^2} + \cdots + b_{n-m+1}\frac{d^{m-1}r(t)}{dt^{m-1}} + b_{n-m}\frac{d^mr(t)}{dt} \\
&\quad = -a_1\left[x_n(t) + b_{n-1}r(t) + b_{n-2}\frac{dr(t)}{dt} + \cdots + b_{n-m+1}\frac{d^{m-2}r(t)}{dt^{m-2}} + b_{n-m}\frac{d^{m-1}r(t)}{dt^{m-1}}\right] \\
&\qquad - a_2\left[x_{n-1}(t) + b_{n-2}r(t) + b_{n-3}\frac{dr(t)}{dt} + \cdots + b_{n-m+1}\frac{d^{m-3}r(t)}{dt^{m-3}} + b_{n-m}\frac{d^{m-2}r(t)}{dt^{m-2}}\right] \\
&\qquad - \cdots - a_{n-1}x_2(t) - a_nx_1(t) + d_0\frac{d^mr(t)}{dt^m} + d_1\frac{d^{m-1}r(t)}{dt^{m-1}} + \cdots + d_mr(t)
\end{aligned}
\tag{11.157}
$$

The constants $b_{n-m}, b_{n-m+1}, \ldots, b_{n-2}, b_{n-1}$ are determined by equating the coefficients of $d^j r(t)/dt^j$ on both sides of Eq. (11.157). This leads to the following recursive relations

$$\begin{aligned}
b_{n-m} &= d_0 \\
b_{n-m+1} &= d_1 - a_1 b_{n-m} \\
b_{n-m+2} &= d_2 - a_2 b_{n-m} - a_1 b_{n-m+1} \\
&\vdots \\
b_{n-2} &= d_{m-2} - \cdots - a_2 b_{n-4} - a_1 b_{n-3} \\
b_{n-1} &= d_{m-1} - \cdots - a_2 b_{n-3} - a_1 b_{n-2} \\
b_n &= d_m - a_m b_{n-m} - \cdots - a_2 b_{n-2} - a_1 b_{n-1}
\end{aligned} \tag{11.158}$$

In view of Eqs. (11.158), Eqs. (11.155) and (11.157) can be written in the form

$$\begin{aligned}
\dot{x}_1(t) &= x_2(t) \\
\dot{x}_2(t) &= x_3(t) \\
&\vdots \\
\dot{x}_{n-m-1}(t) &= x_{n-m}(t) \\
\dot{x}_{n-m}(t) &= x_{n-m+1}(t) + b_{n-m} r(t) \\
\dot{x}_{n-m+1}(t) &= x_{n-m+2}(t) + b_{n-m+1} r(t) \\
&\vdots \\
\dot{x}_{n-1}(t) &= x_n(t) + b_{n-1} r(t) \\
\dot{x}_n(t) &= -a_n x_1(t) - a_{n-1} x_2(t) - \cdots - a_2 x_{n-1}(t) - a_1 x_n(t) + b_n r(t)
\end{aligned} \tag{11.159}$$

which are the desired state equations. They can be expressed in the matrix form

$$\dot{\mathbf{x}}(t) = A\mathbf{x}(t) + \mathbf{b}r(t) \tag{11.160}$$

where $\mathbf{x}(t) = [x_1(t)\ x_2(t) \ldots x_n(t)]^{\mathrm{T}}$ is an n-dimensional state vector,

$$A = \begin{bmatrix}
0 & 1 & 0 & 0 & \cdots & 0 & 0 \\
0 & 0 & 1 & 0 & \cdots & 0 & 0 \\
0 & 0 & 0 & 1 & \cdots & 0 & 0 \\
\vdots & \vdots & \vdots & \vdots & & \vdots & \vdots \\
0 & 0 & 0 & 0 & \cdots & 0 & 1 \\
-a_n & -a_{n-1} & -a_{n-2} & -a_{n-3} & \cdots & -a_2 & -a_1
\end{bmatrix} \tag{11.161}$$

is the coefficient matrix and

$$\mathbf{b} = [0 \quad 0 \quad \cdots \quad 0 \quad b_{n-m} \quad b_{n-m+1} \quad \cdots \quad b_{n-1} \quad b_n]^{\mathrm{T}} \tag{11.162}$$

is an n-dimensional vector whose components represent a measure of the input to each of the state equations. Note that the first $n-m-1$ components of $\mathbf{b}$ are zero.

Equation (11.160) is in a typical state form. Its solution by means of the transition matrix was discussed in Sections 10.2 and 10.3.

Example 11.7
Derive the state equations for the compensated system described by Eq. (11.142). Then, solve the equations for $r(t)$ in the form of a unit step function, plot the response $c(t)=x_1(t)$ as a function of time, and compare the system performance with that of the uncompensated system of Section 11.7.

From Eq. (11.142), the open-loop transfer function of the compensated system is

$$G_c(s)G(s)=\frac{1}{2}\frac{1+2s}{s(1+0.5s)^2(1+s)}=\frac{1}{4}\frac{s+0.5}{s(s+1)(s+2)^2} \tag{a}$$

Hence, the closed-loop transfer function is

$$M(s)=\frac{G_c(s)G(s)}{1+G_c(s)G(s)}=\frac{1}{4}\frac{s+0.5}{s(s+1)(s+2)^2+0.25(s+0.5)}$$
$$=\frac{0.25s+0.125}{s^4+5s^3+8s^2+4.25s+0.125} \tag{b}$$

so that, comparing Eq. (b) with Eqs. (11.151) and (11.152), we conclude that

$$a_1=5,\quad a_2=8,\quad a_3=4.25,\quad a_4=0.125,\quad d_0=0.25,\quad d_1=0.125 \tag{c}$$

Recognizing that $n=4$ and $m=1$, Eqs. (11.158) yield the components

$$b_3=d_0=0.25$$
$$b_4=d_1-a_1b_3=0.125-5\times 0.25=-1.125 \tag{d}$$

Finally, inserting Eqs. (c) and (d) into Eqs. (11.161) and (11.162), we can write the state equations in the matrix form

$$\begin{bmatrix}\dot{x}_1\\ \dot{x}_2\\ \dot{x}_3\\ \dot{x}_4\end{bmatrix}=\begin{bmatrix}0 & 1 & 0 & 0\\ 0 & 0 & 1 & 0\\ 0 & 0 & 0 & 1\\ -0.125 & -4.25 & -8 & -5\end{bmatrix}\begin{bmatrix}x_1\\ x_2\\ x_3\\ x_4\end{bmatrix}+\begin{bmatrix}0\\ 0\\ 0.25\\ -1.125\end{bmatrix}r(t) \tag{e}$$

The step response is obtained by letting the reference input $r(t)$ in Eq. (e) be equal to the unit step function $u(t)$. The step response, obtained by the approach based on the transition matrix, is displayed in Fig. 11.43. Note that, according to the first of Eqs. (11.155), the output $c(t)$ is merely the first component of the state vector.

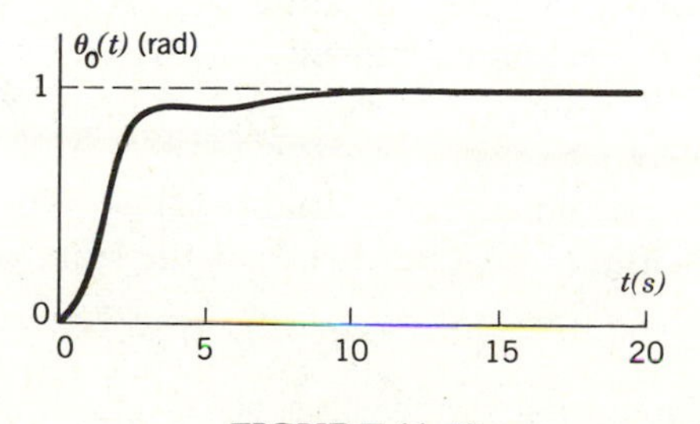

FIGURE 11.43

Comparing the response of the compensated system of Fig. 11.43 with the response of the uncompensated system shown in Fig. 11.31, we conclude that the lead compensator with the transfer function given by Eq. (11.137) and with $a=1/2$ and $b=2$ does an excellent job in causing the position servomechanism to follow the unit step function reference input. The only improvement that can be made is inducing a faster rise time.

11.11 FEEDBACK CONTROL OF MULTIVARIABLE SYSTEMS

The control systems discussed until now have one thing in common. They are all characterized by a single input and a single output. This is true even in the case of the fourth-order system of Example 11.7, in which the input was the reference input $r(t)$ and the output was the first component of the state vector, $c(t)=x_1(t)$. Such systems are referred to as single input, single output (SISO) systems. More complicated systems, such as multi-degree-of-freedom systems, can be treated at times as SISO systems. Quite often, however, it becomes necessary to treat them as multiple input, multiple output (MIMO) systems. In attempting to treat MIMO systems by techniques based on the concept of transfer function, as discussed in earlier sections, one difficulty becomes apparent immediately: We must consider a transfer function for every input–output pair. For two inputs and two outputs, we must consider four transfer functions, and in general the number of transfer functions is equal to the product of inputs and outputs. As the number of inputs and outputs increases, the amount of work involved in the analysis and design of controls becomes unmanageable, so that new approaches are desirable. In the case of SISO control systems, the input and output are scalar quantities, and the analysis and design is carried out either in the Laplace domain or in the frequency domain. In contrast, in MIMO systems the inputs and outputs are represented by vectors, and the analysis and design are carried out more conveniently in the time domain. It is customary to refer to the scalar Laplace-domain, or frequency-domain approach as *classical control* and to the vectorial time-domain approach as *modern control*. Classical control has more of a physical approach, and modern control has more of a mathematical approach. For the most part, classical and modern control complement each other and can be used together. In fact, in Example 11.7 we computed the step response of a compensated position control system by a time-domain technique, namely, one based on the transition matrix. In the remaining part of this chapter, we provide a brief introduction to modern control.

Let us consider a general linear multivariable system. The block diagram of the system is displayed in Fig. 11.44, and it reflects the state equation

$$\dot{\mathbf{x}}(t)=A\mathbf{x}(t)+B\mathbf{u}(t) \tag{11.163}$$

where $\mathbf{x}(t)$ is the m-dimensional state vector, $\mathbf{u}(t)$ is an l-dimensional input vector, and A and B are $m\times m$ and $m\times l$ matrices of coefficients, respectively. Moreover,

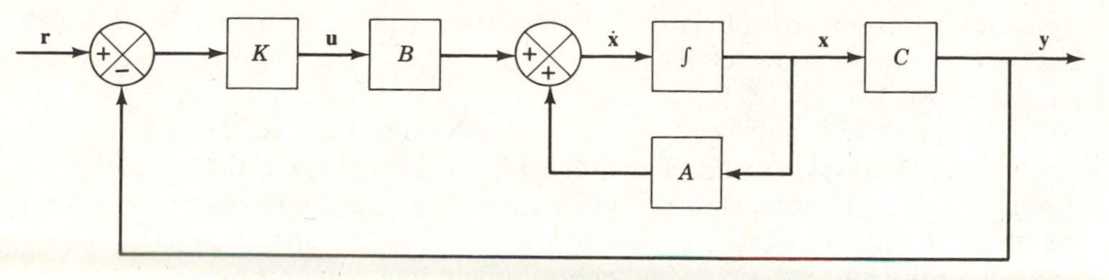

FIGURE 11.44

it is implied in the block diagram that the output vector $\mathbf{y}(t)$ is different from the state vector, where the relation between the two is

$$\mathbf{y}(t) = C\mathbf{x}(t) \tag{11.164}$$

in which C is a rectangular matrix

The relation between the reference input $\mathbf{r}(t)$ and output $\mathbf{y}(t)$ and the input $\mathbf{u}(t)$ is assumed to have the linear form

$$\mathbf{u}(t) = K[\mathbf{r}(t) - \mathbf{y}(t)] \tag{11.165}$$

where K is a *matrix of control gains*. In the case in which the reference input is zero, Eq. (11.165) reduces to

$$\mathbf{u}(t) = -K\mathbf{y}(t) \tag{11.166}$$

Inserting Eq. (11.164) into Eq. (11.166), we can write

$$\mathbf{u}(t) = -KC\mathbf{x}(t) \tag{11.167}$$

Equation (11.167) relates the control input to the state and is known as the *control law*. Introducing Eq. (11.167) into Eq. (11.163), we obtain the *closed-loop equation*

$$\dot{\mathbf{x}} = (A - BKC)\mathbf{x}(t) \tag{11.168}$$

which is a homogeneous equation of order m. The problem of feedback control design consists of deriving a matrix $A - BKC$ so as to cause the state $\mathbf{x}(t)$ to tend to the final state $\mathbf{x}(t_f)$ as $t \to t_f$, where t_f denotes the final time. Quite often the final state is taken as zero, so that the control task amounts to driving the state $\mathbf{x}(t)$ to the origin of the state space. This is consistent with a null reference input and can be ensured by determining the gain matrix K so that the eigenvalues of the matrix $A - BKC$ have negative real parts. Hence, the object of feedback control is to transform a merely stable or an unstable system into one exhibiting asymptotic stability. The net effect is to change the system poles to either real and negative, or complex with negative real parts, or a combination of the two. In general, in the process of changing the system eigenvalues, the eigenvectors also change.

In implementing the control law, we do not use the actual state $\mathbf{x}(t)$ but an estimated state $\hat{\mathbf{x}}(t)$. The estimated state can be obtained by means of an *observer*, known as a Luenberger observer, which has the form of a dynamical system similar in structure to the actual system and driven by the system output. The problem of designing an observer is obviated if the full state $\mathbf{x}(t)$ can be measured, in which case the matrix C becomes the identity matrix. We are concerned exclusively with this case.

There are several approaches to the computation of the gain matrix K. One approach, known as *pole allocation*, consists of first selecting closed-loop poles guaranteeing asymptotic stability and then computing the control gains corresponding to the preselected closed-loop poles. Another approach is known as *optimal control* and consists of first choosing and then minimizing a performance index. A commonly used performance index has the form of a time integral with the integrand comprising two quadratic expressions, the first representing a measure of the distance of the system state from the origin of the state space and the second a measure of the control effort. The minimization process reduces to the solution of a so-called matrix Riccati equation, which in turn permits the calculation of the control gain matrix. Both pole allocation and optimal control lie beyond the scope of this text.

The problem of designing a feedback control system becomes appreciably simpler when the control vector has the same dimension as the state vector, as shown in the next section.

11.12 MODAL CONTROL

Let us consider an undamped multi-degree-of-freedom linear dynamical system and write the differential equations of motion in the matrix form (see Section 8.3)

$$M\ddot{\mathbf{q}}(t)+K\mathbf{q}(t)=\mathbf{Q}(t) \tag{11.169}$$

where $\mathbf{q}(t)$ and $\mathbf{Q}(t)$ are n-dimensional configuration and force vectors, respectively, and M and K are $n\times n$ symmetric mass and stiffness matrices, respectively. The second-order configuration equations can be replaced by twice the number of first-order state equations. This is not necessary, however, as the type of control to be discussed in this section can be better presented in terms of configuration equations.

As shown in Section 8.3, the eigenvalue problem for the system has the form

$$K\mathbf{u}_r=\omega_r^2 M\mathbf{u}_r, \qquad r=1,2,\ldots,n \tag{11.170}$$

where ω_r^2 and $\mathbf{u}_r$ $(r=1,2,\ldots,n)$ are the eigenvalues and eigenvectors, respectively, in which ω_r are the natural frequencies. The eigenvectors, or modal vectors, are orthogonal and can be normalized so as to satisfy

$$\mathbf{u}_s^{\mathrm{T}}M\mathbf{u}_r=\delta_{rs}, \qquad \mathbf{u}_s^{\mathrm{T}}K\mathbf{u}_r=\omega_r^2\,\delta_{rs}, \qquad r,s=1,2,\ldots,n \tag{11.171}$$

where δ_{rs} is the Kronecker delta. Introducing the coordinate transformation

$$\mathbf{q}(t)=\sum_{r=1}^{n}\mathbf{u}_r\eta_r(t) \tag{11.172}$$

into Eq. (11.169), multiplying both sides of the equation on the left by $\mathbf{u}_s^{\mathrm{T}}$, and using

Eqs. (11.171), we obtain the *modal equations of motion*

$$\ddot{\eta}_r(t)+\omega_r^2\eta_r(t)=N_r(t), \qquad r=1, 2, \ldots, n \tag{11.173}$$

in which $\eta_r(t)$ $(r=1, 2, \ldots, n)$ are *modal coordinates* and

$$N_r(t)=\mathbf{u}_r^T\mathbf{Q}(t), \qquad r=1, 2, \ldots, n \tag{11.174}$$

are *modal forces.* In the case of open-loop control, that is, when the force vector $\mathbf{Q}(t)$ is an explicit function of time and does not depend on the displacement vector $\mathbf{q}(t)$ or the velocity vector $\dot{\mathbf{q}}(t)$, Eqs. (11.173) are independent and the modal coordinates $\eta_r(t)$ are referred to as *natural coordinates.*

Next, let us consider feedback control of the system. Assuming that the modal forces, generally known as *modal control forces*, have the form

$$N_r(t)=-\sum_{s=1}^{n}[g_{rs}\eta_s(t)+h_{rs}\dot{\eta}_s(t)] \tag{11.175}$$

where g_{rs} and h_{rs} are *modal control gains*, we conclude that the modal equations become

$$\ddot{\eta}_r(t)+\sum_{s=1}^{n}h_{rs}\dot{\eta}_s(t)+\sum_{s=1}^{n}(g_{rs}+\omega_r^2\,\delta_{rs})\eta_s(t)=0, \qquad r=1, 2, \ldots, n \tag{11.176}$$

Equations (11.176) are recognized as the *closed-loop equations of motion.* The process of using Eqs. (11.173) to control the system is known as *modal control.* Hence, the process reduces to selecting the gains g_{rs} and h_{rs} $(r, s=1, 2, \ldots, n)$ so that the closed-loop eigenvalues are either real and negative or complex with negative real parts. Note that the actual control vector can be synthesized from the modal control forces by writing

$$\mathbf{Q}(t)=\sum_{r=1}^{n}M\mathbf{u}_rN_r(t)=-\sum_{r=1}^{n}\sum_{s=1}^{n}M\mathbf{u}_r[g_{rs}\eta_s(t)+h_{rs}\dot{\eta}_s(t)] \tag{11.177}$$

As pointed out in Section 11.11, the process of producing the gains for given closed-loop poles can be difficult. The difficulty can be traced to the form assumed for the modal controls, Eq. (11.175). Indeed, this form causes recoupling of the equations of motion, so that the closed-loop equations are no longer independent. As a result, the modal coordinates are no longer natural coordinates. We refer to this type of control as *coupled control.* Note that coupled control is the only form possible when the control vector $\mathbf{Q}(t)$ has one or more components equal to zero, which implies that the control is to be carried out with fewer than n forces. In coupled control the control vector $\mathbf{Q}(t)$ is determined directly, without determining the modal controls first.

Substantial simplification of the control task can be achieved if control is to be implemented by means of a control vector $\mathbf{Q}(t)$ of full dimension n. In this case, the modal forces can be assumed to have the form

$$N_r(t)=-g_r\eta_r(t)-h_r\dot{\eta}_r(t), \qquad r=1, 2, \ldots, n \tag{11.178}$$

and the closed-loop modal equations reduce to

$$\ddot{\eta}_r(t)+h_r\dot{\eta}_r(t)+(g_r+\omega_r^2)\eta_r(t)=0, \qquad r=1, 2, \ldots, n \tag{11.179}$$

By contrast with Eqs. (11.176), Eqs. (11.179) are independent, so that the modal coordinates $\eta_r(t)$ $(r=1, 2, \ldots, n)$ remain natural coordinates even in the presence of feedback forces. For this reason, independent control of the type (11.178) is called *natural control*. The characteristic equations associated with the closed-loop equations are

$$s_r^2+h_rs_r+g_r+\omega_r^2=0, \qquad r=1, 2, \ldots, n \tag{11.180}$$

which have the roots

$$s_{1r}=-\alpha_r+i\beta_r, \qquad s_{2r}=-\alpha_r-i\beta_r, \qquad r=1, 2, \ldots, n \tag{11.181}$$

Hence, the pole allocation method reduces simply to the selection of the constants α_r and β_r for every mode. Then, the modal gains are obtained from

$$g_r=\alpha_r^2+\beta_r^2-\omega_r^2, \qquad h_r=2\alpha_r, \qquad r=1, 2, \ldots, n \tag{11.182}$$

and the control vector is given by

$$\mathbf{Q}(t)=\sum_{r=1}^{n} M\mathbf{u}_rN_r(t)=-\sum_{r=1}^{n} M\mathbf{u}_r[g_r\eta_r(t)+h_r\dot{\eta}_r(t)] \tag{11.183}$$

Equations (11.182) can be verified by inserting Eqs. (11.181) into Eqs. (11.180). A block diagram of natural control is shown in Fig. 11.45. Note that the modal control forces can be computed in parallel.

The modal gains g_r and h_r $(r=1, 2, \ldots, n)$ can be given a physical interpretation

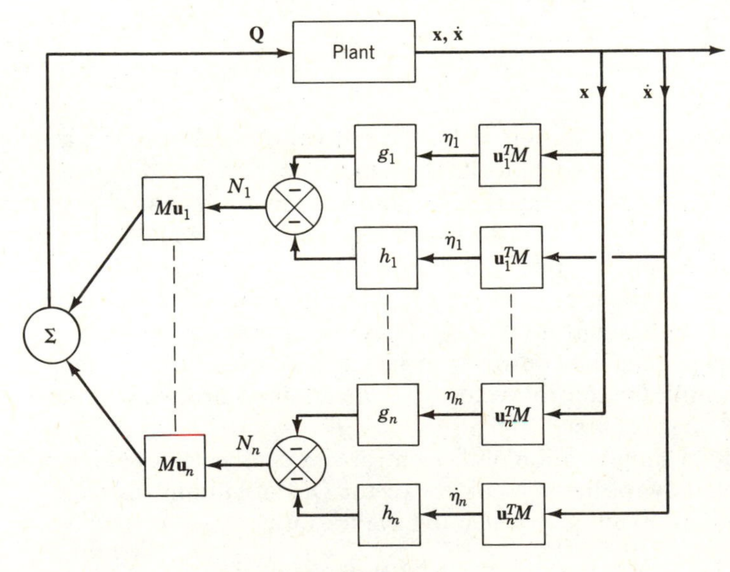

FIGURE 11.45

by observing that $-g_r\eta_r(t)$ represents a modal restoring force and $-h_r\dot{\eta}_r(t)$ a modal viscous force. Hence, g_r can be regarded as a modal stiffness coefficient and h_r as a modal damping coefficient.

Another interesting aspect of natural control is that the closed-loop eigenvectors are the same as the open-loop eigenvectors. This statement can be verified easily by observing that the same transformation (11.172) decouples both the open-loop and the closed-loop equations of motion. This is clearly not the case in coupled control, in which the closed-loop eigenvectors are not the same as the open-loop eigenvectors. In fact, it can be shown that symmetry of the coefficients is lost in coupled control because $g_{sr} \neq g_{rs}$ and $h_{sr} \neq h_{rs}$, so that the closed-loop system has both right and left eigenvectors. Because the velocity terms in the control vectors $\mathbf{Q}(t)$ can be identified as damping terms, we conclude that natural control generates proportional damping (see Section 8.6) in addition to proportional enhanced stiffness.

If the system is unrestrained, then it admits rigid-body motions, such as rigid-body translations and rotations. Rigid-body modes are characterized by zero natural frequencies, so that the closed-loop equation corresponding to a rigid-body mode is

$$\ddot{\eta}_r(t) + h_r\dot{\eta}_r(t) + g_r\eta_r(t) = 0 \qquad (11.184)$$

For velocity feedback alone, $g_r = 0$, the modal coordinate cannot be controlled, and for displacement feedback alone, $h_r = 0$, the modal coordinate is merely stable. Hence, to achieve asymptotic stability, the modal feedback must contain both a displacement and a velocity term, and velocity feedback alone is not sufficient. This is not necessary when $\omega_r \neq 0$, in which case the mode can be controlled even when g_r is zero, as can be concluded from Eq. (11.179).

Example 11.8

Design a three-dimensional control vector $\mathbf{T}(t)$ to control the motion of the three-degree-of-freedom system shown in Fig. 11.46, and plot the response of the system to an impulsive torque of magnitude $\hat{T}$ applied to the left disk.

First, let us derive the equations of motion. To this end, we use the Lagrangian

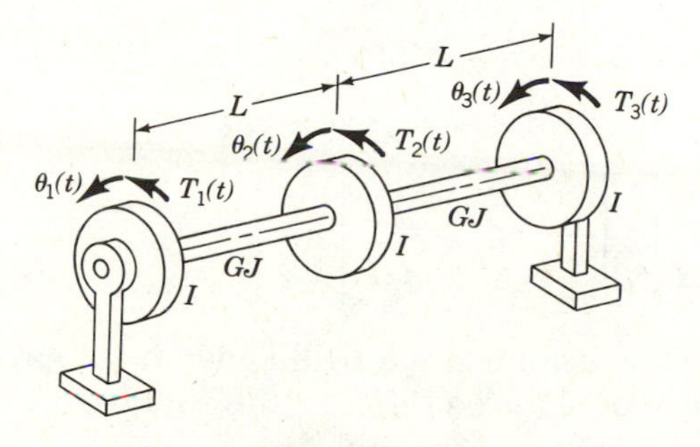

FIGURE 11.46

approach of Chapter 7. Referring to Fig. 11.46, we write the kinetic energy

$$T=\tfrac{1}{2}(I\dot{\theta}_1^2+I\dot{\theta}_2^2+I\dot{\theta}_3^2) \tag{a}$$

where I is the mass moment of inertia of the rigid disks about the axis of rotation, and the potential energy

$$V=\frac{1}{2}\left[\frac{GJ}{L}(\theta_2-\theta_1)^2+\frac{GJ}{L}(\theta_3-\theta_2)^2\right] \tag{b}$$

where GJ/L is the equivalent spring constant of the massless shafts connecting any pair of adjacent disks. Clearly, the generalized coordinates for this problem are the rotations of the disks, $q_j(t)=\theta_j(t)$, and the generalized forces are the torques, $Q_j(t)=T_j(t)$. Introducing Eqs. (a) and (b) into Eqs. (7.44), we obtain the equations of motion

$$\begin{aligned} I\ddot{\theta}_1(t)+\frac{GJ}{L}\theta_1(t)-\frac{GJ}{L}\theta_2(t)&=T_1(t)\\ I\ddot{\theta}_2(t)-\frac{GJ}{L}\theta_1(t)+\frac{2GJ}{L}\theta_2(t)-\frac{GJ}{L}\theta_3(t)&=T_2(t)\\ I\ddot{\theta}_3(t)-\frac{GJ}{L}\theta_2(t)+\frac{GJ}{L}\theta_3(t)&=T_3(t) \end{aligned} \tag{c}$$

The equations can be written in the matrix form

$$M\ddot{\boldsymbol{\theta}}(t)+K\boldsymbol{\theta}(t)=\mathbf{T}(t) \tag{d}$$

in which

$$\boldsymbol{\theta}(t)=[\theta_1(t)\quad \theta_2(t)\quad \theta_3(t)]^{\mathrm{T}},\qquad \mathbf{T}(t)=[T_1(t)\quad T_2(t)\quad T_3(t)]^{\mathrm{T}} \tag{e}$$

are the configuration and control vectors, respectively, and

$$M=I\begin{bmatrix}1&0&0\\0&1&0\\0&0&1\end{bmatrix},\qquad K=\frac{GJ}{L}\begin{bmatrix}1&-1&0\\-1&2&-1\\0&-1&1\end{bmatrix} \tag{f}$$

are the mass and stiffness matrices, respectively.

The natural frequencies can be shown to be

$$\omega_1=0,\qquad \omega_2=\sqrt{GJ/IL},\qquad \omega_3=\sqrt{3GJ/IL} \tag{g}$$

and the associated normalized modal vectors are

$$\boldsymbol{\theta}_1=\frac{1}{\sqrt{3I}}\begin{bmatrix}1\\1\\1\end{bmatrix},\qquad \boldsymbol{\theta}_2=\frac{1}{\sqrt{2I}}\begin{bmatrix}1\\0\\-1\end{bmatrix},\qquad \boldsymbol{\theta}_3=\frac{1}{\sqrt{6I}}\begin{bmatrix}1\\-2\\1\end{bmatrix} \tag{h}$$

and we observe that the first mode is a rigid-body mode representing rigid-body rotation. Using the procedure outlined in this section, we obtain the modal equations

$$\ddot{\eta}_1(t)=N_1(t)$$
$$\ddot{\eta}_2(t)+\omega_2^2\eta_2(t)=N_2(t) \tag{i}$$
$$\ddot{\eta}_3(t)+\omega_3^2\eta_3(t)=N_3(t)$$

The design of feedback control for the system amounts to designing the modal control forces $N_1(t)$, $N_2(t)$, and $N_3(t)$. To this end, we seek some guidance from damped single-degree-of-freedom systems. For the rigid-body mode, we propose aperiodic damping; for the two elastic modes, we wish to provide damping without altering the natural frequencies. Hence, we choose the modal control forces in the form

$$N_1(t)=-\Omega_1^2\eta_1(t)-2\Omega_1\dot{\eta}_1(t)$$
$$N_2(t)=-(\zeta_2\omega_2)^2\eta_2(t)-2\zeta_2\omega_2\dot{\eta}_2(t) \tag{j}$$
$$N_3(t)=-(\zeta_3\omega_3)^2\eta_3(t)-2\zeta_3\omega_3\dot{\eta}_3(t)$$

Substituting Eqs. (j) into Eqs. (i), we obtain the closed-loop equations

$$\ddot{\eta}_1(t)+2\Omega_1\dot{\eta}_1(t)+\Omega_1^2\eta_1(t)=0$$
$$\ddot{\eta}_2(t)+2\zeta_2\omega_2\dot{\eta}_2(t)+(1+\zeta_2^2)\omega_2^2\eta_2(t)=0 \tag{k}$$
$$\ddot{\eta}_3(t)+2\zeta_3\omega_3\dot{\eta}_3(t)+(1+\zeta_3^2)\omega_3^2\eta_3(t)=0$$

so that the closed-loop poles are

$$s_{11}=s_{21}=-\Omega_1, \qquad s_{12}=-\zeta_2\omega_2+i\omega_2, \qquad s_{22}=-\zeta_2\omega_2-i\omega_2 \tag{l}$$
$$s_{13}=-\zeta_3\omega_3+i\omega_3, \qquad s_{23}=-\zeta_3\omega_3-i\omega_3$$

The implication of the first of Eqs. (1) is that the rigid-body mode is subjected to critical damping ($\zeta_1=1$). For the sake of this example, we use

$$\Omega_1=0.25\sqrt{GJ/IL} \tag{m}$$

Moreover, to ensure that the elastic modes are controlled without excessive oscillation, we choose

$$\zeta_2=\zeta_3=0.5 \tag{n}$$

To solve Eqs. (k), we must produce the initial modal coordinates and velocities. The impulsive torque $\hat{T}$ is equivalent to the initial conditions

$$\theta_i(0)=0, \qquad i=1, 2, 3, \qquad \dot{\theta}_1(0)=\frac{\hat{T}}{I}, \qquad \dot{\theta}_2(0)=\dot{\theta}_3(0)=0 \tag{o}$$

so that, from Eqs. (11.172) and (h), we obtain

$$\eta_r(0)=0, \qquad r=1, 2, 3, \qquad \dot{\eta}_1(0)=\hat{T}\sqrt{\frac{I}{3}}, \qquad \dot{\eta}_2(0)=\hat{T}\sqrt{\frac{I}{2}}, \qquad \dot{\eta}_3(0)=\hat{T}\sqrt{\frac{I}{6}} \tag{p}$$

The solution of Eqs. (k) can be shown to be

$$\begin{aligned}
\eta_1(t) &= \hat{T}\sqrt{\frac{I}{3}}\, te^{-\Omega_1 t}\boldsymbol{u}(t) \\
\eta_2(t) &= \hat{T}\sqrt{\frac{I}{2}}\frac{1}{\omega_2}e^{-0.5\omega_2 t}\sin\omega_2 t\,\boldsymbol{u}(t) \\
\eta_3(t) &= \hat{T}\sqrt{\frac{I}{6}}\frac{1}{\omega_3}e^{-0.5\omega_3 t}\sin\omega_3 t\,\boldsymbol{u}(t)
\end{aligned}\tag{q}$$

so that, using Eq. (11.172), we can write the time response of the system

$$\theta(t) = \hat{T}\left\{\frac{t}{3}\begin{bmatrix}1\\1\\1\end{bmatrix}e^{-\Omega_1 t} + \frac{1}{2\omega_2}\begin{bmatrix}1\\0\\-1\end{bmatrix}e^{-0.5\omega_2 t}\sin\omega_2 t + \frac{1}{6\omega_3}\begin{bmatrix}1\\-2\\1\end{bmatrix}e^{-0.5\omega_3 t}\sin\omega_3 t\right\}\boldsymbol{u}(t)\tag{r}$$

where Ω_1, ω_2, and ω_3 are given by Eqs. (g) and (m). To complete the problem, we must show how the control is implemented. To this end, we insert Eqs. (f), (h), and (j) into Eq. (11.183) and obtain the control force vector

$$\begin{aligned}
\mathbf{T}(t) = &-\sqrt{\frac{I}{3}}\begin{bmatrix}1\\1\\1\end{bmatrix}[\Omega_1^2\eta_1(t)+2\Omega_1\dot{\eta}_1(t)] - \sqrt{\frac{I}{2}}\begin{bmatrix}1\\0\\-1\end{bmatrix}[0.25\omega_2^2\eta_2(t)+\omega_2\dot{\eta}_2(t)] \\
&-\sqrt{\frac{I}{6}}\begin{bmatrix}1\\-2\\1\end{bmatrix}[0.25\omega_3^2\eta_3(t)+\omega_3\dot{\eta}_3(t)]
\end{aligned}\tag{s}$$

The response of the system is plotted in Fig. 11.47 for the values $\hat{T}=1$ and $GJ/IL=1$. Moreover, the control torques are plotted in Fig. 11.48.

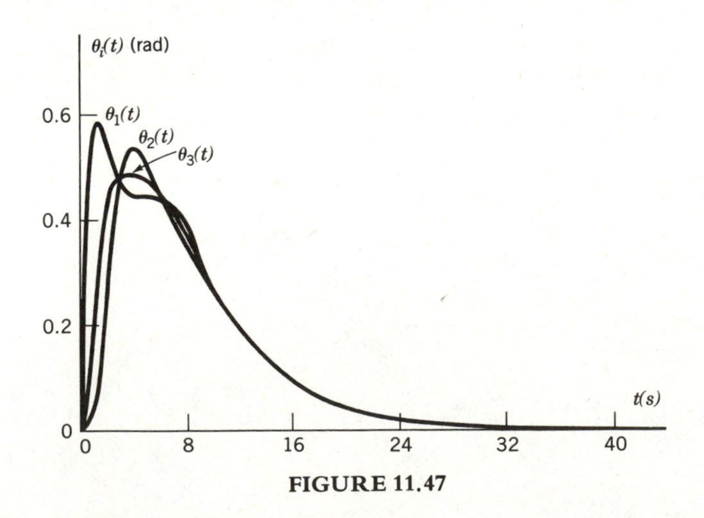

FIGURE 11.47

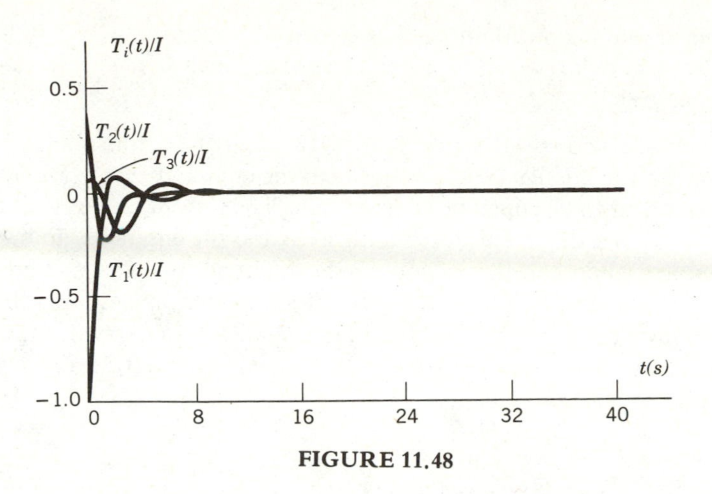

FIGURE 11.48

PROBLEMS

11.1 Figure 11.49 depicts a single-axis gyro attached to an airframe through a torsional spring and a sleeve acting as a viscous damper, where $\theta_o(t)$ represents the output angle and $\omega_i(t)$ the input angular velocity. The gyro can be used to measure angular motions of the airframe. Show that the differential equation of motion can be written in the form

$$I_T\ddot{\theta}_o + c\dot{\theta}_o + K\theta_o = -H\omega_i$$

where I_T is the mass moment of inertia of the rotor about a transverse axis, c the coefficient of viscous damping in torsion, K the spring torsional stiffness, and H the moment of momentum of the gyro about the spin axis. Determine the transfer function for the two cases:

1. The spring is omitted, $K=0$, in which case the gyro can be used to sense angular displacements of the airframe. Such a device is called an *integrating gyro*.

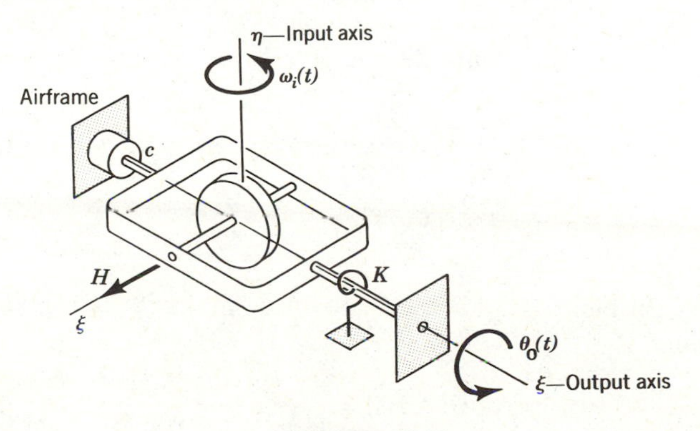

FIGURE 11.49

2. The spring is not omitted, $K \neq 0$, in which case the gyro can be used to sense angular velocities of the airframe. Such a device is known as a *rate gyro*.

11.2 The missile shown in Fig. 11.50 flies through the atmosphere. The forces acting on the missile are the drag D and the engine thrust T, where D and T are assumed to be constant in magnitude. The output angle $\theta(t)$ is measured by means of a single-axis integrating gyro and the engine is gimballed by an angle $\delta(t)$, where the gimbal angle is the difference between a reference angle $r(t)$ and the gyro output angle $\theta_o(t)$. All angles can be assumed to be small. Show how the gyro must be placed in the missile to carry out the intended task, draw a block diagram of the system, and derive the closed-loop transfer function.

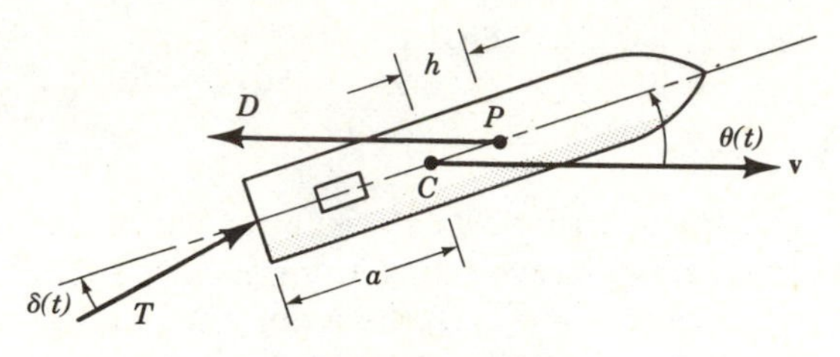

FIGURE 11.50

11.3 Use the final-value theorem (Section A.5) to determine a general expression for the steady-state error of the system described by Eq. (11.7) in which the reference input is the unit step function.

11.4 Repeat Problem 11.3 for the case in which the reference input is the unit ramp function.

11.5 Use the rules given in Section 11.4 to determine the main features of the root loci for the feedback system with the open-loop transfer function

$$G(s)H(s)=\frac{K}{s(s+2)(s+3)}$$

Sketch the root loci, and determine the value of K for which the system becomes unstable.

11.6 Repeat Problem 11.5 for the feedback system with the open-loop transfer function

$$G(s)H(s)=\frac{K}{s(s^2+2s+2)}$$

11.7 Repeat Problem 11.5 for the feedback system with the open-loop transfer function

$$G(s)H(s)=\frac{K(s+0.5)}{s(s+1)(s^2+4s+8)}$$

11.8 Use root-locus techniques to find the roots of the polynomial

$$s^3+6s^2+11s+6=0$$

11.9 Sketch the Nyquist plot for the system of Problem 11.5 for $K=10$ and determine the system stability.

11.10 Sketch the Nyquist plots for the system of Problem 11.6 for the two values: (1) $K=2$ and (2) $K=6$. Determine the system stability in each case.

11.11 Sketch the Nyquist plot for the system of Problem 11.7 for $K=20$ and determine the system stability.

11.12 Construct the Bode plots for the system of Problem 11.9.

11.13 Construct the Bode plots for the two cases in Problem 11.10.

11.14 Construct the Bode plots for the system of Problem 11.11

11.15 Determine the step response of the system of Problem 11.10, Case 1. Assume that $H(s)=1$. Check the maximum overshoot, the settling time, and the steady-state error.

11.16 Obtain the ramp response of the system of Problem 11.15, and determine the steady-state error.

11.17 Derive the transfer function for the lag compensator of Fig. 11.41, as given by Eq. (11.148).

11.18 Derive the transfer function for the lead-lag compensator of Fig. 11.42, as given by Eq. (11.149).

11.19 Sketch Bode plots for the lag compensator of Problem 11.17 for three pairs of parameters, and summarize the effects of the compensator.

11.20 Sketch Bode plots for the lead-lag compensator of Problem 11.18 for a case in which $a_1>b_2$, and summarize the effects of the compensator.

11.21 A system has the open-loop transfer function

$$G(s)H(s)=\frac{24}{s(s+2)(s+3)}$$

where $H(s)=1$. Use Nyquist plots to design a lead compensator so that the compensated system has a gain margin exceeding 12 dB and a phase margin exceeding 60°. Determine the steady-state error of the compensated system for (1) a step function reference input and (2) a unit ramp function reference input.

11.22 Repeat Problem 11.21, but design a lag compensator instead of a lead compensator.

11.23 Repeat Problem 11.21, but use Bode plots to carry out the design of the compensator.

11.24 Repeat Problem 11.22, but use Bode plots to design the compensator.

11.25 Derive the state equations for the uncompensated and lead-compensated system of Problem 11.21, and obtain the response to a unit step function reference input by the transition matrix approach.

11.26 Derive the state equations for the uncompensated and lag-compensated system of Problem 11.22 and obtain the response to a unit ramp function by the transition matrix approach.

11.27 Derive the equations of motion for the system shown in Fig. 11.51. Then, design a feedback control force F so that the closed-loop poles have the values s_1, s_2, s_3, and s_4. *Hint:* Assume that $F=-a\dot{x}-bx-c\dot{\theta}-d\theta$, and solve for the gains a, b, c, and d in terms of the closed-loop poles.

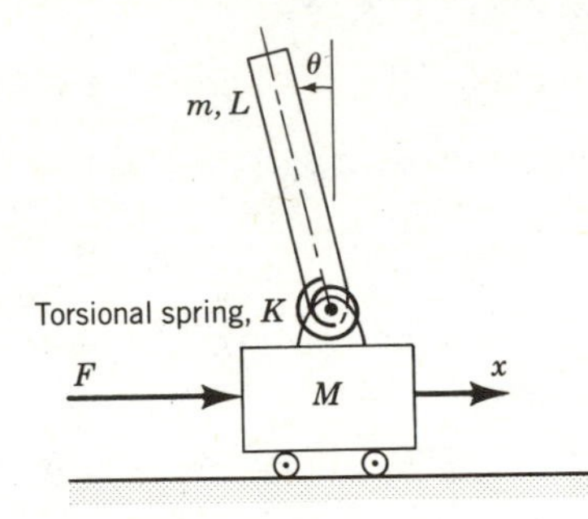

FIGURE 11.51

11.28 Use the modal approach of Section 11.12 to design control forces for the system of Fig. 11.52 so that closed-loop poles reflect 25% damping ($\zeta_1=\zeta_2=\zeta_3=0.25$) and the closed-loop frequencies remain the same as the open-loop frequencies. Determine the response to a unit step function applied to m_3, and examine the performance of the control.

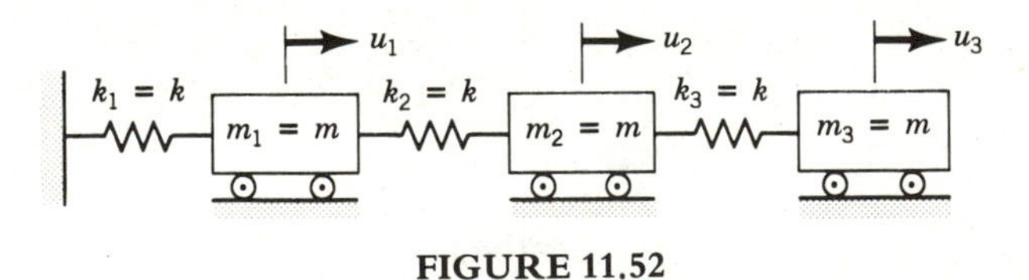

FIGURE 11.52

APPENDIX

Elements of the Laplace Transformation

A.1 DEFINITIONS

The Laplace transformation is a mathematical technique for solving linear differential equations with constant coefficients. The general idea of a transformation is to transform a relatively complicated problem into a simpler one, solve the simpler problem, and perform an inverse transformation on the transformed solution to recover the solution to the original problem.

Let us consider an arbitrary function $f(t)$ and introduce the definition of the Laplace transformation of $f(t)$ in the form

$$F(s) = \mathscr{L}f(t) = \int_0^\infty e^{-st} f(t)\, dt \tag{A.1}$$

Clearly, the result of the integration is a function of s, where s is known as a *subsidiary variable*, generally a complex number. The complex plane defined by the subsidiary variable is called the *s-plane*, or the *Laplace plane*. For the Laplace transformation $F(s)$ to exist, the function $f(t)$ must satisfy certain conditions. Because our interest in the Laplace transformation is merely as a mathematical tool, we shall not go into many details, but simply give one of the most basic conditions, namely,

$$\lim_{t\to\infty} e^{-st} f(t) = 0 \tag{A.2}$$

If this condition is satisfied then the Laplace transformation $F(s)$ can be assumed to exist. This is indeed the case with all the functions of interest in our study.

A.2 TRANSFORMATION OF DERIVATIVES

We shall be interested not only in the Laplace transformation of functions but also in the transformation of derivatives of functions. Hence, let us consider the Laplace transformation of the nth derivative of $f(t)$ with respect to time. Replacing $f(t)$ by $d^n f(t)/dt^n$ in Eq. (A.1) and integrating by parts, we obtain

$$\mathscr{L}\frac{d^n f(t)}{dt^n} = \int_0^\infty e^{-st}\frac{d^n f(t)}{dt^n}\, dt = e^{-st}\frac{d^{n-1} f(t)}{dt^{n-1}}\bigg|_0^\infty + s\int_0^\infty e^{-st}\frac{d^{n-1} f(t)}{dt^{n-1}}\, dt \tag{A.3}$$

But, consistent with Eq. (A.2), we can write

$$\lim_{t\to\infty} e^{-st}\frac{d^{n-1}f(t)}{dt^{n-1}}=0 \tag{A.4}$$

so that Eq. (A.4) can be rewritten in the form

$$\mathscr{L}\frac{d^{n}f(t)}{dt^{n}}=-\frac{d^{n-1}f(0)}{dt^{n-1}}+s\mathscr{L}\frac{d^{n-1}f(t)}{dt^{n-1}} \tag{A.5}$$

where $d^{n-1}f(0)/dt^{n-1}$ is the value of $d^{n-1}f(t)/dt^{n-1}$ evaluated at $t=0$. Equation (A.5) can be used as the basis of a process of induction that lowers the order of the derivatives by one with every application. Repeated use of this induction process yields

$$\mathscr{L}\frac{d^{n}f(t)}{dt^{n}}=-\frac{d^{n-1}f(0)}{dt^{n-1}}-s\frac{d^{n-2}f(0)}{dt^{n-2}}-\cdots-s^{n-2}\frac{df(0)}{dt}-s^{n-1}f(0)+s^{n}f(s) \tag{A.6}$$

where the notation is obvious.

A.3 THE INVERSE LAPLACE TRANSFORMATION

The rigorous definition of the inverse Laplace transformation involves a line integral in the s-plane. For our purposes, however, we need not go into line integrations, but define the inverse Laplace transformation simply as a function known to have the given Laplace transformation. For example, if $f(t)$ has the Laplace transformation $F(s)$, then the inverse Laplace transformation of $F(s)$ is denoted symbolically by

$$f(t)=\mathscr{L}^{-1}F(s) \tag{A.7}$$

In Section A.7 there is a table of Laplace transformation pairs, which permits us to go from $f(t)$ to $F(s)$ and vice versa for a number of simple functions of special interest. More complicated functions can often be decomposed into simple ones by the method of partial fractions, as described in the following.

Let us consider the inverse Laplace transformation $F(s)$ and assume that it can be written in the form

$$F(s)=\frac{A(s)}{B(s)} \tag{A.8}$$

where both $A(s)$ and $B(s)$ are polynomials in s, with the degree of $B(s)$ being higher than that of s. If $s=a_k$ $(k=1,2,\ldots,n)$ represent the roots of $B(s)$, then the polynomial can be written as the product of the factors, or

$$B(s)=(s-a_1)(s-a_2)\cdots(s-a_k)\cdots(s-a_n)=\prod_{k=1}^{n}(s-a_k) \tag{A.9}$$

where $\prod$ is the product symbol. The roots $s=a_k$ are called *simple poles* of $F(s)$.

Under these circumstances, the function $F(s)$ can be expanded into the sum of partial fractions

$$F(s)=\frac{c_1}{s-a_1}+\frac{c_2}{s-a_2}+\cdots+\frac{c_k}{s-a_k}+\cdots+\frac{c_n}{s-a_n}=\sum_{k=1}^{n}\frac{c_k}{s-a_k} \tag{A.10}$$

where the constant coefficients c_k are given by

$$c_k=\lim_{s\to a_k}[(s-a_k)F(s)]=\left.\frac{A(s)}{B'(s)}\right|_{s=a_k} \tag{A.11}$$

in which $B'(s)=dB(s)/ds$.

It is easy to verify that

$$\mathscr{L}e^{a_k t}=\frac{1}{s-a_k} \tag{A.12}$$

so that

$$\mathscr{L}^{-1}\frac{1}{s-a_k}=e^{a_k t} \tag{A.13}$$

Equations (A.10), (A.11), and (A.13) permit us to write the inverse Laplace transformation of $F(s)$ in the form

$$\begin{aligned} f(t)&=\mathscr{L}^{-1}F(s)=\mathscr{L}^{-1}\sum_{k=1}^{n}\frac{c_k}{s-a_k}\\ &=\sum_{k=1}^{n}\left.\frac{A(s)}{B'(s)}\right|_{s=a_k}e^{a_k t}=\sum_{k=1}^{n}\left.\frac{A(s)}{B'(s)}e^{st}\right|_{s=a_k} \end{aligned} \tag{A.14}$$

Next, let us consider the case in which $B(s)$ has a multiple root of degree r. In this case, $F(s)$ is said to possess a *pole of degree r* in addition to the simple poles examined earlier. Hence, let us consider

$$B(s)=(s-a_1)^r(s-a_2)(s-a_3)\cdots(s-a_n) \tag{A.15}$$

The partial fractions expansion in this case has the form

$$\begin{aligned} F(s)=\frac{A(s)}{B(s)}&=\frac{c_{11}}{(s-a_1)^r}+\frac{c_{12}}{(s-a_1)^{r-1}}+\cdots+\frac{c_{1r}}{s-a_1}\\ &+\frac{c_2}{s-a_2}+\frac{c_3}{s-a_3}+\cdots+\frac{c_n}{s-a_n} \end{aligned} \tag{A.16}$$

It can be easily verified that the coefficients corresponding to the repeated root are given by

$$c_{1j}=\left.\frac{1}{(j-1)!}\frac{d^{j-1}}{ds^{j-1}}[(s-a_1)^rF(s)]\right|_{s=a_1}, \qquad j=1,2,\ldots,r \tag{A.17}$$

Moreover, by first calculating the transform of $t^{j-1}e^{a_1t}$, it is not difficult to show that

$$\mathscr{L}^{-1}\frac{1}{(s-a_1)^j}=\frac{t^{j-1}}{(j-1)!}e^{a_1t} \tag{A.18}$$

Hence, the inverse transform of Eq. (A.16) can be written in the form

$$f(t)=\mathscr{L}^{-1}F(s)=\left[c_{11}\frac{t^{r-1}}{(r-1)!}+c_{12}\frac{t^{r-2}}{(r-2)!}+\cdots+c_{1r}\right]e^{a_1 t}$$
$$+c_2e^{a_2t}+c_3e^{a_3t}+\cdots+c_ne^{a_nt} \tag{A.19}$$

which can be shown to be equal to

$$f(t)=\frac{1}{(r-1)!}\frac{d^{r-1}}{ds^{r-1}}[(s-a_1)^rF(s)e^{st}]\bigg|_{s=a_1}$$
$$+\sum_{k=2}^{n}[(s-a_k)F(s)e^{st}]\bigg|_{s=ak} \tag{A.20}$$

Example A.1
Obtain the inverse Laplace transformation of the function

$$F(s)=\frac{1}{(s-a)(s-b)} \tag{a}$$

First, let us write $F(s)$ in the form of the partial fraction expansion

$$F(s)=\frac{c_1}{s-a}+\frac{c_2}{s-b} \tag{b}$$

where, from Eq. (A.11),

$$c_1=\lim_{s\to a}[(s-a)F(s)]=\frac{1}{s-b}\bigg|_{s=a}=\frac{1}{a-b}$$
$$c_2=\lim_{s\to b}[(s-b)F(s)]=\frac{1}{s-a}\bigg|_{s=b}=-\frac{1}{a-b} \tag{c}$$

Introducing Eqs. (c) into Eq. (b), we obtain the expansion

$$F(s)=\frac{1}{a-b}\left(\frac{1}{s-a}-\frac{1}{s-b}\right) \tag{d}$$

Hence, using the table of Laplace transform pairs of Section (A.7), we obtain the inverse transformation

$$f(t)=\frac{1}{a-b}(e^{at}-e^{bt}) \tag{e}$$

Example A.2
Calculate the inverse Laplace transform of the function

$$F(s)=\frac{2\omega^2s}{(s+\omega^2)^2} \tag{a}$$

It is easy to see that $F(s)$ has two double poles, $s=a_1=i\omega$ and $s=a_2=-i\omega$. Hence, we can write the partial fractions expansion

$$F(s)=\frac{c_{11}}{(s-i\omega)^2}+\frac{c_{12}}{s-i\omega}+\frac{c_{21}}{(s+i\omega)^2}+\frac{c_{22}}{s+i\omega} \tag{b}$$

Using formula (A.17), we obtain

$$c_{11}=-c_{21}=\frac{\omega}{2i}, \qquad c_{12}=c_{22}=0 \tag{c}$$

Moreover, using Eq. (A.18), we can write

$$\begin{aligned} f(t)&=c_{11}\mathscr{L}^{-1}\frac{1}{(s-i\omega)^2}+c_{21}\mathscr{L}^{-1}\frac{1}{(s+i\omega)^2}=\frac{\omega}{2i}(te^{i\omega t}-te^{-i\omega t}) \\ &=\omega t \sin \omega t \end{aligned} \tag{d}$$

A.4 FIRST AND SECOND SHIFTING THEOREMS

The evaluation of the Laplace transform can at times be expedited by a shift in the complex plane. Let us consider the Laplace transform of the function $f(t)e^{at}$, where a is in general a complex number. Using Eq. (A.1), we can write

$$\mathscr{L}[f(t)e^{at}]=\int_0^\infty [f(t)e^{at}]e^{-st}\,dt=\int_0^\infty f(t)e^{-(s-a)t}\,dt=F(s-a) \tag{A.21}$$

Hence, the effect of multiplying $f(t)$ by e^{at} in the time domain is to shift the transform $F(s)$ by an amount a in the complex s-domain. Equation (A.21) is known as the *first shifting theorem*, or the *complex shifting theorem*.

Next, consider

$$F(s)=\int_0^\infty e^{-s\lambda}f(\lambda)\,d\lambda \tag{A.22}$$

and let

$$\lambda=t-a, \qquad t>a \tag{A.23}$$

so that

$$F(s)=\int_a^\infty e^{-s(t-a)}f(t-a)\,dt=e^{as}\int_a^\infty e^{-st}f(t-a)\,dt \tag{A.24}$$

or

$$e^{-as}F(s)-\begin{cases}\displaystyle\int_0^\infty e^{-st}f(t-a)\,dt, & t>a \\ 0, & t<a\end{cases} \tag{A.25}$$

It follows that

$$\mathscr{L}^{-1}e^{-as}F(s)=f(t-a)\mathscr{u}(t-a) \tag{A.26}$$

where $\mathscr{u}(t-a)$ is the unit step function with the jump at $t=a$. Equation (A.26) is known as the *second shifting theorem*.

A.5 INITIAL- AND FINAL-VALUE THEOREMS

Let us consider Eq. (A.6) with $n=1$, so that

$$\mathscr{L}\dot{f}(t)=sF(s)-f(0)=\int_0^\infty e^{-st}\dot{f}(t)\,dt=\lim_{t\to\infty}\int_0^t e^{-s\lambda}\dot{f}(\lambda)\,d\lambda \tag{A.27}$$

Now, if we let $s\to\infty$, then Eq. (A.27) yields

$$f(0)=\lim_{s\to\infty} sF(s) \tag{A.28}$$

Equation (A.28) is known as the *initial-value theorem* and can be used to determine the initial value of a function without carrying out the inverse transformation. On the other hand, if we let $s\to 0$, then we can use Eq. (A.27) and write

$$\lim_{s\to 0} sF(s)-f(0)=\lim_{t\to\infty}\int_0^t \dot{f}(\lambda)\,d\lambda=\lim_{t\to\infty}[f(t)-f(0)] \tag{A.29}$$

which reduces to

$$\lim_{t\to\infty} f(t)=\lim_{s\to 0} sF(s) \tag{A.30}$$

Equation (A.30) is known as the *final-value theorem* and can be used to determine the final value of a function without the inversion process. The theorem is often useful in determining the steady-state response of a system to transient excitation.

Example A.3

Determine the steady-state step response of a mass–damper–spring system.

Considering Eqs. (1.32) and (4.157), we can show that, in the absence of initial displacement and velocity, the Laplace transform of the response $c(t)$ of a mass–damper–spring system to the excitation $r(t)$ has the form

$$C(s)=\frac{1}{ms^2+cs+k}R(s) \tag{a}$$

where $R(s)$ is the transform of the excitation. If $r(t)$ is the unit step function, then

$$R(s)=\int_0^\infty e^{-st}\,dt=\frac{e^{-st}}{-s}\bigg|_0^\infty=\frac{1}{s} \tag{b}$$

so that

$$C(s)=\frac{1}{s(ms^2+cs+k)} \tag{c}$$

To obtain the final value of the response, we use Eq. (A.30) to write

$$\lim_{t\to\infty} c(t)=\lim_{s\to 0} sC(s)=\lim_{s\to 0}\frac{1}{ms^2+cs+k}=\frac{1}{k} \tag{d}$$

A.6 THE CONVOLUTION INTEGRAL

Consider two functions $f_1(t)$ and $f_2(t)$, both defined for $t>0$, and denote their Laplace transforms by $F_1(s)$ and $F_2(s)$, respectively. Then, consider the integral

$$x(t)=\int_0^t f_1(\tau)f_2(t-\tau)\,d\tau=\int_0^\infty f_1(\tau)f_2(t-\tau)\,d\tau \tag{A.31}$$

Note that the value of the integral is unaffected by a change in limit, because $f_2(t-\tau)=0$ for $\tau>t$, or $t-\tau<0$. Transforming both sides of Eq. (A.31) and changing the integration order, we obtain

$$\begin{aligned}X(s)&=\int_0^\infty e^{-st}\left[\int_0^\infty f_1(\tau)f_2(t-\tau)\,dt\right]d\tau=\int_0^\infty f_1(\tau)\left[\int_0^\infty e^{-st}f_2(t-\tau)\,dt\right]d\tau\\&=\int_0^\infty f_1(\tau)\left[\int_\tau^\infty e^{-st}f_2(t-\tau)\,dt\right]d\tau\end{aligned} \tag{A.32}$$

where the lower limit in the second integral could be changed without affecting the result because $f_2(t-\tau)=0$ for $t<\tau$. Next, let us introduce the notation $t-\tau=\lambda$ in the second integral in (A.32), note that $\lambda=0$ when $t=\tau$, and write

$$\begin{aligned}X(s)&=\int_0^\infty f_1(\tau)\left[\int_\tau^\infty e^{-st}f_2(t-\tau)\,dt\right]d\tau=\int_0^\infty f_1(\tau)\left[\int_0^\infty e^{-s(\tau+\lambda)}f_2(\lambda)\,d\lambda\right]d\tau\\&=\int_0^\infty e^{-s\tau}f_1(\tau)\,d\tau\int_0^\infty e^{-s\lambda}f_2(\lambda)\,d\lambda=F_1(s)F_2(s)\end{aligned} \tag{A.33}$$

Now, let us consider a function $x(t)$ whose Laplace transform $X(s)$ has the form of the product of the Laplace transforms of the functions $f_1(t)$ and $f_2(t)$, that is,

$$X(s)=F_1(s)F_2(s) \tag{A.34}$$

Then, according to Eqs. (A.31)–(A.33) the function $x(t)$ can be obtained from

$$\begin{aligned}x(t)&=\mathscr{L}^{-1}X(s)=\mathscr{L}F_1(s)F_2(s)\\&=\int_0^t f_1(\tau)f_2(t-\tau)\,d\tau=\int_0^t f_1(t-\tau)f_2(\tau)\,d\tau\end{aligned} \tag{A.35}$$

The second integral has been added in Eq. (A.35) because it does not matter which of the functions $f_1(t)$ and $f_2(t)$ is shifted. Equation (A.35) is known as the *convolution theorem*, and the integrals in (A.35) are known as *convolution integrals.*

Example A.4

Derive a general expression for the response of a mass–damper–spring system by means of the convolution theorem.

The inverse Laplace transform $C(s)$ of the response $c(t)$ is shown in Section 1.7 to have the form

$$C(s)=G(s)R(s) \tag{A.36}$$

where $G(s)$ is the system transfer function and $R(s)$ is the transform of the excitation

$r(t)$. Hence, using Eq. (A.35), we can write

$$c(t) = \int_0^t g(\tau) r(t - \tau) \, d\tau = \int_0^t g(t - \tau) r(\tau) \, d\tau \tag{A.37}$$

where

$$g(t) = \mathscr{L}^{-1} G(s) \tag{A.38}$$

is the impulse response.

A.7 TABLE OF LAPLACE TRANSFORM PAIRS

$f(t)$	$F(s)$
$\delta(t)$ (Dirac delta function)	1
(t) (unit step function)	$\dfrac{1}{s}$
$t^n,\ n=1, 2, \ldots$	$\dfrac{n!}{s^{n+1}}$
$e^{-\omega t}$	$\dfrac{1}{s+\omega}$
$t^n e^{-\omega t}$	$\dfrac{n!}{(s+\omega)^{n+1}}$
$\cos \omega t$	$\dfrac{s}{s^2+\omega^2}$
$\sin \omega t$	$\dfrac{\omega}{s^2+\omega^2}$
$\cosh \omega t$	$\dfrac{s}{s^2-\omega^2}$
$\sinh \omega t$	$\dfrac{\omega}{s^2-\omega^2}$
$1-e^{-at}$	$\dfrac{\omega}{s(s+\omega)}$
$1-\cos \omega t$	$\dfrac{\omega^2}{s(s^2+\omega^2)}$
$\omega t-\sin \omega t$	$\dfrac{\omega^3}{s^2(s^2+\omega^2)}$
$\omega t \cos \omega t$	$\dfrac{\omega(s^2-\omega^2)}{(s^2+\omega^2)^2}$
$\omega t \sin \omega t$	$\dfrac{2\omega^2 s}{(s^2+\omega^2)^2}$
$\dfrac{1}{(1-\zeta^2)^{1/2}\omega} e^{-\zeta\omega t} \sin (1-\zeta^2)^{1/2}\omega t$	$\dfrac{1}{s^2+2\zeta\omega s+\omega^2}$
$e^{-\zeta\omega t}\left[\cos (1-\zeta^2)^{1/2}\omega t+\dfrac{\zeta}{(1-\zeta^2)^{1/2}} \sin (1-\zeta^2)^{1/2}\omega t\right]$	$\dfrac{s+2\zeta\omega}{s^2+2\zeta\omega s+\omega^2}$

Bibliography

1. J. A. Aseltine, *Transform Method in Linear System Analysis*, McGraw-Hill, New York, 1958.
2. W. L. Brogan, *Modern Control Theory*, Quantum Publishers, New York, 1974.
3. B. Carnahan, H. A. Luther, and J. O. Wilkes, *Applied Numerical Methods*, John Wiley, New York, 1969.
4. J. J. DiStefano, A. R. Stubberud, and I. J. Williams, *Theory and Problems of Feedback and Control Systems*, Schaum Publishing, New York, 1967.
5. R. C. Dorf, *Modern Control Systems*, Second edition, Addison-Wesley, Reading, Massachusetts, 1974.
6. B. C. Kuo, *Automatic Control Systems*, Third edition, Prentice-Hall, Englewood Cliffs, New Jersey, 1975.
7. L. Meirovitch, *Methods of Analytical Dynamics*, McGraw-Hill, New York, 1970.
8. L. Meirovitch, *Elements of Vibration Analysis*, McGraw-Hill, New York, 1975.
9. L. Meirovitch, *Computational Methods in Structural Dynamics*, Sijthoff-Noordhoff, The Netherlands, 1980.
10. J. Ll. Morris, *Computational Methods in Elementary Numerical Analysis*, John Wiley, New York, 1983.
11. A. Ralston, *A First Course in Numerical Analysis*, McGraw-Hill, New York, 1965.
12. Y. Takahashi, M. J. Rabins, and D. M. Auslander, *Control and Dynamic Systems*, Addison-Wesley, Reading, Massachusetts, 1970.
13. H. F. VanLandingham, *Introduction to Digital Control*, Macmillan, New York, 1985.

Index